TABLE OF THE ELEMENTS

Name	Symbol	Atomic Number	Atomic Mass	Name	Symbol	Atomic Number	Atomic Mass
Actinium	Ac	89	227[a]	Molybdenum	Mo	42	95.94
Aluminum	Al	13	26.98154	Neodymium	Nd	60	144.24
Americium	Am	95	243[a]	Neon	Ne	10	20.179
Antimony	Sb	51	121.75	Neptunium	Np	93	237.0482
Argon	Ar	18	39.948	Nickel	Ni	28	58.70
Arsenic	As	33	74.9216	Niobium	Nb	41	92.9064
Astatine	At	85	210[a]	Nitrogen	N	7	14.0067
Barium	Ba	56	137.33	Nobelium	No	102	259[a]
Berkelium	Bk	97	247[a]	Osmium	Os	76	190.2
Beryllium	Be	4	9.01218	Oxygen	O	8	15.9994
Bismuth	Bi	83	208.9804	Palladium	Pd	46	106.4
Boron	B	5	10.81	Phosphorus	P	15	30.97376
Bromine	Br	35	79.904	Platinum	Pt	78	195.09
Cadmium	Cd	48	112.41	Plutonium	Pu	94	244[a]
Calcium	Ca	20	40.08	Polonium	Po	84	209[a]
Californium	Cf	98	251[a]	Potassium	K	19	39.0983
Carbon	C	6	12.011	Praeseodymium	Pr	59	140.9077
Cerium	Ce	58	140.12	Promethium	Pm	61	145[a]
Cesium	Cs	55	132.9054	Protactinium	Pa	91	231.0359
Chlorine	Cl	17	35.453	Radium	Ra	88	226.0254
Chromium	Cr	24	51.996	Radon	Rn	86	222[a]
Cobalt	Co	27	58.9332	Rhenium	Re	75	186.207
Copper	Cu	29	63.546	Rhodium	Rh	45	102.9055
Curium	Cm	96	247[a]	Rubidium	Rb	37	85.4678
Dysprosium	Dy	66	162.50	Ruthenium	Ru	44	101.07
Einsteinium	Es	99	252[a]	Samarium	Sm	62	150.4
Erbium	Er	68	167.26	Scandium	Sc	21	44.9559
Europium	Eu	63	151.96	Selenium	Se	34	78.96
Fermium	Fm	100	257[a]	Silicon	Si	14	28.0855
Fluorine	F	9	18.998403	Silver	Ag	47	107.868
Francium	Fr	87	223[a]	Sodium	Na	11	22.98977
Gadolinium	Gd	64	157.25	Strontium	Sr	38	87.62
Gallium	Ga	31	69.72	Sulfur	S	16	32.06
Germanium	Ge	32	72.59	Tantalum	Ta	73	180.9479
Gold	Au	79	196.9665	Technetium	Tc	43	98[a]
Hafnium	Hf	72	178.49	Tellurium	Te	52	127.60
Helium	He	2	4.00260	Terbium	Tb	65	158.9254
Holmium	Ho	67	164.9304	Thallium	Tl	81	204.37
Hydrogen	H	1	1.0079	Thorium	Th	90	232.0381
Indium	In	49	114.82	Thulium	Tm	69	168.9342
Iodine	I	53	126.9045	Tin	Sn	50	118.69
Iridium	Ir	77	192.22	Titanium	Ti	22	47.90
Iron	Fe	26	55.847	Tungsten	W	74	183.85
Krypton	Kr	36	83.80	Unnilhexium	Unh	106	263[a]
Lanthanum	La	57	138.9055	Unnilpentium	Unp	105	262[a]
Lawrencium	Lr	103	260[a]	Unnilquadium	Unq	104	261[a]
Lead	Pb	82	207.2	Uranium	U	92	238.029
Lithium	Li	3	6.941	Vanadium	V	23	50.9415
Lutetium	Lu	71	174.967	Xenon	Xe	54	131.30
Magnesium	Mg	12	24.305	Ytterbium	Yb	70	173.04
Manganese	Mn	25	54.9380	Yttrium	Y	39	88.9059
Mendelevium	Md	101	258[a]	Zinc	Zn	30	65.38
Mercury	Hg	80	200.59	Zirconium	Zr	40	91.22

[a]Not naturally occurring

Introduction to College Chemistry

Drew H. Wolfe
Hillsborough Community College

Introduction to College Chemistry

McGraw-Hill
Book Company

New York • St. Louis • San Francisco • Auckland
Bogotá • Hamburg • Johannesburg • London
Madrid • Mexico • Montreal • New Delhi
Panama • Paris • São Paulo • Singapore
Sydney • Tokyo • Toronto

Introduction to College Chemistry

1 2 3 4 5 6 7 8 9 0   VNHVNH   8 9 8 7 6 5 4

ISBN 0-07-071410-X

This book was set in Electra by York Graphic Services, Inc.
The editors were Stephen Zlotnick, Kathleen M. Civetta, and Jo Satloff;
the designer was Jo Jones;
the production supervisor was Phil Galea.
The photo editor was Randy Matusow.
The cover photographs were taken by Fundamental Photographs, N. Y.
The drawings were done by Fine Line Illustrations, Inc.
Von Hoffmann Press, Inc., was printer and binder.

Library of Congress Cataloging in Publication Data

Wolfe, Drew H.
    Introduction to college chemistry

    Includes index.
    1. Chemistry.   I. Title.
QD33.W84   1984   540   83-17554
ISBN 0-07-071410-X

To my mother, Cynthia, and Natasha

· CONTENTS ·

· PREFACE ·

Introduction to College Chemistry is a comprehensive beginning chemistry textbook primarily designed for use in courses that prepare students to take general college chemistry. It can also be used in the first semester of an allied health chemistry course or in a terminal chemistry course for nonscience majors.

A major emphasis of *Introduction to College Chemistry* is chemical problem solving. Chapter 2, which is devoted to this topic, includes a section on scientific problem solving and then shows students how they can follow a similar systematic procedure to solve chemistry problems. This procedure requires four steps and incorporates the factor-label method (sometimes called the unit-conversion method). When appropriate, various steps and the final answer are thoroughly discussed to encourage students to think through the problem.

Principles and ideas developed in Chapter 2 are applied throughout the book. Many sample problems worked out in detail are included in all chapters that contain numerical problems. Diagrams and flowcharts are also used in conjunction with the sample problems. Stepwise procedures are presented to write dot formulas of compounds, to name compounds, and to balance equations. Additionally, each chapter has between 50 and 100 questions and problems. After each section in a chapter there are *Review Problems*. In this section, both questions and problems are presented to help students understand the material within that section. At the end of the chapter, the *Questions and Problems* cover all the material in the chapter. Answers to problems and selected questions can be found in Appendix 6.

Introduction to College Chemistry follows a fairly traditional order of presentation. Chapter 3 covers the fundamentals of chemical measurement. Chapter 4 looks at the basic concepts of matter and energy. Chapters 5 and 6 consider the structure and properties of atoms and elements. Before proceeding to molecules and compounds, Chapter 7 introduces the mole concept. An early introduction of the mole concept helps students to better learn this central topic of chemistry and supports the laboratory program. Should the instructor prefer, this chapter can be delayed until after Chapter 10 without loss of continuity. Chapter 8 discusses compounds, molecules, and chemical bonding, and Chapter 9 is an overview of inorganic nomenclature. Chapter 10 deals with chemical reactions and chemical equations, and Chapter 11 is a discussion of stoichiometry. Chapters 3 to 11 present the core material for most introductory courses.

Depending on the nature of the course, various combinations of the remaining chapters might be selected. Chapter 12, Gases, and Chapter 13, Liquids, Solids, and State Changes, are an introduction to the physical states of matter. Chapter 14, Descriptive Inorganic Chemistry, is a discussion of selected substances that emphasizes the principles learned in the early chapters. Descriptive chemistry is presented as an application of chemical theory and principles, not solely as factual material to be memorized for no apparent reason. Should the instructor wish, Chapter 14 could be used effectively to conclude the course. Relevant descriptive inorganic chemistry also is integrated into other chapters. Chapter 15 considers solutions and their properties. Should time allow, Chapters 16 through 20 can be used to introduce kinetics, equilibrium, acids and bases, redox, nuclear chemistry, and organic and biological chemistry.

Each chapter contains the following pedagogical aids:

1. *Study guidelines* that list what the students should be able to do after successfully completing the chapter.
2. *Margin notes* that amplify and expand topics in the body of the text. Sometimes the margin notes cite historical information or chemical trivia. These short notes help to make the task of learning chemistry more enjoyable.
3. Many *diagrams and photographs* that illustrate abstract chemical principles, procedures for solving problems, and chemical structures.
4. *Sample problems* containing careful explanations of most steps. Significant figures are emphasized in each sample problem.
5. *Chapter summaries* of the important ideas and concepts within the chapter.
6. *Questions and problems* ranging from trivial to quite challenging. Questions and problems are located after each major section (Review Problems) and at the end of the chapter (Questions and Problems). In all, there are more than 1200 questions and problems in the volume.

The appendixes include:

1. *Review of Math Skills* includes basic algebra, exponential and scientific notation, and graphing skills. Each section of this appendix contains sample problems and practice problems.
2. *Chemistry Calculations Using Calculators* discusses both algebraic notation and reverse polish notation.
3. *Physical Constants and Conversion Factors*
4. *Table of Ions and Their Formulas*
5. *Logarithms to Base 10* table with an explanation of its use.
6. *Answers to Problems and Selected Questions*

The volume concludes with a Glossary, which gives succinct definitions and descriptions of the most important words and terms in the book, and a carefully prepared Index.

A number of ancillary materials accompany the text: A *Laboratory Manual* written by Dr. Alan J. Pribula, Towson State University, contains laboratory experiments that parallel the material in the textbook. Each experiment includes an introduction, procedures, data tables, and summary questions. The *Study Guide/Solutions Manual*, written by Ms. Elsie Gross, Hillsborough Community College, exactly follows the textbook. Added explanations of difficult topics are included, as well as additional sample problems. Practice tests help students prepare for their exams. The *Instructor's Manual* contains detailed solutions to most problems and the answers to more important questions. A test bank is also included in the Instructor's Manual and is available on floppy disks for popular microcomputers.

Acknowledgments

I would like to thank each of the following reviewers whose comments, suggestions, and criticisms helped immensely in the development of this book.

David L. Adams, North Shore Community College, Beverly, MA
Robert Buckley, Edison Community College, Ft. Myers, FL
David S. Byrd, Northeast Louisiana University, Monroe, LA
Terry L. Eyrich, Merced College, Merced, CA
Marjorie Gardner, University of Maryland, College Park, MD
John W. Gilje, University of Hawaii at Manoa, Honolulu, HA
Julian J. Hamerski, Eastern Illinois University, Charleston, IL
Henry Heikkinen, University of Maryland, College Park, MD
F. Axtell Kramer, St. Louis Community College, St. Louis, MO
Patricia W. Lee, Bakersfield College, Bakersfield, CA
Roland Loeffler, Modesto Junior College, Modesto, CA
John P. Lowe, The Pennsylvania State University, University Park, PA
William H. McMahan, Mississippi State University, Mississippi State, MS
Gordon A. Parker, University of Toledo, Toledo, OH
Nancy C. Reitz, American River College, Sacramento, CA
B. L. Stump, Virginia Commonwealth University, Richmond, VA
Joseph F. Testa, Essex Community College, Baltimore County, MD
James Tortorelli, University of Wisconsin, Madison, WI
John P. Warriner, Central Michigan University, Mount Pleasant, MI
James A. Weiss, Pennsylvania State University, Worthington Scranton Campus, PA

I would also like to thank all those in the college division of McGraw-Hill Book Company who have supported and assisted me over the last three years. This project was initiated and carried through the formative stages by the former chemistry editor Jay Ricci. His knowledge and guidance helped me immeasurably. David Horvath helped to complete the project as chemistry editor. Special thanks go to Kathleen Civetta who spent innumerable hours working to refine and improve the manuscript and get it ready for production. The editing supervisor, Jo Satloff, and the production supervisor, Phil Galea, coordinated the myriad of tasks that are required to change a manuscript into a textbook. Also, I would like to thank Jo Jones, the designer, Marie Dumbra, the copy editor, and Randy Matusow, the photographic researcher.

Finally, I want to express my sincerest appreciation for the help and support that I received from my wife Cynthia and my daughter Natasha. Without their love and understanding this book could not have been written.

Drew H. Wolfe

Many students think that chemistry is studied simply to acquire chemistry facts; this is not correct. Learning the facts of chemistry is a relatively minor outcome of completing a chemistry course, considering that a rather large percentage of the facts learned in an introductory course are soon forgotten. Instead, studying chemistry should help you to develop long-lasting skills and abilities that will be useful to you both in your future studies, and in many aspects of your day-to-day life.

Some of the important outcomes of a careful study of chemistry are the ability to solve scientific problems systematically, the development of greater abstract reasoning skills, the ability to organize and grasp a large quantity of factual information, and a better understanding of the language spoken by chemists. These outcomes are in addition to those objectives that are customarily thought of as the primary chemistry course goals: the development of an appreciation of chemistry and what chemists do; an understanding of the most important laws, principles, concepts, and facts of chemistry; and the attainment of some special skills unique to the field of chemistry.

One of the primary goals of all education is the development of problem-solving ability. A person who can solve problems is much more effective, and is usually a more valuable employee, than one who only can recite facts. Students who develop good problem solving skills can apply these skills to their future studies, and eventually to their professions. Studying chemistry is an excellent way to expand your problem solving skills. This text presents a general problem-solving method that can be used to solve many different types of chemistry problems.

Chemistry involves many abstract concepts because it deals, for the most part, with things we cannot see or experience directly. For this reason, many students think of chemistry as "hard." However, throughout our lives we are faced with many abstract ideas such as the concepts of good, evil, existence, right, wrong, and many others. Studying chemistry can help you to expand your abstract reasoning skills, and these skills will allow you to make more intelligent decisions in all areas of your life.

Often, students confronted with a large body of facts or ideas, such as those encountered in a chemistry course, are overwhelmed by the magnitude of what is to be learned. This task can become more managable if related facts and ideas are grouped together. The material to be learned is thus organized into managable portions. Chemists organize information about substances in this way, identifying regularities which are used to link a large number of substances together. Acquiring methods of systematically organizing facts and ideas is a desirable outcome of studying chemistry.

Your study of chemistry will also give you an understanding of the specific language of the chemist. Chemists and other physical scientists use words carefully, and often use the fewest words possible to express a concept. Chemists generally use words that have exact meanings, and the interchanging of terms is rare. When possible, words are not used at all. Instead, chemists write chemical and mathematical expressions. These are the most concise means for expressing chemical relationships. Knowledge of the language of chemistry can help you to make better decisions about the medicines, chemicals, and foods you put into your body, and will help greatly in any future study of chemistry or the other sciences.

Drew H. Wolfe

· CHAPTER ·

1

Introduction to Chemistry

Study Guidelines

After completing Chapter 1, you should be able to:
1. Develop your own successful strategy for learning chemistry
2. Write a simple definition of chemistry
3. Define and give examples of matter
4. Identify and describe the main divisions of chemistry
5. Explain how the discovery of new materials like bronze and iron changed the development of civilization
6. Discuss the contributions of the Greek philosophers to the development of a better understanding of the nature of matter
7. Identify the scientific contributions of Boyle, Stahl, Priestley, and Lavoisier
8. List contributions by chemists to modern society

In order to be most successful in your study of chemistry, you should have some sort of learning strategy or plan. Consider the following steps as a model for developing your own strategy.

Be positive. **1.** Approach the study of chemistry in a positive way. If you think chemistry is impossible, and that you are going to fail—you will! Instead, try to regard chemistry as an important, central subject that will not only assist you in obtaining your educational goals but will give you a deeper understanding of yourself and your environment.

Read and stay ahead. **2.** Begin your study of each new chemistry topic by quickly reading the material in the textbook before attending the lecture. It is not important to memorize new terms or solve problems at this time. Read the study guidelines as a guide to identify the most significant topics in the chapter.

Reread. **3.** After attending the lecture and obtaining a good set of lecture notes—the highlights of what your professor thinks is important, reread the chapter, paying careful attention to those topics that were discussed in the lecture. During the second reading, answer all review questions and problems that are in each chapter section. Strive to understand as much as possible from one section of the textbook before going on to the next section. Before proceeding, you might want to answer appropriate questions at the end of the chapter.

Solve all problems. Answer all questions. **4.** After you have reread the chapter and answered the review problems, you must complete all assigned questions and problems at the end of each chapter. **The only way to learn chemistry is to do chemistry by solving problems.** If you can solve the problems at the end of the chapter, then, and only then, will you have an adequate knowledge of the material in the chapter. Some helpful hints to consider when working problems are:

(a) Read each problem carefully and determine if you understand what the problem is asking. If not, reread the appropriate section in the chapter or refer to your lecture notes.

(b) Determine what is to be found, and write it on paper.

(c) Extract from the problem all relevant information, and write it below what is to be found.

(d) Use the factor-label method (discussed in Chapter 2), when appropriate, to solve the problem. **Never forget units.** Knowing the correct unit associated with a number is usually more important than knowing the number itself.

(e) Check to see if your answer corresponds to the correct answer given in the Appendix.

(f) If the answer is incorrect, check for arithmetic errors; if none are found, compare the numbers used with those given, and carefully examine the units for possible errors.

(g) Instead of wasting a lot of time on a problem that you are unable to solve at the moment, leave the problem and try others. Come back to the unsolved problem later—in a few days perhaps. After returning to the problem, you may find that you can solve it with ease (brains work in mysterious ways).

(h) Seek help any time that you are unable to solve a problem. Do not be afraid to ask questions; that is an effective way to learn chemistry.

5. To determine how well you understand the material presented in the chapter, carefully go over the study guidelines listed at the beginning of the chapter to see if you have mastered them completely. Alternatively, find the practice test located in the study guide that accompanies the textbook, and take the test without consulting your notes or textbook. After grading the test, try to resolve any problems that you were unable to complete under the test conditions. Return to the chapter section and review those topics that you have not mastered.

The important thing is to not stop questioning.
—Albert Einstein

Pretest yourself before an exam.

The above list is only a suggested learning strategy. Although it is not necessary to follow each step exactly, it is necessary to develop a successful strategy for learning chemistry, whatever that strategy may be. If you do not have a better method, use the one presented until you have developed your own plan.

A few words concerning your lecture notes are in order. Lecture notes are very important because they give an added perspective (your professor's) to the textbook. But, in your zeal to obtain a complete set of notes, do not forget to listen to what the professor is saying! Shortly after each lecture, rewrite your lecture notes. Rewriting your notes will allow you to rethink what was said, correct errors, clear up undecipherable passages, and add any extra thoughts.

Listening is important.

Repetition is another key to learning chemistry efficiently. The amount of study time necessary for learning chemistry is normally greater than that needed to learn many other subjects. A large proportion of your study time should be spent working and reworking problems and transferring thoughts from your brain to paper. Correctly solving a problem once does not guarantee that you have totally learned how to solve problems of this type. Work many similar problems before proceeding to new problems. When learning new ideas, definitions, concepts, or rules, write them on paper over and over again. Chemistry is best learned through repetition!

Finally, to maximize your achievement in chemistry, plan to spend time each day studying chemistry. Cramming at the last minute rarely works in chemistry. Shorter, less intense periods of study spread out over a longer time interval are most effective. Using this technique, you will not be frustrated by trying to accomplish too much at the last moment.

Last-minute cramming is usually not successful in chemistry.

1.2 THE SCOPE OF CHEMISTRY

Chemistry is concerned with matter and its interactions.

Chemistry is the study of the material of the universe—matter—and its interactions. A **chemist** is a person who studies the composition, structure, and properties of matter and seeks to explain the changes that matter undergoes.

Matter is anything that has mass and occupies space.

But what is matter? Put simply, **matter** is anything that has mass and occupies space. The earth, and everything on it, is composed of matter. We use terms such as substances, materials, objects, and bodies when referring to matter. As we shall see throughout this textbook, matter is closely associated with energy,

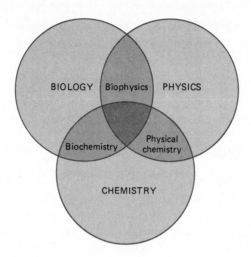

FIGURE 1.1
Science is divided into three
major overlapping disciplines:
chemistry, physics, and biology.

and in some instances it cannot be easily distinguished from energy. At this time
we will defer our discussion of matter and energy to complete our brief look at
the scope of chemistry.

Chemistry overlaps with and is an integral part of the other sciences (see
Figure 1.1). Biology, the study of living systems, applies chemical principles to
help understand cells, the basic units of life. Geology, the study of the earth,
incorporates chemical observations to comprehend the processes that occur on
earth.

Physics and chemistry, both physical sciences, overlap to a large degree
because physics also deals with matter, energy, and the interaction of the two.
The difference between physics and chemistry is that physicists are more interested in the fundamental components and regularities of nature and how they fit
together to yield our universe. Chemists apply the same fundamental laws, often
first elucidated by physicists, to gain a better understanding of the properties and
behavior of matter.

The study of chemistry can be separated into artificial divisions or branches
that categorize the most significant areas of study. These overlapping divisions
include (1) analytical, (2) inorganic, (3) organic, (4) biological, (5) physical, and
(6) geological chemistry. An insight into each division is gained by looking at
what various types of chemists study.

Analytical chemists examine matter to find the identity and amount of its
components. Most chemists use analytical procedures and techniques. Inorganic
chemists study the properties, structures, and reactions of all elementary substances. As we shall see, one of these substances, carbon, has a special set of
properties. Therefore, a division is devoted entirely to this vast topic. It is called
organic chemistry. Organic chemists investigate substances containing carbon
and attempt to produce new carbon compounds. Biological chemists, also called
biochemists, study the compounds that make up living things. Physical chemists

Chemistry is sometimes called
the central science.

Physics is the natural science
that deals with subjects such as
light, heat, motion, electricity,
optics, and the basic structure
of matter.

apply the concept of physics to better understand the behavior of matter. Lastly, geochemists investigate the structures and properties of substances found in the earth's crust, atmosphere, and oceans.

REVIEW PROBLEMS

1.1 Using a dictionary, find the definition of (a) chemistry, (b) physics, (c) biology, and (d) geology.

1.2 List and describe the major divisions of chemistry.

1.3 EARLY HISTORY OF CHEMISTRY

Ancient Peoples

The birth of chemistry coincides with the first time people became aware that they could improve upon what nature offered. By observing lightning, fire, decay, and other natural phenomena, primitive people eventually discovered that the properties of objects could be altered.

After harnessing fire, ancient people solved day-to-day problems more efficiently. Pottery or bricks were formed from baked clay. Through trial and error, ceramics, glazes, and glasses were discovered. Advances in primeval "chemistry" helped people develop the foundation for civilization.

Our word "metal" is thought to be derived from a Greek word meaning "to search for."

During this early period of history, now called the stone age, people saw that their lives were simplified through the development of new materials. The search for new substances to replace their stone tools led to the discovery of metals.

Metals known to the ancients included gold, silver, lead, tin, and copper.

Metals added a new dimension to life. Unlike stone, metals could be hammered and shaped into a multitude of forms. Weapons created from metals remained sharper longer than stone counterparts, and metal weapons could be resharpened. Copper, gold, and tin were the first metals used—a copper cooking pan was found in an Egyptian tomb dating back to 3200 B.C.

Bronze: 70–90% copper and 10–30% tin.

Sometime around 3000 B.C. a startling discovery was made. If copper and tin ores were heated together, a new metal (an alloy) was obtained which was much harder than either copper or tin. The metal, bronze, ushered in a new era—the bronze age.

A thousand years elapsed before the bronze age ended. It was common knowledge that a superior metal, iron, existed. But it was rare, and no method was readily available to extract the metal from its ore. In approximately 1500 B.C. the Hittites, a group of people who lived in Asia Minor, found a means for liberating iron from its ore via a method that is a forerunner of our present-day smelting process. By chance, the Hittites heated iron ore in a charcoal (a form of carbon) smelter, producing an iron-carbon mixture resembling steel. Thus the world was thrust into the iron age about 1000 B.C.

Steel is an alloy of iron containing other metals and <0.5% carbon.

During the iron age, other practical chemical advances were made. In Egypt, chemicals were incorporated into all aspects of life and death. The Egyptians produced alcoholic beverages by fermentation, concocted embalming fluids to preserve the dead, and developed pigments, dyes, and paints that have lasted to modern times.

The Greeks

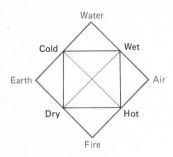

FIGURE 1.2
The ancient Greeks thought that all matter was composed of four "elements:" earth, air, fire, and water.

Aristotle (*Burndy Library*.)

The Greeks, about 600 B.C., were the first people to pose questions about why matter acted as it did. They wanted to know why certain metals, when heated, were changed into new substances; whether there were certain basic forms of matter; and if any metal could be changed into any other metal. Greek scholars sought to determine the composition of the universe.

Thales of Miletus (640–546 B.C.) was one of the first Greek philosophers to conceive the idea of "elements." Thales suggested that elements were the fundamental form of matter. Other Greek thinkers looked to nature and speculated that the entire universe was composed of four elements: earth, air, fire, and water (see Figure 1.2).

Aristotle (384–322 B.C.) accepted and advanced the four-element theory. He suggested that in addition to the four basic elements there could be two pairs of opposite qualities: hot–cold and wet–dry. Aristotle also believed that each element had its own set of properties—earth should fall and fire should rise. With each modification to the four-element theory, a greater number of plausible answers were proposed to questions and problems that had puzzled humankind for centuries.

Today, some people might think of these ideas as humorous or simple-minded solutions to complex problems. Nevertheless, the four-element theory, in a variety of forms, lasted for almost 2000 years. The longevity of the theory is attributed, in part, to its simplicity. Many problems of nature are effortlessly explained in terms of earth (solid), air (gas), fire (energy), and water (liquid).

Greek philosophers also attempted to understand the problem of breaking down matter into smaller parts. If a stone was broken or crushed, could the fragments of stone be further subdivided? If so, was there a point at which the fragment could no longer be divided? Leucippus and his disciple Democritus (about 450 B.C.) proposed the idea of the atom. The word *atom* is derived from the Greek term for "indivisible." Democritus suggested that the smallest particle of matter was an atom—a unit of matter that was indivisible.

Democritus also speculated that atoms in different substances varied with respect to size and shape. This was an incredible proposition, considering that Democritus was a philosopher—one who proposes ideas about nature through logic. He was not a true scientist—one who conducts controlled systematic experiments based on observable facts.

Alchemists

What is accomplished by fire is alchemy, whether in the furnace or kitchen stove.
—Paracelsus

From approximately A.D. 300 to A.D. 1100 the Dark Ages prevailed in Europe and chemical advancement came to a standstill. However, in another part of the world, Arab cultures continued to make significant chemical contributions during the Dark Ages. A small group of Arabs tried to find a way to convert (transmute) cheap, abundant metals to gold. This period in chemical history is now known for these dedicated men who searched for gold—the alchemists (see Figure 1.3).

As part of their quest to change metals to gold, alchemists attempted to find the magic elixir of life, or as it is better known, the philosopher's stone. It was thought that the philosopher's stone could rid one's body of disease and was the key to eternal life. Thousands of people searched in vain for gold and the magic

FIGURE 1.3
An alchemist and his assistant working in a "laboratory." Alchemists developed many different types of glassware, equipment, and procedures in their quest for gold. (*New York Public Library, Picture Collection.*)

Many of the discoveries of alchemists outside of Europe were lost and later rediscovered by European alchemists.

elixir. Even though the alchemists never achieved their goals, they did uncover a vast amount of chemical knowledge. Various contemporary laboratory techniques and instruments are traced to them. Historians believe that the term "chemistry" is derived from the alchemists' term for mixing chemicals.

Alchemy lasted over 2000 years, from the period before the birth of Christ until the eighteenth century. Alchemy died and chemistry emerged because curious people started to ask more probing questions about matter: What explains the behavior of matter? Was all matter composed of only earth, air, fire, and water? Do all substances act in a predictable, regular manner?

Early Scientists

Robert Boyle, an Irishman, saw the shortcomings of alchemy, and decided to apply what is now known as scientific reasoning to the study of chemistry. Boyle (1627–1691) followed the lead of other great scientific investigators of his time: Galileo Galilei (1564–1642), Jan Baptista Van Helmont (1577–1644), Evangelista Torricelli (1608–1647), and Otto von Guericke (1602–1686).

Robert Boyle (*National Portrait Gallery, Smithsonian Institution, Washington, DC.*)

Boyle's exacting studies of gases and their properties supported the idea proposed 1000 years before by certain Greek philosophers—that matter is composed of atoms. In his famous book *The Sceptical Chymist*, published in 1661, Boyle attacked the old idea that matter is composed of only four elements. Instead, Boyle proposed that if a substance thought to be an element is capable of being broken down into simpler forms, then it is not an element. One of the most significant outcomes of Boyle's work is the idea of experimentation—that any propositions regarding matter must be supported by reproducible observations.

Other scientists of the seventeenth century were concerned with the nature of energy, which they called "fire." Their interest was spurred by the invention of

Joseph Priestley (*National Portrait Gallery, Smithsonian Institution, Washington, DC.*) Priestley became friends with Benjamin Franklin and Thomas Jefferson in the United States after leaving England because of religious persecution. He was a Unitarian minister.

the steam engine and by the possibility of developing more efficient engines capable of performing heavy work. Scientists wanted to know why certain substances burn while others do not, how heat is transferred from one object to another, and the nature of the combustion process.

A German physician and chemist, Georg Ernest Stahl (1660–1734), proposed the phlogiston theory to help answer some of these questions about "fire." The term "phlogiston" was derived from the Greek term meaning "to set on fire." Stahl's phlogiston theory described combustible objects as those containing a large quantity of phlogiston. As an object burned, Stahl suggested that phlogiston flowed from the object and the object stopped burning when all of the phlogiston was released. A log burned because it contained phlogiston. The resulting ashes lacked phlogiston; hence, ashes were noncombustible.

Joseph Priestley's (1733–1804) discovery of oxygen as a component of air, and its ability to support combustion (burning), brought about the end of the phlogiston theory. Priestley informed Antoine Laurent Lavoisier (1743–1794) of his discovery of oxygen. Lavoisier immediately repeated Priestley's experiments and found that oxygen was truly formed. But Lavoisier saw something much more important—mathematics could explain the decomposition of matter. When dealing with matter, Lavoisier found that the whole was equal to the sum of its parts. He then conducted a classic experiment: he heated mercury and oxygen to show conclusively that oxygen and not phlogiston supported combustion.

Lavoisier is considered the father of modern chemistry. His textbook *Elementary Treatise on Chemistry*, published in 1789, indicated to the world that chemistry was a science based on theories supported by reproducible experiments. In the *Treatise* he discussed 33 elements known at that time. To his credit, all but two, caloric and light, are considered elements today. Lavoisier's contributions to chemistry are equated to those of Isaac Newton in physics.

Antoine Laurent Lavoisier (*New York Public Library, Picture Collection.*) Lavoisier was arrested and tried for crimes against the people during the French Revolution. He pleaded that he was a scientist, and not an aristocrat. The reply from the rebels was "the Republic has no need for scientists." He was guillotined on May 8, 1794.

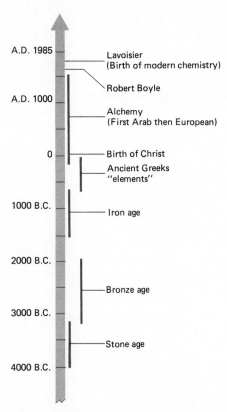

A.D. 1985 — Lavoisier
(Birth of modern chemistry)

Robert Boyle

A.D. 1000 —

Alchemy
(First Arab then European)

0 — Birth of Christ

Ancient Greeks
"elements"

1000 B.C. — Iron age

2000 B.C. —

Bronze age

3000 B.C. —

Stone age

4000 B.C. —

FIGURE 1.4
Time line of chemical history

Lavoisier possessed a rare talent found in few people who have ever lived—he correctly organized and interpreted a large body of facts, yielding a completely new area of human concern, that of modern chemistry. Figure 1.4 illustrates a time line of chemical history.

REVIEW PROBLEMS

1.3 Write a brief explanation of how the following materials changed people's lives: (a) stone during the stone age, (b) bronze during the bronze age, (c) iron during the iron age.

1.4 What contributions did the Greeks make to the advancement of people's understanding of matter?

1.5 Write the names of the scientists responsible for the following: (a) phlogiston theory, (b) discovery of oxygen, (c) discovery that matter was not composed solely of the four elements.

1.6 What were the main goals of the alchemists?

1.4 CHEMISTRY IN TODAY'S WORLD

Chemical advancements contribute greatly to our modern lifestyle. Today, deadly diseases are controlled by using potent antimicrobial agents, like tetracy-

FIGURE 1.5
Today, people rely on medicines and drugs from birth to death. (Chemical Manufacturers Association)

Paul Ehrlich, in 1910, was the first to use antimicrobial agents. Penicillin was discovered by Sir Alexander Fleming in 1929. Sulfa drugs were first identified by Domagk in 1935.

clines, sulfa drugs, and penicillins (see Figure 1.5). We take for granted that almost any illness we contract can be treated with a drug.

We see chemistry in action daily. For example, when we cook meals, wash clothes, wax furniture, or eliminate pests, we are involved with chemistry. The "art" of cooking employs chemical techniques in frying, baking, roasting, and boiling. Cleaning involves selecting the soap, detergent, or cleaning fluid that best removes dirt and stains. Waxes are rubbed or sprayed on surfaces to protect and beautify. Insecticides are used to rid our houses and gardens of insect pests.

Home recreation incorporates many advances in chemical technology. Color television sets contain substances called phosphors that glow various colors when hit by an electron beam. Magnetic recording tape is manufactured by affixing various metallic substances to a plastic tape that is strong and does not stretch or shrink. Advances in the chemical processes used to manufacture transistors and integrated circuits (ICs) have resulted in personal computers, videotape recorders, giant TV screens, electronic games, and superb high-fidelity stereophonic sound systems.

Modern means of transportation also rely on the chemical industry. Passenger jets are able to fly faster and carry more people as a result of strong new metals used in the bodies of aircraft. In addition, new jets have an ever-increasing number of tough plastic parts. Plastic parts are superior in many ways to the heavier, more costly metal parts they replace. The same is true for automobiles. Each year the number of pounds of plastic per car increases. In part, the increased gas mileage of new cars is due to the decreased weight of the cars.

Modern society is becoming increasingly dependent upon the advances of chemistry. To a degree, people expect chemists to rescue them from the clutches of life's problems and annoyances. But, unfortunately, with each new advance we must pay some price. Environmental problems result from by-products of the chemical industry. New health problems are caused by life-saving drugs. Pesticides used to protect our crops accumulate in our bodies, causing serious health problems.

There are no "free lunches" in this world.

REVIEW PROBLEMS

1.7 List three ways in which your life would change if all plastic goods were no longer available.

1.8 Write a short paragraph describing the role of chemists in the modern world.

SUMMARY

Chemistry is the study of matter and its interactions. Chemistry contains six major divisions: (1) analytical chemistry, (2) inorganic chemistry, (3) organic chemistry, (4) physical chemistry, (5) biochemistry, and (6) geochemistry. Each area is concerned with a special aspect of how matter behaves. Within science, chemistry is sometimes considered the central science because all other sciences, to a degree, deal with matter. Chemistry's domain extends into every aspect of life, from birth to death.

Chemistry began in ancient times, when people saw that matter could be changed and used to improve the quality of life. Discoveries at this time were made through trial and error, leading people through the stone, bronze, and iron ages. The Egyptians and Greeks were among the first civilizations to question why matter behaved as it did. Early Greek philosophers proposed explanations for the composition and structure of matter. Around 450 B.C. Democritus suggested that matter was composed of tiny particles called atoms.

Modern chemistry grew out of the pseudoscience called alchemy. Alchemists searched for methods to convert cheap metals to gold. Many discoveries and techniques of alchemists have survived to this day. Boyle was the first scientist to suggest that ideas and thoughts about matter must be supported by reproducible experiments. Lavoisier is credited with being the father of modern chemistry as a result of his pioneering experiments on the properties of matter.

QUESTIONS AND PROBLEMS

1.9 What are 10 different types of matter found in your house?

1.10 List two different areas that each of the following scien-

tists might study: (a) chemists, (b) biologists, (c) physicists, (d) geologists.

1.11 Explain what each of the following chemists investigates:

(a) analytical, (b) inorganic, (c) organic, (d) biological, (e) physical, (f) geological.

1.12 What type of chemist would study: (a) rocks and minerals, (b) synthesis of a new carbon compound, (c) antibiotics, (d) structure of metals, (e) amount of pollution in the air?

1.13 List five household consumer products that help simplify the task of living.

History of Chemistry

1.14 What effect did the harnessing of fire by early peoples have on the development of civilization?

1.15 For prehistoric peoples, what advantages were afforded by the discovery of metals to replace stone objects?

1.16 What chemical advances were made during the (a) bronze age, (b) iron age?

1.17 What are the four elements of matter suggested by the Greeks?

1.18 How did Aristotle modify the four-element theory?

1.19 The four-element theory lasted for thousands of years. List three plausible reasons for its longevity.

1.20 How do philosophers differ from scientists when approaching and solving problems?

1.21 Who is credited with the idea that matter is composed of small particles called atoms?

1.22 List accomplishments of the alchemists that contributed to modern chemistry.

1.23 Propose a reason to explain why Boyle entitled his book *The Sceptical Chymist*.

1.24 Who investigated the following problems: (a) mathematics applied to chemical changes, (b) release of phlogiston when objects burned (c) divisibility of particles, (d) discovery of oxygen, (e) properties of gases?

1.25 Why is Antoine Lavoisier considered the father of modern chemistry?

1.26 If human beings still exist in A.D. 3000, what material age (stone, bronze, iron, . . .) might they label the middle to late twentieth century?

· C H A P T E R ·

2

Problem Solving in Chemistry

Study Guidelines

After completing Chapter 2, you should be able to:
1. List and explain the six steps of the scientific method
2. Define: fact, data, hypothesis, theory, and law
3. Distinguish among facts, theories, and laws
4. Discuss limitations of the scientific method
5. Develop a personal strategy for approaching and solving chemistry problems
6. List and apply all steps required to solve a given chemistry problem
7. Write conversion factors, given equalities
8. Apply the factor-label method in solving problems

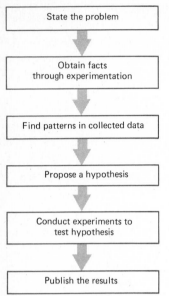

FIGURE 2-1
Typical steps followed when investigating scientific problems.

Today, a systematic, logical procedure called the **scientific method** is used to solve scientific problems. In reality, there is no one method that applies in all cases. The scientific method is a general set of rules that guide scientists in pursuing the solution to a problem. Not all of the rules are followed all of the time, but in general the steps are as follows:

1. State the problem precisely in terms of the most relevant variable factors.
2. Obtain facts regarding the problem through carefully controlled experiments.
3. Organize, analyze, and evaluate the collected facts considering the problem being solved. Try to find a pattern. If no pattern exists, reevaluate the problem; possibly the wrong problem is being investigated, or the problem is not stated clearly.
4. Propose an explanation (hypothesis) to account for the pattern found within the data.
5. Conduct experiments to determine if the proposed hypothesis applies in similar situations.
6. If the experiments support the hypothesis, publish the findings to inform the rest of the world; however, if the experiments do not support the hypothesis, then modify the hypothesis, experimental procedures, or problem and start again.

Let's look at each step of the scientific method individually (see Figure 2.1). The first step is to state the problem precisely in terms of what it is you are trying to solve. An exact statement of the problem should indicate a direction to follow in solving the problem. A large quantity of information (facts, laws, and theories) is collected. This information is organized, evaluated, and analyzed prior to stating the problem.

After defining the problem, an experiment is performed to collect additional facts or data to help solve the problem. A *fact* is an accepted truth—something that everyone accepts as correct. *Data* are facts collected during an experiment; from the data, the problem is solved.

A fact is known to be true.

When collecting data, experimenters are careful to manipulate, or change, one variable quantity at a time. In other words, they do a **controlled experiment.** The variable of interest, called the independent variable, is changed, and the effect on the outcome, or dependent, variable is observed. All other potentially variable quantities are kept constant and not allowed to change. Performing a controlled experiment allows the investigator to find relationships that exist between two variables.

Variable quantities do not have fixed numerical values.

After all of the data are collected, they are analyzed to find regularities and patterns. Experimenters then ask themselves what accounts for the regularity in the data. After answering this question, they can formulate a **hypothesis**—a tentative guess that explains the patterns found in the data. Outcomes are then predicted for new experiments to test the hypothesis. New experiments are con-

Think of an experiment as a method of discovering a cause-and-effect relationship.

ducted and analyzed to determine if the hypothesis is supported or not. If it is supported, the hypothesis is labeled valid; if not, the hypothesis is modified.

A hypothesis that is supported through tight, controlled experiments can be elevated to the level of a **theory**—an experimentally verified hypothesis. Theories are generally regarded as useful or not useful, rather than right or wrong. Theories, once they are published, are subject to scientific scrutiny and criticism. Good theories stand up under the most severe testing; theories that are not so good fall apart under similar conditions. Useful theories have fewer assumptions and exceptions than inferior theories. In chemistry, many phenomena— atoms, molecules, physical states, and chemical changes—are understood through application of generally accepted theories.

> No amount of experimentation can ever prove me right; a single experiment can prove me wrong.
> —Albert Einstein

Theories also explain scientific laws. A **scientific law** is a statement that explains the occurrence of a process or event in the universe given a particular set of conditions. Only a few basic laws exist; each law explains some consistency of behavior in the natural world. For instance, the law of conservation of matter states that matter cannot be created or destroyed under normal conditions. When applying a scientific law, one is confident of its universality—laws have few, if any, exceptions.

> Nature's laws affirm instead of prohibit. If you violate her laws you are your own prosecuting attorney, judge, jury, and hangman.
> —Luther Burbank

Even though the scientific method has passed the test of time and is based on a sound logical foundation, it has limitations and flaws. The scientific method assumes that nature acts in a consistent, rational, and understandable way. But in some cases the laws of nature apply only to a subset of matter—some laws work only for normal-size objects, but fail miserably when explaining the behavior of extremely small or extremely large objects. Certain "why" questions regarding the universe are unanswerable, and no matter what problem-solving method is used, they will remain unanswered. Finally, the scientific method is based on a foundation of logic that has inherent limitations; hence, scientific investigators must be careful not to overstep the limits of logical thought or they may logically fabricate incorrect results with a high level of confidence!

REVIEW PROBLEMS

2.1 List the six basic steps in the scientific method.

2.2 Define each term: (a) fact, (b) data, (c) hypothesis, (d) theory, (e) law, (f) variable quantity, (g) constant quantity.

2.3 What are the limitations of the scientific method?

2.2 STUDENT GUIDE TO PROBLEM SOLVING IN CHEMISTRY

> Haphazard thrusts at solving problems are generally unproductive.

Solving chemistry problems requires the application of many of the same procedures employed by research scientists. Haphazard attempts to solve problems generally are unproductive whether they are undertaken in research laboratories or in the classroom. The "key" that unlocks the mysteries of how to succeed in chemistry is to use a systematic procedure for approaching and solving chemistry problems.

Before plunging into the details of learning how to solve chemistry problems, let us consider a simple problem faced by most people everyday: How does one go from one place to another? Solving this problem is easy: (1) Determine where you are initially located; (2) pinpoint the exact location of your destination; (3) using a map or prior knowledge, plan a pathway to follow; and (4) go! Few people would jump into their cars and drive endlessly until they found their destination. Instead, they follow a logical plan.

If you think about this simple problem, it is evident that the problem cannot be solved if you do not know either your location or your destination. How can you go to an unknown place? How can you plan a trip if you do not know where you are starting from? Even if you know where you are located and where you are headed, it is impossible to plan a trip if you lack appropriate knowledge of what lies between the two points. If the destination is a familiar place, then you can search through your brain for the route that has the smallest number of obstacles such as traffic jams, traffic lights, and stop signs. If you aren't familiar with the destination, you will probably use a map to plan an expedient route. When reading a map, most people normally look for major highways and the shortest path possible, although some trips require the use of back roads and alternate routes. The situation dictates what pathway is needed.

Solving the problem of going from one place to another is similar to solving chemistry problems. Chemistry students must also know (1) where they are, (2) where they are going, and (3) what pathway they intend to follow before they can solve chemistry problems. Too often, in a rush to solve a problem, students forget about following a logical route and, in effect, jump into their cars without a map, to search endlessly for their destinations. Random problem-solving methods usually end in frustration.

Determine what is unknown in the problem you are trying to solve.

Chemistry problem solving begins with a careful reading of the problem to determine what is unknown. What are you trying to find? If the answer is not apparent, reread the problem and list any words or terms which are unfamiliar. Determine the exact meanings of the unfamiliar terms by referring to the textbook. If you still cannot decide what is unknown, go back to the chapter and read appropriate sections, paying attention to the example problems. After you have determined what is unknown, write it down. Now you know where you are headed!

List all given information, and gather other appropriate information.

Continue the chemistry problem-solving process by listing all information (data) that is given or known. Write numbers with their labels or units. A number is meaningless without a label (unless it's unitless). Units, or labels, are words describing the number. For example, 6 dogs, 10 houses, 2 days, and 5 seconds are examples of labeled numbers. In these examples, dogs, houses, days, and seconds describe the numbers to which they are attached. Frequently, besides the data found in the problem, additional data are needed. Missing data are found in tables or charts located in the chapter and appendixes. After completing this second step in chemistry problem solving, you know where you are located!

Develop a logical plan to solve the problem.

Now you need to develop a logical plan to travel from your location to your destination. The logical plan comes from an understanding of the chemistry principles that pertain to the problem. Ask yourself, What is the connection between what is known and what is unknown? One of the most important procedures that is applied in this situation is the **factor-label** method (sometimes called the unit-conversion method).

The factor-label method is an orderly procedure in which known labeled numbers are converted to new numbers with new labels: 14 days is changed to 2 weeks by knowing that 1 week equals 7 days, or 3 dozen donuts is changed to 36 donuts by knowing that 12 donuts are found in a dozen. The specifics of the factor-label method are discussed and illustrated in the next section. At this point, it is important to realize that some systematic procedure like the factor-label method is required to solve chemistry problems. A systematic procedure is the "map" that guides us to the destination.

Perform the indicated math operations.

A chemistry problem is finally solved by performing the indicated mathematical operations. This final step is purely mechanical. All numbers and their associated units in the factor-label setup are added, subtracted, multiplied, or divided to yield the final numerical answer with its units.

You have not completed the problem until you check to see if the answer is correct.

When solving the problems in this book, check your answer with the correct answer located in the Appendix. If you have successfully solved the problem, your answer should agree with the correct answer. If your answer does not agree, check to see if you made an arithmetic error. If no arithmetic error is detected, assume you have made an error in logic. Analyze the reasoning that you originally used, or attempt to work back from the correct answer. A word of caution is in order: Do not totally rely on having the correct answer available. It is not available during quizzes and exams!

In summary, you should usually follow four steps when solving chemistry problems:

1. Carefully read the problem and determine what it asks. On paper, write down exactly what you are trying to find—the unknown quantity.
2. Extract from the problem all information that is given, and obtain any other information that is necessary to solve the problem.
3. Apply appropriate chemistry principles and the factor-label method to convert the known information to what is desired.
4. Perform the indicated mathematical operations and find the answer to the problem. Check to see that the answer is correct. If the answer is incorrect, repeat any steps as required.

Before proceeding, it is extremely important to understand and learn all four chemistry problem-solving steps. **Possibly the most important thing you can learn in a beginning chemistry course is how to approach and solve problems.**

2.3 THE FACTOR-LABEL METHOD

Numerous problems encountered in chemistry are conveniently solved using the factor-label method. In the factor-label method, one or more conversion factors are used to change the given units to the desired units. A conversion factor is an exact relationship between two quantities expressed as a fraction. For instance, 1 dozen objects is defined as 12 objects.

$$1 \text{ dozen objects} = 12 \text{ objects}$$

The correct way to express this equality as a conversion factor is

$$\frac{12 \text{ objects}}{1 \text{ dozen objects}} \quad \text{or} \quad \frac{1 \text{ dozen objects}}{12 \text{ objects}}$$

In a conversion factor, the fraction line is read as "per." So the above expressions are read as "12 objects per 1 dozen objects or 1 dozen objects per 12 objects."

Mathematically, conversion factors are obtained by dividing both sides of the equality by one of the quantities. Dividing both sides of the equality, 1 dozen objects = 12 objects, by 1 dozen objects yields

$$\frac{1 \text{ dozen objects}}{1 \text{ dozen objects}} = \frac{12 \text{ objects}}{1 \text{ dozen objects}}$$

and canceling the "1 dozen objects" results in the conversion factor

$$1 = \frac{12 \text{ objects}}{1 \text{ dozen objects}}$$

To obtain the inverted form of the conversion factor divide the equality by "12 objects,"

$$\frac{1 \text{ dozen objects}}{12 \text{ objects}} = \frac{12 \text{ objects}}{12 \text{ objects}}$$

and then cancel the "12 objects."

$$\frac{1 \text{ dozen objects}}{12 \text{ objects}} = 1$$

$3 \times 1 = 3$
$10 \times 1 = 10$
$85 \times 1 = 85$

Note that a conversion factor always equals 1. Therefore, if a quantity is multiplied by a conversion factor, the value of the quantity is unchanged. Multiplying 1 times any number does not alter the value of the number.

Conversion factors are used to change the units associated with a number to another set of units. This is accomplished by multiplying the conversion factor times the given quantity, so that the given unit cancels and yields the desired unit.

$$\cancel{\text{given unit}} \times \frac{\text{desired unit}}{\cancel{\text{given unit}}} = \text{desired unit}$$

Conversion Factor

For example, how many dozen eggs is 120 eggs?

$$120 \ \cancel{\text{eggs}} \times \frac{1 \text{ dozen eggs}}{12 \ \cancel{\text{eggs}}} = 10 \text{ dozen eggs}$$

In this case, we see that the given unit, eggs, is canceled by the unit eggs in the denominator of the conversion factor, yielding dozen eggs, the unit in the numerator of the conversion factor. Thus, 120 eggs is equal to 10 dozen eggs.

Study the following examples of factor-label conversions.

Example Problem 2.1

Fifty houses containing 10 rooms each are constructed. How many rooms are contained in all 50 houses?

Solution

1. *What is unknown?*   number of rooms in 50 houses

2. *What is known?* $\dfrac{10 \text{ rooms}}{1 \text{ house}}$ and 50 houses

3. Apply the factor-label method.

$$\cancel{\text{houses}} \times \frac{10 \text{ rooms}}{\cancel{\text{house}}} = ? \text{ rooms}$$

Since the problem asks for the number of rooms, and we know the number of houses and the number of rooms per house, we write the conversion factor with rooms in the numerator (top) and houses in the denominator (bottom). When houses are multiplied by rooms per house, the houses cancel, leaving the number of rooms.

4. Perform indicated math operations.

$$50 \ \cancel{\text{houses}} \times \frac{10 \text{ rooms}}{\cancel{\text{house}}} = 500 \text{ rooms}$$

In this example, we have applied the basic principles of problem solving: (1) identifying what is unknown, (2) identifying what is known, (3) applying the factor-label method to find the desired units, and (4) performing the indicated arithmetic operations.

Example Problem 2.2

A soup company packages 30 cans of soup per box. How many boxes are needed to hold 8700 soup cans?

Solution

1. *What is unknown?*   Number of boxes that hold 8700 soup cans

2. *What is known?* $\dfrac{30 \text{ soup cans}}{\text{box}}$ and 8700 soup cans

3. Apply the factor-label method.

$$\text{soup cans} \times \frac{1 \text{ box}}{30 \text{ soup cans}} = ? \text{ boxes}$$

To obtain boxes and cancel soup cans, we invert the conversion factor, placing the desired unit boxes in the numerator and the unit soup cans in the denominator.

4. Perform the indicated math operations.

$$8700 \text{ soup cans} \times \frac{1 \text{ box}}{30 \text{ soup cans}} = 290 \text{ boxes}$$

The factor-label setup shows that 30 is divided into 8700 to obtain the correct answer of 290 boxes.

Example Problems 2.1 and 2.2 are simple, almost trivial, nonchemical examples that illustrate the general procedure for solving chemistry problems. Most chemistry problems are as easy to solve as these, once the techniques of problem solving and the factor-label method are learned.

REVIEW PROBLEMS   Use the factor-label method to solve the following problems.

2.4 An orange crate holds 60 oranges. (a) How many crates are needed to hold 2200 oranges? (b) How many oranges are contained in 290 full crates?

2.5 One brand of gasoline costs $1.75 per gallon. How many gallons are purchased with exactly $30?

In many instances, more than one conversion factor is utilized to solve a problem. Let's consider the problem of converting a given number of years to hours. Most people do not know an exact relationship between years and hours, and such a relationship is generally not found in a table. Consequently, application of our rules regarding problem solving and conversion factors quickly gives the correct answer.

Knowing that we want to find the number of hours, given the number of years, and that there are 365 days per year and 24 hours per day, we write:

$$\text{years} \times \frac{365 \text{ days}}{\text{year}} \times \frac{24 \text{ hours}}{\text{day}} = \text{hours}$$

The number of years is first multiplied by the conversion factor relating days to

FIGURE 2.2
To convert years to hours, multiply years by the conversion factor that relates years to days, and then multiply by the conversion factor that relates days to hours.

years. The years cancel, resulting in the number of days. Days are then converted to hours by multiplying by the conversion factor equating hours and days. In a similar manner the days cancel, yielding the number of hours.

Figure 2.2 illustrates the pathway followed to find the number of hours in a given number of years. Since a direct path is not available, an indirect route, obtaining the number of days, is followed.

Conversion factors are introduced successively until the desired unit is reached. Any number of conversion factors can be used to solve a problem.

The following example problems illustrate time conversions using more than one conversion factor.

Example Problem 2.3

How many seconds elapse in exactly 4 hours?

Solution

1. *What is unknown?* Seconds

2. *What is known?* 4 hours, $\dfrac{60 \text{ minutes}}{\text{hour}}$, and $\dfrac{60 \text{ seconds}}{\text{hour}}$

3. Apply the factor-label method.

$$\text{hours} \times \frac{60 \text{ minutes}}{\text{hour}} \times \frac{60 \text{ seconds}}{\text{minute}} = ? \text{ seconds}$$

Hours are first converted to minutes, using the conversion factor of 60 minutes per hour. In a similar manner, minutes are converted to seconds, given 60 seconds per minute. The first conversion factor cancels hours, yielding minutes, and the next conversion factor cancels minutes, giving seconds, the answer. This is represented diagrammatically in Figure 2.3.

4. Perform the indicated math operations.

FIGURE 2.3
To change hours to seconds, multiply hours by the conversion factor that relates hours to minutes, and then multiply by the conversion factor that relates minutes to seconds.

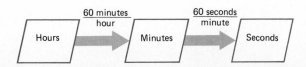

$$4 \text{ hours} \times \frac{60 \text{ minutes}}{\text{hour}} \times \frac{60 \text{ seconds}}{\text{minute}} = 14{,}400 \text{ seconds}$$

Example Problem 2.4

Convert 3600 minutes to days.

Solution

1. *What is unknown?* Number of days

2. *What is known?* 3600 minutes, $\dfrac{60 \text{ minutes}}{\text{hour}}$, and $\dfrac{24 \text{ hours}}{\text{day}}$

3. Apply the factor-label method.

$$\text{minutes} \times \frac{1 \text{ hour}}{60 \text{ minutes}} \times \frac{1 \text{ day}}{24 \text{ hours}} = ? \text{ days}$$

Starting with minutes requires us to invert the conversion factor between minutes and hours so that minutes cancel when we multiply. Hours are converted to days by inverting the conversion factor between hours and days, and multiplying.

4. Perform the indicated math operations.

$$3600 \text{ minutes} \times \frac{1 \text{ hour}}{60 \text{ minutes}} \times \frac{1 \text{ day}}{24 \text{ hours}} = 2.5 \text{ days}$$

The number of minutes, 3600, is first divided by 60 and then divided by 24 to obtain the correct answer of 2.5 days.

A common error when performing the arithmetic operations in this type of conversion is forgetting that both conversion factors require division—both 60 minutes and 24 hours are divided into 3600 minutes.

After completing a conversion using the factor-label method, look at the answer to see if it is reasonable or not. When units small in size are converted to units of larger size, the numerical value decreases. Conversely, when changing large units to smaller units, the magnitude of the number increases. For example, if you change years, a larger time unit, to seconds, a smaller time unit, the numerical value for the number of seconds is significantly larger. As we saw in Example Problem 2.4, 3600 minutes equals 2.5 days. Thus changing a unit of smaller magnitude to one of larger magnitude results in a decrease in the size of the number. Do not let the factor-label method become a mechanical operation. Apply the factor-label method prudently, considering where you are coming from and where you are heading.

REVIEW PROBLEMS
2.6 Convert 6.8 days to (a) hours, (b) minutes, (c) seconds, and (d) years.
2.7 Convert 5 centuries to decades (100 years = 1 century, and 10 years = 1 decade).

SUMMARY

A systematic set of procedures, called the scientific method, is more or less followed in modern scientific research. The heart of the scientific method is collecting facts and data in order to propose a hypothesis, a suggested explanation of the problem. A controlled experiment is then conducted to determine if the hypothesis is valid or not. Experiments provide evidence to support hypotheses.

Studying chemistry is best accomplished by following a systematic plan. Solving problems and answering questions is the most effective way to learn chemistry. You must know what you are looking for and what information is available before mapping out a strategy for solving a problem. Once the problem is solved, check to see if you obtained the correct answer. If not, go back and try again.

Many chemistry problems can be solved using the factor-label method. This is a systematic procedure for converting a number with one unit to a number with another unit using conversion factors, i.e., fractions expressing an equality. Conversion factors are multiplied in such a way that the given unit is canceled and the desired unit is retained.

QUESTIONS AND PROBLEMS

Scientific Method

2.8 Why is it useless to conduct an uncontrolled experiment?
2.9 If you were conducting the following experiments, what variables would you control to produce the most valid results?
(a) Which of two toothpastes is most effective in preventing tooth decay in children?
(b) Which of two automobiles is most fuel-efficient?
(c) What pain reliever is most effective at controlling headaches?
2.10 Classify each of the following as a fact, a theory, or a law:

(a) Grass is green.
(b) Whatever goes up must come down.
(c) Human beings evolved from apes.
(d) Earth revolves around the sun.
(e) Matter is composed of atoms.
(f) Diamonds are forever.
(g) The harder an object is pushed, the faster it accelerates.
2.11 Why can't an experiment be conducted to measure "love"?
2.12 Outline a complete experiment (all six steps) that might be used to determine the effectiveness of a newly synthesized antimicrobial agent, i.e., a drug that fights disease.

Factor-Label Method

2.13 A small economy car, when cruising on a highway, can travel 47 miles on 1 gallon of gasoline. How many miles can it travel on 10 gallons?

2.14 Perform the following conversions:
(a) 7 minutes = ? seconds
(b) 155 days = ? years
(c) 6 decades = ? centuries
(d) 8600 apples = ? dozen
(e) 140 hours = ? days
(f) 775 pencils = ? gross (1 gross = 144)

2.15 How many hours would it take to travel 1000 miles if a constant speed of 55 miles per hour were maintained?

2.16 How many seconds elapse during 4.5 hours?

2.17 What is the age of a 21-year-old person in days?

2.18 Perform the following time conversions:
(a) 570 hours = ? years
(b) 2500 days = ? decades
(c) 9 centuries = ? days
(d) 15,000 seconds = ? days
(e) 100,000 days = ? millennia (1 millennium = 1000 years)

2.19 Excellent long-distance runners can run 25 miles in 2 hours.
(a) Assuming they run at the same pace throughout the race, how many miles will they travel in 0.5 hour?
(b) How many feet are traveled? (5280 ft = 1 mile.)

2.20 If nine bananas weigh 2.7 pounds, and bananas cost 30 cents per pound, what is the cost of four bananas?

2.21 On a trip from New York to Miami, 1317 miles, an automobile consumed exactly 46 gallons of gasoline costing $1.55 per gallon.
(a) What was the average mileage per gallon?
(b) How many gallons of gasoline were consumed per mile?
(c) What was the cost of gasoline per 100 miles?
(d) What was the total cost of gasoline for the trip?

2.22 A box containing 100 paper clips costs $0.89; each paper clip weighs 0.0015 pound.
(a) What is the cost of 750 paper clips?
(b) What is the cost of a single paper clip?
(c) What is the cost of paper clips per pound?
(d) How many boxes must be emptied to produce 3 pounds of paper clips?
(e) How many pounds of paper clips are purchased with $25?

2.23 Convert 490,000 seconds to centuries.

2.24 The approximate life span of an American is 75 years. Convert the life span to seconds.

2.25 A furlong is a unit of length equal to 660 feet. If a rod, another unit of length, is 198 inches, determine the number of rods per furlong.

2.26 Light travels at 186,000 miles per second. Light from the sun takes about 0.23 days to travel to the planet Pluto. What is the distance from the sun to Pluto?

· C H A P T E R ·

Chemical Measurement

Study Guidelines

After completing Chapter 3, you should be able to:
1. List the seven base SI units and explain what they measure
2. List common prefixes used in SI and their meaning
3. Explain the difference between base and derived SI units
4. Convert the base SI units of length and mass to common multiples
5. Convert the base SI units of length and mass to non-SI unit equivalents
6. Distinguish between mass and weight
7. Define temperature and briefly explain how it is different from heat
8. Convert Kelvin to Celsius and vice versa
9. Convert both Kelvin and Celsius to Fahrenheit
10. Convert an SI unit of volume to multiples and non-SI units
11. Calculate the density of a substance given the mass and volume
12. Calculate either the mass or volume of an object given its density
13. List other commonly used derived SI units
14. Distinguish between precision and accuracy of measurements
15. Distinguish between systematic and random errors in measurement
16. Define significant figures
17. Determine whether or not a zero digit in a measured quantity is significant
18. Add, subtract, multiply, and divide measured amounts and express the answer to the correct number of significant figures

The measurement system utilized by scientists and by people in virtually all countries in the world is the metric system. It was first developed by the French Academy of Sciences in 1790 as a response to a request by the French National Assembly for a simple and organized system of weights and measures.

The metric system has evolved since 1790, but its basic structure has remained the same. The metric system is a decimal system, one that requires only the movement of the decimal point to change larger to smaller units or vice versa.

Today, scientists use the International System (SI) of units.

A conference held in 1960 made significant modifications to the metric system, enough that the name "metric system" was dropped. The revised system is called Système International d'Unités or **International System of Units (SI).** Even though we commonly speak of the metric system, in actuality we are usually referring to the International System.

Throughout this book, SI units are used along with some non-SI units that are still employed by the scientific community.

The seven base units of the International System are:

1. Meter (m)—unit of length
2. Kilogram (kg)—unit of mass
3. Second (s)—unit of time
4. Kelvin (K)—unit of temperature
5. Mole (mol)—unit of amount of substance
6. Ampere (A)—unit of electric current
7. Candela (cd)—unit of light intensity

All other SI units are created from combinations of these base units. Examples of derived SI units include m^2 (area), m^3 (volume), kg/m^3 (density), and m/s (velocity).

Prefixes are placed in front of the base SI unit to change the size of the unit. We will consider the meter (m) as an example; the meter is a unit of length approximately equal to 39.4 inches (in.). While it is convenient to measure the dimensions of a room in meters, it is unwieldy to express distances between cities in meters. To magnify the meter 1000 times, it is only necessary to add the prefix *kilo*, meaning 1000 times, in front of *meter*, producing the unit *kilometer* (km).

1 mi = 1.609 km

$$1 \text{ kilometer} = 1000 \text{ meters}$$
$$1 \text{ km} = 1000 \text{ m}$$

For measuring small distances, the meter is an awkward unit. Thus, a prefix is added to specify a unit smaller than the meter. The two most common prefixes are *centi*, $\frac{1}{100}$, and *milli*, $\frac{1}{1000}$.

1 in. = 2.54 cm

$$\frac{1}{100} \text{ meter} = 1 \text{ centimeter}$$

TABLE 3.1   SI PREFIXES

Prefix	Symbol	Meaning
exa	E	10^{18}
peta	P	10^{15}
tera	T	10^{12}
giga	G	10^{9}
mega	M	10^{6}
kilo	k	10^{3}
hecto	h	10^{2}
deca	da	10^{1}
—	—	10^{0} (=1)
deci	d	10^{-1}
centi	c	10^{-2}
milli	m	10^{-3}
micro	μ	10^{-6}
nano	n	10^{-9}
pico	p	10^{-12}
femto	f	10^{-15}
atto	a	10^{-18}

$$100 \times \frac{1}{100} \text{ meter} = 100 \times 1 \text{ centimeter}$$

$$1 \text{ meter} = 100 \text{ centimeters}$$

$$\frac{1}{1000} \text{ meter} = 1 \text{ millimeter}$$

$$1000 \times \frac{1}{1000} \text{ meter} = 1000 \times 1 \text{ millimeter}$$

$$1 \text{ meter} = 1000 \text{ millimeters}$$

Common prefixes encountered in chemistry are:

$$\text{mega (M)} = 1,000,000 = 10^6$$
$$\text{kilo (k)} = 1000 = 10^3$$
$$\text{deci (d)} = 0.1 = 10^{-1}$$
$$\text{centi (c)} = 0.01 = 10^{-2}$$
$$\text{milli (m)} = 0.001 = 10^{-3}$$
$$\text{micro } (\mu) = 0.000001 = 10^{-6}$$

Since these prefixes are constantly used in chemistry, you should learn them immediately. Table 3.1 provides a complete listing of SI prefixes. Exponential notation is used in Table 3.1. If you are not familiar with or need a review of exponential and scientific notation, refer to the Review of Essential Math Skills, Appendix 1 in the back of the book.

REVIEW PROBLEMS
3.1 What advantages does SI have over non-SI systems of measurement?
3.2 List the seven base SI units and what they measure.
3.3 Distinguish between base and derived SI units.
3.4 List six common prefixes used in SI and their meaning.

3.2 BASE SI UNITS: METER, KILOGRAM, AND KELVIN

Length

Many countries spell "meter" differently—"metre."

The base SI unit of length is the **meter.** Before 1960, the meter was defined as the distance between two marks on a platinum-iridium rod located in France. These etched marks were separated by one ten-millionth the distance from the equator to the north pole on a line going through Sèvres, France. After 1960, the meter was redefined in terms of a more universal standard. Today, a meter is 1,650,763.73 wavelengths of a certain light emitted by the gas ^{86}Kr (krypton, an element). Just why the meter was defined in terms of light emitted from krypton gas is beyond the scope of this discussion, but note that this standard is more reproducible than the old standard. Light given off by krypton is not affected by temperature or pressure changes, and its wavelength can be measured anywhere with the proper equipment.

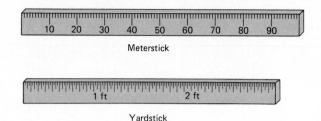

A meter is equal to 39.3701 in., slightly longer (9%) than a yard (36 in.) (see Figure 3.1). While the meter is a convenient day-to-day measurement, it is generally too large to use for length measurements in chemistry because chemists deal with infinitesimal (minute) objects like atoms and molecules. Therefore, various prefixes are added to the meter to give more useful units of length.

1 m = 39.37 in.

A centimeter (cm) is 0.01 m, and a millimeter (mm) is 0.001 m. Even these units are gigantic when measuring dimensions of objects as small as atoms. A commonly used prefix for the meter when considering atomic dimensions is *nano*, or 10^{-9}. A nanometer (nm) is equal to 0.000000001 m. To give an idea of how small a nanometer is, we can apply the factor-label method and scientific notation to determine what fraction of an inch is a nanometer. We know that a nanometer is 10^{-9} m and that there are 39.37 in. per meter:

The measurement system used in the United States is called the U. S. Customary System of units (USCS). It was developed from the old English system of measurement.

$$nm = 1 \, \cancel{nm} \times \frac{10^{-9} \, \cancel{m}}{\cancel{nm}} \times \frac{39.37 \, in.}{\cancel{m}} = 3.937 \times 10^{-8} \, in.$$

Some multiples of the meter are shown in Figure 3.2.

Mass

The **kilogram** (kg) is the SI unit of mass. **Mass** is the quantity of matter contained in an object. The terms "mass" and "weight" are frequently confused. **Weight** is the gravitational force of attraction acting on a mass. Mass is constant no matter where it's located, whereas weight is variable depending on where the object is situated in the universe.

Mass is the quantity of matter in an object.

If a man travels to the moon, his mass remains the same but his weight is approximately one-sixth of his weight on earth because the gravitational attraction on the moon is one-sixth that on earth. Even on earth there is a nonuniform gravitational field: closer to the poles, the gravitational force is greater than near

Weight depends on gravity. Mass does not.

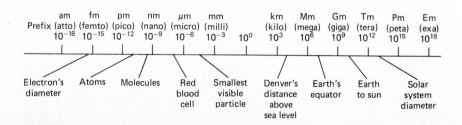

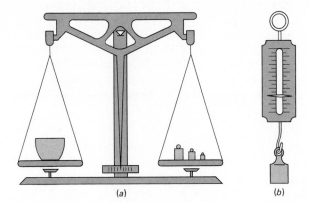

FIGURE 3.3
(a) A double-pan balance is an instrument for measuring the mass of an object by comparing with known masses. (b) A spring scale is an instrument for measuring the weight of an object.

the equator. Hence, a person's weight changes as she or he travels from the equator to the north pole. The person's mass, however, remains the same.

Mass and weight measurements require different instruments. Mass is measured with a balance; weight is measured with a scale. Since a balance operates in a gravitational field, there must be a way to cancel the effects of gravity. Consider a double-pan balance (see Figure 3.3): An unknown mass is placed on one pan of the balance, and known masses are successively added to the other pan until the pointer returns to its original setting. Gravity is canceled because it pulls equally on both sides of the balance. Thus, the unknown mass is equal to the sum of the known masses.

In contrast, a scale—the device for measuring weight—is sensitive to gravitational changes. Consider a common spring scale (see Figure 3.3). A spring scale has a hook attached to a spring, and the spring has a pointer affixed that indicates the weight. When an object is hung on the hook, the spring expands with the pointer, indicating the degree that the spring has expanded.

When we discuss the measurement of mass or weight, we usually use the verb "to weigh."

The standard kilogram, a block of platinum-iridium alloy from which all mass measurements are compared, is located in France. A kilogram is equal to 2.2046 pounds (lb). The kilogram is too massive for routine chemical laboratory measurements; consequently, smaller multiples of the kilogram are used. Grams (g), milligrams (mg), and micrograms (μg) are the most common units of mass in chemistry labs. Inexpensive triple beam balances measure masses within 0.01 g, or 1 centigram (cg), and expensive analytical balances easily determine masses to the nearest 0.0001 g, or $\frac{1}{10}$ mg (100 μg) (see Figure 3.4).

1 kg = 2.2046 lb

FIGURE 3.4
Triple beam balances (left) and electronic balances (right) are most commonly used in chemistry laboratories. Most triple beam balances measure the mass of objects to within ±0.01 g, and electronic balances measure the mass of objects to within ±0.001 or ±0.0001 g. (Fisher Scientific Company)

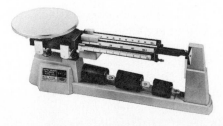

Even with expensive balances, the small mass of the individual atoms and molecules, which make up matter, is undetectable. A staggering quantity of atoms is needed to give a measurable mass on a sensitive analytical balance. For instance, it can be calculated that 3.0×10^{17} gold atoms (rather large atoms, as atoms go) is the smallest number of gold atoms detectable on a balance that has a sensitivity of 0.1 mg.

Temperature

Temperature is the relative hotness of an object. Temperature and heat are frequently confused. Heat and temperature are related, but they are totally separate measurements.

We will discuss heat energy in more detail in Chapter 4. For now, recognize that **heat** is the amount of energy associated with matter and depends on mass. Temperature is independent of the mass of an object. Temperature is the property of matter which determines heat flow. Everyone knows that if a hot object is touched, you are burned. Why? Heat flows from the hot object to you, a colder (less hot) object (see Figure 3.5).

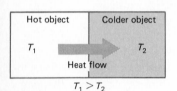

$T_1 > T_2$

FIGURE 3.5
Heat always flows spontaneously from hotter to colder objects.

Celsius and **Kelvin** temperature scales are utilized by chemists. In 1742, Anders Celsius, a Swedish astronomer, developed the Celsius scale. He designed his scale so that the freezing point of water was assigned 0 degrees, and the normal boiling point of water was given the value of 100 degrees — 100 divisions separate the freezing and boiling points of water. On the Celsius scale room temperature is 25°C and average human body temperature is approximately 37°C.

The SI unit of temperature was named after Lord Kelvin, the title given to William Thomson (1824–1907), a brilliant British physicist. The Kelvin temperature scale contains units called kelvins that are of the same magnitude as Celsius degrees but are displaced by 273 degrees (actually 273.15 degrees). Thus the zero point on the Kelvin scale, called absolute zero, is equivalent to −273.15°C. There are no negative values in the Kelvin scale.

Lord Kelvin entered Glasgow University when he was 11 years old. He published his first paper in his teens. His scientific work ranks with the greatest. He is buried in Westminster Abbey next to Sir Isaac Newton.

To change Celsius degrees to kelvins it is only necessary to add 273 to the Celsius temperature:

$$K = °C + 273$$

On the Kelvin scale the freezing and boiling point of water are 273 K and 373 K, respectively.

$$\text{Freezing point}_{H_2O}\ K = 0°C + 273 = 273\ K$$
$$\text{Boiling point}_{H_2O}\ K = 100°C + 273 = 373\ K$$

William Thomson, Lord Kelvin
(*Burndy Library*)

While °C and K are universally used in science, in the United States the Fahrenheit scale predominates among nonscientists. The Fahrenheit temperature scale was proposed by a German scientist, Gabriel Fahrenheit, in 1714. He somewhat arbitrarily decided on 180 divisions between the freezing and boiling points of water. He placed the zero point at the coldest temperature that he could attain in his laboratory using salt solutions (salts lower the freezing point of water). The resulting scale fixed the freezing point of pure water at 32°F and the boiling point at 212°F (exactly 180 degrees above the freezing point).

FIGURE 3.6
Comparison of the Celsius,
Kelvin, and Fahrenheit
temperature scales.

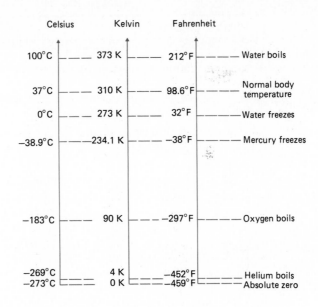

FIGURE 3.7
Mercury thermometers are the
most common temperature-
measuring devices in chemistry
laboratories.

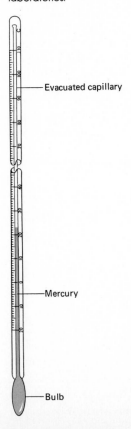

Figure 3.6 shows the relative magnitudes of the freezing and boiling points of common substances on the three temperature scales: K, °C, and °F.

Thermometers are used for most temperature measurements. While many different types of thermometers exist, the mercury thermometer is the laboratory "work horse." A mercury thermometer contains a glass bulb which houses the mercury, a liquid metal. Attached to the bulb is a thin glass tube from which the air has been removed (see Figure 3.7). As the bulb is heated, the mercury expands into the tube. One reason mercury is frequently placed in thermometers is that it expands nearly uniformly when heated.

Sometimes it is necessary to convert temperatures from Fahrenheit to Celsius and then to Kelvin or vice versa. Temperature conversions are efficiently made by using algebraic conversion formulas. To convert a Fahrenheit temperature to a Celsius temperature, the following formula is used:

$$°C = \tfrac{5}{9}(°F - 32)$$

Thus, given a Fahrenheit temperature we would first subtract 32 from it and then multiply by $\tfrac{5}{9}$ to obtain the equivalent Celsius temperature.

For example, when 212°F is substituted into the equation, 100°C is obtained; and when 32°F is substituted, 0°C results.

$$°C = \tfrac{5}{9}(212°F - 32) = 100°C$$
$$°C = \tfrac{5}{9}(32°F - 32) = 0°C$$

Example Problem 3.1 is an illustration of a temperature conversion.

Temp. °C	
Star interior	40,000,000
Mercury lamp	14,000
Tungsten, mp	3,410
Hottest climate (Tripoli)	58
Coldest climate (Alaska)	−63
Liquid air	−195
Absolute zero	−273

Example Problem 3.1

What is the corresponding Celsius temperature for −100°F?

Solution

The answer to the problem is found by substituting the Fahrenheit temperature into the equation $°C = \frac{5}{9}(°F - 32)$.

$$°C = \frac{5}{9}(-100°F - 32)$$
$$°C = \frac{5}{9}(-132)$$
$$°C = -73.3°C$$

We find that −100°F is −73.3°C.

To change a Celsius temperature to Fahrenheit, rearrange the equation and solve it for °F. First multiply both sides by $\frac{9}{5}$ to eliminate $\frac{5}{9}$ from the right side of the equation:

$$\frac{9}{5}°C = \frac{5}{9}(°F - 32)\frac{9}{5}$$

Add 32 to both sides to isolate °F on the right side:

$$32 + \frac{9}{5}°C = °F - 32 + 32$$

This yields

$$(\frac{9}{5}°C) + 32 = °F \quad \text{or} \quad °F = \frac{9}{5}°C + 32$$

Example Problem 3.2 shows how this equation is used.

Example Problem 3.2

Change 200°C to °F.

Solution

Again, we just substitute into the newly derived equation.

$$°F = (\frac{9}{5} \times 200°C) + 32$$
$$°F = 360 + 32 = 392°F$$

Hence, 200°C is 392°F.

The three temperature conversion equations:

$$K = °C + 273$$
$$°C = \tfrac{5}{9}(°F - 32)$$
$$°F = \tfrac{9}{5}°C + 32$$

are used to convert any given temperature on one scale to a temperature on any other scale.

Example Problem 3.3

Convert 300 K to °F.

Solution

First convert the K temperature to °C, and then convert to °F.

1. Change 300 K to °C.

$$K = °C + 273 \quad \text{or} \quad °C = K - 273$$
$$°C = 300 - 273 = 27°C$$

2. Change 27°C to °F.

$$°F = \tfrac{9}{5}°C + 32$$
$$°F = \tfrac{9}{5}(27°C) + 32 = 81°F$$

The correct answer, 81°F, is equivalent to 300 K.

REVIEW PROBLEMS

3.5 What are the base SI units for length, mass, and temperature?

3.6 What standards are used for the base units of length and mass?

3.7 Use the factor-label method and scientific notation to perform the following conversions:
(a) 350 m = ? mm,
(b) 17 yards (yd) = ? cm,
(c) 1.05 g = ? lb,
(d) 7.6 kg = ? ng

3.8 What is the difference between mass and weight?

3.9 Perform the following temperature conversions.
(a) 125°C = ? K,
(b) −224°C = ? K,
(c) 51°F = ? °C,
(d) −85.6°C = ? °F

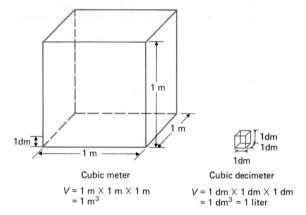

FIGURE 3.8
One cubic meter $(1 m^3)$ is a very large unit of volume; it contains $1000 dm^3$. One cubic decimeter is equal to the volume unit called the liter. Both the cubic meter and the cubic decimeter are SI units of volume. The liter is not an SI unit of volume.

Cubic meter

$V = 1 m \times 1 m \times 1 m$
$= 1 m^3$

Cubic decimeter

$V = 1 dm \times 1 dm \times 1 dm$
$= 1 dm^3 = 1$ liter

3.3 DERIVED SI UNITS OF MEASUREMENT: VOLUME AND DENSITY

Derived SI units are combinations of the base SI units. While many derived units exist, only volume and density units are discussed here because they are very important in measuring matter.

Volume

Volume is the amount of space occupied by matter. All matter takes up some space and therefore has volume. The amount of space is measured in cubic units, or a length unit to the third power. For instance, the volume of a rectangular object is computed by multiplying the object's length times its width times its height, as shown in Figure 3.8.

$$V = lwh$$

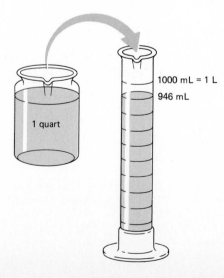

1000 mL = 1 L
946 mL

1 quart

FIGURE 3.9
One quart is 0.946 L, and 1 liter is 1.06 qt.

The base unit of length is the meter; thus, one SI unit of volume is m × m × m or m^3 (cubic meter). One cubic meter is approximately 264 gallons (gal), and for most measurements in chemistry the cubic meter is much too large.

Instead of working with such an enormous unit as the m^3, chemists normally use a fraction of this unit, the cubic decimeter, dm^3. *Deci* is the prefix meaning $\frac{1}{10}$, so a decimeter (dm) is 0.1 m (3.937 in.). The cubic decimeter is exactly the same volume as the non-SI unit, the liter (L). Both the cubic decimeter and the liter are $0.001\ m^3$. Cubic decimeters are less frequently encountered because of the popularity of the liter. Therefore, the liter is utilized throughout this textbook. Just remember that the liter and cubic decimeter are different units for the same volume, $0.001\ m^3$.

A liter is approximately the same volume as a quart (qt) (see Fig. 3.9). One liter is 1.057 qt, or slightly larger than the quart.

$1\ dm^3$ is the same volume as 1 L.

1 L = 1.057 qt
1 qt = 946 mL

Example Problem 3.4

Some gasoline stations now sell gasoline by the liter. How many liters of gasoline must be purchased to fill a 15.0-gal gas tank?

Solution

1. *What is unknown?* Liters

2. *What is known?* 15.0 gal, $\dfrac{1.057\ qt}{1\ L}$, $\dfrac{4\ qt}{1\ gal}$

3. Apply the factor-label method.

$$15.0\ \text{gal} \times \frac{4\ \text{qt}}{1\ \text{gal}} \times \frac{1\ L}{1.057\ \text{qt}} = ?\ L$$

4. Perform the indicated math operations.

$$15.0\ \text{gal} \times \frac{4\ \text{qt}}{1\ \text{gal}} \times \frac{1\ L}{1.057\ \text{qt}} = 57.7\ L$$

A 15.0-gal tank holds 57.7 L gasoline.

INSTRUMENTS THAT MEASURE
VOLUME
Pipets
Burets
Graduated cylinders
Graduated beakers
Volumetric flasks
Syringes

Even the liter is too large for most laboratory volume measurements so the milliliter, mL, or the cubic centimeter, cm^3, is regularly used in chemistry. Milliliters and cubic centimeters are measurements for the same volume, 0.001 L.

$$1\ mL = 1\ cm^3 = 0.001\ L$$

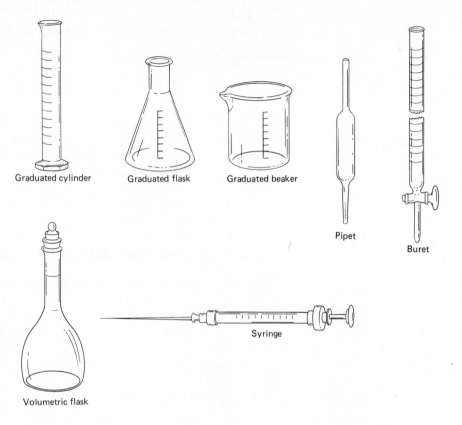

FIGURE 3.10
If only a rough estimate of volume is required, a graduated beaker or graduated flask is used. For more precise volume measurements, graduate cylinders, volumetric flasks, burets, pipets, and syringes are used.

In some instances, very small volumes are required; in such cases the microliter (μL) is used. Example Problem 3.5 shows how to determine the number of microliters in exactly 1 m^3.

Example Problem 3.5

How many microliters are contained in exactly 1 m^3?

Solution

1. *What is unknown?*   Microliters
2. *What is known?*   1 m^3, 1 dm^3 = 0.001 m^3, 1 dm^3 = 1 L, 1 μL = 1 $\times$ 10^{-6} L
3. Apply the factor-label method.

$$1 \, \cancel{m^3} \times \frac{1 \, \cancel{dm^3}}{0.001 \, m^3} \times \frac{1 \, \cancel{L}}{1 \, \cancel{dm^3}} \times \frac{1 \, \mu L}{1 \times 10^{-6} \, \cancel{L}} = ? \, \mu L$$

4. Perform the indicated math operations.

$$1 \, m^3 \times \frac{1 \, dm^3}{0.001 \, m^3} \times \frac{1 \, L}{1 \, dm^3} \times \frac{1 \, \mu L}{1 \times 10^{-6} \, L} = 1 \times 10^9 \, \mu L$$

Thus $1 \, m^3$ contains $1 \times 10^9 \, \mu L$.

Density Density is an important property of matter. A substance's **density** describes how much mass is contained in a unit volume.

$$\text{Density} = \frac{\text{mass}}{\text{volume}}$$

Substances with larger densities have more mass packed into an equivalent volume than do lower-density substances. For example, 1.0 mL of gold has a mass of 19 g; in contrast, 1.0 mL of water has a mass of 1.0 g. Comparing equal masses of gold and water, the gold would occupy $\frac{1}{19}$ the volume of water.

Densities of substances are measured by finding the mass of a known volume of the substance. Mass is measured in grams, and volume is measured in liters and milliliters. Thus the units for density are grams per milliliter (g/mL) and grams per liter (g/L).

It is common practice to express the densities of solids in the units g/cm³. In this textbook, densities of solids are expressed in g/mL.

$$d = \frac{g}{mL} \qquad \text{or} \qquad d = \frac{g}{L}$$

Density is a conversion factor that relates mass to volume.

For more compact forms of matter, liquids and solids, densities are expressed as grams per milliliter. Less dense forms of matter (gases) have densities expressed in grams per liter.

Each substance has a characteristic density; for instance, water at 4°C has a density of 1.00 g/mL. This means that 1.00 g of water at 4°C occupies a volume of 1.00 mL. See Table 3.2 for a listing of common materials and their densities.

TABLE 3.2   DENSITIES OF SELECTED MATERIALS

Material	Density g/mL (25°C)	Material	Density g/L (1 atm, 0°C)
Liquids and solids		Gases	
Grain alcohol	0.785	Air	1.29
Battery acid		Oxygen	1.43
(38% sulfuric acid)	1.285	Nitrogen	1.251
Gold	19.3	Carbon dioxide	1.977
Aluminum	2.70	Xenon	5.85
Mercury	13.6	Chlorine	3.21
Uranium	18.9	Krypton	3.71
Potassium	0.86	Hydrogen	0.090
Benzene	0.860		
Osmium	22.57		
Lead	11.4		
Iron	7.86		

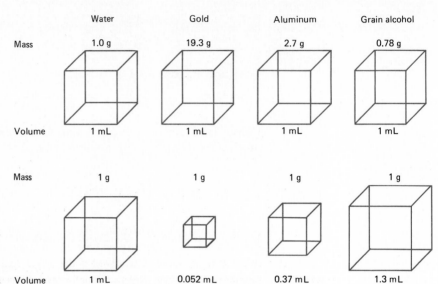

	Water	Gold	Aluminum	Grain alcohol
Mass	1.0 g	19.3 g	2.7 g	0.78 g
Volume	1 mL	1 mL	1 mL	1 mL

Mass	1 g	1 g	1 g	1 g
Volume	1 mL	0.052 mL	0.37 mL	1.3 mL

FIGURE 3.11
Considering equal volumes of matter, substances with higher densities have greater masses than those with smaller densities. Considering equal masses of matter, substances with higher densities have smaller volumes than those with smaller densities.

Density varies with temperature, because most substances either expand or contract when heated.

Altering the temperature of matter results in a volume change. The density of water is 1.000 g/mL at 4°C only. At room temperature, 25°C or 298 K, water's density is 0.997 g/mL. At 80°C, the density of water drops to 0.971 g/mL.

Densities are used to compare masses of substances which have the same volume. It is incorrect to say "gold is a heavier metal than iron." Stated properly, one would say "gold is more dense than iron." Table 3.2 shows that the density of gold is 19.3 g/mL, which is greater than iron's density, 7.86 g/mL. Relative volumes of substances with the same mass are shown in Figure 3.11.

In the laboratory, densities are measured by finding the mass of a known volume. Liquid densities are easily obtained. A volumetric instrument (graduated cylinder or volumetric flask) is weighed, then filled to a specific volume level and reweighed. If we subtract the mass of the container from the mass of the container plus liquid, we get the liquid's mass (see Figure 3.12).

Mass of liquid = (mass of container + liquid) − mass of container

FIGURE 3.12
To obtain the mass of a liquid, measure the mass of the container that holds the liquid, and then measure the mass of the container plus the liquid. The mass of the liquid is the difference between these two masses.

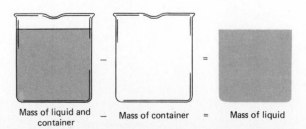

Mass of liquid and container − Mass of container = Mass of liquid

To find the liquid's density, the volume is divided into the mass. A typical liquid density determination is illustrated in Example Problem 3.6.

Example Problem 3.6

An empty 10-mL graduated cylinder has a mass of 53.24 g. When 10.0 mL of an unknown liquid is poured into the cylinder, the total mass is 63.12 g. Calculate the density of the unknown liquid.

Solution

1. *What is unknown?* Density of the unknown liquid, $\dfrac{g}{mL}$

2. *What is known?* Liquid volume = 10.0 mL, mass of empty graduated cylinder = 53.24 g, and mass of liquid plus graduated cylinder = 63.12 g

$$d_{liq} = \frac{\text{mass of liquid}}{\text{volume of liquid}}$$

3. Finding the density of a substance requires that the mass of a known volume is measured. The mass of the unknown liquid is determined by subtracting the mass of the graduated cylinder from the mass of the graduated cylinder plus liquid.

$$\text{Mass of liquid} = 63.12 \text{ g} - 53.24 \text{ g} = 9.88 \text{ g}$$

4. Divide the mass of the liquid by its volume.

$$d = \frac{\text{mass}}{\text{volume}} = \frac{9.88 \text{ g}}{10.0 \text{ mL}}$$

$$d = 0.988 \ \frac{g}{mL}$$

The density of the unknown liquid is 0.988 g/mL.

Densities are helpful quantities when working in the laboratory. Sometimes it is inconvenient to measure the mass of a substance directly. If the volume and density of the substance are known, the mass can be calculated. Think of the density of an object as another conversion factor, mass/volume. Example Problem 3.7 shows how density is applied as a conversion factor.

Example Problem 3.7

A student pours 25.0 mL of decane into a graduated cylinder. If the density of decane is 0.730 g/mL, what is the mass of the decane in the cylinder?

Solution

1. *What is unknown?*   Mass of decane, g decane
2. *What is known?*   Volume of decane = 25.0 mL, density of decane = 0.730 g/mL
3. Apply the factor-label method.

$$25.0 \; \text{mL} \times \frac{0.730 \; \text{g}}{\text{mL}} = ? \; \text{g decane}$$

4. Perform the indicated math operations.

$$25.0 \; \text{mL} \times \frac{0.730 \; \text{g}}{\text{mL}} = 18.3 \; \text{g decane}$$

Example Problem 3.7 is another application of the factor-label method— when working with density, treat it as any other conversion factor.

Other Derived SI Units A number of other derived units of measurement are utilized in chemistry. They include such measurements as (1) energy, (2) force, (3) power, (4) quantity of electric charge, and (5) pressure. We shall investigate some of these in later chapters.

REVIEW PROBLEMS **3.10** (a) What is an SI unit of volume? (b) What other volume units are used in chemistry?

3.11 Perform the following conversions:
(a) $12.5 \text{ L} = ? \text{ mL} = ? \text{ cm}^3 = ? \text{ dm}^3 = ? \text{ m}^3$
(b) $5.6 \times 10^5 \; \mu\text{L} = ? \text{ m}^3 = ? \text{ L}$.

3.12 A barrel is a unit of volume used when measuring crude oil; 1 barrel contains 44 gal. How many liters of oil are found in a barrel?

3.13 An unknown solid is found to have a mass of 133.1 g and a volume of 155.93 mL. (a) What is the density of the solid? (b) If the unknown solid is one of those listed in Table 3.2, what is it?

3.14 What is the volume of 1233 g osmium? Osmium's density is given in Table 3.2.

3.15 An empty 25.0-mL graduated cylinder has a mass of 82.14 g. If 25.0 mL of an unknown liquid is poured into the cylinder, the total mass of the cylinder and liqud is 104.52 g. Determine the density of the liquid.

3.4 UNCERTAINTY IN MEASUREMENTS

Whenever chemical measurements are made, both precision and accuracy are considered. The **accuracy** of a measurement is how close the obtained value is to a standard or "true" value. A more accurate measurement is one that is closer to the standard than a less accurate measurement. Accuracy is measured in terms of deviation of the measurement(s) (called the error) from the "true" value.

Precision, on the other hand, is how closely repeated measurements are grouped together or how reproducible the measurements are. The smaller the range of values obtained when measuring the same quantity, the greater the precision. Normally, good precision is an indication of high accuracy, but not always, as we shall see.

The game of darts provides a good analogy to help understand the terms "accuracy" and "precision." A dart player attempts to "hit the bull's-eye," just as a chemist tries to "hit" the true value when measuring matter. Consider Figure 3.13a; the first dart landed far from the bull's-eye—not an accurate throw. In Figure 3.13b, the dart is closer to the bull's-eye—a more accurate throw. Figure 3.13c shows a dart board after six throws; notice that the six throws are grouped closely together, which is an indication of good precision, and they are near the bull's eye, which shows good accuracy. In contrast, the dart distribution in Figure 3.13d indicates good precision but poor accuracy; and Figure 3.13e illustrates poor accuracy and poor precision.

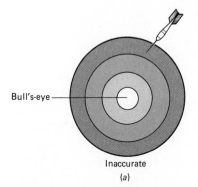

Bull's-eye

Inaccurate
(a)

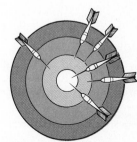

Accurate
(b)

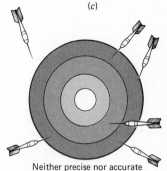

Highly precise and accurate
(c)

FIGURE 3.13
The distribution of darts on a dart board can be used as an analogy to understand the terms "precision" and "accuracy." Precise measurements are closely grouped together. Less precise measurements are more spread out. The accuracy of a measurement is related to how close a measurement is to the actual value.

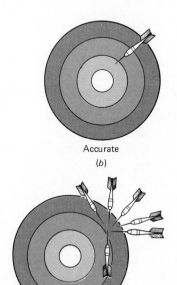

Precise, but not accurate
(d)

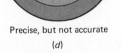

Neither precise nor accurate
(e)

Measurement errors account for obtaining varying values for the same thing. Two types of errors are generally found in chemical measurements: **systematic** and **random** errors.

Systematic errors can be eliminated, if they are identified.

Systematic errors result from (1) poor procedures and methods, (2) malfunctioning and uncalibrated instruments, (3) human error, and (4) some unrecognized factor that is influencing the results. For example, suppose that you measure the mass of an object on a balance but fail to adjust the balance to the zero point prior to weighing the object. The measurement will be high or low, depending on the initial incorrect setting of the balance. Systematic errors are reduced by finding their causes and eliminating them.

Random errors cannot be corrected.

Random errors are found in all chemical measurements. Even if every precaution is taken to avoid systematic errors, small deviations or random errors arise that are unavoidable and not identifiable. Random errors, by definition, are impossible to illustrate. If random errors could be identified, they would be corrected, and thus would not be considered random errors.

Collectively, systematic and random errors introduce uncertainty—or lack of confidence—in all measured values. Thus, all reported measurements should indicate the degree of certainty and uncertainty of the measurement. In chemistry, this is accomplished through the use of significant figures, which is described in the following section.

REVIEW PROBLEMS

3.16 Define (a) precision and (b) accuracy.

3.17 What can be said about 10 measurements that are known to be highly precise?

3.18 The true value for the mass of an object is 239.1 g. What statement can be made about the precision of the following mass measurements of the object: 240.9 g, 241.1 g, 240.8 g, and 241.0 g?

3.19 Explain the difference between systematic and random measurement errors.

3.5 SIGNIFICANT FIGURES

Significant figures are measured digits plus one uncertain or estimated digit.

Significant figures are the measured digits in a number that are known with certainty plus one uncertain digit. Stated differently, significant figures are all digits measured with certainty plus the first doubtful or estimated digit.

Exact numbers are not uncertain.

Note that significant figures apply only to *measured* values, not to exact numbers. For example, if you correctly count the number of students in a classroom, the value obtained is absolutely accurate (an exact number) with no uncertainty. When we discussed mass, we defined the kilogram as 1000 grams (1 kg = 1000 g), an exact relationship. Significant figures are applied only to measurements that are uncertain.

In Figure 3.14a, observe the level of the liquid relative to the scale etched on the cylinder. This indicates the liquid's volume. The volume indicated is 35.5 mL, since the liquid level (always read the bottom of the meniscus, the concave upper surface of a liquid in a small container) is about halfway between the 35- and 36-mL marks. This measurement, 35.5 mL, represents three significant figures—the "35" is directly measured (digits are certain) and the ".5" is a

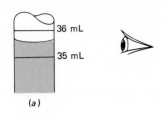

(a)

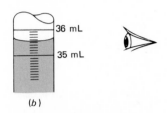

(b)

FIGURE 3.14
(a) With 1 mL graduations, the volume can be estimated to within ±0.1 mL. In this case the volume is reported as 35.5 mL; this is three significant figures. (b) With 0.1 mL graduations, the volume can be estimated to within ±0.01 mL. Using this more precise graduated cylinder, the volume is reported as 35.58 mL; this is four significant figures.

good estimate obtained from a careful reading of the graduated cylinder (first uncertain digit).

Normally, the last significant figure is thought to be uncertain by ±1. In the above example, by stating the volume as 35.5 mL, we are indicating that the measured volume is at most 35.6 mL (+0.1) and at least 35.4 mL (−0.1). If the same liquid is totally transferred to a more precise volumetric instrument—let's say one that has a scale with 0.1-mL marks accurately etched on the side—it's possible to obtain another significant figure. In Figure 3.14b we can read the volume to one-hundredth of a mL, giving 35.58 mL. Three digits are certain (35.5) with the last digit (8) as the uncertain digit. Once again, the range of uncertainty is ±1 in the last significant figure—35.59 to 35.57 mL.

In Figure 3.14b if you read the volume as 35.5872389 mL, you are overstepping the limits of the measuring device. There is no way to be certain of the second decimal place; it is impossible to know the third or fourth decimal places—they are meaningless numbers, especially if you consider that the uncertainty is ±0.01. Never report any more significant figures than the measuring device is capable of providing.

Whenever you encounter a measurement, always keep in mind that besides the numerical value and units, the number also indicates significant figures or the precision with which the measurement was made. Observe the following measurements and the indicated number of significant figures.

1.25 m indicates three significant figures (1, 2, 5)
434.56 K indicates five significant figures (4, 3, 4, 5, 6)
3 g indicates one significant figure (3)
8.913477 cm indicates seven significant figures (8, 9, 1, 3, 4, 7, 7)

When writing measured quantities, always ask yourself: Did I report the correct number of significant figures?

Zero digits in measurements pose a special problem; a zero digit acting as a placeholder is not significant. Placeholders are not measured quantities; therefore, by definition, they are not significant.

The following rules summarize and illustrate all possible cases where zero digits are found in measurements:

Rule 1. Zeros located in the middle of a number. In all cases, zeros in the middle of a number are significant. In each of the following, the zero is counted as a significant figure:

10.004 g indicates five significant figures (1, 0, 0, 0, 4)
47,000.15 mL indicates seven significant figures (4, 7, 0, 0, 0, 1, 5)
103 cm^2 indicates three significant figures (1, 0, 3)

Zeros in the middle of measurements are digits that have been measured and are not placeholders; thus, they are significant in all cases.

Rule 2. Zeros located in front of a number. Zero digits in front of numbers are usually to the right of the decimal point. These zeros act as placeholders (they are not measured), so they are not significant figures:

.0005 L indicates one significant figure (5)
.0000477 m indicates three significant figures (4, 7, 7)
.000000091 kg indicates two significant figures (9, 1)

Sometimes a zero is placed in front of the decimal point to show that no other digit is present. Similarly, this zero is not significant.

0.00831 mm indicates three significant figures (8, 3, 1)
0.1 mL indicates one significant figure (1)

When you convert measurements from one set of units to another, the number of significant figures does not change if the conversion factor used contains at least as many significant figures as the measurement. If a sample is found to have a mass of 26.9 mg (three significant figures), changing the units to grams, 0.0269 g, has no effect on the number of significant figures. Any added zero digits that are placed in front of the measured quantity are placeholders.

Rule 3. Zeros located after a number to the right of the decimal point. For this specific case, the zero digit is either a measured quantity (certain) or a good estimate (first uncertain digit); consequently, zeros after a number and to the right of the decimal point are all significant.

.650 dm indicates three significant figures (6, 5, 0)
.178300 ms indicates six significant figures (1, 7, 8, 3, 0, 0)
.3550000 km indicates seven significant figures (3, 5, 5, 0, 0, 0, 0)

For each of the above cases, the zero digits were measured by some device. All of the zeros were measured with certainty except the last zero, which is uncertain but still significant.

Rule 4. Zeros located after a number to the left of the decimal point. Zero digits found after a number and to the left of the decimal point are significant if they are measured, and are not significant if they are placeholders. An object with a measured mass of 800 g has a questionable number of significant figures; more information is required to determine the correct number. The measurement, 800 g, contains three significant figures only if the second zero (units place) is the first uncertain figure. It contains two significant figures if the first zero (tens place) is the first uncertain figure. Lastly, the 8 could be the digit that is uncertain; if this is the case, the number would only have one significant figure.

To avoid the confusion generated by the ambiguous nature of zero digits to the left of decimal points, scientists commonly express such measurements in **scientific notation,** where the decimal factor represents the correct number of significant figures. So, 800 g is expressed as 8×10^2 (one significant figure) or

8.0×10^2 (two significant figures) or 8.00×10^2 (three significant figures), depending on the proper number of significant figures.

As another example, the measurement 45,000 mm is expressed as two, three, four, or five significant figures as follows:

4.5×10^4 mm represents two significant figures (4, 5)
4.50×10^4 mm represents three significant figures (4, 5, 0)
4.500×10^4 mm represents four significant figures (4, 5, 0, 0)
4.5000×10^4 mm represents five significant figures (4, 5, 0, 0, 0)

REVIEW PROBLEMS   **3.20** What is a significant figure?
3.21 Write the correct number of significant figures indicated by each measurement:
(a) 1977 g   (b) 0.98 cm
(c) 5000.02 L   (d) 0.00500 kg
(e) 6.010×10^8 mg   (f) 0.00000000000030 m
(g) 9,111,000.00000030 nm   (h) 0.6 s

Normally, after measurements are obtained they are used in subsequent calculations. Specific rules are followed so that results of the calculations also have the proper number of significant figures. Two different rules are followed, depending upon the arithmetic operation performed. The rules for handling significant figures when adding and subtracting are different from those used when multiplying and dividing. Attempt to learn each rule, and try not to confuse one with the other.

Adding and Subtracting Significant Figures

When adding and subtracting measured quantities, the answer can have no more digits to the right of the decimal point than does the measured quantity with the least number of decimal places. If, for instance, we add the masses 2.0965 g and 1.41 g, the answer can have only two decimal places.

First obtain the sum of the two numbers:

$$\begin{array}{r} 2.0965 \text{ g} \\ +1.41 \quad\ \text{ g} \\ \hline 3.5065 \text{ g} \end{array}$$

Then round off to the correct number of decimal places. Since the second mass has only two decimal places, the answer should have only two decimal places.

Another commonly used system for rounding off is the same as stated in the text except for the case where 5 is the first nonsignificant figure and it is followed by zeros or nothing. In this case, if the least significant figure is an even number then it is retained; all other figures are dropped. However, if it is odd, then 1 is added to make it even.

When rounding off, look at the first nonsignificant figure—6 in our example—one place to the right of the least significant figure (0). The least significant figure is the last figure in the number that is retained after rounding off. If the value of the first nonsignificant figure is equal to or greater than 5, add 1 to the least significant figure and drop all numbers to the right. But, if the value of the first nonsignificant figure is less than 5, retain the least significant figure and drop all digits to the right. In our example, the first nonsignificant figure is 6, so 1 is added to 0 giving 1 as the second decimal place. The final answer is expressed as 3.51 g. Any other answer is incorrect.

$$3.5065 \text{ g}$$

Least significant figure————First nonsignificant figure

$$3.5065 \text{ g}$$

Add 1————Drop

Example Problem 3.8 is another illustration of adding measurements.

Example Problem 3.8

What is the sum of 10.0043 mL + 5.3 mL + 9.230 mL?

Solution

$$
\begin{array}{ll}
10.0043 \text{ mL} & \text{(four decimal places)} \\
5.3 \quad\;\; \text{mL} & \text{(one decimal place)} \\
+\;9.230 \;\; \text{mL} & \text{(three decimal places)} \\
\hline
24.5343 \text{ mL} & \text{(round to one decimal place)}
\end{array}
$$

Least significant figure ————First nonsignificant figure

We must round off our answer, 24.5443 mL, to one decimal place because the second measured quantity, 5.3 mL, contains only one decimal place. Since the first nonsignificant figure is 3, we drop all nonsignificant digits and retain the least significant figure (5), giving a final answer of **24.5 mL**.

You may wonder why such a rule is employed when adding and subtracting measurements. Always, when dealing with significant figures, the last number is uncertain; thus, if it's added or subtracted from a figure that is certain, the resultant quantity is rendered uncertain. For example, if 3.46 m is added to 1.9 m, the 6 is uncertain in 3.46 m and the 9 is uncertain in 1.9 m. Within the answer, 5.36 m, both the 3 and 6 are uncertain.

$$
\begin{array}{ll}
3.46 \text{ m} & \text{uncertain} \\
+1.9 \;\; \text{m} & \text{uncertain} \\
\hline
5.36 \text{ m} & \text{uncertain}
\end{array}
$$

Least significant figure————

After adding, we see that the new least significant figure is 3; consequently, the number must be rounded off to 5.4 m.

Multiplying and Dividing Significant Figures

When we multiply and divide measurements, the answer can have no more significant figures than the measurement with the least number of significant figures. If two numbers are multiplied together, one with six and the other with three significant figures, the answer can have only three significant figures. Observe the following example:

$$
\begin{array}{r}
5.82131 \text{ cm} \quad \text{(six significant figures)} \\
\times \quad 4.11 \text{ cm} \quad \text{(three significant figures)} \\
\hline
23.\underbrace{9255841}_{\text{drop}} \text{ cm}^2
\end{array}
$$

The first nonsignificant figure, 2, is less than 5. Therefore, it and all other figures to the right are dropped, and the answer is rounded off to 23.9 cm² (three significant figures).

The multiplication and division rule results from the fact that, when an uncertain figure—the last figure—is multiplied or divided, it produces numbers that are uncertain. An answer can have only one uncertain figure, so all other uncertain figures are dropped. Let's consider another multiplication problem:

$$2.41 \text{ m} \times 2.1 \text{ m} = ?$$

$$
\begin{array}{r}
2.4\underline{1} \text{ m} \quad \text{uncertain} \\
\times \ 2.\underline{1} \text{ m} \quad \text{uncertain} \\
\hline
24\underline{1} \quad \text{uncertain} \\
48\underline{2} \quad \text{uncertain} \\
\hline
5.\underline{061} \text{ m}^2 \quad \text{uncertain}
\end{array}
$$

Least significant figure ⟶

When 2.41 is multiplied by the 1 (uncertain figure) in 2.1, all resulting numbers are rendered uncertain. When these uncertain figures are added to the product of 2.41 and 2, the least significant figure becomes the first decimal place; thus, the final answer can have only two significant figures. After rounding off, the answer is reported as 5.1 m².

Example Problem 3.9

Perform the indicated arithmetic operations and express the answer to the correct number of significant figures.

$$\frac{7.290 \text{ m} \times 2.0400 \text{ m}}{0.95 \text{ m}} =$$

Solution

Notice that the denominator contains a measurement with only two significant figures; this limits the answer to two significant figures. Perform the indicated math operations and round off the resulting answer to two significant figures.

$$\frac{7.290 \text{ m} \times 2.0400 \text{ m}}{0.95 \text{ m}} = 15.\underset{\uparrow\uparrow}{65}431579 \text{ m} = 16 \text{ m}$$

Least significant figure ⌐ ⌐First nonsignificant figure

The first nonsignificant figure is 6, and we round off by adding 1 to 5 and dropping the numbers to the right, leaving 16 m as the answer.

Don't forget significant figures in all chemistry problems.

It is not uncommon to perform calculations where both addition and multiplication are required. Example Problem 3.10 is a model of such problems.

Example Problem 3.10

Perform the indicated arithmetic operations and express the answer to the correct number of significant figures.

$$(11.2050 \text{ mm} - 10.322 \text{ mm}) \times 6.030000 \text{ mm} =$$

Solution

Both rules regarding significant figures apply in this example; first, after subtracting we can only have three decimal places in the answer because 10.322 mm contains only three decimal places.

$$11.2050 \text{ mm} - 10.322 \text{ mm} = 0.8830 \text{ mm} = 0.883 \text{ mm}$$

Apply the multiplication rule when multiplying 0.883 mm (three significant figures) by 6.030000 mm (seven significant figures). Three significant figures is the maximum number allowed in the answer.

$$0.883 \text{ mm} \times 6.030000 \text{ mm} = 5.32449 \text{ mm}^2 = 5.32 \text{ mm}^2$$

Subtracting values that have magnitudes close to each other usually results in a loss of significant figures. In this problem, a measurement with five significant figures is subtracted from a measurement with six significant figures and gives an answer with only three significant figures.

REVIEW PROBLEMS   **3.22** Perform the indicated additions and subtractions and express the answers to the correct number of significant figures:
(a) 300.55 g − 1.9 g
(b) 0.01122 mL + 0.183 mL
(c) 504 m + 46.884 m

(d) $34.000000 \text{ s} - 5.011328 \text{ s} + 195.0887552 \text{ s}$

(e) $6.55 \times 10^{23} + 4.9 \times 10^{21}$

3.23 Perform the indicated multiplications and divisions and express the answers to the correct number of significant figures:

(a) $\dfrac{4.900 \text{ g}}{9.2 \text{ mL}}$

(b) $39.0030 \text{ m} \times 9 \text{ m}$

(c) $0.822 \text{ cm} \times 1.332 \text{ cm} \times 2.1925 \text{ cm}$

(d) $\dfrac{327 \text{ L}}{0.001444 \text{ L}}$

(e) $64,329.0 \text{ mg} \times 8.7 \text{ mg}$

3.24 Perform the indicated arithmetic operations and express the answers to the correct number of significant figures:

(a) $(4.99 \text{ m} - 3.255 \text{ m}) \times 2 \text{ m}$

(b) $\dfrac{124.04 \text{ g} - 124.0285 \text{ g}}{6.916 \text{ s} + 35.5913 \text{ s}}$

(c) $\dfrac{(9.122 \times 10^5 \text{ cm}) \times (4.2 \times 10^{-3} \text{ cm})}{(8.0000 \times 10^9 \text{ cm}) \times (4.921 \times 10^{-30} \text{ cm})}$

SUMMARY

Today, the International System of Units (SI) is the measuring system used by scientists throughout the world. All SI units of measurement are classified as either base or derived units. Seven base units exist in SI; they are the meter, kilogram, kelvin, second, mole, ampere, and candela. All other units are derived from these seven base units; thus, they are called derived SI units. To increase or decrease the magnitude of SI units, it is only necessary to attach the appropriate prefix to the unit. Frequently encountered prefixes are kilo ($10^3 \times$), centi ($10^{-2} \times$), milli ($10^{-3} \times$), and micro ($10^{-6} \times$).

Mass is the quantity of matter in an object and is measured in kilograms, grams, and milligrams. No matter where an object is located, mass remains constant. Weight, in contrast, varies with the gravitational force of attraction on an object. Kelvin (K) is the SI unit of temperature. The zero point on the Kelvin scale is absolute zero, the lowest possible temperature. Kelvin degrees are of the same magnitude as Celsius degrees.

Volume is measured in cubic meters, cubic decimeters, or cubic centimeters in SI. They are derived from the base units of length—meters, decimeters, and centimeters. Liters and milliliters, two non-SI volume units, are commonly used in chemistry. Density is the ratio of mass to volume and is most frequently measured in grams per milliliter or grams per liter.

Whenever measurements are made, there is a degree of uncertainty that depends on measurement errors. Two general types of errors are found in all measurements: systematic and random. Systematic errors result from improper techniques, uncalibrated equipment, or human error: they are correctable. Random errors are those that cannot be identified but are always present.

Chemists use significant figures to express the degree of certainty for each measurement. Significant figures are measured digits plus one estimated digit. Measurements with a greater number of significant figures are more certain than those with less. Generally, more expensive instruments give more significant figures. When we use significant figures in calculations, we must be careful to follow a specific set of rules so that the results of the calculations have only one uncertain digit.

QUESTIONS AND PROBLEMS

3.25 Define each of the following terms: SI, metric system, base unit, derived unit, kilogram, mass, weight, balance, scale, absolute zero, volume, density, systematic error, random error, measurement uncertainty, and significant figure.

Base SI Units

3.26 Write the base SI unit that corresponds to the following measurements: (a) time, (b) amount of substance, (c) length, (d) light intensity, (e) mass, (f) temperature.

3.27 (a) Use a dictionary to find the meaning of the following measurement units: dram, scruple, minim, and slug. (b) What difficulties are encountered using units that are not encountered in the International System?

3.28 Write the prefix that is used for each of the following: (a) $0.01\times$, (b) $1,000,000\times$, (c) $10^{-6}\times$, (d) $0.1\times$, (e) $1000\times$, (f) $0.001\times$, (g) $10^{-9}\times$.

3.29 State the meaning of each prefix: (a) milli, (b) deca, (c) giga, (d) pico, (e) nano, (f) deci, (g) kilo, (h) centi, (i) micro, (j) hecto.

3.30 Perform the following length conversions: (a) 12 mm = ? m, (b) 509 cm = ? mm, (c) 1.2 km = ? cm, (d) 6.8 km = ? μm, (e) 8.5 mm = ? nm, (f) 4.6 Mm = ? km.

3.31* Change each to the designated unit: (a) 1 m = ? yd, (b) 1 m = ? ft, (c) 100 km = ? mi, (d) 50 cm = ? in., (e) 0.5 km = ? fathoms.

3.32* Change each to the designated SI unit:
(a) 10 mi = ? km
(b) 36 in. = ? mm

(c) 6 ft 1 in. = ? m
(d) 100 yd = ? m
(e) 0.015 in. = ? nm
(f) 5.1 mi = ? mm.

3.33 The distance from the earth to the sun is 9.3×10^7 mi. Express this distance in (a) kilometers, (b) meters, and (c) millimeters.

3.34 What is the modern standard of length in SI, and why did it replace the old standard?

3.35 A person's body measurements are as follows: weight = 130 lb, height = 5 ft 7 in., chest = 37 in., waist = 30 in., and hips = 35 in. Convert the weight measurement to kilograms and the length measurements to centimeters.

3.36 Assuming that you could travel to Jupiter and make a quick weight measurement, how would your weight there compare with your weight on earth? (Jupiter is the largest planet in our solar system, with a mass many times that of earth.)

3.37 (a) What instruments are used to measure mass and weight? (b) How do they differ?

3.38 Perform the following mass conversions:
(a) 5 kg = ? mg
(b) 2.7 mg = ? g
(c) 8.571×10^{10} μg = ? kg
(d) 0.009 kg = ? pg
(e) 7.5×10^{-5} g = ? μg
(f) 5.5×10^{12} mg = ? Mg
(g) 0.00053 kg = ? dg
(h) 3.147×10^3 g = ? cg

3.39* Change each metric unit to the designated unit:
(a) 505.6 kg = ? lb
(b) 9.111 g = ? oz
(c) 6.3×10^{12} g = ? tons
(d) 204 mg = ? lb

---

* See Appendix 3 for needed conversion factors.

(e) 2703 ng = ? grains

(f) 1 metric ton = ? tons

3.40* Change each to the designated SI unit:

(a) 10 lb = ? g

(b) 925 oz = ? mg

(c) 4.2 tons = ? Mg

(d) 0.033 lb = ? cg

(e) 6.7×10^{-8} lb = ? ng

(f) 1.4×10^{10} oz = ? dg

Temperature Conversions

3.41 Change each Celsius temperature to Kelvin: (a) 91°C, (b) −35°C, (c) −265.6°C, (d) 273°C.

3.42 Change each Kelvin temperature to Celsius: (a) 25 K, (b) 100 K, (c) 400.3 K, (d) 298.9 K.

3.43 Change each Fahrenheit temperature to Celsius: (a) 195°F, (b) −195°F, (c) 1000°F, (d) −459°F.

3.44 Change each Celsius temperature to Fahrenheit: (a) 195°C, (b) −195°C, (c) 1000°C, (d) −35°C.

3.45 Convert each Kelvin temperature to Fahrenheit: (a) 195 K, (b) 98.6 K, (c) 1000.0 K, (d) 310.8 K.

3.46 One temperature has exactly the same numerical value in both the Celsius and Fahrenheit temperature scales. Use algebra to find this temperature.

3.47 Cesium is a highly reactive metal that melts at 28.7°C. (a) What is cesium's melting point in Fahrenheit? (b) If you could hold cesium in your hand, what would happen to it?

3.48 Hydrogen's melting and boiling points are −269.7°C and −268.9°C, respectively. Determine hydrogen's melting and boiling points in Kelvin.

Derived SI Units

3.49 A block of wood is 1.7 cm long, 0.39 cm wide, and 2.3 cm high. What is the volume of the block in (a) cubic centimeters, (b) milliliters, (c) liters?

3.50 Compute the volume in liters of a metal block that is 19,340 mm long, 890 cm wide, and 0.0083 m high.

3.51 Perform the indicated volume conversions:

(a) 1 dm^3 = ? cm^3 (b) 24 mL = ? m^3

(c) 834.22 mL = ? μL   (d) 0.44 m^3 = ? mL

(e) 840,000 L = ? m^3

(f) 9.14×10^{14} μL = ? m^3

3.52 Change each to the indicated unit:

(a) 55.0 mL = ? qt   (b) 1 m^3 = ? in.^3

(c) 100 mL = ? qt   (d) 7000 cm^3 = ? ft^3

3.53 Change each to the indicated SI units:

(a) 100.0 gal = ? m^3 (b) 95 pt = ? cm^3

(c) 0.0041 qt = ? dm^3 (d) 1.1 gal = ? mm^3

3.54 Calculate the density of each substance in grams per milliliter, given mass and volume:

(a) mass = 110 g and volume = 39.3 mL

(b) mass = 9.37 g and volume = 11.6 mL

(c) mass = 0.588 kg and volume = 0.602 L

(d) mass = 6.24 lb and volume = 0.1 ft^3

3.55 Vanadium, an element, has a density of 6.1 g/mL. Calculate the mass of vanadium contained in the following volumes: (a) 10 mL, (b) 174 mL, (c) 1.5 L, (d) 0.28 cm^3, (e) 1 qt, and (f) 0.871 m^3.

3.56 Pure silicon's density is 2.33 g/mL. Calculate the volume of the following masses of silicon: (a) 48 g, (b) 125 mg, (c) 0.110 kg, (d) 10.69 lb, and (e) 1 metric ton (1000 kg).

3.57 Air has a density of 1.29 g/L. Express air's density in: (a) g/ft^3, (b) kg/m^3, and (c) lb/ft^3.

3.58 The metal platinum has a density of 21.4 g/mL. If you are given a metal sample weighing 3.6 g that occupies 0.168 mL, how could you quickly determine if the sample is pure platinum or not? Is it?

3.59 A graduated beaker is used to determine the density of an unknown liquid. The beaker's mass is 40.1 g. What is the density of the unknown liquid if the beaker and 20.1 mL of the unknown liquid weigh 67.3 g?

3.60 A quart bottle is used to store liquid mercury (density = 13.6 g/mL). What is the mass of mercury in the bottle in (a) grams and (b) pounds?

3.61 A 47.355-g graduated cylinder is used to measure the density of a liquid. After 15.0 mL of this liquid is poured into the cylinder, the combined mass of the cylinder and liquid is 68.452 g. Calculate the density of the liquid.

3.62 What is the mass of air in a room measuring 5 m × 5 m × 3 m?

3.63 A bar of osmium metal has the following dimensions: 1.45 m × 5.93 cm × 83.1 mm. What is the mass of the osmium bar?

3.64 List five measurements where derived SI units are used. Do not include volume or density.

Uncertainty in Measurement

3.65 How is the accuracy of a measurement determined?

3.66 What factors can decrease the accuracy of measuring your weight on a bathroom scale?

---

*See Appendix 3 for needed conversion factors.

3.67 What could be said about the precision of the following set of measurements of the same volume: 144 mL, 163 mL, 155 mL, 182 mL, 159 mL, and 149 mL?

3.68 Select the more precise measuring device from the following pairs: (a) centigram balance or analytical balance, (b) pipet or calibrated beaker, (c) large graduated cylinder or small graduated cylinder, and (d) volumetric flask or graduated cylinder.

3.69 What systematic errors might influence the measurement of air temperature on an outdoor thermometer?

3.70 Why can't random errors be eliminated from measurements?

Significant Figures

3.71 How many significant figures are there in each of the following measured quantities?
(a) 204.841 (b) 3.0070
(c) 1970.0 (d) 0.0000120000
(e) 9 (f) 0.000000808
(g) 2000.0000001 (h) 0.0029
(i) 7.5×10^4 (j) 1.38×10^{-4}

3.72 For each measurement listed in Question 3.71, write the least significant figure.

3.73 For each measurement listed in Question 3.71, express each with one significant figure.

3.74 Round off each of the following to four significant figures:
(a) 61.645 (b) 1.9998
(c) 641.1439 (d) 0.999119999
(e) 799,533.9 (f) 199,954,832,000

3.75 Use scientific notation to express 70,000 with: (a) one, (b) two, and (c) three significant figures.

3.76 Add each of the following and express the answer to the correct number of significant figures:
(a) 2.345 g + 2.05 g
(b) 12.0030 mL + 11 mL
(c) 4.5 g + 3.2388 g + 9.18 g
(d) 94.14591 m + 71.449 m + 61.12 m

3.77 Subtract each of the following and express the answer to the correct number of significant figures:
(a) 53.236 mL − 36.9 mL
(b) 1.003278 cm − 0.939 cm
(c) 41.21 g − 40.977 g
(d) 100.00122 mm − 58 mm − 31.55 mm

3.78 Multiply each of the following and express the answer to the correct number of significant figures:
(a) 0.024 g × 3 g

(b) 100.0 s × 6.25 s
(c) 3.195 cm × 9 cm
(d) 0.000147 kg × 3587.0 kg

3.79 Divide each of the following and express the answer to the correct number of significant figures:
(a) $\dfrac{5.83 \text{ g}}{1.8 \text{ mL}}$ (b) $\dfrac{421.8 \text{ m}}{0.669514 \text{ m}}$
(c) $\dfrac{1.43 \times 10^{12} \text{ m}}{9 \times 10^{-4} \text{ s}}$ (d) $\dfrac{0.003000 \text{ g}}{0.10 \text{ cm}^3}$

3.80 Perform the indicated arithmetic operations and express the answer to the correct number of significant figures:
(a) $\dfrac{51 \times 234.99}{1.1 + 0.9743}$
(b) $(54.7325 - 54.6356) \times 4.913322$
(c) $\dfrac{\dfrac{0.000500}{0.166} - 0.000172 + 0.004}{1.119}$

Additional Problems

3.81 Commercial airplanes regularly travel in excess of 500 mi/hr. Convert this speed to (a) km/hr, (b) m/s, and (c) mm/ns.

3.82 A cylinder is measured and is found to weigh 103.44 g. Its height is 3.22 cm, and its diameter is 1.73 cm. The formula for the volume of a cylinder is $V = \pi r^2 h$. Calculate the density of the cylinder.

3.83 What size cube—length of each side—could be formed from 435 g of chromium metal? Chromium's density is 7.2 g/cm^3. (*Hint:* An equation to find the volume of a cube is $V_{cube} = l^3$, where l is the length of each side.)

3.84 As part of the U.S. program to change to the metric system, auto makers now express the engine displacement in liters. What is the displacement of a 2.5-L engine in cubic inches?

3.85 (a) What Fahrenheit temperature is numerically equal to twice the Celsius temperature?
(b) What Celsius temperature is numerically equal to twice the Fahrenheit temperature?

3.86 A 32.326-g container is filled with water. The mass of the container and water is found to be 57.205 g. The water is removed, and the container is filled with an unknown liquid. The mass of the container and the unknown liquid is 55.223 g. Assume that the density of water is 1.000 g/mL and calculate the density of the unknown liquid.

· C H A P T E R ·

4

Matter and Energy

Study Guidelines

After completing Chapter 4, you should be able to:
1. Distinguish between the composition and structure of matter
2. Define and give examples of physical and chemical properties
3. Distinguish between and give examples of physical and chemical changes
4. List fundamental properties of solids, liquids, and gases
5. Explain trends in fluidity, viscosity, average density, volume, and compressibility among the three states of matter
6. Identify what state change occurs during melting, freezing, boiling, and subliming
7. List and explain three criteria that are used to classify pure substances
8. Distinguish between elements and compounds
9. Write symbols for various simple elements, and write names when given symbols of elements
10. Read and interpret formulas of simple compounds
11. Give examples of simple compounds
12. Briefly discuss the smallest unit of a compound—the molecule
13. List and discuss the three criteria used to classify mixtures
14. Give examples of simple mixtures
15. Distinguish a homogeneous mixture from a heterogeneous mixture
16. Give examples of solutions and heterogeneous mixtures
17. Explain how different types of mixtures can be separated
18. Define and explain "energy" in terms of "work"
19. Discuss the different forms of potential and kinetic energy
20. Explain how energy can be interconverted
21. Explain the direction of heat flow
22. Explain the difference between heat and temperature
23. Identify the two units utilized for measuring heat energy
24. Convert calories to joules and vice versa
25. Define and calculate the specific heat of a substance
26. Calculate heat gain or loss of substances, given the specific heat, mass, and temperature
27. State the laws of conservation of matter and energy

As we discussed in Chapter 3, matter is anything that has mass and occupies space. Chemists, when studying matter, are usually interested in matter's composition and structure. **Composition** refers to the identity and quantity of the components (ingredients) of matter. The **structure** of matter is the physical arrangement of the components within matter. Collectively, the composition and structure determine the **properties,** or characteristic traits, of matter.

Each type of matter has its own unique set of properties. Thus, different types of matter are distinguished by their properties, just as people are distinguished by their physical appearance and personality traits. There are two types of properties, physical properties and chemical properties.

Physical Properties and Changes

The physical properties of a substance can be measured without referring to any other substance.

Physical properties of matter are characteristics of individual substances that can be determined without changing a substance's composition. A physical property of matter that we have already discussed thoroughly is density. Density is the mass of a unit volume. When measuring density, it's only necessary to find the mass and volume of the substance—no other substance is involved.

Examples of physical properties include (1) color, (2) hardness, (3) electric conductivity, (4) heat conductivity, (5) physical state, (6) melting point, (7) boiling point, and (8) tensile strength. Physical properties are measured by observing what happens when matter interacts with heat, light, electricity, and other forms of energy, or when matter is subjected to various stresses and forces.

Tensile strength is the resistance to pulling, and is measured by finding the breaking stress.

A substance has a unique set of physical properties that distinguishes it from all other substances. If two substances have exactly the same set of physical properties, the most plausible conclusion is that they are the same substance with the same composition and structure.

To illustrate, let's investigate the physical properties of pure gold. Gold is the only bright yellow metal. Gold melts at 1063°C and boils at 2966°C. Its density is 19.3 g/mL, a very high density for metals. On a hardness scale of 1 to 10, gold is 2.8—quite soft. It shares the common physical property of metals called **malleability,** i.e., the ability of a substance to be hammered into different shapes or thin foils (called leaf). Gold leaf, 1.3×10^{-5} cm thick, is used in expensive decorations. Gold is one of the best conductors of electricity and is an excellent heat conductor.

When physical changes occur the composition remains the same.

When the physical properties of a substance are altered but the composition remains the same, we say that a **physical change** has occurred. No new substance is formed in a physical change. Examples of physical changes are changes in (1) size, (2) state, (3) density, (4) shape, (5) magnetic properties, and (6) conductivity (see Figure 4.1). After a physical change, the starting substance is still present, but in a modified state. For example, after a rock is crushed, it has the same composition; only the particle size has changed. When ice melts, it changes from the solid to the liquid state. When iron is magnetized, it is still iron. When gems such as rubies, emeralds, and diamonds are cut, only their shape changes.

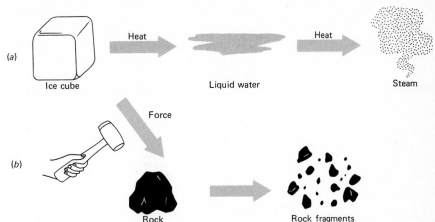

FIGURE 4.1
(a) After undergoing a physical change, a substance's composition remains the same. Addition of heat to ice, which is water in the solid state, changes it to liquid water. Addition of heat to liquid water changes it to steam, which is gaseous water. (b) Only the shape of the rock changes when the rock is broken into pieces. A change in shape is a physical change.

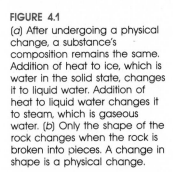

Chemical Properties and Changes

FIGURE 4.2
Decomposing water by using an electric current is called electrolysis. During electrolysis, water is chemically changed to hydrogen gas and oxygen gas.

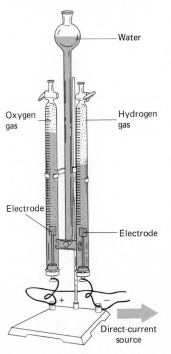

Chemical properties are those properties that describe how the composition of a substance changes or does not change when the substance interacts with other substances or energy forms. Terms used to describe chemical properties are "reactive," "inert," and "combustible." Chemical properties are observed when a substance changes composition. Paper burns in air, iron rusts, silver tarnishes, TNT explodes—these are all illustrations of chemical properties. In each case a new substance is formed, and a **chemical change** has occurred.

Chemical changes or chemical reactions are the result of the chemical properties of matter. After a chemical change, the composition is no longer the same. For each chemical change, we can write a chemical equation that indicates what the original substance, or substances, ultimately changes into. The starting materials, called reactants, undergo a chemical change and produce the products.

$$\text{Reactants} \longrightarrow \text{Products}$$

The arrow is a symbol meaning "yield" or "produce." In chemical reactions, the reactants combine to yield the products. If only one reactant is initially present, we say the reactant "decomposes" or "rearranges" to yield the products.

Chemical changes are monitored by first observing the physical properties of the reactants, and then looking at the physical properties of the products. If a chemical change has truly taken place, the physical properties of the products are different from those of the reactants.

Some everyday examples of chemical changes are (1) iron rusting, (2) food cooking, (3) gasoline igniting, (4) dynamite exploding, and (5) wood burning. Each of these examples is fairly complex and requires elaborate equations to explain what happens to the reactants. Let us consider a less complex chemical change, the decomposition or breakdown of water using electricity.

When an electric current is passed through water (see Figure 4.2), the water is changed to new substances, hydrogen and oxygen. A word equation for water's decomposition is as follows:

(4.1)   Water $\xrightarrow{\text{electricity}}$ hydrogen + oxygen

The decomposition of water by electricity is called electrolysis.

This equation states: "Water is decomposed by electricity to yield hydrogen and oxygen." How do we know that a chemical change rather than a physical change has taken place? First we consider the physical properties of water:

1. It is a liquid.
2. It has a boiling point of 100°C.
3. It has a melting point of 0°C.
4. It flows easily.
5. It has a high specific heat.
6. It is a poor heat conductor.
7. Its density is 1 g/mL.

These hardly resemble the physical properties of the products, hydrogen and oxygen. Both are gases with boiling and melting points below 0°C, and their densities are quite low relative to water.

Example Problem 4.1

Classify each as a physical or chemical property. (a) Silver will tarnish upon standing. (b) Ice floats on water. (c) Copper can be drawn into thin wires.

Solution

(a) "Silver will tarnish upon standing" is a chemical property. Tarnishing describes a chemical change that silver undergoes with oxygen in the air, resulting in a change of composition. We observe that silver is tarnished by the blackish coating on the surface of the silver. Silver is not black!

(b) "Ice floats on water" is a physical property. Floating involves no change in physical properties, which immediately eliminates the possibility of a chemical change. Ice floats on water because ice has a lower density than water. Density is a physical property. A substance always floats on the surface of a fluid of higher density (assuming it does not mix with the fluid).

(c) "Copper can be drawn into thin wires" is a physical property. Copper's shape is changed when drawn into wires; its composition does not change.

When classifying physical and chemical properties, ask yourself the following: Has the composition changed? Can an equation be written to indicate reactants and products? If the answer is "yes" to both of these questions, the property being described is chemical. Solving problems and answering questions by eliminating possibilities is an effective technique. Sort through all given information and discard what you can.

Example Problem 4.2

Classify the following as physical or chemical changes. (a) Steam is condensed to liquid water. (b) Fermenting grapes produce alcohol. (c) A candle is burned.

Solution

(a) "Steam is condensed to liquid water" is a physical change. Steam is water in the gaseous state; hence, the composition is the same before and after the change, a physical change.

(b) "Fermenting grapes" is an example of a chemical change. Fermentation is the process whereby yeast is added to crushed fruits or grains to produce alcohol, the drinking variety. Alcohol, a new substance, is formed after fermentation, indicating a chemical change.

(c) A burning candle illustrates a chemical change. Whenever a substance is burned, the initial substance, in this case the candle, is changed to new substances with new compositions; accordingly, the change is chemical.

REVIEW PROBLEMS

4.1 Classify each property as physical or chemical: (a) Sulfur is bright yellow. (b) Silicon is a relatively hard substance. (c) Cadmium is corroded by acids. (d) Hydrogen explodes when ignited in the air.

4.2 Classify each as a chemical or physical change: (a) Paper ignites when placed in a flame. (b) Sand and water are separated when passed through a filter. (c) Charcoal burns, leaving ashes. (d) Dry ice forms a vapor "cloud" upon heating. (e) Sugar dissolves in water.

4.2 STATES OF MATTER

All matter on earth exists in three physical states: **solids, liquids,** and **gases.** Various properties distinguish these three states of matter. The properties most frequently considered when identifying physical states of matter are (1) shape, (2) volume, (3) average density, (4) structure, (5) fluidity, and (6) compressibility. (See Table 4.1.) Shape, volume, and density have been discussed before; the last two require explanations.

Fluidity is the measure of a substance's ability to flow. Resistance to flow, the opposite of fluidity, is called **viscosity.** Water is more fluid than molasses; the viscosity of water, consequently, is less than that of molasses. **Compressibility** is the measure of the decrease in volume of a substance with an applied pressure. A substance is deemed compressible if a force exerted on its surface results in a compacting of the substance.

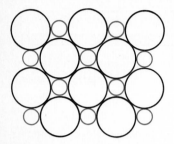

FIGURE 4.3
Most solids are composed of a regular array of closely packed particles. Particles within solids are more organized and packed more tightly than the particles within liquids and gases.

TABLE 4.1 PROPERTIES OF THE THREE PHYSICAL STATES

Property	Solid	Liquid	Gas
Shape	Constant	Variable	Variable
Volume	Constant	Constant	Variable
Density	High	Moderately high	Low
Structure	Organized	Semiorganized	Random
Fluidity	Low	High	Very high
Compressibility	Incompressible	Incompressible	Compressible

Let's consider each physical state individually, starting with the solid state and proceeding to the liquid and gaseous states. A substance's physical state depends on temperature and pressure. Unless otherwise noted, room conditions of 25°C (298 K) and normal atmospheric pressure (1 atmosphere, a unit of gas pressure) are assumed.

Solids

Solids have a fixed shape and cannot be compressed.

Solids have fixed shapes that are independent of their containers. The volume of a solid is also fixed, and does not change when pressure is exerted. Solids are **incompressible.** Solids that seem to be compressible, like styrofoam or corrugated paper, actually are solids that contain holes, or empty regions, throughout. When "compressed," the solid structure fills into the empty regions: the solid itself is not compressed.

Of the three states of matter, solids have the highest average density. Densities in excess of 1 g/mL are the norm for solids, whereas liquids infrequently have densities of this magnitude, and gases never have a density anywhere near 1 g/mL. A high average density reflects the fact that structurally the particles within solids are packed together closer than are the particles in liquids or gases. The tightly packed particles of solids are also highly organized (Figure 4.3); regular patterns of particles are found in solids that are not detected in liquids and gases.

Solids have practically no ability to flow because the particles of a solid are very tightly held together. Stated in another way, solids have very high viscosities.

Liquids

Liquids have a constant volume and take the shape of their container's bottom.

Liquids are quite different from solids in many respects, but there are some common characteristics. Like solids, liquids are incompressible. Pressure exerted on liquids generally produces little, if any, change in volume. When placed into a container, liquids assume the shape of the container's bottom (see Figure 4.4).

As previously mentioned, the average density of liquids is less than that of solids and greater than that of gases. Liquid particles are not held together as strongly as solids, and they are less orderly—more random. Both of these factors tend to increase the average volume of liquids relative to solids. Thus, for equal masses of an average solid and average liquid, the liquid's volume is usually larger, resulting in a lower density.

Liquids have a lower average density than solids, and their structure is more random.

Fluidity is a special property of liquids. Generally, people immediately

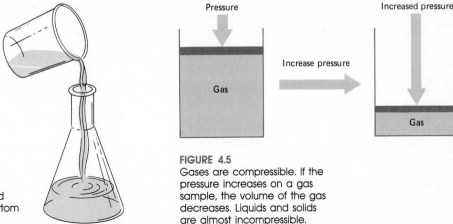

FIGURE 4.4
A liquid completely fills and takes the shape of the bottom of its container.

FIGURE 4.5
Gases are compressible. If the pressure increases on a gas sample, the volume of the gas decreases. Liquids and solids are almost incompressible.

associate fluidity with liquids. Liquids are more fluid than solids, but less fluid than gases. Viscosities of liquids vary over a broad range.

Gases

Gases have an indefinite shape and volume. They have low average density and are the most fluid physical state.

Gases bear little resemblance to the more condensed states of matter. To a degree, gases' properties are the opposite of those of solids. Gases completely fill the volume of their containers, approach infinite compressibility (see Figure 4.5), have a disorganized structure, possess the lowest average density, and are more fluid (have lower viscosities) than liquids and solids. See Table 4.2 for a list of common solids, liquids, and gases.

TABLE 4.2 COMMON SOLIDS, LIQUIDS, AND GASES (25°C and 1 atm)

Solids	Liquids	Gases
Ice	Water	Steam
Iron	Mercury	Neon
Steel	Alcohol	Carbon dioxide
Sugar	Oils	Helium
Carbon	Kerosene	Nitrogen
Salt	Ether	Methane

Changes of State

Solid and liquid states coexist at the melting point.

Matter can change from one physical state to another. For example, solids, when heated, change to liquids. The characteristic temperature at which a solid changes to a liquid is called the **melting point**. At the melting point, both the solid and liquid forms of the substance can exist together. Liquids, in turn, change to solids as they are cooled. The temperature at which a liquid becomes a solid is called the **freezing point**. Freezing and melting occur at the same temperature. In one case, the solid is changed into a liquid, it melts; in the other

$$\text{Solid} \; \underset{\text{freezing}}{\overset{\text{melting}}{\rightleftharpoons}} \; \text{liquid}$$

TABLE 4.3   MELTING AND BOILING POINTS OF COMMON SUBSTANCES

Substance	Melting point, °C	Boiling point, °C
Hydrogen	−259.2	−252.7
Helium	−269.7	−268.9
Oxygen	−218.8	−183
Sodium	97.8	892
Mercury	−38.4	357
Iron	1536	3000
Aluminum	660	2450
Water	0	100

direction a liquid is changed into a solid, it freezes. Water normally freezes or melts at 0°C. Table 4.3 lists melting (freezing) points of a number of substances.

On the other end of the scale, liquids can change to gases. The transition temperature is termed the **boiling point.** Just as at the melting point, at the boiling point two states coexist. These are liquid and gas. When a gas changes to liquid, the change is called **condensation.** Hence, when a gas is cooled, it changes back to a liquid at the condensation point (this term is rarely used). Once again, refer to Table 4.3 for examples of normal boiling points.

Liquid and gaseous states coexist at the boiling point.

$$\text{Liquid} \underset{\text{condensation}}{\overset{\text{boiling}}{\rightleftarrows}} \text{gas}$$

Numerous solids change directly to gases without going through the liquid state. This state change is called **sublimation,** and as with other transitions, it occurs at a specific temperature and pressure. At this point, called the subliming point, the substance exists in both the solid and gaseous states. A good example of a solid that sublimes is "dry ice" or solid carbon dioxide. At −78°C (195 K), solid carbon dioxide and gaseous carbon dioxide exist together.

Changing from a solid directly to a gas is called subliming.

REVIEW PROBLEMS

4.3 Define fluidity and viscosity. What trend exists in fluidity when considering the solid, liquid, and gaseous states?

4.4 What properties distinguish an average solid from an average liquid?

4.5 (a) What physical state has the most unorganized structure?
(b) What accounts for the unorganized structure?

4.6 What states coexist at the following transition points: (a) melting point, (b) subliming point, (c) freezing point, (d) boiling point?

4.3 CLASSIFICATION OF MATTER

Matter is found in many different forms. Every year, thousands of new types of matter are synthesized. When dealing with such a gigantic variety of substances, it is best to divide this enormous group into a series of smaller categories containing similar matter forms.

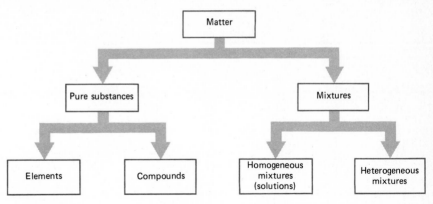

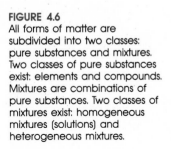

FIGURE 4.6
All forms of matter are subdivided into two classes: pure substances and mixtures. Two classes of pure substances exist: elements and compounds. Mixtures are combinations of pure substances. Two classes of mixtures exist: homogeneous mixtures (solutions) and heterogeneous mixtures.

Matter is first divided into two broad categories—pure substances and mixtures. (See Figure 4.6.)

Pure Substances A substance is classified as pure if it meets the following criteria:

1. It has the same composition throughout the sample.
2. Its components are inseparable using physical methods.
3. Changes of state occur at a constant temperature.

Analysis of a pure substance reveals the same composition throughout the sample. All parts of a pure substance contain the same percentage of each component. Water is a pure substance—all water samples contain 11% hydrogen and 89% oxygen by mass. If some other percentage is found for hydrogen and oxygen, then the substance is not pure water or it may not be water at all.

Pure substances cannot be separated by physical methods. Pure substances cannot be separated by physical methods. If they are capable of being separated (not all are), chemical means are needed. For example, one way of separating water is by passing an electric current through it. This results in the production of hydrogen and oxygen (see Equation 4.1). This is a chemical method. Physical methods, such as filtering or heating at normal temperatures, have no effect on water with respect to separating it into its components. Water passes unchanged through filters, and heat changes water to steam (gaseous water).

Pure substances undergo state changes at a specific temperature. All pure substances undergo state changes at a certain fixed temperature. In contrast, state changes of mixtures occur over a wider temperature range. Chemists can use this property to assess the purity of a substance; a sharper melting point (smaller observable temperature range) indicates a higher level of purity. A broad melting temperature range is characteristic of an impure sample, a mixture.

Pure substances are subdivided into two groups, elements and compounds. Elements are pure substances that cannot be decomposed by chemical methods. Compounds are pure substances that can be chemically decomposed to elements.

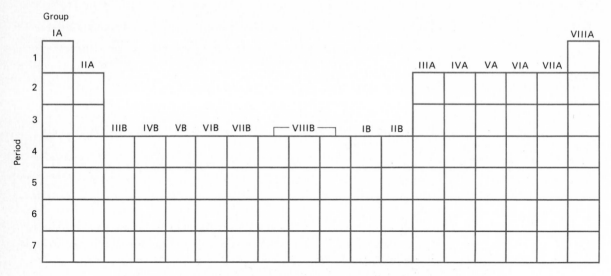

Group

FIGURE 4.7 All the elements are listed in the periodic table. Each element is a member of a chemical group, which is shown as a vertical column of elements, and a period, shown as a horizontal row of elements. Each chemical group is assigned a roman numeral and a letter (A or B). Each period is denoted by a number.

Elements

The periodic table lists symbols and other valuable information about the elements.

ELEMENTS

Hardest:
    carbon (diamond)
Highest density:
    osmium (22.6 g/mL)
Highest mp:
    tungsten (3410°C)
Lowest mp:
    helium (−270°C)
Best electric conductor:
    silver

Elements are the basic units of matter. All matter contains elements. Today, we know of approximately 106 different elements. Whenever the number of elements is stated, it's necessary to introduce some doubt. New elements are occasionally discovered, and they are added to the list. About 92 elements occur in nature, and the rest are synthetic. At 25°C, 93 elements are solids, 2 are liquids, and 11 are gases.

All of the known elements are listed in the **periodic table,** probably the most important table in chemistry (see inside front cover). Each element is located in a horizontal row called a *period* and in a vertical column called a *group* (sometimes called a *family*). Each period is numbered consecutively from 1 to 7; each group of elements is assigned a roman numeral and a letter (see Figure 4.7). An alphabetical listing of the elements is also provided inside the front cover.

Chemists normally use symbols to represent elements. Representing elements by symbols dates back to the Greeks, who originally suggested that matter was composed of elements. Table 4.4 lists some ancient symbols for four of the earliest known elements.

Modern symbols are usually derived from the first one or two letters of the element's name. Twelve elements have one-letter symbols which correspond to the first letter of the element's name. They are hydrogen, H; boron, B; carbon,C; nitrogen, N; oxygen, O; fluorine, F; phosphorus, P; sulfur, S; vanadium, V; yttrium, Y; iodine, I; and uranium, U.

Two other elements possess one-letter symbols, but the symbols are not the first letter of the element's English name. K is the symbol for the element potassium. Why K? K is the first letter of the old Latin name for potassium, *kalium,*

TABLE 4.4   ANCIENT SYMBOLS OF THE ELEMENTS

Source	Sulfur	Gold	Copper	Lead
Ancient Greeks	ω	♂	♀	♄
Alchemists	♁	⊙	♀	♄
John Dalton (1808)	⊕	Ⓖ	Ⓒ	Ⓛ
Modern	S	Au	Cu	Pb

meaning ashes. It's convenient to use K instead of P or Po since there are other elements with these symbols. Similarly, the symbol for tungsten, W, is derived from an old name—wolfram.

The first letter of an element's symbol is capitalized. The second letter is always lowercase.

Most of the remaining elements are assigned two-letter symbols. The first letter is always a capital letter, and the second is a lowercase letter. Cobalt's symbol is Co, not CO. CO is not a symbol, it is the formula of a compound. Some symbols are the first two letters of the English name, others are two letters of an old name, and the remainder contain the first letter plus some other letter in the name.

Ytterby, a town in Sweden, has four elements named after it: yttrium (Y), terbium (Tb), erbium (Er), and ytterbium (Yb).

Origins of the modern names of elements are interesting. Elements are named after geographical locations: Ge, germanium (Germany); Po, polonium (Poland); Fr, francium (France); Eu, europium (Europe); Am, americium (America); Bk, berkelium (Berkeley, CA); Cf, californium (California); Sr, strontium (Strontia, Scotland); etc. Others are named for important scientists: Es, einsteinium (Albert Einstein); Cm, curium (Marie and Pierre Curie); Fm, fermium (Enrico Fermi); No, nobelium (Alfred Nobel); etc. Still other elements are named for mythological gods or astronomical bodies: U, uranium (Uranus); Pu, plutonium (Pluto); Th, thorium (Thor); Np, neptunium (Neptune); Pd, palladium (an asteroid called Pallas); Se, selenium (Greek *selene*, "moon"), and He, helium (Greek *helios*, "sun").

An atom is the smallest unit of an element.

Atoms are the smallest particles that retain the chemical properties of elements. Atoms are extremely small; a gram of carbon, C, contains 5×10^{22} C atoms. Placed end to end, approximately 2×10^8 C atoms are needed to span an inch. Each element is composed of similar atoms. Both chemical and physical properties of an element are directly related to the structure and properties of its atoms. In Chapter 5 we shall investigate this central topic in chemistry.

Compounds

Compounds make up the other class of pure substances. Compounds are more complex than elements. Elements undergo chemical reactions to form compounds; thus, **compounds** are chemical combinations of elements. Consequently, compounds can only be separated into elements by chemical means. The smallest subdivision of a compound is a **molecule**—a chemical combination of atoms.

Formulas are used to represent compounds. Each formula shows the specific composition of a compound. Most people are familiar with the chemical formula of water, H_2O. What information is conveyed by water's formula? It indicates that all water is composed of two parts hydrogen and one part oxygen. If water is decomposed electrically, two volumes of hydrogen are recovered for each volume of oxygen.

$$\text{Water} \longrightarrow 2 \text{ hydrogens} + 1 \text{ oxygen}$$

Additionally, the formula of a compound gives the ratio of atoms within one molecule of that compound. For example, water molecules are particles that contain two atoms of hydrogen and one atom of oxygen.

In each chemical formula the atoms are listed, using their symbols, with a subscript—a number written to the right and below the symbol—indicating the total number of atoms within the molecule. If only one atom of a given element is found in the molecule, the subscript 1 is not written; it is understood to be 1. In the formula for the water molecule, the subscript 2 is placed next to hydrogen, but no subscript is written next to oxygen since there is only one oxygen atom per molecule. Let's take carbon dioxide as another example.

$$CO_2 \quad \text{(carbon dioxide)}$$

1 carbon⌐ ⌐2 oxygens

In the CO_2 molecule there are one carbon atom and two oxygen atoms. When reading the formula, we say "see-oh-two."

Other examples of simple compounds are: NaCl, sodium chloride; NH_3, ammonia; CH_4, methane; and $C_{12}H_{22}O_{11}$, sucrose. NaCl is common table salt and has a wide variety of uses. All living systems maintain a balance between the NaCl and fluids within their cells. NH_3, ammonia, is an ingredient of many common household cleaners. Ammonia is also found in some fertilizers and is chemically changed to other helpful substances. CH_4, methane, is the gas found in largest concentration in what is called natural gas, a fuel. Sucrose, $C_{12}H_{22}O_{11}$, is table sugar, the sweet compound that we place in food and drinks.

In more complex formulas, parentheses are used to group repeating units. For example, the formula of calcium phosphate is $Ca_3(PO_4)_2$. In calcium phosphate there are two PO_4 (phosphate) groups. Instead of writing PO_4 twice, we enclose it in parentheses and write 2 as a subscript to the right. The formula for calcium phosphate indicates that there are three calcium atoms, two phosphorus atoms, and eight oxygen atoms in calcium phosphate.

Only 106 elements are now positively identified. Each element can form thousands or even millions of compounds; hence, the number of compounds that can exist is limitless.

REVIEW PROBLEMS

4.7 What are the three criteria used to identify pure substances?

4.8 What are the two classes of pure substances?

4.9 What are the names of elements for which the following are symbols: (a) B, (b) Be, (c) Ba, (d) Br, (e) Bi?

4.10 Write the symbols for and name 10 elements whose first letter is C.

4.11 List five elements where the first letter of the symbol is different than the name.

4.12 Given the formula of a compound, what information is known?

4.13 Write the name of each atom and tell how many there are in each of the following molecules: (a) KF, (b) $BeCl_2$, (c) NiS, (d) $Fe(OH)_3$, (e) PI_3, (f) $Ba(NO_3)_2$, (g) $NaMnO_4$, (h) $(NH_4)_2C_2O_4$.

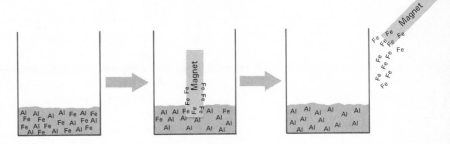

FIGURE 4.8
Mixtures are separated by physical methods. A mixture of iron and aluminum is separated by placing a magnet in the mixture. Aluminum is not attracted by a magnet, but iron is strongly attracted.

Mixtures
Mixtures are the second major division of matter. They are more complex than pure substances. **Mixtures** are composed of two or more pure substances that are physically associated. Three criteria are used to classify a mixture:

1. Its composition is variable.
2. Its components are separated by using physical methods.
3. Changes of state occur over a range of temperatures.

Many possibilities are open when preparing a mixture. Two substances can be mixed in virtually any proportion. For example, a mixture of sugar and water might contain 1 g of sugar in 100 mL of water, or 15 g of sugar in 80 mL of water, or many other possible combinations.

Unlike pure substances, mixtures are separable by physical methods. Sugar and water are separated by evaporating the water, leaving the sugar behind. A mixture of iron and aluminum is separated with a magnet. Iron is attracted to the magnet, leaving the aluminum behind (Figure 4.8).

When a mixture undergoes a change of state, the observed temperature is not constant. Usually, a fairly broad range of temperatures is observed—from the point where the mixture begins to change state until the entire mixture is in the new state. Since mixtures are composed of two or more pure substances,

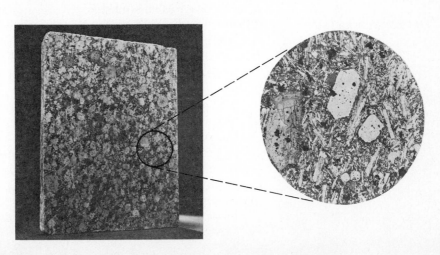

FIGURE 4.9
Granite is an example of a mixture. It is composed of quartz, feldspars, and mica. A close look at granite reveals these different minerals within the rock. (*Photograph by Ron Testa; Field Museum of Natural History, Chicago*)

each component changes state at a different temperature, producing the wide range of temperature.

Milk, gasoline, asphalt, ocean water, granite, and air are some examples of mixtures. Milk's components include such compounds as water, proteins, fats, and vitamins. Gasoline is a mixture of organic compounds, chiefly carbon-hydrogen compounds. Asphalt contains tarry substances blended with sand, gravel, glass, and stones. Ocean water is primarily water with a large number of dissolved substances: minerals, salts, and gases. Granite is a rock made of quartz, feldspar, and mica—three common minerals (see Figure 4.9). Lastly, air is a mixture of gases, mainly nitrogen and oxygen, with smaller amounts of carbon dioxide, water vapor, argon, and others.

Homogeneous and Heterogeneous Mixtures

Like pure substances, mixtures are divided into two classes, homogeneous and heterogeneous mixtures. "Homogeneous" is a word derived from *homo*, meaning the "same" or "equal," and *genus*, which means "kind" or "structure." "Hetero" is a prefix meaning "different."

A heterogeneous mixture is identified by observing more than one phase.

A **heterogeneous mixture** is one exhibiting different (more than one) phases. A phase is an observable region of matter with a different composition than the surrounding regions. Each phase can be distinguished from bordering regions by its properties (see Figure 4.10). For example, when sand is added to water, the sand does not dissolve; it just falls to the bottom of the water. When observing sand and water, we can see the solid sand phase and the liquid water phase. Sand and water, oil and water, salt and sand, and granite, are all examples of heterogeneous mixtures.

Solutions are homogeneous mixtures of pure substances.

Homogeneous mixtures are also called **solutions.** When observing a solution, we see only one phase. Sugar water is an example of a solution. It is prepared by adding solid sugar to liquid water. After the sugar dissolves, a homogeneous mixture results. Looking at sugar water, we cannot tell if it is pure water or a solution.

Other examples of solutions are (1) alcohol and water, (2) air, and (3) alloys. Grain alcohol and water is a popular beverage throughout the world. Alcohol content of drinks can range from a few percent to as high as 60% alcohol by volume. Air is a gaseous solution. All gaseous mixtures exhibit only one phase; therefore, all are solutions. An alloy is a solution of metals. Most commonly used metals are alloys. Sterling silver is 92.5% Ag (silver) and 7.5% Cu (copper). Brass, another alloy, is 67% Cu (copper) and 33% Zn (zinc).

As we shall see later, the properties of solutions are different from those of the pure substances that make up the solution.

Mixtures, whether heterogeneous or homogeneous, are separated by physical methods. A solution of salt and water can be separated by heating, for example. During heating, water changes to a vapor, leaving the salt behind. This is the basis of the process called **distillation.** Figure 4.11 illustrates a laboratory distillation setup. Crude oil, after being pumped out of the ground, is separated in a similar manner. A tall building is required to separate efficiently such a complex mixture as crude oil.

Filtration is a method that is utilized to separate some heterogeneous mix-

FIGURE 4.10
In a heterogeneous mixture, two or more phases are observed. A phase is an observable region of matter with a different composition than the surrounding regions. This diagram shows a three-phase system. There is an interface (boundary) between one phase and another phase.

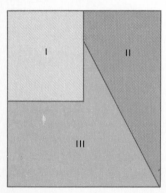

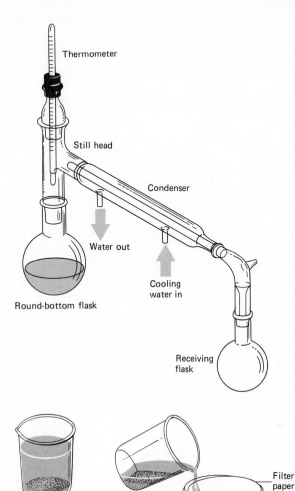

FIGURE 4.11
Some mixtures can be separated by distillation. A mixture is placed in a round-bottom flask. The mixture is then carefully heated to vaporize the component with the lowest boiling point. These vapors are condensed in a water-cooled condenser, and the liquid drips into a receiving flask.

FIGURE 4.12
A sand-and-water mixture is separated by gravity filtration. The mixture is poured into a funnel containing filter paper. Water passes through the filter paper. Sand particles are trapped in the filter paper.

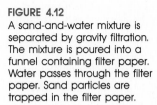

Wine is filtered to remove particulate matter before it is bottled.

tures. Such mixtures are poured into a funnel containing filter paper, which is paper containing small, uniform openings or pores. The pores block large particles and allow small particles to pass through. Sand and water are separated by pouring the mixture into a filter. Water, composed of small water molecules, easily passes through the paper. Particles of sand are too large to travel through the filter paper's pores, and thus they remain behind (see Figure 4.12).

> ## Example Problem 4.3

How could a mixture of sand and salt be separated?

Solution

A combination of sand and salt is a heterogeneous mixture. One contrasting property of each, their abilities to dissolve in water, provides a convenient method of separation. Salt dissolves in water, whereas sand does not. Thus, if water is added to the mixture, the salt dissolves leaving the sand undissolved. The salt water solution that results is filtered. Sand is trapped by the filter paper while the salt water passes through. To recover the salt, the water is evaporated. Thus to separate a mixture of sand and salt:

1. Add water to dissolve salt. The sand is unaffected.
2. Filter resulting salt water solution from the sand.
3. Evaporate water to recover the salt.

REVIEW PROBLEMS

4.14 What are the criteria for classifying mixtures?
4.15 Distinguish between heterogeneous and homogeneous mixtures.
4.16 Give two examples of solutions commonly used around the house.
4.17 How are the following mixtures separated: (a) oil and water, (b) carbonated water (carbon dioxide, a gas, dissolved in water), and (c) crushed rock and salt?

4.4 ENERGY

Work is done when a force applied to matter causes the matter to change its position.

Energy is defined as the capacity to do work. What does this mean? In science, the term "work" does not mean the 9 A.M. to 5 P.M. type of work. **Work** is done when matter is moved by applying a force—a push or pull. Lifting a book off a table or pushing a stalled car requires work. In science, something must be moved to do work. Energy, therefore, is the capacity to move or effect changes in matter.

Two general classes of energy exist, potential and kinetic energy. **Potential energy** is stored energy. This potential or stored energy results from an object's position, condition, or composition. A boulder moved from the ground to the top of a cliff has potential energy of position. The boulder can fall off the cliff and crush objects below. A compressed spring has potential energy of condition; spontaneously, the spring can expand and do work (push something). A vial of nitroglycerin possesses potential energy of composition, or chemical potential energy. You do not want to drop a bottle of nitroglycerin and release this stored energy!

Potential energy is stored energy.

Kinetic energy is the energy associated with matter in motion. Whenever an

FIGURE 4.13
(*a*) A boulder on the ground has no kinetic energy since it is not moving, and it has no potential energy with respect to the ground. (*b*) Energy is required to move the boulder from the ground to the top of the cliff. On top of the cliff the boulder has no kinetic energy, but it has potential energy with respect to the ground. The greater the distance above the ground (*h*), the greater its potential energy. (*c*) If the boulder falls off the cliff, it possesses both kinetic energy and potential energy until it reaches the ground.

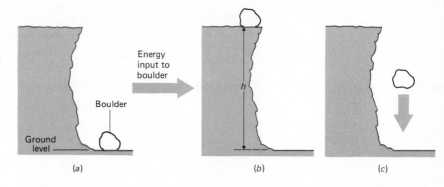

(a)   (b)   (c)

Kinetic energy is energy in motion.

Velocity is the speed of an object in a particular direction.

object is moving, it possesses kinetic energy. Potential energy is released in the form of kinetic energy. For example, a boulder at the top of a cliff has potential energy only. But as soon as it falls from the cliff, the boulder's potential energy is changed to energy of motion, or kinetic energy. When the boulder hits the ground, it stops moving; it no longer has kinetic energy. (See Figure 4.13.)

The kinetic energy of an object is proportional to its mass and velocity squared. The more massive an object and the faster the object is moving, the greater the kinetic energy it possesses.

Energy is encountered in a wide variety of forms: (1) mechanical energy, (2) electric energy, (3) nuclear energy, (4) light, (5) heat, and (6) sound. (See Figure 4.14.) All of these examples are energy forms because each has the capacity to produce changes in matter.

Energy can be converted from one form to another. For example, gasoline is burned in an engine and transformed into mechanical energy. The chemical potential energy of the gasoline is released, causing a piston to move which ultimately moves the wheels of an automobile. Along with the mechanical en-

FIGURE 4.14
Lightning is an electric discharge that produces light energy, heat energy, and sound energy. (*U.S. Department of Agriculture/Lyle Orwig*)

FIGURE 4.15
A nuclear power plant differs from a conventional power plant primarily in the way it produces heat energy. In a nuclear power plant, energy released from splitting atoms (nuclear energy) is converted into heat energy. The liberated heat energy changes liquid water into steam. High-pressure steam turns a turbine that is connected to an electric generator. (*Atomic Industrial Forum*)

ergy, heat is also released, an unavoidable transformation. Electric potential energy stored in a battery is transformed into light (radiant) energy inside a flashlight.

Power plants burn fuels to release heat energy. This heat is transferred to water, which is converted to high-pressure steam, which turns a turbine. Turbines are connected to electric generators that change the turbine's mechanical energy into electricity. Finally, the electricity is sent to consumers who use it to perform innumerable tasks. In this illustration, chemical potential energy is converted to heat energy, mechanical energy, and finally electric energy.

Heat Energy

Heat energy is especially important to chemists. Why? All other forms of energy can be transformed into heat, and heat changes accompany chemical reactions. Heat is a form of kinetic energy; it can never be classified as potential energy. As we have already discussed, heat brings about both chemical and physical changes. Solids melt, liquids evaporate, and some substances decompose or undergo chemical changes when heated.

When heat energy is transferred to matter, it increases the potential energy, or the kinetic energy, or both. What happens when a substance is heated? It becomes hotter, you may say. In some cases that is correct. Substances become hotter because the heat transferred increases the motion of the molecules—a kinetic energy increase. We observe this increase in kinetic energy by measuring the substance's temperature. Temperature is a measure of the average velocity of atoms and molecules. Higher temperatures of pure substances indicate faster-moving molecules.

Temperature is proportional to the average velocity of molecules.

But heat transferred to ice at 0°C gives a different result. As long as ice is present, added heat does not increase the temperature. State changes of pure substances occur at a constant temperature. Added heat increases the potential energy of the ice to the point where it becomes a liquid. Only after all of the ice

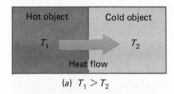

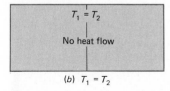

FIGURE 4.16
(a) If two objects at different temperature ($T_1 > T_2$) contact each other, heat flows spontaneously from the hot object at T_1 to the colder object at T_2. Initially T_1 decreases and T_2 increases until T_1 equals T_2; at that time, heat flow ceases. (b) If two objects at the same temperature ($T_1 = T_2$) contact each other, no heat is transferred between the objects.

is melted does the temperature increase; this temperature increase indicates an increase in the average kinetic energy of the molecules.

Heat energy is detected only when it moves from one body to another. In our world heat travels down a one-way street. Heat is always transferred from a hotter object to a colder object; the reverse never happens spontaneously (see Figure 4.16).

When two objects at different temperatures touch each other, heat is transferred to the object with the lower temperature. Faster-moving molecules, within the hotter object, collide with the slower-moving molecules in the colder object. During the molecular collisions energy is transferred, decreasing the average kinetic energy of the faster-moving molecules and increasing that of the slower-moving molecules. When their average energies are the same, heat flow stops.

Consider what happens when a cup of hot coffee is allowed to sit for a period of time in a room. As time passes, the temperature of the coffee drops. If the coffee is not consumed, its temperature will continue to drop until it is the same as the temperature of the room. What happens to the heat? Heat flows from the coffee and heats the room. But, compared with the total amount of heat in the room, the small quantity of heat released by the coffee is insignificant.

To summarize, heat is a form of energy that always flows from warmer to colder objects. Temperature is a measure of how fast the particles that compose matter are moving. In other words, temperature is a measure of the average kinetic energy of atoms and molecules. Knowing the temperature of two objects allows us to predict the direction of heat flow.

REVIEW PROBLEMS

4.18 If work is done on an object, what happens to that object?
4.19 Distinguish between kinetic and potential energy.
4.20 List four forms of energy.
4.21 What energy transformations take place within a television set?
4.22 What is the difference between heat and temperature?
4.23 Describe the heat flow and temperature change in a cup of hot coffee in an empty room.

4.5 MEASUREMENT OF ENERGY

Joules and Calories

Joule was the son of a wealthy brewer. He was self-educated, and he constructed his laboratory in his house. Joule befriended William Thomson (Lord Kelvin) when Thomson was a young student.

Energy is measured using the derived SI unit called the joule (J). Technically, a joule is a unit of mechanical energy. The joule was named after James P. Joule (1818–1889), an English scientist who performed significant early experiments relating energy forms.

Prior to SI, joules were used exclusively for expressing mechanical equivalents of energy. A second unit, the calorie (cal), was employed to express heat energy. Today, both are used to measure energy, and so we shall work with both units, although the calorie is non-SI.

How does the magnitude of the calorie compare with the joule? The calorie

James P. Joule (*National Portrait Gallery, Smithsonian Institution, Washington, DC.*)

is over 4 times larger than the joule. By definition, one calorie is equal to 4.184 J—an exact conversion factor.

$$1 \text{ cal} = 4.184 \text{ J}$$

As you can see, the joule is a relatively small unit of energy.

In the past, a calorie was defined as the amount of heat necessary to raise the temperature of 1 g of pure water from 14.5°C to 15.5°C. Since calories are a rather small unit of heat, kilocalories are commonly encountered. A kilocalorie (kcal) is equal to 1000 cal. If 1 kcal of heat is transferred to 1 kg of water at 14.5°C, the added heat increases the temperature to 15.5°C. Example Problem 4.4 shows how kilocalories can be converted to joules.

Example Problem 4.4

SOME CALORIC VALUES

Food	Heat content, kcal
Apple	55
Egg	75
10 potato chips	100
Orange juice, ½ glass	50
10 peanuts	50
1 soda	100
Coffee, plain	0
Beer	200
Cheesecake, 1 slice	275
Chocolate bar with nuts	250

Convert 2.9 kcal to joules.

Solution

1. *What is unknown?*   Joules, J
2. *What is known?*   2.9 kcal, 1000 cal = 1 kcal, 1 cal = 4.184 J
3. Apply the factor-label method.

$$2.9 \text{ k̶c̶a̶l̶} \times \frac{1000 \text{ c̶a̶l̶}}{1 \text{ k̶c̶a̶l̶}} \times \frac{4.184 \text{ J}}{1 \text{ c̶a̶l̶}} = ? \text{ J}$$

4. Perform the indicated math operations.

$$2.9 \text{ k̶c̶a̶l̶} \times \frac{1000 \text{ c̶a̶l̶}}{1 \text{ k̶c̶a̶l̶}} \times \frac{4.184 \text{ J}}{1 \text{ c̶a̶l̶}} = 1.2 \times 10^4 \text{ J}$$

Specific Heat

Water's high specific heat makes it an excellent coolant.

Copper, a metal with a low specific heat, is placed on the bottom of some cookware to maximize the heat transfer from the heat source to the food.

When substances that are not undergoing state changes are heated, they increase in temperature. **Specific heat** (c) is the amount of heat required to increase the temperature of 1 g of substance by 1°C. Each substance absorbs heat at a different rate. It requires 1 cal to increase the temperature of 1 g of water by 1°C. Only 0.1 cal is needed to increase 1 g of copper by 1°C. Copper has a lower specific heat than water; i.e., less heat is needed to raise the temperature of an equal quantity of copper compared with water.

The specific heat of a substance is calculated with the following equation:

$$\text{Specific heat } (c) = \frac{\text{cal (or J)}}{g \cdot \Delta T}$$

TABLE 4.5   SPECIFIC HEAT OF SELECTED SUBSTANCES

Substance	Specific heat	
	cal/(g·°C)	J/(g·°C)
Water (*l*)	1.000	4.184
Copper (*s*)	0.092	0.39
Hydrogen (*g*)	3.45	14.4
Silver (*s*)	0.056	0.23
Gold (*s*)	0.031	0.13
Helium (*g*)	1.25	5.23
Boron (*s*)	0.309	1.29
Sugar (*s*)	0.30	1.3
Sand (*s*)	0.19	0.80

where g is the mass of the substance in grams, ΔT is the change in temperature $(T_2 - T_1)$, T_2 is the final temperature, and T_1 is the starting temperature. The units of heat energy can be either calories or joules.

You should add specific heat to your ever-growing list of conversion factors. Think of specific heat as the conversion factor that relates heat transfer to the mass and temperature change of a substance.

Table 4.5 lists the specific heats of some common substances. Specific heats of substances are determined by applying a known quantity of heat to a known mass, and then observing the increase in temperature. Example Problem 4.5 illustrates the calculations involved.

Example Problem 4.5

Calculate the specific heat of a solid in cal/(g·°C) and J/(g·°C) if it is found that 346 cal is needed to raise the temperature of 215 g of the solid from 25.3°C to 44.9°C.

Solution

1. *What is unknown?* c, specific heat in $\dfrac{\text{cal}}{\text{g·°C}}$ and $\dfrac{\text{J}}{\text{g·°C}}$

2. *What is known?*   Heat needed = 346 cal, mass = 215 g, initial T = 25.3°C, and final T = 44.9°C

3. Apply the appropriate equation.

$$c = \frac{\text{cal}}{g \cdot \Delta T}$$

$$c = \frac{346 \text{ cal}}{215 \text{ g} \times (44.9°C - 25.3°C)}$$

4. Perform the indicated math operations.

$$c = \frac{346 \text{ cal}}{215 \text{ g} \times 19.6°C} = 0.0821 \text{ cal}/(g·°C)$$

5. Convert cal/(g·°C) to J/(g·°C) using the appropriate conversion factor.

$$1 \text{ cal} = 4.184 \text{ J}$$

$$\frac{0.0821 \text{ cal}}{g·°C} \times \frac{4.184 \text{ J}}{\text{cal}} = 0.344 \text{ J/g·°C}$$

The solid's specific heat is 0.0821 cal/(g·°C) or 0.344 J/(g·°C).

Knowing the specific heat, mass, and change of temperature of a substance allows us to calculate the quantity of heat added or removed. Rearranging the specific heat equation to solve for heat, we get

$$c = \frac{\text{cal}}{g·\Delta T}$$

Multiplying both sides of the equation by g·Δt, we obtain

$$(g·\Delta T)c = g·\Delta T \frac{\text{cal}}{g·\Delta T}$$

or

$$\text{cal} = g \times \Delta T \times c$$

Thus, to find the amount of heat transferred, the mass of substance is multiplied by its change in temperature and its specific heat.

Example Problem 4.6

How many calories are required to increase the temperature of 1.0 kg of silver, Ag, from 22.1°C to 71.9°C?

Solution

1. *What is unknown?*   calories, cal
2. *What is known?*   Mass = 1.0 kg Ag, $\Delta T = 71.9°C - 22.1°C$,
   and $c_{Ag} = \dfrac{0.056 \text{ cal}}{g·°C}$ (from Table 4.5)
3. Apply the equation:

$$\text{cal} = g \times \Delta T \times c$$

$$\text{cal} = 1.0 \, \cancel{kg} \times \frac{1000 \, g}{\cancel{kg}} \times (71.9 - 22.1)°\cancel{C} \times \frac{0.056 \, \text{cal}}{g \cdot °\cancel{C}}$$

Mass in kilograms is converted to grams so the units will cancel when dividing by grams in the specific heat.

4. Perform the indicated math operations.

$$\text{cal} = 1000 \, g \times 49.8°\cancel{C} \times \frac{0.056 \, \text{cal}}{g \cdot °\cancel{C}}$$

$$\text{cal} = 2788.8 \, \text{cal, rounded to } 2.8 \times 10^3 \, \text{cal}$$

The correct answer, 2.8×10^3 cal, is expressed to two significant figures because the c_{Ag} and mass are only known to two significant figures. In multiplication, the answer can have no more significant figures then the smallest number of significant figures in the measured quantities.

Example Problem 4.7

What is the final temperature of sand when 855 cal is added to 375 g of sand that is originally at 29.4°C?

Solution

1. *What is unknown?*   Final temperature, T_2
2. *What is known?*   Heat added = 855 cal, mass = 375 g, initial temperature (T_1) = 29.4°C, and specific heat (c_{sand}) = 0.19 cal/(g·°C) (from Table 4.5)
3. Apply the equation.

$$\text{cal} = g \times (T_2 - T_1) \times c_{sand}$$

$$855 \, \text{cal} = 375 \, g \times (T_2 - 29.4°\text{C}) \times \frac{0.19 \, \text{cal}}{g \cdot °\text{C}}$$

4. Perform the indicated math operations.

Divide both sides by $\dfrac{375 \, g \times 0.19 \, \text{cal}}{g \cdot °\text{C}}$, and then add 29.4°C to both sides to obtain the final answer.

$$12°\text{C} = T_2 - 29.4°\text{C}$$
$$T_2 = 41°\text{C}$$

The sand reaches a final temperature T_2 of 41°C.

REVIEW PROBLEMS

4.24 (a) What is the SI unit of energy? (b) What other energy units are commonly used in chemistry?

4.25 Perform the following conversions: (a) 9.3×10^4 J to calories, (b) 0.033 cal to joules, (c) 1.44×10^5 J to kilocalories.

4.26 What do you know about a substance when you know its specific heat?

4.27 (a) Calculate the specific heat of a substance that requires 134 cal to raise 93.4 g from 15.9°C to 36.1°C. (b) What is the substance's specific heat expressed in $J/(g \cdot °C)$?

4.28 Determine the amount of heat needed to raise 75.4 g of water from 10°C to 90°C.

4.29 If 439 J of heat is added to 50.0 g of water at 12.3°C, what is the water's final temperature?

4.6 CONSERVATION OF MATTER AND ENERGY

One of the most fundamental laws of nature is applied when dealing with matter and energy. This is the **law of conservation of matter/energy.**

At one time this was stated separately as two laws: the law of conservation of matter and the law of conservation of energy. However, these laws were merged by Albert Einstein, who discovered that matter and energy are equivalent. Einstein summarized his discovery in the universally known equation:

$$\Delta E = \Delta mc^2$$

In 1905, while working in a patent office in Switzerland, Albert Einstein published three papers. Each paper was worthy of a Nobel prize. In 1921, he was awarded the Nobel prize in physics for one of these works.

where ΔE is the change in energy in joules, Δm is the change in mass in kilograms, and c is the velocity of light, 3.00×10^8 m/s.

This equation tells us that matter can be converted to energy and energy can be converted to matter. A tiny amount of matter is converted into a huge quantity of energy in nuclear bombs or nuclear power plants. For example, if 1 g of uranium, U, could be totally converted to energy, approximately 9×10^{13} J or 2×10^{13} cal would be released. Stated another way, the energy contained in 1 g U is enough to heat 2×10^5 tons of water from 0°C to 100°C.

Albert Einstein (*The American Friends of the Hebrew University*)

As it turns out for most chemical reactions, which are nonnuclear changes, the amount of matter converted to energy is too small to measure on a balance, and it is safe to say that both matter and energy are conserved. Chemists are primarily concerned with matter-energy conversions in nuclear reactions (see Chapter 19).

The **law of conservation of matter** states that matter cannot be created or destroyed during "normal" chemical changes. Stated another way: The total mass of reactants lost equals the total mass of products formed. Let's illustrate by considering what happens when a log is burned in a fireplace. The mass of all gaseous products, soot, dust particles, and ash equals the initial mass of the log burned. If for some reason we find that the masses are not equal, we can be perfectly sure we overlooked something. Matter cannot vanish. If matter is "lost" in one part of the universe, then it is residing somewhere else; it is not really lost.

The **law of conservation of energy** states that energy cannot be created or destroyed under "normal" conditions. As we have discussed, energy is converted from one form to another. During these conversions no energy is lost. If there is an apparent energy loss, we have not looked hard enough; the energy is somewhere else in the universe. Generally, energy that cannot be accounted for has escaped as heat.

Heat is unavoidably given off in energy transformations. Only a small quantity of the electricity which is pumped into the filament of a light bulb is changed into light; most is given off as heat. An efficient engine only converts 10 to 20% of the energy stored in the fuel into mechanical energy. Most of the energy released heats the engine and surrounding areas.

REVIEW PROBLEMS **4.30** How many joules would be released if 1.0 μg of matter could be totally converted to energy?
4.31 State the law of conservation of matter. What does the law imply?
4.32 State the law of conservation of energy. What does the law imply?

SUMMARY

Our universe is composed entirely of matter and energy. Matter is anything that has mass and occupies space. Energy is the capacity to do work.

Physical and chemical properties are utilized to identify different types of matter. Physical properties are those that can be measured without referring to any other substance. Examples of physical properties are melting point, tensile strength, color, and shape. Chemical properties are observed when a substance is changing its composition—becoming a new substance. Inertness, reactivity, and combustibility are chemical properties.

Solids, liquids, and gases are the three states in which matter is found. Solids are the most dense and least fluid of all the physical states. In contrast, gases are the least dense and most fluid. Solids have a fixed shape and volume; gases expand and take the full shape of their containers, and have a variable volume—depending on applied pressure. Liquids resemble solids more closely than gases. Liquids have a structure somewhat similar to that of solids, with relatively high average density and fixed volume, and they are incompressible. However, liquids share with gases the common property of high fluidity.

Matter is subdivided into two general classes, pure substances and mixtures. Pure substances have a constant composition, cannot be separated by physical means, and undergo state changes at a constant temperature. In contrast, mixtures have a variable composition, can be separated using physical means, and undergo state changes over a wide temperature range.

Pure substances are further subdivided into elements and compounds. Elements are the most fundamental units of matter and are found in all other matter classifications. Approximately 106 elements are known. Compounds are produced when elements are chemically combined. Millions of compounds exist because there are many ways to combine elements. Elements are composed of small particles called atoms, whereas compounds are made up of molecules, which are combinations of atoms.

Mixtures can be either homogeneous or heterogeneous. A homogeneous mixture is a combination of pure substances that can be separated by physical means, has a variable composition, undergoes state changes over a temperature range, and exhibits only one phase. Homogeneous mixtures are called solutions. Heterogeneous mixtures differ from homogeneous mixtures in that they possess two or more distinct phases.

Energy is classified as being potential or kinetic. Potential energy is stored energy, and results from an object's position, condition, or chemical composition. Kinetic energy is the energy of motion. All things that are moving have kinetic energy.

Energy can be interconverted; for instance, electric energy is changed to heat in an electric stove. whenever energy is interconverted, some of the energy is usually lost as heat.

The SI unit of energy is the joule (J), but calories are frequently used as a measure of heat energy. One calorie is equal to 4.184 J. The quantity of heat flow depends on the temperature of an object, and the temperature of what it is in contact with. Heat always flows from a hotter object to a cooler one.

Matter and energy are interconvertible—matter can be changed to energy or vice versa. However, in normal chemical changes, both matter and energy are conserved. The same quantity of matter is present after a chemical reaction as was originally present. Matter cannot be created or destroyed. Likewise, energy cannot be created or destroyed.

QUESTIONS AND PROBLEMS

4.33 Define each of the following terms: matter, energy, property of matter, physical property, chemical property, reactant, product, physical change, chemical change, decomposition, physical state, fluidity, viscosity, change of state, melting point, boiling point, freezing point, subliming point, pure substance, mixture, element, compound, chemical symbol, chemical formula, homogeneous, heterogeneous, solution, alloy, work, kinetic energy, potential energy, calorie, joule, and specific heat.

Physical and Chemical Properties

4.34 Classify each as either a physical or chemical property: (a) existence in the solid state, (b) magnetic properties, (c) explosiveness, (d) combustibility, (e) flammability, (f) boiling point, (g) rusting, (h) density, (i) specific heat, (j) decay, (k) viscosity, (l) hardness, (m) tensile strength, (n) reactivity, and (o) inertness.

4.35 Classify each as a physical or chemical change: (a) formation of an ice cube from liquid water, (b) frying of an egg, (c) fizzing of an Alka-Seltzer tablet, (d) evaporation of gasoline, (e) distillation of alcohol, (f) cutting of a piece of paper, (g) digesting of food, (h) corrosion of a metal, and (i) shaping of steel.

4.36 Consider the following properties of diamond (a pure form of C): (a) electric insulator, (b) density = 3.51 g/mL, (c) chemically inert, (d) extremely hard, and (e) burns in oxygen to produce carbon dioxide. Classify each of the listed properties of diamond as physical or chemical.

4.37 Sulfur is a yellow solid which burns in air to yield poisonous sulfur oxides. On heating, sulfur discolors and turns dark brown at 180°C. Identify all stated chemical and physical properties of sulfur.

Physical States

4.38 List four general properties of (a) solids, (b) liquids, (c) gases.

4.39 From each of the following pairs, determine the substance that has the greater viscosity: (a) water or vegetable oil, (b) motor oil or antifreeze, (c) pudding or soft drink, (d) shaving cream or molasses.

4.40 Identify the state(s) of matter with the following properties: (a) lowest average density, (b) most fluid, (c) intermediate densities, (d) fixed volume, (e) takes volume of the bottom of its container, (f) most orderly structure, (g) strongest forces, (h) greatest viscosity, (i) random structure, (j) highest specific heat.

4.41 What accounts for the fact that various substances cannot exist in all three states of matter?

4.42 What physical state of matter is most commonly found under the following conditions: (a) very high temperatures and low pressures, (b) very low temperatures and high pressures?

4.43 What type(s) of matter possesses the following properties: (a) has a variable composition with one phase, (b) is inseparable by chemical means, (c) exhibits two or more phases, (d) changes state at constant temperature and is separable chemically, (e) is composed of uncombined atoms?

Classification of Matter

4.44 Classify the following as pure substances or mixtures: (a) beer, (b) beef, (c) gold bars at Fort Knox, (d) tap water, (e) charcoal, (f) baking soda, (g) sugar cube, (h) paint, (i) steam, (j) penny.

4.45 Name the following elements: (a) Fe, (b) Li, (c) Se, (d) Ne, (e) Zr, (f) Mg, (g) Be, (h) F, (i) Ce, (j) Ca.

4.46 Give the symbols for the following elements: (a) nickel, (b) nitrogen, (c) neodymium, (d) neon, (e) niobium, (f) nobelium, (g) neptunium.

4.47 Write the names and symbols for all seven elements whose name begins with A.

4.48 Write the names and symbols for all eight elements in the second period of the periodic table.

4.49 What is the difference between a chemical formula and a chemical symbol?

4.50 State the name and number of each atom in the following formulas: (a) N_2O, (b) NO_2, (c) N_2O_4, (d) Na_2CO_3, (e) KNO_3, (f) LiH_2PO_4, (g) $Fe(OH)_2$, (h) $(NH_4)_2CrO_4$, (i) $XePtCl_6$, (j) CCl_2F_2.

4.51 Classify each of the following as an element or a compound from the given information:
(a) Melts at 120°C, boils at 228°C, and decomposes to Si and I
(b) A white substance, it melts at 44°C and combines with oxygen to form P_2O_5
(c) A soft, silvery metal, it reacts violently with water.

4.52 Explain the statement, "mixtures have variable composition."

4.53 Classify the following as being homogeneous or heterogeneous mixtures:
(a) silver dollar, (b) coffee, (c) cement, (d) motor oil, (e) cotton, (f) paper, (g) oil and vinegar, (h) smog.

4.54 How could each of the following mixtures be separated: (a) sand and salt, (b) alcohol and water, (c) sand and water, (d) oil and water?

4.55 How many phases would be observed in the following (not including the container and air): (a) glass of iced tea, (b) bottle of seawater and oil residue, (c) aquarium with four different colors of sand, (d) glass of soda water with ice cubes?

Energy

4.56 What are the three ways potential energy is stored?

4.57 What two factors are directly related to an object's kinetic energy?

4.58 What types of potential energy are possessed by the following: (a) TNT, (b) apple on a tree, (c) wound watch mainspring, (d) stretched rubber band, (e) water at the top of a dam, (f) hamburger?

4.59 For each of the following pairs, determine which possesses the most heat: (a) match flame or bunsen burner flame, (b) cup of water at 90°C or bathtub filled with 90°C water, (c) ice cube at 0°C or a large block of ice at 0°C, (d) teaspoon of boiling water or gallon of water at 50°C, (e) two identical wooden blocks in contact with each other.

4.60 Perform the indicated energy conversions:
(a) 333 cal = ? J   (b) 4391 J = ? cal
(c) 7.8×10^5 J = ? kcal (d) 9.93×10^{10} cal = ? kJ
(e) 0.147 cal = ? kJ

Specific Heat

4.61 Calculate the specific heat of a 60.0-g sample that requires 512 J to raise the temperature 3.9°C.

4.62 Calculate the specific heat of a 125-g sample where the addition of 193 cal increases its temperature from 19.5°C to 23.4°C.

4.63 How much heat is required to raise the temperature of 37.6 g of water from 41.0°C to 100.0°C?

4.64 How many calories of heat energy are released when 112 g of water at 50.0°C cool to 25.0°C?

4.65 Calculate and compare the amount of heat given off when 1.00 kg of water and 1.00 g of water, both at 100.0°C, cool to 25.0°C.

4.66 If all of the energy from the food that we consume in a day could be transferred to water and if a person eats a total of 3000 kcal per day: (a) What mass of cold water at 0°C could be heated to 100°C? (b) How many liters of water would this be (assume water's density to be 1.0 g/mL)?

4.67 Gold's specific heat is 0.031 cal/(g·°C). (a) How much heat is required to increase 400 g Au from 25.0°C to 50.0°C? (b) Compare this quantity of heat with the amount of heat that would be required for an equal mass of water to be increased the same temperature range.

Conservation Laws

4.68 (a) If exactly 12 g of carbon is combined with 32 g of oxygen, how many grams of carbon dioxide form? (b) What law does this illustrate?

4.69 What energy transformations occur when electricity is produced through hydroelectric generation?

4.70 Calculate the amount of energy released if 1 μg of matter is converted to energy.

Atoms

Study Guidelines

1. List the main principles of Dalton's atomic theory, and explain how Dalton's model of the atom differs from the modern model
2. List the properties of the fundamental components of an atom
3. Discuss the overall organization of an atom
4. Determine the composition of an atom, given the atomic and mass numbers
5. Write the symbol of an atom, given the number of protons and neutrons
6. Identify isotopes of an element.
7. Calculate the atomic mass of an element, given the masses of its isotopes and their natural abundances
8. Explain the derivation of the unified atomic mass scale
9. Describe what is meant by an electron orbital
10. Determine the total number of electrons in an atom
11. Distinguish among electron energy levels, sublevels, and orbitals
12. List the maximum number of electrons found in levels, sublevels, and orbitals
13. Write electronic configurations for a given atom
14. Write the outer (valence) electronic configuration for an atom
15. Write electron dot formulas for atoms

Our contemporary theory regarding atoms has evolved over more than 100 years. Theory development is an ever-changing process. Whenever new solid evidence is uncovered that contradicts a theory, the theory must be flexible enough to incorporate the new data. Theories are not facts; they are the best explanations we can offer in light of known information.

The historical development of the modern atomic theory is an unusually interesting story with an exciting plot and extraordinary characters. Most chemical historians would begin this saga with John Dalton (1766–1844), an English chemist. He is credited with developing the first scientific theory of atoms. Dalton, unlike the Greek philosophers (see Section 1.3), based his theory on approximately 150 years of scientific investigations.

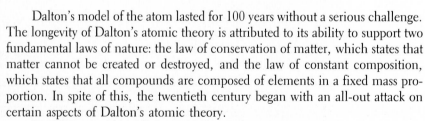

Dalton was also a meteorologist. He wrote a book entitled *Meteorological Observations and Essays.* Dalton also discovered color blindness, an affliction he had.

Dalton's atomic theory stated:
1. Matter is composed of small particles called atoms.
2. Atoms within an element have the same properties, but differ from all other types of atoms found in other elements.
3. Atoms cannot be subdivided or changed to other atoms.
4. Atoms combine together in simple whole-number ratios.
5. Chemical changes involve linking and separation of atoms.

Dalton's model of the atom lasted for 100 years without a serious challenge. The longevity of Dalton's atomic theory is attributed to its ability to support two fundamental laws of nature: the law of conservation of matter, which states that matter cannot be created or destroyed, and the law of constant composition, which states that all compounds are composed of elements in a fixed mass proportion. In spite of this, the twentieth century began with an all-out attack on certain aspects of Dalton's atomic theory.

By 1903, enough evidence was compiled to suggest that the atom possessed a substructure: atoms were made up of smaller particles. At this time physicists unearthed startling facts about matter that totally changed our concepts concerning particles and how they behave. For instance, small particles seemingly obey a different set of laws than those that large particles follow. By the early 1930s, chemists and physicists succeeded in producing what we now call the modern atomic theory. Figure 5.1 diagrammatically illustrates the evolution of scientists' theoretical conceptions of the atom.

Most of the work after 1930 was directed toward the composition of the subatomic particles. Little has changed in the minds of scientists regarding overall atomic structure since the 1930s, suggesting that our atomic model is consistent with the real world.

John Dalton *(National Portrait Gallery, Smithsonian Institution, Washington, DC)*

REVIEW PROBLEMS

5.1 Why is an atomic theory required to describe atoms?
5.2 List the most significant points of Dalton's atomic theory.
5.3 Why is Dalton credited with the discovery of the first scientific atomic model when Democritus proposed an atomic theory many years before?

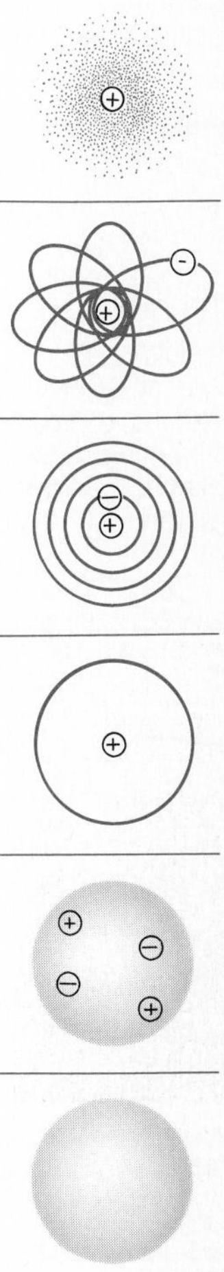

Quantum Mechanical Model (Current Model)

In 1923, Schrödinger proposed a wave equation from which atomic orbitals are derived. In 1932, Chadwick discovered the neutron, a second major particle in the nucleus.

Bohr-Sommerfeld Model (1916)

Modification of the Bohr model, placing electrons in ellipitical orbits.

Bohr Model (1913)

First quantum model of the atom with the electrons following circular orbits around the nucleus.

Nuclear Model (1911)

Rutherford discovered that the atom possessed a small dense core called the *nucleus*.

Thomson Model (1903)

"Plum pudding" model of the atom, with electrons and protons as the "plums" in a matter "pudding."

Dalton Model (1803)

Atoms as solid indestructible spheres.

FIGURE 5.1
Development of the atomic theory.

5.2 ATOMIC SUBSTRUCTURE

Protons, Neutrons, and Electrons

In terms of our current theories, atoms are composed of three fundamental particles—protons, neutrons, and electrons. These particles are characterized by their mass and electric charge (see Table 5.1). Protons and neutrons have approximately the same mass, about 1.674×10^{-24} g. An electron has a mass of 9.110×10^{-28} g, about $\frac{1}{1837}$ that of a proton (or neutron); in other words, 1837 electrons are needed to equal the mass of one proton.

Masses of subatomic particles are frequently expressed using a relative unit, termed a *unified atomic mass unit*, u. Later, we shall see from where this unit is

TABLE 5.1   PROPERTIES OF SUBATOMIC PARTICLES

Particle	Symbol	Mass, g	Mass, u	Relative charge
Proton	p^+	1.673×10^{-24}	1.007276	1+
Neutron	n^0	1.675×10^{-24}	1.008665	0
Electron	e^-	9.110×10^{-28}	0.000544	1−

$1u = 1.6606 \times 10^{-24}$ g

derived. For now, one unified atomic mass unit, 1 u, is approximately the mass of a proton (or neutron). An electron's mass in this scale is only 0.000544 u.

A property of matter that has not been discussed is electric charge. Bodies can have a positive (+), a negative (−), or no (neutral) net charge. Electrons and protons have the smallest elementary unit of charge found in matter. Electrons are negatively charged (1−); protons possess the same magnitude of charge as electrons, but their charge is positive (1+). Neutrons, as the name implies, are electrically neutral.

FIGURE 5.2
Two particles with the same electric charge repel each other. Two particles with unlike electric charges attract each other.

From physics, we find that particles with the same electric charge repel each other. Objects with unlike charges attract each other. Two electrons or two protons in close proximity repel each other, or push each other apart. Unlike charged particles (+ and −), when brought close together, attract each other.

The force of electric attraction or repulsion is inversely related to the square of the distance separating the particles (Coulomb's law),

Repulsion of like charges

$$F = k\frac{q_1 q_2}{r^2}$$

where q_1 and q_2 are the magnitudes of the charges, r is the distance between particles, and k is a constant. Neutral particles do not interact with charged particles (see Figure 5.2).

Attraction of unlike charges

Nucleus

Protons and neutrons are located in a very small region of the atom called the *nucleus* (plural, *nuclei*). Most nuclei have diameters of roughly 10^{-6} nm (1 nm is 1×10^{-9} m). Diameters of whole atoms are many times (100,000×) larger than nuclei, ranging from 0.1 to 0.5 nm. The electrons are found in the relatively vast space outside the nucleus.

Ernest Rutherford (1871–1937), a New Zealander who lived in England, discovered the nucleus in the now famous "gold leaf" experiment, when he bombarded thin gold foil (leaf) with alpha particles. His assistant Hans Geiger invented the radiation-detecting device that bears his name.

To summarize atomic structure, the atom consists of a small dense core, the nucleus, surrounded by minute particles, electrons, which occupy an immense, mainly empty, region of space. Atoms are not solid forms of matter; they are sparsely populated with matter.

Nuclear density is incredibly large considering that virtually the entire mass of an atom (protons and neutrons) is concentrated in such an infinitesimal volume. Nuclear densities, roughly, are 100,000,000 tons/mL!

Nuclear Properties of Atoms

Chemists are concerned with the number of protons and neutrons located in the nucleus. Two values are needed to determine the composition of the nucleus. One is called the atomic number, and the other is the mass number. An atom's

H. G. J. Moseley (1887–1915), a
student of Rutherford, was the
first to measure the charge on
the nucleus—its atomic number.
He accomplished this using an
x-ray technique.

atomic number is the number of positive charges found in the nucleus, or stated another way, the number of protons in the nucleus of an atom.

$$\text{Atomic number} = \text{number of protons in the nucleus of an atom}$$

Hydrogen, the simplest atom, has an atomic number of 1. Given its atomic number, we know that hydrogen has one proton in the nucleus. Helium's atomic number is 2; there are two protons in the nucleus of all helium atoms. A lithium nucleus has three protons; hence, lithium's atomic number is 3. Following the periodic table, you can see that the integer (whole number) in each element's block is the atomic number. Atoms are arranged in the periodic table in order of increasing atomic number.

In addition to the atomic number, the **mass number** of an atom is required to find the composition of an atom's nucleus. An atom's mass number is equal to the total number of protons plus neutrons in the nucleus.

$$\text{Mass number} = \text{number of protons} + \text{neutrons}$$

Note that mass numbers are just "numbers," and not really masses.

To express the atomic number and mass number of an atom, atomic symbols are written as follows:

$$^A_Z\text{X}$$

where X is the symbol of the atom, A is the mass number, and Z is the atomic number.

Let us write the atomic symbol for a krypton atom that has 36 protons and 48 neutrons in its nucleus. First determine the atomic number and mass number of krypton. Krypton's atomic number is 36 because it has 36 protons in its nucleus. Its mass number is equal to 84, the sum of the protons plus neutrons ($36\ p^+ + 48\ n^0 = 84$). After writing the symbol for krypton (Kr), we write 36 as a subscript to the left of the symbol and 84 as a superscript, also to the left of the symbol.

$$^{84}_{36}\text{Kr}$$

To find the number of neutrons in a nucleus, given the mass number and atomic number, we subtract the atomic number (number of p^+) from the mass number (number of p^+ and n^0).

$$\text{Number of } n^0 = \text{mass number} - \text{atomic number}$$
$$= (\text{number of } p^+ + n^0) - \text{number of } p^+$$

Mass numbers are not found in the periodic table, so they will be given whenever they are needed.

Let's consider the hydrogen atom, and find the complete composition of its nucleus. Most hydrogen atoms have a mass number equal to 1 and an atomic number also equal to 1, $_1^1H$.

$$\text{Number of } n^0 = \text{mass number} - \text{atomic number}$$
$$\text{Number of } n^0 \text{ in } _1^1H = 1 - 1 = 0$$

Consequently, $_1^1H$ atoms have one proton and no neutrons in their nuclei. $_1^1H$ is the only atom that lacks a neutron in the nucleus.

What is the nuclear composition of $_{26}^{56}Fe$ atoms? From the symbol, we obtain the atomic number 26; hence, there are 26 protons in this atom. To find the number of neutrons, subtract the atomic number, 26, from the mass number, 56, to obtain 30 neutrons.

$$\text{Number of } n^0 = \text{mass number} - \text{atomic number}$$
$$= 56 - 26 = 30 \ n^0$$

$_{26}^{56}Fe$ atoms have 26 protons and 30 neutrons in their nuclei. Example Problem 5.1 is another illustration of determining an atom's nuclear composition.

Example Problem 5.1

Given the following atom, determine the composition of the nucleus:

$$_{88}^{226}Ra$$

Solution

Marie Curie first isolated radium in 1911. Radium metal glows in the dark as a result of its high level of radiation. It is quite dangerous if inhaled or ingested.

In our example, the mass number A is 226 and the atomic number Z is 88. Mass number is the number of protons and neutrons, and the atomic number is the number of protons. Therefore, there are 88 protons in the nucleus. The number of neutrons is calculated by subtracting the atomic number from the mass number.

$$\text{Number of } n^0 = \text{mass number} - \text{atomic number}$$
$$= 226 - 88$$
$$= 138$$

$_{88}^{226}Ra$ contains 88 protons and 138 neutrons in its nucleus.

Isotopes Whenever atoms have the same atomic number but different mass numbers, they are referred to as **isotopes.** Stated differently, isotopes are atoms with the

FIGURE 5.3
Nuclei of protium (^{1}H) contain one proton and no neutrons. Nuclei of deuterium (^{2}H) contain one proton and one neutron. Tritium atoms, the heaviest isotope of hydrogen (^{3}H), have nuclei containing one proton and two neutrons.

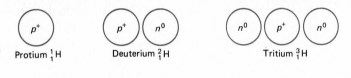

Protium $^1_1$H   Deuterium $^2_1$H   Tritium $^3_1$H

same number of protons and varying numbers of neutrons in their nuclei.

In addition to $^1_1$H, a small percentage of hydrogen atoms have a different nuclear composition. These hydrogen atoms have a mass number of 2 and an atomic number of 1.

$$\text{Number of neutrons in } {}^2_1\text{H} = 2 - 1 = 1 \; n^0$$

Each of these atoms has one proton and one neutron in its nucleus. A third hydrogen species exists with a mass number of 3. Its nuclei possess two neutrons and a single proton. Consequently, $^1_1$H, $^2_1$H, and $^3_1$H are all said to be isotopes of hydrogen.

Hydrogen's three isotopes are expressed as follows:

$$^1_1\text{H} \qquad ^2_1\text{H} \qquad ^3_1\text{H}$$

Protium   Deuterium   Tritium

$^1_1$H is called regular hydrogen or protium and is the most abundant type of hydrogen. $^2_1$H is called deuterium; it has a mass approximately twice that of regular $^1_1$H because it has both a proton and neutron in the nucleus. Tritium, $^3_1$H, the heaviest form of hydrogen, is not found in naturally occurring hydrogen samples. Figure 5.3 shows the nuclear composition of the hydrogen isotopes.

A large percentage of elements are composed of mixtures of different isotopes. For example, three isotopes are found in naturally occurring samples of uranium: $^{234}_{92}$U, $^{235}_{92}$U, and $^{238}_{92}$U. Each of these uranium isotopes contains 92

TABLE 5.2   NATURAL ISOTOPIC COMPOSITION OF SELECTED ELEMENTS

Atomic number	Isotope	Number of protons	Number of neutrons	Natural abundance, %
6	^{12}C	6	6	98.89
	^{13}C	6	7	1.11
8	^{16}O	8	8	99.76
	^{17}O	8	9	0.04
	^{18}O	8	10	0.20
29	^{63}Cu	29	34	69.09
	^{65}Cu	29	36	30.91
32	^{70}Ge	32	38	20.51
	^{72}Ge	32	40	27.43
	^{73}Ge	32	41	7.76
	^{74}Ge	32	42	36.54
	^{76}Ge	32	44	7.76

protons in the nucleus. They differ with respect to the number of neutrons in the nucleus. $^{234}_{92}U$ nuclei contain 142 neutrons, $^{235}_{92}U$ nuclei contain 143 neutrons, and $^{238}_{92}U$ nuclei contain 146 neutrons. Of the three, $^{238}_{92}U$ is most abundant in natural samples, representing 99% of the total atoms. Consequently, the other two isotopes represent less than 1% of the atoms. See Table 5.2 for added examples of natural isotopic mixtures.

Masses of Individual Atoms

An atom has a very small mass. A hydrogen atom, for example, has a mass of only 1.67×10^{-24} g. To avoid the inconvenience of working with such small numbers, chemists use a relative scale for the mass of individual atoms. In this scale, masses of all atoms are expressed relative to the mass of one $^{12}_{6}C$ atom.

By definition the mass of a $^{12}_{6}C$ atom is equal to exactly 12 unified atomic mass units (u). Thus, one unified atomic mass unit (1 u) is $\frac{1}{12}$ the mass of the $^{12}_{6}C$ atom. Since, $^{12}_{6}C$ has six protons and six neutrons in its nucleus, 1 u is about the average mass of a proton or a neutron.

The masses of other atoms are determined relative to the standard, $^{12}_{6}C$. For example, if an atom is found to have a mass 3 times the mass of $^{12}_{6}C$, its relative mass is 36 u, that is, 3 times 12 u. An atom with a mass one-fourth the mass of $^{12}_{6}C$ is assigned a mass of 3 u, one-fourth of 12 u. When the mass of an individual atom is given, it should be thought of relative to the mass of a $^{12}_{6}C$ atom.

Atomic Mass (Atomic Weight)

Berzelius produced one of the first accurate tables of atomic masses in 1828. Most of his values agree with those we use today. He also proposed the modern symbols for elements.

Some elements are not isotopic mixtures in nature; they are composed of a single isotope. Examples include ^9Be, ^{23}Na, ^{27}Al, ^{31}P, ^{45}Sc, and ^{55}Mn. Thus their atomic masses are not averages.

Atomic masses are the numbers listed in each block in the periodic table along with an element's symbol and atomic number. Notice that the atomic masses of all the elements are decimal numbers (Al, 26.9815; S, 32.06; V, 50.942; etc.); none are integers like the atomic number. Why is this?

A large percentage of the naturally occurring elements exist as a mixture of isotopes, each isotope with a different mass. Since chemists work with a large quantity of atoms, they usually find it convenient to use the average mass of an element's isotopes, called its atomic mass. **Atomic mass** is the average mass of the naturally occurring isotopes relative to $^{12}_{6}C$.

Consider the atomic mass of carbon, 12.011 u. If you refer to Table 5.2, you will find that carbon is primarily composed of two isotopes, $^{12}_{6}C$, the one we just referred to as the standard for the atomic mass scale, and $^{13}_{6}C$. In nature, approximately 9889 out of 10,000 carbon atoms are $^{12}_{6}C$, and 111 are the heavier $^{13}_{6}C$ isotopes. If we average 9889 particles with a mass of 12 u and 111 particles with a mass of 13 u, the average is 12.011 u. How is this number calculated? First we must consider how a *weighted average* is calculated.

Weighted averages are calculated in a similar way to any other average: add up the values and divide by the total number of values. For instance, if a test is given to 100 students, and 60 students achieve the score of 80 points and 40 students obtain 95 points, the average score is 86 points. The weighted average is obtained as follows: First, multiply 60, the number of students with 80 points, times their score, 80 points. Second, multiply 40, the number of students with 95 points, times their score, 95. Finally, add these two products together and divide by the total number of students.

$$\text{Average} = \frac{(60 \text{ students} \times 80 \text{ pts}) + (40 \text{ students} \times 95 \text{ pts})}{100}$$

$$= 86 \text{ points}$$

Atomic number

→1
H
→1.008

Atomic mass

Periodic tables differ with respect to where the atomic number and atomic mass are listed within each element's block. Remember that the atomic number is always an integer, and the decimal number is the atomic mass.

A similar calculation allows us to find the atomic mass of chlorine from the natural abundance of the two main isotopes, $^{35}_{17}\text{Cl}$ and $^{37}_{17}\text{Cl}$. In natural samples of the element chlorine, 75.53% is $^{35}_{17}\text{Cl}$ with a mass of 34.969 u. The remaining 24.47% is $^{37}_{17}\text{Cl}$, which has a mass of 36.966 u. To obtain chlorine's atomic mass, the weighted average of the isotopes' masses, assume that there are 100 atoms. Of those 100 total atoms, 75.53 have a mass of 34.969 u and 24.47 have a mass of 36.966 u. Therefore, multiply 75.53 times 34.969 u, add that to the product of 24.47 times 36.966 u, and divide by the total number of atoms, 100.

$$\text{Atomic mass of Cl} = \frac{(75.53 \times 34.969 \text{ u}) + (24.47 \times 36.966 \text{ u})}{100}$$

$$= 35.46 \text{ u}$$

Chlorine's atomic mass is 35.46 u.

Example Problem 5.2 is another illustration of how to calculate the atomic mass of an element.

Example Problem 5.2

Calculate the atomic mass of boron, using the following data:

Isotope	Relative mass	Percent abundance
$^{10}_{5}\text{B}$	10.013 u	19.70%
$^{11}_{5}\text{B}$	11.009 u	80.30%

Solution

Atomic mass is found by computing the average mass of the naturally occurring isotopes:

$$\text{Atomic mass} = \frac{(10.013 \text{ u} \times 19.70) + (11.009 \text{ u} \times 80.30)}{100}$$

$$= 10.81 \text{ u}$$

Percents indicate the number of parts per 100 total parts. In this calculation, for every 100 B atoms, 19.70 possess a mass of 10.013 u and 80.30 possess a mass of 11.009 u. Thus, 19.70 is multiplied by 10.013 u and 80.30 is multiplied by 11.009 u. The resulting two quantities are added, and the sum is divided by 100, giving the average, 10.81 u.

TABLE 5.3   RADIOACTIVE PARTICLES

Name	Symbol	Mass, u	Charge	Penetrating power
Alpha	α	4	2+	Low
Beta	β	$\frac{1}{1837}$	1−	Intermediate
Gamma	γ	0	0	High

Radioactivity

Henri Becquerel, in 1896, accidentally discovered radioactivity when he found that a uranium salt emitted something that fogged his photographic plates.

Nuclei of certain atoms are not stable. Instead they undergo spontaneous changes that result in a particle or ray being emitted from the nucleus at high speed. Such nuclei are said to be radioactive: they release radioactive "particles." There are three common types of nuclear emissions: alpha (α) particles, beta (β) particles, and gamma (γ) rays.

Alpha particles are the most massive of the three, with a mass of 4 u. Electrically, alpha particles have a charge of 2+, twice that of a proton. Beta

FIGURE 5.4
Alpha radiation is the least penetrating form of radiation and is blocked by a thin metallic foil. Beta radiation can penetrate more matter than alpha radiation, but beta particles are stopped by a heavy metallic foil. Gamma radiation is the most penetrating, and can be stopped only by thick walls of dense materials such as lead.

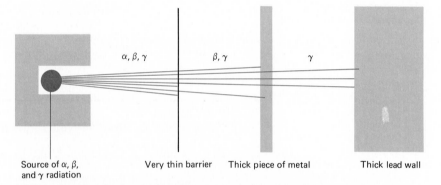

Source of α, β, and γ radiation   Very thin barrier   Thick piece of metal   Thick lead wall

α, β, γ   β, γ   γ

FIGURE 5.5
Tracks of radioactive particles can be observed in a device called a bubble chamber. As radioactive particles travel through liquid hydrogen they produce ions (charged atoms), which cause small bubbles of vapor to form. A bubble track shows the path of a particle, not the particle itself. In this picture, a collision occurs with a proton, producing three new particles. *(Argonne National Laboratory)*

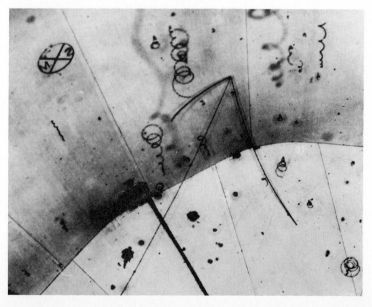

Electromagnetic radiations include radio waves, microwaves, infrared, visible light, ultraviolet, x-rays, and gamma rays.

particles are electrons that are ejected from the nucleus at high velocity. Gamma rays are not actual particles; they resemble energy forms like x-rays or ultraviolet light (electromagnetic radiations). Hence, they have no measurable mass or charge. Table 5.3 summarizes the properties of these three forms of radioactive emissions.

Another distinguishing characteristic of the three types of radioactivity is their ability to penetrate matter. Gamma rays have the greatest capacity to penetrate matter. Most gamma rays have little trouble passing through wood and various metals. Thick walls of lead or other dense types of matter are needed to totally block out gamma rays. Beta particles are less penetrating than gamma rays. A 0.5-cm-thick metal barrier is generally effective in stopping most beta particles. Alpha particles are the least penetrating, and most can be stopped by a piece of paper or other thin sample of matter (see Figures 5.4 and 5.5).

The greater the mass and charge of a radioactive particle, the more it interacts with matter, and the less it penetrates.

REVIEW PROBLEMS

5.4 What happens when particles with (a) the same and (b) different charges interact with each other?

5.5 What can be determined about an atom given its (a) atomic number, (b) mass number, (c) atomic mass, (d) atomic number and mass number?

5.6 How many protons and neutrons are found in the nucleus of (a) $^{31}_{15}P$, (b) $^{78}_{34}Se$, (c) $^{227}_{89}Ac$?

5.7 (a) What are isotopes?
(b) List the naturally occurring isotopes of germanium, Ge (Table 5.2).
(c) Which isotope of germanium is the most abundant in nature?

5.8 (a) Define atomic mass.
(b) How is the atomic mass of a element determined?

5.9 Calculate the atomic mass of copper using 62.93 as the mass of $^{63}_{29}Cu$ and 64.93 as the mass of $^{65}_{29}Cu$ and given that the natural abundances of $^{63}_{29}Cu$ and $^{65}_{29}Cu$ are 69.09% and 30.91%, respectively.

5.10 List three types of radioactivity and their characteristic properties.

5.3 ELECTRONIC CONFIGURATION

Electrons So far we have seen that electrons have a tiny mass and a negative charge and are found somewhere outside the nucleus. All atoms are electrically neutral; hence, the number of electrons equals the number of protons (atomic number).

$$\text{Number of } e^- = \text{number of } p^+ = \text{atomic number}$$

It is only necessary to look at the periodic table to find the number of electrons in an atom: H, 1 e^-; C, 6 e^-; Ne, 10 e^-; and U, 92 e^-.

Orbitals An **orbital** is a region of space where there is a higher probability of finding electrons. An orbital can contain no more than two electrons: orbitals can be empty (no electrons), half-filled (one electron), or filled (two electrons).

Electrons are elusive particles, ones that cannot be directly observed. A German physicist, Werner Heisenberg, first proposed what is now called the

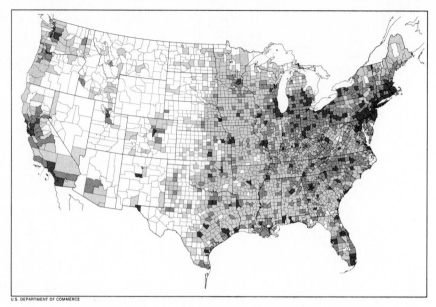

FIGURE 5.6
A population density map of the United States shows the regions where there is a high probability of locating people. Regions that are shaded darkly indicate a higher probability of locating a person. Unshaded regions are places where there is a low probability of locating a person.

Werner Heisenberg (1901–1975), a student of Bohr, proposed the uncertainty principle in 1927. This radical idea, and what it implies, was ultimately accepted by the scientific community, but it was never totally acceptable to Albert Einstein.

FIGURE 5.7
Electrons behave as if they are spinning on an axis. An electron can spin in either one direction or the opposite direction.

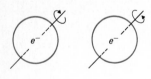

Electron Energy Levels

uncertainty principle, which states that it is impossible to accurately determine the exact position and velocity of an electron at the same time. In simpler words, an electron's position in space cannot be pinpointed at a particular instant. Hence, scientists no longer worry about identifying what path an electron takes as its travels around the nucleus. Instead, they identify the regions, or *orbitals,* where electrons are most likely located.

A chemist looks at orbitals as a geographer looks at a population density map. A population density map shows an area like the United States with regions shaded where there is a high probability of finding a person (see Figure 5.6). Darkly shaded areas indicate regions where there is a good chance of finding a person, and lightly shaded areas indicate regions where there is a lower probability. Actually going to a region that is darkly shaded does not ensure that you will find a person; there is just a better chance of finding a person in this region.

Within an orbital, electrons act as if they are spinning on an axis. An electron can spin in either one direction or the other, as a spinning top. If two electrons are contained in an orbital, one spins in one direction and the other spins in the opposite direction (see Figure 5.7). The idea that two electrons in an orbital have opposite spins was proposed by Wolfgang Pauli (1900–1958) and is referred to as *Pauli's exclusion principle.*

Orbitals are organized into energy levels. A collection of orbitals at approximately the same average distance from the nucleus is referred to as an **energy level.** Electrons closer to the nucleus are in lower energy levels. Higher energy levels are located, on the average, succeedingly farther from the nucleus. The first energy level is closest to the nucleus; the second energy level is located beyond the first level. Electrons in the third level are even farther from the nucleus. See Figure 5.8.

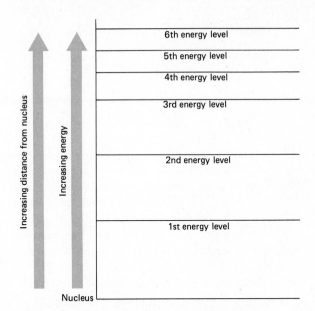

FIGURE 5.8
The first electron energy level is closest to the nucleus. Higher energy levels are farther from the nucleus.

The first energy level is the lowest energy level. It requires energy to move an electron from a lower level to a higher one. Why? Protons in the nucleus attract electrons and hold them in their lowest energy levels, in a manner similar to the gravitational attraction of objects to the earth, where the ground is usually the lowest energy level.

Each energy level contains a fixed number of orbitals and electrons. Lower energy levels are closer to the nucleus, where there is less room to house electrons. Mutual repulsion of electrons limits the number of electrons in a region. Succeedingly higher energy levels contain more space for electrons to populate, diminishing the repulsion of electrons. Table 5.4 presents the theoretical maximum number of orbitals and electrons in each energy level.

Each energy level is divided into one or more sublevels. A **sublevel** is made up of one or more orbitals within an energy level that has the same shape and characteristics. The number of sublevels within an energy level corresponds to the number given to the energy level, the principal quantum number. Thus, the first energy level contains only one sublevel—they are one and the same. The

> The theoretical maximum number of electrons in an energy level is $2n^2$, where n is the energy level.

TABLE 5.4 POPULATION OF ELECTRON ENERGY LEVELS

Energy level	Theoretical maximum number of	
	Orbitals	Electrons
1	1	2
2	4	8
3	9	18
4	16	32
5	25	50

TABLE 5.5   ELECTRON SUBLEVEL POPULATIONS

Sublevel	Number of orbitals	Maximum number of electrons
s	1	2
p	3	6
d	5	10
f	7	14

Electron energy levels are composed of sublevels, and sublevels are divided into orbitals.

SUBLEVELS

s
p
d
f

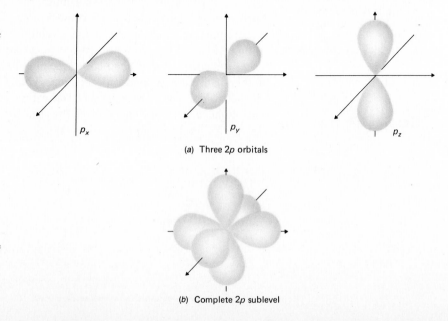

1s

FIGURE 5.9
The 1s sublevel is the lowest energy sublevel. Only one orbital makes up an s sublevel; therefore, the 1s sublevel and 1s orbital are the same region of space. An s orbital has the shape of a sphere.

second level has two sublevels, the third level has three sublevels, and so on. We shall consider the four lowest energy sublevels. Sublevels are denoted by a letter: s, the lowest energy sublevel, p, next highest energy sublevel, d, higher still, and f, the highest energy sublevel of the four.

Each sublevel contains a maximum population of orbitals and of electrons that they can hold. Table 5.5 lists the maximum number of orbitals and electrons in each sublevel. All s sublevels hold a maximum of two electrons because the s sublevel only contains a single orbital. A p sublevel is composed of three orbitals, so six electrons are the maximum population of the p sublevel. Five and seven orbitals are found in the d and f sublevels, which hold 10 and 14 electrons, respectively.

Sublevels are distinguished by the shape of the region where electrons are most probably found. An s sublevel distribution is shown in Figure 5.9. Notice that the shading is darkest near the nucleus and becomes lighter farther from the nucleus, indicating that an electron in an s orbital has a greater chance of being found closer to the nucleus than farther away.

A more complex distribution of electrons is found for the p sublevel. Each of the three p orbitals is shown in Figure 5.10a. Each p orbital is distributed along

p_x p_y p_z

(a) Three 2p orbitals

FIGURE 5.10
The 2p sublevel is composed of three 2p orbitals. Each 2p orbital is located on a different axis and has a "dumbbell" shape.

(b) Complete 2p sublevel

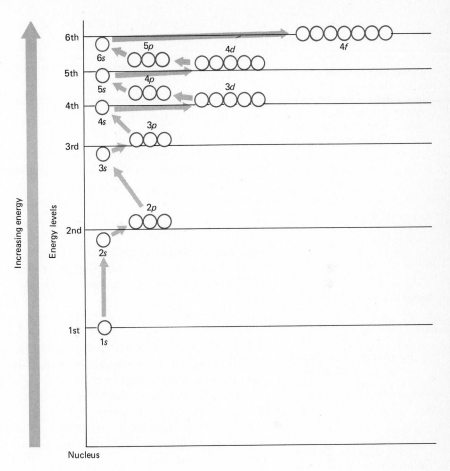

FIGURE 5.11

Each orbital is represented by a circle. Two electrons fill each orbital, starting with the lowest energy orbital and proceeding to the next higher energy orbital until all the electrons of an atom are accounted for.

one of the three axes in a three-dimensional coordinate system. An entire *p* sublevel, three *p* orbitals, is shown in Figure 5.10*b*. Both *d* and *f* sublevels are more complex distributions, and will not be considered.

Electron Arrangements   So far, we have seen that each energy level is divided into smaller regions called sublevels, and each sublevel is divided into orbitals. Each orbital contains a maximum of two electrons. Figure 5.11 shows the first, second, third, and fourth levels with their sublevels and orbitals. Electrons fill these levels starting from the lowest energy orbital (first energy level), and proceed, one electron at a time, filling each lower energy orbital before filling a higher energy orbital.

We will now begin our study of the arrangement of electrons in atoms by considering where hydrogen's one electron is found. Then, we shall proceed to helium, which has two electrons. After helium, we will add another electron to the two electrons that are found in helium to find lithium's electronic configuration. In a similar manner, we shall proceed along the periodic table adding one more electron to those that are found in the previous atom.

In all atoms, the lowest energy orbital, the one closest to the nucleus, is the $1s$ orbital.

$$1s$$
Energy level ⌐↑↑⌐ Sublevel

The number 1 refers to the energy level where the electron is located, and the s identifies the sublevel.

Hydrogen is the simplest atom, containing only one electron. We express hydrogen's electronic configuration—representation of occupied orbitals—as

⌐Number of $e-$ in the sublevel

H $1s^1$

Hydrogen's electron resides in the s orbital of the lowest energy level.

Helium has two electrons. Since there is room for another electron in the $1s$ orbital, both electrons populate this orbital. Helium's electronic configuration is

He $1s^2$

A 2 is written as a superscript above the s to indicate that two electrons are contained in the $1s$ orbital; the orbital is now filled. No more than two electrons can occupy the first energy level.

Lithium (atomic number = 3) is the first element to possess an electron in the second energy level. Three electrons are found in Li atoms; the first two electrons occupy the lower energy level, $1s$, and the remaining electron is found in the next higher energy $2s$ orbital; see Figure 5.12. We write lithium's electronic configuration as

Li $1s^2 2s^1$

Even though the second energy level contains two sublevels, s and p, the s sublevel is lower in energy (see Figure 5.11). Hence, the $2s$ sublevel fills before an electron enters the $2p$ sublevel.

Beryllium's (atomic number = 4) two lowest energy electrons occupy the $1s$ orbital, the He configuration. The two outer electrons are located in the $2s$ sublevel, filling it.

Be $1s^2 2s^2$

In boron, which has five electrons, the first four electrons occupy the same orbitals as beryllium's four electrons, $1s^2$ and $2s^2$. The fifth electron enters the higher energy $2p$ sublevel. Boron is the first element to possess an electron in the $2p$ sublevel; boron's electronic configuration is

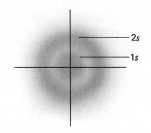

FIGURE 5.12
Electrons that occupy the 1s orbital are on an average closer to the nucleus than electrons in the 2s orbital. The 2s orbital distribution is similar to the 1s except for being farther from the nucleus.

$$\text{B} \quad 1s^2 2s^2 2p^1$$

Since the p sublevel contains three orbitals and thus has the capacity to hold six electrons (see Table 5.5), the next five atoms on the periodic table, C through Ne, fill the $2p$ sublevel. Carbon atoms have one more electron than B; therefore, carbon's electronic configuration is

$$\text{C} \quad 1s^2 2s^2 2p^2$$

In nitrogen, the next element on the periodic table, the $2p$ sublevel is half-filled.

$$\text{N} \quad 1s^2 2s^2 2p^3$$

In the atoms O, F, and Ne, the $2p$ sublevel fills.

$$\text{O} \quad 1s^2 2s^2 2p^4$$
$$\text{F} \quad 1s^2 2s^2 2p^5$$
$$\text{Ne} \quad 1s^2 2s^2 2p^6$$

Neon's $2p$ sublevel contains six electrons, the maximum that the $2p$ sublevel can hold. Therefore, a neon atom has a completely filled second energy level. All atoms beyond neon on the periodic table have more than 10 electrons and thus have outer electrons in higher energy levels.

Sodium, the atom with 11 electrons, has the same electronic configuration as neon for its inner electrons ($1s^2 2s^2 2p^6$); the remaining electron occupies the lowest energy sublevel in the third energy level, $3s^1$. Sodium's configuration is

$$\text{Na} \quad 1s^2 2s^2 2p^6 3s^1$$

Note that sodium's outer electronic configuration is the same as that of H and Li. Each of these atoms has one outer electron in an s sublevel.

$$\text{H} \quad 1s^1$$
$$\text{Li} \quad 2s^1$$
$$\text{Na} \quad 3s^1$$

Each atom listed in group IA of the periodic table possesses the same outer electronic configuration. All group IA atoms have one outer-level s electron (s^1).

The outer electronic configuration of an atom is also called the **valence** electronic configuration. Both of these terms refer to the highest occupied electron energy level in an atom.

TABLE 5.6   OUTER ELECTRONIC CONFIGURATIONS OF REPRESENTATIVE ELEMENTS

Group number in periodic table	Number of outer electrons	Electronic configuration
IA	1	s^1
IIA	2	s^2
IIIA	3	s^2p^1
IVA	4	s^2p^2
VA	5	s^2p^3
VIA	6	s^2p^4
VIIA	7	s^2p^5
VIIIA	8	s^2p^{6*}

*Except He, which is s^2 only.

Magnesium, the next atom on the periodic table, has the electronic configuration

$$\text{Mg} \quad 1s^2 2s^2 2p^6 3s^2$$

Magnesium's outer electronic configuration is the same as other members of its group (IIA) on the periodic table. All atoms in group IIA have two outer electrons in the s sublevel.

$$
\begin{array}{ll}
\text{Be} & 2s^2 \\
\text{Mg} & 3s^2 \\
\text{Ca} & 4s^2 \\
\text{Sr} & 5s^2 \\
\text{Ba} & 6s^2 \\
\text{Ra} & 7s^2 \\
\end{array}
$$

In all cases, atoms listed within a group on the periodic table have the same number of outer electrons. Table 5.6 lists the outer electronic configurations of the **representative elements**, those in groups IA through VIIIA. From Table 5.6, we see that Al, an atom in group IIIA with 13 electrons, has three outer electrons, two in the $3s$ sublevel and one in the $3p$ sublevel. Aluminum's complete electronic configuration is:

$$\text{Al} \quad 1s^2 2s^2 2p^6 3s^2 3p^1$$

Each succeeding atom after aluminum has one more electron in the $3p$ sublevel, ending with Ar with a completed $3p$ sublevel.

$$\text{Si} \quad 1s^2 2s^2 2p^6 3s^2 3p^2$$
$$\text{P} \quad 1s^2 2s^2 2p^6 3s^2 3p^3$$
$$\text{S} \quad 1s^2 2s^2 2p^6 3s^2 3p^4$$
$$\text{Cl} \quad 1s^2 2s^2 2p^6 3s^2 3p^5$$
$$\text{Ar} \quad 1s^2 2s^2 2p^6 3s^2 3p^6$$

Up to this point, the electrons have filled the orbitals in an orderly fashion. Despite the fact that the third energy level has three sublevels, s, p, and d, the $3d$ sublevel is slightly higher in energy than the $4s$ sublevel. Hence, the $4s$ sublevel fills before the $3d$ level. Thus, the electronic configurations for potassium and calcium are

$$\text{K} \quad 1s^2 2s^2 2p^6 3s^2 3p^6 4s^1$$
$$\text{Ca} \quad 1s^2 2s^2 2p^6 3s^2 3p^6 4s^2$$

Scandium, Sc, is the first atom to have an electron in the $3d$ orbital.

$$\text{Sc} \quad 1s^2 2s^2 2p^6 3s^2 3p^6 3d^1 4s^2$$

From Sc to Zn, the $3d$ sublevel fills somewhat irregularly. Zinc is the first element to have a complete $3d$ sublevel. Its electronic configuration is

$$\text{Zn} \quad 1s^2 2s^2 2p^6 3s^2 3p^6 3d^{10} 4s^2$$

After the $3d$ is full, the $4p$ fills. Gallium, Ga, is the first element with a $4p$ electron.

$$\text{Ga} \quad 1s^2 2s^2 2p^6 3s^2 3p^6 3d^{10} 4s^2 4p^1$$

Table 5.7 lists the electronic configurations for all of the elements.

It is not necessary to memorize the order of filling of sublevels. Instead recognize that the periodic table is organized according to the electronic configurations of atoms. In Figure 5.13, the periodic table is marked to indicate the order in which electrons fill sublevels. All elements in groups IA and IIA have s electrons in their outer energy level. Elements in groups IIIA through VIIIA have both s and p electrons in their outer levels. Elements in groups IB through VIIIB, the transition metals, have electrons filling d and f sublevels.

TABLE 5.7 ELECTRONIC CONFIGURATION OF ATOMS

I. Atoms with atomic numbers 1 to 54

Atomic Number	Symbol	1s	2s	2p	3s	3p	3d	4s	4p	4d	4f	5s	5p
1	H	1											
2	He	2											
3	Li	2	1										
4	Be	2	2										
5	B	2	2	1									
6	C	2	2	2									
7	N	2	2	3									
8	O	2	2	4									
9	F	2	2	5									
10	Ne	2	2	6									
11	Na	2	2	6	1								
12	Mg	2	2	6	2								
13	Al	2	2	6	2	1							
14	Si	2	2	6	2	2							
15	P	2	2	6	2	3							
16	S	2	2	6	2	4							
17	Cl	2	2	6	2	5							
18	Ar	2	2	6	2	6							
19	K	2	2	6	2	6		1					
20	Ca	2	2	6	2	6		2					
21	Sc	2	2	6	2	6	1	2					
22	Ti	2	2	6	2	6	2	2					
23	V	2	2	6	2	6	3	2					
24	Cr	2	2	6	2	6	5	1					
25	Mn	2	2	6	2	6	5	2					
26	Fe	2	2	6	2	6	6	2					
27	Co	2	2	6	2	6	7	2					
28	Ni	2	2	6	2	6	8	2					
29	Cu	2	2	6	2	6	10	1					
30	Zn	2	2	6	2	6	10	2					
31	Ga	2	2	6	2	6	10	2	1				
32	Ge	2	2	6	2	6	10	2	2				
33	As	2	2	6	2	6	10	2	3				
34	Se	2	2	6	2	6	10	2	4				
35	Br	2	2	6	2	6	10	2	5				
36	Kr	2	2	6	2	6	10	2	6				
37	Rb	2	2	6	2	6	10	2	6			1	
38	Sr	2	2	6	2	6	10	2	6			2	
39	Y	2	2	6	2	6	10	2	6	1		2	
40	Zr	2	2	6	2	6	10	2	6	2		2	
41	Nb	2	2	6	2	6	10	2	6	4		1	
42	Mo	2	2	6	2	6	10	2	6	5		1	
43	Tc	2	2	6	2	6	10	2	6	6		1	
44	Ru	2	2	6	2	6	10	2	6	7		1	
45	Rh	2	2	6	2	6	10	2	6	8		1	
46	Pd	2	2	6	2	6	10	2	6	10			
47	Ag	2	2	6	2	6	10	2	6	10		1	
48	Cd	2	2	6	2	6	10	2	6	10		2	
49	In	2	2	6	2	6	10	2	6	10		2	1
50	Sn	2	2	6	2	6	10	2	6	10		2	2
51	Sb	2	2	6	2	6	10	2	6	10		2	3
52	Te	2	2	6	2	6	10	2	6	10		2	4
53	I	2	2	6	2	6	10	2	6	10		2	5
54	Xe	2	2	6	2	6	10	2	6	10		2	6

Atomic number	Symbol	[Xe]*	4f	5d	5f	6s	6p	6d	6f	7s
55	Cs					1				
56	Ba					2				
57	La			1		2				
58	Ce		2			2				
59	Pr		3			2				
60	Nd		4			2				
61	Pm		5			2				
62	Sm		6			2				
63	Eu		7			2				
64	Gd		7	1		2				
65	Tb		9			2				
66	Dy		10			2				
67	Ho		11			2				
68	Er		12			2				
69	Tm		13			2				
70	Yb		14			2				
71	Lu		14	1		2				
72	Hf		14	2		2				
73	Ta		14	3		2				
74	W		14	4		2				
75	Re		14	5		2				
76	Os		14	6		2				
77	Ir		14	7		2				
78	Pt		14	9		1				
79	Au		14	10		1				
80	Hg		14	10		2				
81	Tl		14	10		2	1			
82	Pb		14	10		2	2			
83	Bi		14	10		2	3			
84	Po		14	10		2	4			
85	At		14	10		2	5			
86	Rn		14	10		2	6			
87	Fr		14	10		2	6			1
88	Ra		14	10		2	6			2
89	Ac		14	10		2	6	1		2
90	Th		14	10		2	6	2		2
91	Pa		14	10	2	2	6	1		2
92	U		14	10	3	2	6	1		2
93	Np		14	10	4	2	6	1		2
94	Pu		14	10	6	2	6			2
95	Am		14	10	7	2	6			2
96	Cm		14	10	7	2	6	1		2
97	Bk		14	10	9	2	6			2
98	Cf		14	10	10	2	6			2
99	Es		14	10	11	2	6			2
100	Fm		14	10	12	2	6			2
101	Md		14	10	13	2	6			2
102	No		14	10	14	2	6			2
103	Lr		14	10	14	2	6	1		2
104	Unq		14	10	14	2	6	2		2

*Elements 55 to 104 have the inner electronic configuration of Xe.

s — d — p

IA																	VIIIA
1 H $1s^1$	IIA											IIIA	IVA	VA	VIA	VIIA	2 He $1s^2$
3 Li $2s^1$	4 Be $2s^2$											5 B $2s^2 2p^1$	6 C $2s^2 2p^2$	7 N $2s^2 2p^3$	8 O $2s^2 2p^4$	9 F $2s^2 2p^5$	10 Ne $2s^2 2p^6$
11 Na $3s^1$	12 Mg $3s^2$	IIIB	IVB	VB	VIB	VIIB	⎯VIIIB⎯			IB	IIB	13 Al $3s^2 3p^1$	14 Si $3s^2 3p^2$	15 P $3s^2 3p^3$	16 S $3s^2 3p^4$	17 Cl $3s^2 3p^5$	18 Ar $3s^2 3p^6$
19 K $4s^1$	20 Ca $4s^2$	21 Sc	22 Ti	23 V	24 Cr	25 Mn	26 Fe	27 Co	28 Ni	29 Cu	30 Zn	31 Ga $4s^2 4p^1$	32 Ge $4s^2 4p^2$	33 As $4s^2 4p^3$	34 Se $4s^2 4p^4$	35 Br $4s^2 4p^5$	36 Kr $4s^2 4p^6$
37 Rb $5s^1$	38 Sr $5s^2$	39 Y	40 Zr	41 Nb	42 Mo	43 Tc	44 Ru	45 Rh	46 Pd	47 Ag	48 Cd	49 In $5s^2 5p^1$	50 Sn $5s^2 5p^2$	51 Sb $5s^2 5p^3$	52 Te $5s^2 5p^4$	53 I $5s^2 5p^5$	54 Xe $5s^2 5p^6$
55 Cs $6s^1$	56 Ba $6s^2$	57 La*	72 Hf	73 Ta	74 W	75 Re	76 Os	77 Ir	78 Pt	79 Au	80 Hg	81 Tl $6s^2 6p^1$	82 Pb $6s^2 6p^2$	83 Bi $6s^2 6p^3$	84 Po $6s^2 6p^4$	85 At $6s^2 6p^5$	86 Rn $6s^2 6p^6$
87 Fr $7s^1$	88 Ra $7s^2$	89 Act	104 Unq	105 Unp	106 Unh	107	108										

Period (left margin). 3d, 4d, 5d, 6d designations for the d-block rows.

f

*	58 Ce	59 Pr	60 Nd	61 Pm	62 Sm	63 Eu	64 Gd	65 Tb	66 Dy	67 Ho	68 Er	69 Tm	70 Yb	71 Lu
								—4f—						

†	90 Th	91 Pa	92 U	93 Np	94 Pu	95 Am	96 Cm	97 Bk	98 Cf	99 Es	100 Fm	101 Md	102 No	103 Lw
								—5f—						

FIGURE 5.13

Elements are listed on the periodic table according to the electronic configurations of their atoms. Atoms in the same group have the same outer electronic configurations. Following the periodic table in order of increasing atomic number is an easy way to determine the electronic configuration of an atom.

To write electronic configurations, follow the periodic table in order of increasing atomic number. Numbers that denote periods correspond to electron energy levels, and the number on the top of each vertical column leads to the outer configuration. Example Problem 5.3 illustrates how the periodic table is used to assist in writing electronic configurations.

Example Problem 5.3

Use the periodic table to write the complete electronic configuration for strontium, Sr.

Solution

1. Find Sr on the periodic table.

Sr has 38 electrons, is in the fifth period, and is a member of group IIA. All elements in group IIA have an outer electronic configuration of two electrons in the s orbital, s^2.

2. Follow the periodic table in order of increasing atomic number.

Looking at Figure 5.14 and following the atomic numbers, we see that the

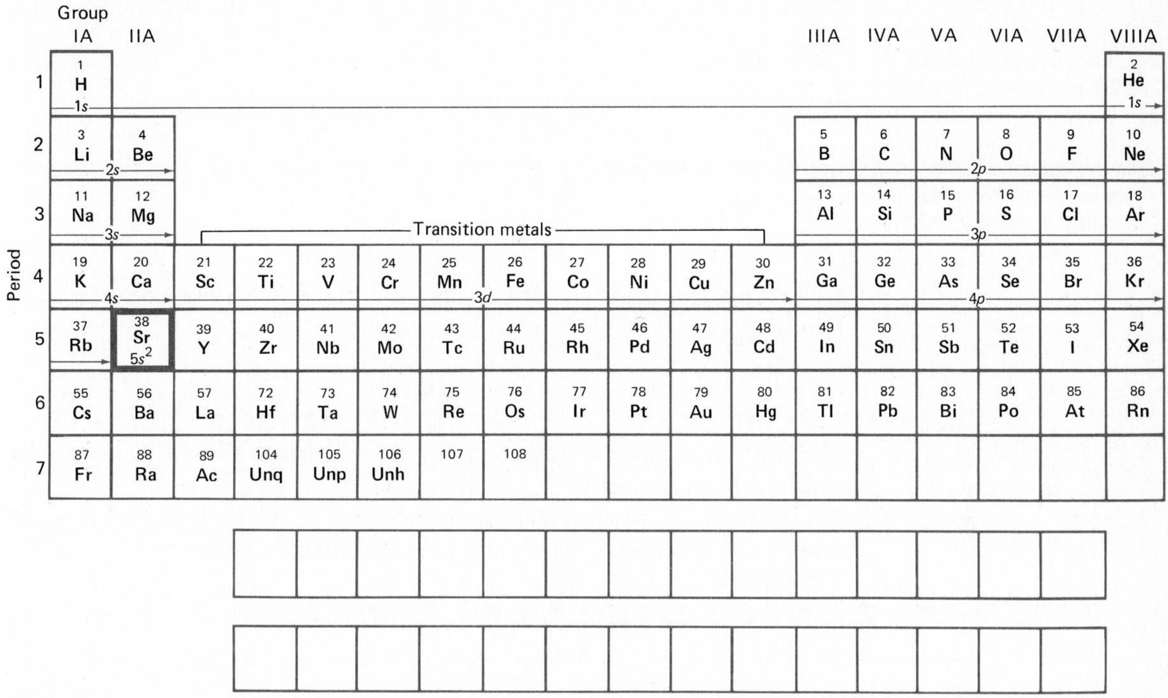

Group IA	IIA											IIIA	IVA	VA	VIA	VIIA	VIIIA
1 H 1s																	2 He 1s
3 Li	4 Be 2s											5 B	6 C	7 N	8 O 2p	9 F	10 Ne
11 Na	12 Mg 3s						Transition metals					13 Al	14 Si	15 P	16 S 3p	17 Cl	18 Ar
19 K	20 Ca 4s	21 Sc	22 Ti	23 V	24 Cr	25 Mn 3d	26 Fe	27 Co	28 Ni	29 Cu	30 Zn	31 Ga	32 Ge	33 As 4p	34 Se	35 Br	36 Kr
37 Rb	38 Sr 5s²	39 Y	40 Zr	41 Nb	42 Mo	43 Tc	44 Ru	45 Rh	46 Pd	47 Ag	48 Cd	49 In	50 Sn	51 Sb	52 Te	53 I	54 Xe
55 Cs	56 Ba	57 La	72 Hf	73 Ta	74 W	75 Re	76 Os	77 Ir	78 Pt	79 Au	80 Hg	81 Tl	82 Pb	83 Bi	84 Po	85 At	86 Rn
87 Fr	88 Ra	89 Ac	104 Unq	105 Unp	106 Unh	107	108										

(Period numbers 1–7 shown at left)

FIGURE 5.14

Start with hydrogen and follow along the periodic table until you reach strontium. As you proceed toward strontium, write the electrons that fill each sublevel in Sr: $1s^2\ 2s^2\ 2p^6\ 3s^2\text{-}3p^6\ 4s^2\ 3d^{10}\ 4p^6\ 5s^2$.

inner electronic configuration of Sr is the same as Kr (atomic number $= 36$). Krypton's electronic configuration is

$$\text{Kr} \quad 1s^2 2s^2 2p^6 3s^2 3p^6 3d^{10} 4s^2 4p^6$$

To this we add the two outer electrons, $5s^2$, to give the complete configuration of strontium:

$$\text{Sr} \quad 1s^2 2s^2 2p^6 3s^2 3p^6 3d^{10} 4s^2 4p^6 5s^2$$

Always check to see that the total number of electrons in the electronic configuration equals the atomic number. In this case, add all of the superscripts together to give 38, the atomic number of Sr.

Dot Formulas

Outer electronic configurations of atoms determine most of the properties of elements, especially chemical properties. Chemists frequently draw dot formulas (sometimes called Lewis electron dot formulas, after G. N. Lewis) as a means of conveniently showing the outer electronic configurations of atoms.

Two rules are followed in writing the dot formula of an atom:

1. Write the symbol of the atom.

G. N. Lewis (1875–1946) was an American who obtained his Ph.D. from Harvard University. In 1933, he prepared a sample of water in which all the hydrogens were the heavy isotope deuterium. This water was called *heavy water* and was used in nuclear reactors.

2. Place one dot around the symbol for each electron in the **outermost** energy level.

Inner-level electrons are never shown in dot formulas.

Since hydrogen has only one electron ($1s^1$), its dot formula shows only one dot next to its symbol:

$$H\cdot$$

Helium has two electrons ($1s^2$), so its dot formula shows two dots next to its symbol.

$$He\text{:}$$

Note that helium's two electrons are written together to show that they are located in the same orbital.

Lithium has an electronic configuration of $1s^2 2s^1$. Only the $2s^1$ is in the outer level, so only one dot is placed next to lithium's symbol (don't count the dot in the i of Li).

$$Li\cdot$$

Beryllium's outer electronic configuration is $2s^2$; therefore, two dots are placed next to beryllium's symbol.

$$Be\text{:}$$

If the atom has p electrons in the outer energy level, a number of different arrangements of dots around the symbol can be written. For example, the electron dot formula of carbon ($1s^2 2s^2 2p^2$) can be written as follows:

$$\cdot \overset{\cdot}{C}\cdot \quad \text{or} \quad \text{:}\overset{\cdot}{C}\cdot$$

In the first dot formula, the dots representing electrons are written symmetrically around the symbol. However, some prefer to write the dots corresponding to how they are paired in orbitals. In carbon, two electrons are paired in the 2s orbital, but each of the two electrons in the $2p$ orbital occupies a different p orbital.

All atoms in a chemical group have the same number of dots around their symbols because they have the same number of outer electrons. For instance, atoms in group VIIA all possess seven outer-level electrons. Thus, the general dot formula for these atoms is

$$\text{:}\overset{\cdot\cdot}{\underset{\cdot\cdot}{X}}\cdot$$

where X is either F, Cl, Br, I, or At.

Figure 5.15 illustrates the dot formulas for all atoms in groups IA through VIIIA in the periodic table.

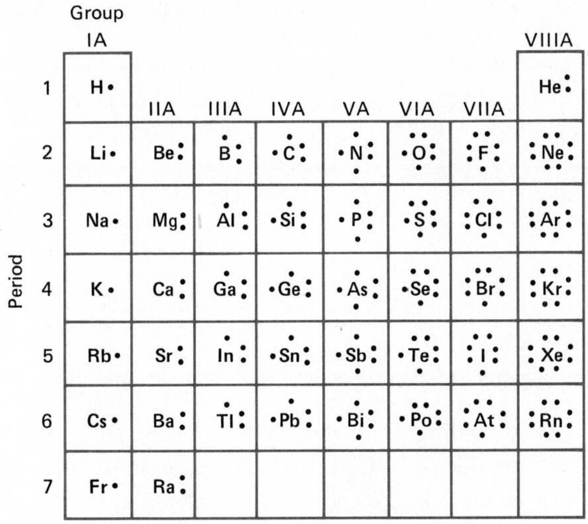

Group

FIGURE 5.15
Electron dot formulas of the elements

Example Problem 5.4

Draw the electron dot formulas for (a) Si, (b) Rb, and (c) Te.

Solution

(a) Silicon belongs to group IVA and has an outer electronic configuration of $3s^2 3p^2$. Each member of group IVA has four outer electrons ($s^2 p^2$); therefore, the dot formula for Si is

$$\cdot \overset{\displaystyle \cdot}{Si} \cdot$$

(b) Rubidium belongs to group IA and has an outer electronic configuration of $5s^1$, one outer electron. Consequently, the dot formula for Rb is

$$Rb\cdot$$

(c) Tellurium belongs to group VI and has an outer electronic configuration of $5s^2 5p^4$; thus, six dots are placed around its symbol:

$$:\overset{\displaystyle \cdot}{\underset{\displaystyle \cdot}{Te}}:$$

REVIEW PROBLEMS

5.11 How many electrons are found in each of the following atoms: (a) Be, (b) P, (c) Mn, (d) Cs, (e) Xe?

5.12 At one time, scientists thought electrons traveled in orbits around the nucleus, similar to the way planets travel around the sun. How is this description different from the modern atomic theory?

5.13 (a) Where is the first energy level in relation to the second level? (b) Explain why electrons are identified according to the energy level in which they reside.

5.14 What is the maximum number of electrons that can populate the following: (a) first energy level, (b) second energy level, (c) an s sublevel, (d) a p sublevel, (e) an orbital?

5.15 Write the complete electronic configurations for: (a) He, (b) B, (c) Ne, (d) Mg, (e) Cl, (f) K.

5.16 Draw the dot formulas for: (a) Na, (b) N, (c) Xe, (d) Ba, (e) Se.

SUMMARY

According to modern atomic theory, the atom is a small particle composed of a very dense nucleus containing protons and neutrons, with electrons sparsely populating the outer regions of the atom. Protons and neutrons have approximately the same mass, 1 u (unified atomic mass unit). Electrons have an extremely tiny mass, only $\frac{1}{1837}$ u. However, electrons possess a full negative charge, equal in magnitude but opposite in charge to the proton. Neutrons have no electric charge.

Each atom has a characteristic atomic number and mass number. Atomic number is equal to the number of protons in the nucleus of an atom. Mass number is the total number of protons and neutrons in the nucleus. If atoms have the same atomic number but different mass numbers, they are called isotopes. When dealing with elements that contain a mixture of isotopes, chemists utilize atomic mass (atomic weight), the average mass of the isotopes compared with the ^{12}C isotope.

Electrons are located in regions of space called orbitals. Orbitals are areas around the nucleus where there is a high probability of finding electrons. Orbitals hold a maximum of two electrons. A set of orbitals with the same shape and nearly the same energy is called a sublevel. Electrons in their lowest energy states are found in four different sublevels: s, p, d, and f. Sublevels with similar energies are grouped into energy levels.

Electrons fill orbitals in atoms, starting with the lowest energy orbital and proceeding to higher energy orbitals. Each different atom has its own specific electronic configuration. Electronic configurations are represented by writing the number corresponding to the energy level next to the letter that designates the sublevel, followed by writing the number of electrons that occupy the sublevel as a superscript above the letter. For example, hydrogen's electronic configuration is $1s^1$.

Atoms in the same group in the periodic table have the same outer electronic configuration. The number at the top of each group corresponds to the total number of electrons in the outer energy level. For example, each element in group IA has one outer electron. Outer electronic configurations are represented by writing Lewis electron dot formulas.

QUESTIONS AND PROBLEMS

5.17 Define the following terms: atomic mass unit, electric charge, nucleus, radioactivity, atomic number, mass number, isotope, atomic mass, orbital, electron energy level, sublevel, and electron dot formula.

Structure of Atoms

5.18 What is the mass, in unified atomic mass units, and charge of (a) a proton, (b) an electron, (c) a neutron?

5.19 What happens when the following charged particles are brought close to each other: (a) two protons, (b) proton and electron, (c) neutron and electron?

5.20 Complete the following table.

Symbol	Atomic number	Mass number	No. of protons	No. of neutrons	No. of electrons
(a) $^4_2$He	____	____	____	____	____
(b) $^{26}_{12}$Mg	____	____	____	____	____
(c) $^{36}_{16}$S	____	____	____	____	____
(d) ____	22	48	____	____	____
(e) ____	34	80	____	____	____
(f) ____	40	90	____	____	____
(g) ____	____	____	44	55	____
(h) ____	____	____	47	61	____
(i) ____	____	____	____	66	52
(j) ____	____	136	____	____	58

5.21 (a) What is a unified atomic mass unit? (b) How is it defined?

5.22 What radioactive particles have the following properties: (a) penetrates relatively dense matter, (b) is an electron moving at high speed, (c) resembles energy more than matter, (d) is the most massive and least penetrating?

Isotopes

5.23 What are the three primary isotopes found in a natural sample of uranium?

5.24 (a) What isotope is used as the standard for the atomic mass system? (b) Could another isotope be used as the standard?

5.25 Europium, Eu, contains two isotopes: $^{151}_{63}$Eu (mass = 150.92 u) and $^{153}_{63}$Eu (mass = 152.92 u). If the natural abundances of $^{151}_{63}$Eu and $^{153}_{63}$Eu are 47.82% and 52.18%, respectively, calculate the atomic mass of Eu.

5.26 Use the following mass and natural abundance data to calculate the atomic mass of thallium, Tl:

Isotope	Mass, u	Percent abundance
$^{203}_{81}$Tl	202.97	29.50%
$^{205}_{81}$Tl	204.97	70.50%

5.27 Determine the atomic mass of germanium, Ge. Consult Table 5.2 for all necessary information.

Electronic Configuration

5.28 Write the maximum number of electrons that can be located in the following: (a) an orbital, (b) the p sublevel, (c) a Li atom, (d) second energy level, (e) the d sublevel, (f) a Co atom, (g) the $3p$ sublevel of S, (h) the $3d$ sublevel of Ti, (i) the $6s$ sublevel of Cs, (j) fourth energy level of As.

5.29 How is the modern explanation for the location of electrons similar to a population density map in geography?

5.30 For each of the following, identify the region closest to the nucleus: (a) fifth, sixth, or seventh energy level, (b) $3s$, $3p$, or $3d$ sublevels, (c) $3s$, $4s$, or $5s$ sublevel.

5.31 Write the complete electronic configuration for each of the following atoms: (a) Be, (b) N, (c) Ne, (d) S, (e) Ga, (f) Br, (g) Rb, (h) Sb, (i) Xe, (j) Ba.

5.32 Write the outer electronic configuration for each of the following atoms: (a) Li, (b) Mg, (c) Al, (d) Se, (e) I, (f) Cs, (g) Sn, (h) In, (i) Pb, (j) Fr.

5.33 Write the symbols for the elements with the following electronic configurations:

(a) $1s^2 2s^2 2p^1$ (b) $1s^2 2s^2 2p^6 3s^1$

(c) $1s^2 2s^2 2p^6 3s^2 3p^4$ (d) $1s^2 2s^2 2p^6 3s^2 3p^6 4s^1$

(e) $1s^2 2s^2 2p^6 3s^2 3p^1$ (f) $1s^2 2s^2 2p^6 3s^2 3p^6 4s^2 3d^{10} 4p^4$

(g) $1s^2 2s^2 2p^6 3s^2 3p^6 4s^2 3d^{10} 4p^6 5s^2$

(h) $1s^2 2s^2 2p^6 3s^2 3p^6 4s^2 3d^{10} 4p^6 5s^2 4d^{10} 5p^6$.

5.34 Draw the electron dot formula for (a) C, (b) Na, (c) As, (d) I, (e) Ar, (f) S, (g) Ga, (h) Se, (i) Br, and (j) In.

5.35 Draw the electron dot formula for (a) Al, (b) O, (c) Mg, (d) Pb, (e) Xe, (f) K, (g) Si, (h) Ra, (i) Rb, and (j) Br.

5.36 Identify the atom or atoms with each of the following characteristics:

(a) First group IA atom to have a $3s$ outer electron

(b) Group VIIA atom(s) with $4p$ electrons

(c) Atom(s) with atomic number less than 47 that possess electrons in $4d$ sublevel

(d) Group IIA atom(s) without occupied d orbitals

(e) Period 3 atoms with more than four outer electrons

(f) Atom(s) with s electrons only.

5.37 What electronic configuration would the following have if they lost or gained the stated number of electrons:

(a) Mg, loses two electrons

(b) N, gains three electrons

(c) F, gains one electron?

· C H A P T E R ·

6

Periodic Properties

Study Guidelines

After completing Chapter 6, you should be able to:
1. State the periodic law
2. State the names of representative groups of elements on the periodic table
3. Distinguish between the properties of metals and nonmetals
4. Define and give examples of cations and anions
5. Predict the charge on ions from their placement in the periodic table
6. Predict a property of an element, given the properties of related elements
7. Predict trends in ionization energy and atomic radius
8. Discuss properties, characteristics, and uses of Na, Cl, and Si as representative of metals, nonmetals, and metalloids, respectively
9. Explain why hydrogen could be placed in a group all by itself

Dimitri Mendeleev (*Bettmann Archive*)

Mendeleev (1834–1907) published his first periodic table in 1869 in Russia. Russian discoveries were not usually translated into Western languages, so the Western world had to rediscover them. Luckily, Mendeleev's periodic table and explanations were translated immediately into German.

Before the electronic configurations of atoms were determined, scientists like Lothar Meyer and Dimitri Mendeleev recognized that if elements were placed in order of atomic mass (they did not know about atomic numbers) certain properties of atoms recurred at regular intervals. Both Meyer and Mendeleev proposed periodic tables that placed elements with similar properties in the same group (see Figure 6.1). Today, we know that these recurring properties result from the fact that the elements in the same chemical group have the same outer electronic configuration.

In the late nineteenth century, Mendeleev gained the attention of the scientific world by using his periodic table to predict the properties of elements which had not yet been discovered. See Table 6.1, which presents Mendeleev's predictions for the undiscovered element that he called "ekasilicon" (literally "under Si" on his table). After its discovery, ekasilicon was named germanium, Ge.

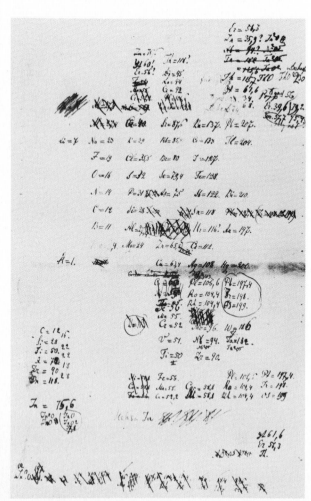

FIGURE 6.1
Mendeleev placed elements in a table in order of their atomic masses and grouped elements with similar properties. This figure shows an early draft of his periodic table that he published in 1869. (*Burndy Library*)

TABLE 6.1   MENDELEEV'S PREDICTIONS FOR EKASILICON (GERMANIUM*)

Property	Modern values	Predicted values (1871)
Atomic mass	72.6	72
Color	Grayish white	Dark gray
Density	5.4 g/mL	5.5 g/mL
Specific heat	0.074 cal/(g·°C)	0.073 cal/(g·°C)
Formula of oxide	GeO_2	ESO_2

*Mendeleev's predictions were made 15 years prior to the discovery of Ge.

Over the years since Mendeleev developed the modern form of the periodic table, groups of elements have been given "family" names, which are summarized in Figure 6.2. Elements in group IA, except H, are called **alkali metals,** while elements located in group IIA are called **alkaline earth metals** (the name used by the alchemists). The next 10 columns on the periodic table (group B elements) are referred to as the **transition metals** and contain commonly encountered metallic elements. All nontransition elements, group A elements, are called **representative elements,** including groups IA through VIIIA.

Group IIIA has no unique name; it is called the **aluminum** or **boron-aluminum group.** Similarly, groups IVA and VA are called the **carbon** and **nitrogen groups,** respectively. An old name for group VIA, the **chalcogens,** is

FIGURE 6.2
Periodic table with the names
of the chemical groups

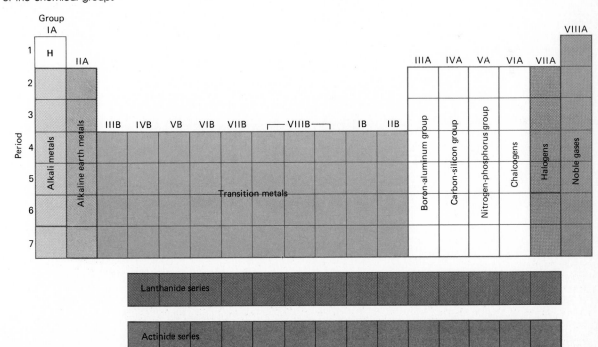

presently being used. Finally, group VIIA contains the **halogens,** and VIIIA contains the **noble gases.**

6.2 PERIODIC LAW

Let's turn our attention to what is now called the periodic law. The **periodic law** states that the properties of the elements are periodic functions of their atomic number. In other words, if the elements are listed in order of their atomic numbers, a regular pattern of chemical and physical properties is observed. To illustrate, we shall look at the two main classes of elements—metals and nonmetals.

Metals and Nonmetals

The largest percentage of elements, over 80 of them, are metals.

Metals are usually solid elements with a silvery gray color; their melting and boiling points are normally quite high. Metals share a set of common properties: (1) they have high densities, (2) they are excellent conductors of heat and electricity, and (3) they are malleable, i.e., able to be hammered into various shapes and foils.

Nonmetals possess properties that, in many cases, are opposite to those of metals. A large percentage of nonmetals are liquids and gases, not solids. On an average, the melting points, boiling points, densities, and electric and heat conductivities of nonmetals are lower than those of metals.

Metals are found on the left side of the periodic table, and nonmetals are on the right side.

In the second period of the periodic table, Li and Be are metals and C, N, O, F, and Ne are nonmetals. Boron has intermediate properties, not exactly metallic or nonmetallic, and is classified as a **metalloid** (or semimetal). Elements have a greater degree of nonmetallic character from left to right across the periodic table.

Metalloids have properties intermediate between those of metals and nonmetals.

Looking beyond Ne to the third period of the periodic table, we observe a similar trend: Na, Mg, and Al possess metallic properties; Si is considered a metalloid; and P, S, Cl, and Ar are nonmetals. Table 6.2 presents selected physical properties of the elements in the third period of the periodic table.

Figure 6.3 shows a periodic table that classifies elements as metals, nonmetals, and metalloids. Notice that the eight metalloids border a zigzag line

TABLE 6.2   SELECTED PROPERTIES OF ELEMENTS IN PERIOD 3

Element	Atomic number	Physical state, 25°	Melting point, °C	Density, g/mL	Type of element
Na	11	Silvery solid	97.8	0.97	Metal
Mg	12	Gray solid	650	1.74	Metal
Al	13	Light gray solid	660	2.70	Metal
Si	14	Grayish solid	1414	2.33	Metalloid
P	15	White or red solid	44	1.83	Nonmetal
S	16	Pale yellow solid	112	2.07	Nonmetal
Cl	17	Green gas	−101	0.00321	Nonmetal
Ar	18	Colorless gas	−189	0.00178	Nonmetal

FIGURE 6.3
Metals are the elements located on the left side of the periodic table. Nonmetals are listed on the right side of the periodic table. Metalloids are located along a zigzag line separating the metals from the nonmetals.

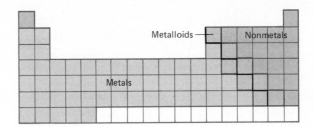

A substance's melting point is the temperature at which it exists in both the solid and liquid states. Higher melting points indicate stronger forces of attraction among atoms in the solid state.

An easy way to remember that anions are negative ions is to break the word *anion* into *a n* (negative) *ion*. Mnemonics (memory aids) such as this can simplify learning chemistry.

starting to the left of boron, B, and ending between Po and At. This line separates metals from nonmetals.

Specifically, if a physical property such as melting point is plotted on a graph, a better view of the periodic nature of properties of metals and nonmetals is obtained. Observe the graph of melting point (°C) versus atomic number in Figure 6.4. Note the semiregular pattern of increasing and decreasing melting points. With the exception of carbon, metals and metalloids are found at the maximum peaks, while the nonmetals occupy the valleys, the minimum values. This repeating pattern of peaks and valleys as atomic number increases is an illustration of the periodic law.

Chemically, metals are different from nonmetals. Metals have a small number of outer-level electrons, and they tend to lose these electrons during chemical changes. In contrast, nonmetals have more complete outer levels and tend to gain electrons (except the noble gases). When atoms either lose or gain electrons, they form ions (see Figure 6.5). An **ion** is a charged atom or, as we will see in a later chapter, a charged group of atoms. Two different types of ions exist: positive ions called **cations** and negative ions called **anions**. Cations are produced when metals give up one or more electrons:

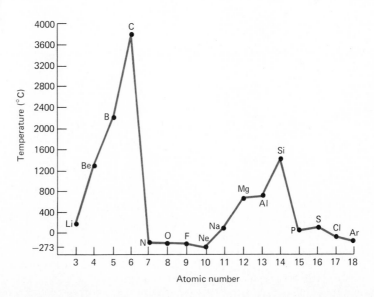

FIGURE 6.4
Graph of melting points versus atomic numbers

FIGURE 6.5
Metal atoms lose one or more electrons, forming cations. Nonmetal atoms gain one or more electrons, forming anions.

Charges on ions are written as superscripts to the right of symbols, with the magnitude of the charge preceding either + or −.

A sodium atom's electronic configuration is $1s^22s^22p^63s^1$. After losing an electron, the sodium ion's electronic configuration is $1s^22s^22p^63s^0$.

$$M \longrightarrow M^+ + e^-$$
$$M \longrightarrow M^{2+} + 2e^-$$

Anions form when nonmetals pick up electrons.

$$X + e^- \longrightarrow X^-$$
$$X + 2e^- \longrightarrow X^{2-}$$

Metals give up their outer-level electrons, resulting in a cation with a noble gas electronic configuration. Nonmetals often take in enough electrons to complete their outer levels, also producing a noble gas configuration. As we shall see in Chapter 8, the noble gas electronic configuration is very stable.

Metals in group IA, the alkali metals, tend to lose one electron when they combine with other substances, producing 1+ cations:

$$Na \longrightarrow Na^+ + e^-$$
$$K \longrightarrow K^+ + e^-$$

Metals in group IIA, alkaline earth metals, tend to lose two electrons when they combine with other elements, yielding 2+ cations:

$$Ca \longrightarrow Ca^{2+} + 2e^-$$
$$Ba \longrightarrow Ba^{2+} + 2e^-$$

In contrast, nonmetals, except noble gases, gain electrons and produce anions. Elements in group VIIA, the halogens, most frequently gain one electron, producing 1− anions:

Group							
IA	IIA	IIIA	IVA	VA	VIA	VIIA	
H^+							
Li^+	Be^{2+}			N^{3-}	O^{2-}	F^-	
Na^+	Mg^{2+}			P^{3-}	S^{2-}	Cl^-	
K^+	Ca^{2+}					Br^-	
Rb^+	Sr^{2+}					I^-	
Cs^+	Ba^{2+}					At^-	
Fr^+	Ra^{2+}						

FIGURE 6.6
Periodic table with selected ions

A chlorine atom's electronic configuration is $1s^22s^22p^63s^23p^5$. After gaining an electron, the chloride ion's electronic configuration is $1s^22s^22p^63s^23p^6$.

$$Cl + e^- \longrightarrow Cl^-$$
$$Br + e^- \longrightarrow Br^-$$

Similarly, group VIA elements, chalcogens, gain two electrons, forming $2-$ anions:

$$O + 2e^- \longrightarrow O^{2-}$$
$$S + 2e^- \longrightarrow S^{2-}$$

A periodic table with selected ions is presented in Figure 6.6.

6.3 PERIODIC PROPERTIES OF ATOMS

Ionization Energy

Periodic trends are observed when studying properties of individual atoms. An important property of atoms that helps explain ion formation is called ionization energy. **Ionization energy** is the minimum amount of energy required to remove the outermost electron from a neutral gaseous atom.

$$A(g) + \text{ionization energy} \longrightarrow A^+(g) + e^-$$

Ionization energy is a measure of the degree to which the nucleus attracts the outermost electrons. A low ionization energy indicates a low degree of attraction and a high ionization energy indicates a high degree of attraction between the nucleus and the outermost electrons.

Figure 6.7 presents a graph of ionization energies versus atomic number for

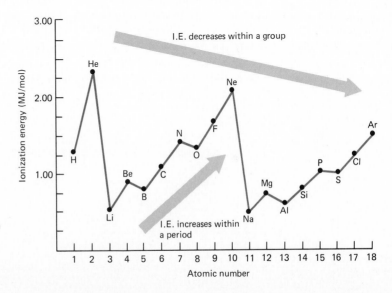

FIGURE 6.7
A graph of the ionization energies of atoms reveals two trends. Within a period, the ionization energy increases with increasing atomic number. Within a group, the ionization energy decreases with increasing atomic number.

Group

	IA	IIA	IIIB	IVB	VB	VIB	VIIB	VIIIB			IB	IIB	IIIA	IVA	VA	VIA	VIIA	VIIIA
1	1 H 1.31																	2 He 2.37
2	3 Li 0.520	4 Be 0.900											5 B 0.800	6 C 1.09	7 N 1.40	8 O 1.31	9 F 1.68	10 Ne 2.08
3	11 Na 0.496	12 Mg 0.738											13 Al 0.578	14 Si 0.787	15 P 1.01	16 S 1.00	17 Cl 1.25	18 Ar 1.52
4	19 K 0.418	20 Ca 0.590	21 Sc 0.631	22 Ti 0.658	23 V 0.650	24 Cr 0.653	25 Mn 0.717	26 Fe 0.759	27 Co 0.758	28 Ni 0.737	29 Cu 0.746	30 Zn 0.906	31 Ga 0.579	32 Ge 0.762	33 As 0.944	34 Se 0.941	35 Br 1.14	36 Kr 1.35
5	37 Rb 0.403	38 Sr 0.550	39 Y 0.616	40 Zr 0.660	41 Nb 0.664	42 Mo 0.685	43 Tc 0.702	44 Ru 0.711	45 Rh 0.720	46 Pd 0.805	47 Ag 0.731	48 Cd 0.868	49 In 0.558	50 Sn 0.709	51 Sb 0.832	52 Te 0.869	53 I 1.01	54 Xe 1.17
6	55 Cs 0.376	56 Ba 0.503	57 La 0.538	72 Hf 0.654	73 Ta 0.761	74 W 0.770	75 Re 0.760	76 Os 0.84	77 Ir 0.88	78 Pt 0.87	79 Au 0.890	80 Hg 1.01	81 Tl 0.589	82 Pb 0.716	83 Bi 0.703	84 Po 0.812	85 At 0.916	86 Rn 1.04

Period

FIGURE 6.8
Periodic table containing
ionization energies

the first three periods, and Figure 6.8 lists the actual ionization energies in megajoules per mole (MJ/mol). Two distinct trends are evident. First, as we proceed from left to right across a period (increasing atomic number within a period), the ionization energy generally increases. More energy is required, on the average, to remove the outermost electron from nonmetals than metals. Alkali metals, group IA elements, have the lowest ionization energies. Noble gases, group VIIIA elements, have the highest ionization energies.

Second, within a group like the noble gases (VIIIA), ionization energies decrease with increasing atomic number. Helium has the largest ionization energy (2.37 MJ/mol) among the noble gases, and Rn has the lowest (1.04 MJ/mol). Similar trends are observed in all chemical groups.

Overall, elements in the lower left corner of the periodic table have the lowest ionization energies; francium, Fr, is the element with the lowest ionization energy. As we move up and to the right on the periodic table, the ionization energies generally increase; helium, He, has the largest ionization energy (see Figure 6.9).

Ionization energies increase from the lower left corner of the periodic table to the upper right corner.

Two factors account for the trends in ionization energy: (1) nuclear attraction of the electrons, and (2) the shielding effect of the inner-level electrons. Greater attraction of the electrons by the nucleus results in greater ionization

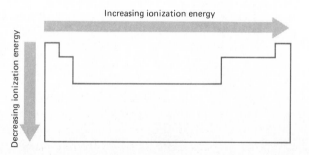

Increasing ionization energy

Decreasing ionization energy

FIGURE 6.9
Periodic table showing trends in
ionization energies

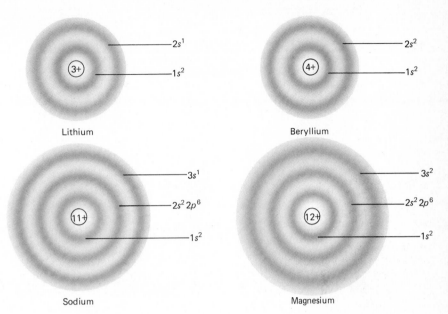

$2s^1$
$1s^2$
(3+)
Lithium

$2s^2$
$1s^2$
(4+)
Beryllium

$3s^1$
$2s^2 2p^6$
$1s^2$
(11+)
Sodium

$3s^2$
$2s^2 2p^6$
$1s^2$
(12+)
Magnesium

FIGURE 6.10
In Li, two of the three nuclear charges are shielded by the inner 1s orbital, leaving only one nuclear charge to hold the outer electron. In Be, two of the four nuclear charges are shielded, leaving two nuclear charges to hold the outer electrons. As a result of having a greater nuclear charge, Be has a greater ionization energy than Li after shielding. A similar situation exists with Na and Mg.

Inner electrons shield a portion of the attractive force of the nucleus from outer-level electrons.

energies. Shielding refers to the blocking effect that inner-level electrons have on the nuclear attraction of the outer electrons. Greater shielding of the nucleus by inner electrons results in lower ionization energies.

The general increasing trend in ionization energy across a period is explained as follows. The shielding effect of the inner electrons remains constant across a period, while the charge on the nucleus increases. Increasing nuclear charge produces a greater attractive force on the outer electrons, resulting in a larger ionization energy.

Decreasing trends in ionization energy within a chemical group are explained in terms of the location of the outer electrons. With each new energy level, the outer electrons are farther from the nucleus, with more levels of inner electrons shielding the nuclear charge (see Figure 6.10). Greater distance from the nucleus and greater shielding effects result in a smaller attractive force on the outer electrons; hence, they are easier to remove.

Trends in Atomic Sizes: Atomic Radius

Atomic radius is a measure of the size of atoms.

Atomic radii decrease from the lower left corner of the periodic table to the upper right corner.

The atomic radius is a measure of an atom's size. It is the distance from the nucleus to a point corresponding to the outermost region of a neutral atom. Atomic radii for the elements in the second and third periods are plotted in Figure 6.11, and the relative sizes of the atoms are illustrated as circles in Figure 6.12.

Once again, two trends are plainly seen. Across a period from left to right, the atomic radii decrease, and within a group, the atomic radii increase. Thus, the largest atoms are located in the lower left corner of the periodic table, and the smallest atoms are located in the upper right corner (see Figure 6.13).

Is this consistent with the trends in ionization energy? Yes! Across a period, the shielding effects of inner electrons are constant and the nuclear charge in-

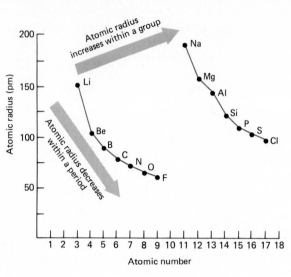

FIGURE 6.11
The covalent radius is used as a measure of the atom's size (atomic radius). Noble gases are not included, since different means are used to measure their radii.

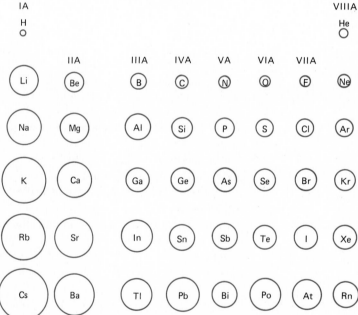

FIGURE 6.12
Periodic table showing relative sizes of atoms

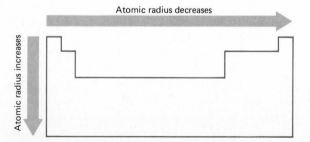

FIGURE 6.13
Periodic table showing trends in atomic radii

creases with the addition of more protons to the nucleus. Consequently, the attractive force of the nucleus for the electrons increases, and the size of the atom decreases as the electrons are "pulled in" toward the nucleus. Within a group, the attractive forces on the outer-level electrons decrease because these electrons are farther from the nucleus and there are greater shielding effects from inner electrons. Thus, the atomic radius increases.

Example Problem 6.1 illustrates predictions of atomic properties from trends in ionization energy and atomic radius.

Example Problem 6.1

Of the four atoms Si, P, S, and Cl, which (a) is the most nonmetallic, (b) has the largest ionization energy, and (c) has the largest atomic radius?

Solution

Note that the four atoms all belong to the third period of the periodic table. Thus, to answer this question, it is only necessary to know the periodic trends when proceeding from left to right across the periodic table.

(a) Metallic character decreases or nonmetallic character increases as we go from left to right across the periodic table. Accordingly, Cl, a halogen, is the most nonmetallic atom of the group.
(b) Ionization increases from left to right across a period; consequently, Cl has the largest ionization energy of the four atoms.
(c) Atomic radius decreases across a period; hence, Si has the largest atomic radius.

REVIEW PROBLEMS **6.1** Write the name of the groups on the periodic table with the following letter and number designations: (a) IA, (b) IIA, (c) IVA, (d) VIA, (e) VIIA, (f) VIIIA.

6.2 State the periodic law, and give an example.

6.3 List three properties that distinguish metals from nonmetals.

6.4 What periodic trends are observed when considering densities within a period on the periodic table?

6.5 Define and give two examples of cations and anions.

6.6 Select the appropriate element from the following set of elements:
(a) Largest ionization energy: C, N, O, F.
(b) Smallest ionization energy: B, C, Al, Si.
(c) Largest atomic radius: Be, B, Mg, Al.
(d) Smallest atomic radius: Li, Be, Na, Mg.
(e) Most metallic: Al, Si, Ga, Ge.

6.4 A REPRESENTATIVE METAL, NONMETAL, AND METALLOID

Before leaving the topic of periodic properties, let's take a more in-depth look at a representative metal, nonmetal, and metalloid to gain a better understanding of elements and their properties.

Sodium

Davy discovered K, Na, Ba, Sr, Ca, and Mg, but his greatest discovery was the man he selected for his assistant, Michael Faraday. Faraday became one of the world's most productive scientists.

Sodium, the most abundant alkali metal, was discovered in 1807 by Sir Humphry Davy (1778–1829). He isolated sodium by passing an electric current through a sodium compound. What he found was a reactive metal with properties unlike those of frequently encountered metals (Fe, Ag, Au, or Cu).

Sodium is a soft, silvery gray solid with a low density, 0.97 g/mL. Most metals have high densities and sink when placed in water, but sodium does not. If a small chunk of sodium is placed in water, it floats. However, this is of secondary importance, for immediately on contacting water, Na reacts violently, liberating hydrogen gas and a large amount of heat energy.

$$2Na(s) + 2H_2O(l) \longrightarrow 2NaOH(aq) + H_2(g) + energy$$

Sodium's melting and boiling points, 97.8°C and 889°C, are both low for metals. While these values are low compared with those of nonalkali metals, only lithium has higher melting and boiling points among the alkali metals. The remaining alkali metals have even lower melting and boiling points (see Table 6.3). On hot days cesium, Cs, is a liquid metal.

Sodium is much too reactive to exist in nature as a separate element. Instead, it exists in various compounds. Many rocks contain sodium compounds that, on weathering, dissolve, ultimately ending up in the oceans as sodium ions, Na^+. Many dissolved sodium compounds are contained in the oceans, most notably, sodium chloride (table salt), NaCl.

Over 2% by mass of the earth's crust and atmosphere is sodium chemically bonded in compounds.

Sodium has a multitude of uses. It acts as a dehydrating agent, removing water from many liquids. Sodium vapor lamps illuminate highways. In industry, sodium is combined with other metals (Hg, K, Sn, and Sb) to form important alloys. The petroleum industry utilizes sodium in the production of antiknock compounds for gasoline. Nuclear power plants use molten (liquid) sodium as a heat transfer agent, i.e., to carry away heat given off by radioactive substances.

TABLE 6.3 MELTING AND BOILING TEMPERATURES OF THE ALKALI METALS

Symbol	Melting point, °C	Boiling point, °C
Li	180	1347
Na	98	889
K	63	757
Rb	39	679
Cs	29	690

TABLE 6.4   MELTING AND BOILING POINTS OF THE HALOGENS

Element	Melting point, °C	Boiling point, °C
Fluorine	−220	−188
Chlorine	−101	−34
Bromine	−7	59
Iodine	114	184

Chlorine   Among his many discoveries, Karl Wilhelm Scheele, (1742–1786) is credited with the discovery of chlorine in 1774. Over 30 years elapsed before it was given its modern name by Davy in 1810. Chlorine is a greenish yellow gas with a density (3.2 g/L) greater than that of air. Chlorine melts at −101°C and boils at −34.1°C. A regular trend in melting and boiling points can also be seen in the halogens. While fluorine and chlorine are gases at room conditions, bromine is a liquid and iodine is a solid (see Table 6.4).

Like sodium, chlorine is reactive and combines with many substances. Because of its extreme reactivity, chlorine is acutely toxic. It is so toxic that, as a precautionary measure, towns are evacuated if a train containing chlorine tank cars derails nearby. During World War I, chlorine was used in trench warfare. Chlorine's toxicity and high density made it an effective killing agent, until gas masks were perfected.

Chlorine compounds, some resembling laundry bleach, are added to swimming pools to destroy bacteria and algae. Chlorine gas, under controlled conditions, is pumped into drinking water and then removed during the purification process. Chlorine destroys most microorganisms that live in the water and removes compounds that are responsible for water's bad taste and unpleasant odor.

Over 1000 chlorine-containing compounds are produced industrially, and they are used in pesticides, refrigerants, anesthetics, cleaning solutions, and plastics.

Like sodium, chlorine is prepared from sodium chloride NaCl. Sodium chloride is obtained from salt mines and ocean water. An electric current is passed through a sodium chloride solution, producing chlorine and another industrially useful substance called sodium hydroxide (lye), NaOH.

Pure chlorine gas is not composed of individual chlorine atoms. Chlorine atoms are so reactive that they chemically combine with other chlorine atoms to form diatomic molecules, i.e., molecules with two atoms each. All halogen molecules share this property and exist as diatomic molecules. Molecules and their formation are discussed in Chapter 8.

Silicon   Silicon is a representative metalloid. It is a brittle, shiny, black-gray element that appears to be metallic but isn't. Structurally, silicon resembles diamond (carbon). It is extremely hard, is capable of scratching glass, melts at 1414°C, and boils at 2327°C. Jons Jakob Berzelius (1779–1848), a Swedish chemist, discovered silicon in 1823.

Silicon is not the same as silicone. Silicon is an element; silicone is a complex compound.

Silicon belongs to a diverse group on the periodic table, group IVA. Carbon, the first member of group IVA, is a nonmetal; both silicon and germanium

TABLE 6.5   SELECTED PROPERTIES OF GROUP IVA ELEMENTS

Symbol	Melting point, °C	Density, g/mL
C	3570	2.25
Si	1414	2.33
Ge	937	5.32
Sn	232	7.30
Pb	328	11.4

are metalloids; and tin and lead are metals. The melting point of silicon is less than that of carbon, but it is higher than those of other group members. Both carbon and silicon have low densities compared with those of the more metallic elements, Sn and Pb. Observe these trends in group IVA properties in Table 6.5.

Silicon is the second most abundant element on earth; oxygen is the most abundant. Approximately 25% of the earth's crust is silicon, in the form of silicon compounds. Quartz (Figure 6.14), sand, agate, jasper, and opal are oxides of silicon (i.e., silicon bonded to oxygen). In many cases, silicon is combined with both oxygen and metals; common examples include talc, mica, asbestos, beryl, and feldspar.

Silicon compounds are commercially important, especially compounds called silicates (silicon–oxygen compounds). Clay, cement, and glass are all silicates. When silicon is combined with carbon, the result is silicon carbide, SiC, a very hard material with numerous industrial uses.

Silicon dioxide (silica), SiO_2, is found in 20 or more different forms in the earth and its inhabitants. Silica dissolves in water, forming silicic acid, $Si(OH)_4$. This acid provides the silica to produce quartz crystals inside geodes, skeletons for diatoms, and petrified wood.

REVIEW PROBLEMS

6.7 What scientists discovered (a) sodium, (b) chlorine, (c) silicon?

6.8 What properties distinguish sodium from metals like Cu, Fe, or Au?

6.9 Compare melting point trends in the alkali metals with that of the halogens.

6.10 (a) How does silicon exist in nature? (b) List three silicon compounds.

FIGURE 6.14
Quartz, SiO_2, is one of the many silicon oxides on earth. Quartz is a hard, colorless, brittle crystalline solid. It is used to control frequencies in electronic devices and is used in optical instruments. (Ron Testa/Field Museum of Natural History, Chicago)

6.5 HYDROGEN: AN ELEMENT AND A GROUP

Hydrogen is usually written at the top of group IA of the periodic table because it shares one common property with the alkali metals; hydrogen and the alkali metals all possess an s^1 valence electronic configuration. Nevertheless, hydrogen is unique in so many other ways that it truly belongs in a group by itself.

Hydrogen was discovered by Henry Cavendish in 1766 and was named by Lavoisier. The term "hydrogen" literally means "water former." Just about all hydrogen on earth is chemically combined with another element. The largest percentage is bonded to oxygen in water, H_2O. In the uncombined state, hydrogen exists diatomically, as H_2. Free, isolated hydrogen atoms, H, do not exist at normal conditions.

Hydrogen is fairly abundant on earth, and it is the most abundant element in the universe. Most of the mass of stars is ionized hydrogen gas.

Hydrogen has the lowest atomic mass and density of all elements. It must be cooled to $-253°C$ before liquefying, and it freezes at $-259°C$, only $6°C$ below its boiling point! Hydrogen is a poor conductor of heat and electricity. Notice the extreme differences in hydrogen's physical properties compared with those of the alkali metals, which are solids, have higher densities, and are good electric conductors. Table 6.6 compares the properties of hydrogen with those of the alkali metal of lowest atomic mass, lithium.

In terms of specific atomic properties, we have seen that the alkali metals have the lowest ionization energies and largest atomic radii within their period. In contrast, hydrogen has a large ionization energy and a small atomic radius.

Chemically, hydrogen is quite reactive. It combines explosively with oxygen, O_2, forming water. Industrially, a large quantity of hydrogen is combined with nitrogen, N_2, producing ammonia, NH_3.

$$2H_2(g) + O_2(g) \longrightarrow 2H_2O(g)$$
$$3H_2(g) + N_2(g) \longrightarrow 2NH_3(g)$$

In the future, hydrogen might occupy the position that fossil fuels hold today as the primary energy source. Stars obtain their energy from hydrogen in a process called nuclear fusion. If scientists could sustain and control nuclear fusion reactions, then the energy released from these reactions could be con-

Henry Cavendish (1731–1810) was truly an eccentric scientist. He was a total "loner"; rarely did he speak, and when he did, it was only to men, never to women. He wrote notes when he had to communicate with women. He spent virtually none of an inheritance worth more than $5 million. He devoted 60 years of his life to scientific research.

Nuclear fusion occurs when the nuclei of low-mass atoms combine to produce an atom of higher mass. During nuclear fusion, a large quantity of energy is released.

TABLE 6.6 COMPARISON OF PHYSICAL PROPERTIES OF HYDROGEN AND LITHIUM

Property	Hydrogen	Lithium
Outer electronic configuration	$1s^1$	$2s^1$
Melting point, °C	-259.2	180.5
Boiling point, °C	-252.7	1347
Density, g/mL	9.0×10^{-5}	0.53
Ionization energy, MJ/mol	1.3	0.52

verted to electricity. One of the technological problems with using hydrogen as a fuel is the enormous temperatures ($>100,000,000°C$) needed to sustain nuclear fusion reactions. Nuclear fusion and other nuclear phenomena are discussed in Chapter 19.

REVIEW PROBLEMS

6.11 What evidence could be presented to show that hydrogen is not an alkali metal?

6.12 All halogens obtain the noble gas configuration by gaining one electron. If H receives one electron, it also obtains the noble gas configuration. What other properties of hydrogen could possibly allow it to be classified as a halogen?

SUMMARY

When atoms are organized according to their atomic number, recurring chemical and physical properties are observed. This is called the periodic law. In our modern periodic table, elements with the same outer electronic configuration are in the same group (vertical column). Group members share many common properties, and within most groups regular trends in properties are seen. Periods (horizontal rows) contain elements that have outer electrons in the same energy level. Likewise, semiregular trends in properties are observed when progressing from one side of the table to the other.

Elements can be classified as metals, nonmetals, or metalloids (semimetals), depending on their properties. Metals occupy the left side of the periodic table, and are usually solids with high densities and high melting and boiling points. Metals are good conductors of heat and electricity and tend to lose electrons in chemical changes. Nonmetals, on the right side of the periodic table, generally possess properties opposite to those of metals. They have lower average densities, melting points, boiling points, and conductivities. In chemical reactions, nonmetals tend to gain electrons and form anions. Metalloids resemble metals but have various nonmetallic properties.

Two properties of atoms that illustrate the periodic law are ionization energy and atomic radius. Ionization energy is the amount of energy required to remove the outermost electron from a neutral gaseous atom. Across a period from left to right, the ionization energy increases: metals have lower ionization energies than nonmetals. Within a chemical group, the ionization energy decreases with increasing atomic number. Overall, elements near the lower left side of the periodic table have the lowest ionization energies, and those near the upper right side have the highest ionization energies.

Atomic radius is a measure of the size of an isolated atom. From left to right across a period on the periodic table, the atomic radius decreases, and within a group the atomic radius increases. Elements at the lower left side of the periodic table are the largest, and those near the upper right are the smallest.

Sodium is a reactive representative metal. Sodium is a soft metal with low melting and boiling points. Like other metals, sodium is a good conductor of heat and electricity. Chemically, it combines with many other elements; in many cases, it gives up an electron to form a $1+$ cation.

Chlorine is a representative nonmetal. Chlorine exists as a gas at room temperature. It posseses the general properties of nonmetals: it has low density, is a poor electric and heat conductor, and forms anions in chemical reactions. Chlorine is a reactive gas that combines with most other elements to produce chlorine compounds.

Hydrogen is unique because its properties are so diverse. It does not readily fit into any group on the periodic table. Hydrogen is best placed in a group by itself. In spite of this, because H has a $1s^1$ electronic configuration, it is usually placed in group IA of the periodic table.

QUESTIONS AND PROBLEMS

6.13 Define the following terms: periodic properties, metal, nonmetal, metalloid, ion, cation, anion, ionization energy, atomic radius, diatomic, and shielding.

Periodic Properties

6.14 Why is the periodic table called by that name?

6.15 Write the name of the group on the periodic table to which each of the following elements belongs: (a) Se, (b) Ar, (c) Sr, (d) K, (e) Co, (f) As, (g) Xe, (h) Ga, (i) Sn, (j) Y.

6.16 How is a representative element distinguished from a transition metal, using a periodic table?

6.17 What is the outer electronic configuration in all (a) alkali metals, (b) halogens, (c) chalcogens, (d) noble gases, (e) alkaline earth metals, (f) nitrogen–phosphorus group elements?

6.18 How do nonmetals differ from metals? Give a specific example.

6.19 Classify each as a metal, nonmetal, or metalloid: (a) Zn, (b) As, (c) Y, (d) Br, (e) H, (f) P, (g) I, (h) Ba, (i) Pd, (j) Ge.

6.20 Consider the following properties of hypothetical elements A and B, and classify each substance as a metal, nonmetal, or metalloid. Fully explain your answer.
   (a) Element A boils at $-195.8°C$, has a density of 1.3 g/L, and is a colorless gas at room temperature.
   (b) Element B boils at $3200°C$, has a density of 10 g/mL, and is a good conducting solid.

6.21 Select the element that is the most metallic in each of the following sets of elements: (a) Be, B, C, and N; (b) C, Si, Ge, Sn; (c) As, Se, Sb, and Te; (d) Mg, Al, Si, and P.

6.22 (a) How many added electrons are required to complete the outer electron energy level of (1) N, (2) S, (3) Br, (4) Se, (5) P, and (6) I? (b) What charge would be found on each of the atoms in part (a) after gaining the indicated number of electrons?

6.23 (a) How many electrons must be removed from each of the following atoms to give it a noble gas electronic configuration: Sr, Li, Al, Sc, Ba, and Si? (b) What charge would be found on each of the atoms in part (a) after the electrons are lost?

6.24 What charge would the most stable ions of the following atoms possess: (a) K, (b) S, (c) F, (d) O, (e) Ca, (f) Te, (g) Rb, (h) P, (i) Br, (j) I?

6.25 Write an equation illustrating what happens when the following atoms form ions: (a) N, (b) Cl, (c) Mg, (d) Li, (e) P, (f) Ca.

6.26 Predict the value for the ionization energy of antimony, Sb, given the values of the ionization energies for all other members of group V. Their ionization energies (kcal) are: N, 335; P, 254; As, 231; and Bi, 185.

6.27 From the following groups of atoms predict the atom with the smallest ionization energy: (a) C, Si, Ge, Sn, (b) As, Se, Br, Kr, (c) K, Ca, Rb, Sr, (d) F, Ne, Cl, Ar.

6.28 For each set of elements in Problem 6.27, select the element with the smallest atomic radius.

6.29 Write the symbol for the element or elements that best fits the following descriptions: (a) noble gas with largest ionization energy, (b) chalcogen with most metallic character, (c) Al group member with the highest density, (d) lowest-melting alkali metal, (e) transition metal with largest atomic radius, (f) period 3 member that forms $3-$ ions.

6.30 The element with atomic number 118 (ununoctium, Uuo) has yet to be discovered. For each of the listed properties, make a prediction about the properties of element 118, and explain your decision: (a) metal, nonmetal, or metalloid; (b) gas, liquid, or solid; (c) good or bad conductor of electricity; (d) colored or colorless; (e) large or small atomic radius compared with its group; (f) large or small ionization energy compared with its group.

6.31 Write the symbol for the element that best fits each description: (a) largest ionization energy on the periodic table, (b) largest atomic radius on the periodic table, (c) smallest halogen, (d) atom that forms a 1− ion with the same arrangement as Xe, (e) metalloid belonging to the chalcogens, (f) group VIA element with the largest density.

Representative Metal, Nonmetal, and Metalloid

6.32 Considering the electronic configurations of sodium and chlorine, what could explain the reactivity of these elements?

6.33 When Na is placed in water, what are the products of the reaction?

6.34 How are (a) Na, (b) Cl, and (c) Si used commercially?

6.35 Study Si's properties, and determine whether it has more of a metallic or nonmetallic character.

6.36 In what type of rocks can silicon compounds be found?

6.37 What are some everyday uses of silicates?

General Questions

6.38 Utilizing data presented in Table 6.2, plot a graph of the density of period 3 elements. Describe density trends in period 3 of the periodic table.

6.39 (a) Graph the ionization energies of the elements in period 4 versus their atomic numbers, using Figure 6.8. (b) Compare and contrast the trends in ionization energy of the third period elements (Figure 6.7) with that of the fourth period.

6.40 Provide a complete explanation for the fact that helium has the highest ionization energy of all atoms. To answer this question consider nuclear charge, shielding, and the size of a helium atom.

6.41 Predict as many properties as possible for the yet to be discovered element 120, unbinilium, Ubn.

6.42 An atom's second ionization energy is the amount of energy required to remove a second electron, after the outermost electron has been removed. Stated differently, an atom's second ionization energy is a measure of the amount of energy needed to remove an electron from a cation with a 1+ charge. What group on the periodic table would be expected to have the largest second ionization energy? Fully explain.

· C H A P T E R ·

7

Mole Concept and Calculations

Study Guidelines

After completing Chapter 7, you should be able to:

1. Explain the meaning of a counting unit
2. Define mole
3. State the number of particles in one mole
4. Calculate the molar mass of atoms, molecules, formula units, and ions
5. Calculate the mass of a substance, given the number of moles
6. Calculate the number of moles of a substance, given the mass
7. Calculate the number of atoms or molecules contained within a substance, given either the mass or number of moles
8. Calculate the number of moles or mass of a substance, given the number of atoms or molecules
9. Determine the molecular mass of a compound
10. Determine the formula masses of compounds that are not composed of discrete molecules
11. Calculate the number of moles of atoms in a compound, given appropriate data
12. State the law of constant composition, and show how it is applied
13. Determine the percent composition of a compound
14. Distinguish between a compound's empirical and molecular formulas
15. Derive the simplest (empirical) formula of a compound, given mass data
16. Find the molecular formula of a compound given mass data and the molecular mass

Amedeo Avogadro *(The Bettmann Archive, Inc.)*

Atoms and molecules are very small particles and cannot be handled on an individual basis. Even minute samples of matter contain a staggering number of particles. We shall study the means that chemists utilize to deal with matter in such a way that macroscopic (large) samples reflect their microscopic (very small) composition.

Earlier we mentioned that the mole is a base SI unit for the amount of substance. What is a mole, and how is it applied in chemistry? A mole, abbreviated mol, is a counting unit, a unit that allows us to keep track of the number and mass of both atoms and molecules. While you may not have heard of the mole, you are familiar with a common counting unit called the "dozen."

A dozen is the counting unit for 12 objects.

$$1 \text{ dozen objects} = 12 \text{ objects}$$

Moles and dozens are alike, except that a mole refers to a different number of objects.

Discrete (separate) objects, like oranges or eggs, are often not sold by their weight, but by their quantity. Here, the dozen can represent a mass of oranges or eggs. Counting units can be used to "weigh things by counting."

Frequently, the dozen is too small, so other counting units are employed. Paper is bought by the ream (480 sheets or 20 quires). We say that a gross of pencils is purchased, rather than 144 pencils. Counting units are used whenever it is not easy to deal with objects as a result of their large number or small size.

Moles

$1 \text{ mol} = 6.022045 \times 10^{23}$ entities

Amedeo Avogadro, Count of Quaregna (1776–1856) was the first person to distinguish between atoms and molecules (a word he coined). Most of his work was overlooked during his lifetime, and it was not until a few years after his death that his discoveries were recognized.

Chemists have produced a counting unit that allows them to keep track of small particles such as atoms and molecules efficiently. This counting unit, the mole, is a fixed number of objects. The number of objects in a **mole** is 6.022×10^{23}, called **Avogadro's number.** Amedeo Avogadro was an Italian scientist who performed pioneering experiments on the properties of gases.

What is the importance of the number 6.022×10^{23}? If Avogadro's number is multiplied by the mass of an individual atom expressed in grams, the product is the atomic mass expressed in grams. For example, an individual hydrogen atom's mass is approximately 1.67×10^{-24} g, and hydrogen's atomic mass (from the periodic table) is 1.01. What number, when multiplied times 1.67×10^{-24} g, gives 1.01 **grams?** You guessed it, 6.022×10^{23}.

Stated in another way, if 6.02×10^{23} atoms of an element are placed on a balance, their mass is the atomic mass expressed in grams. Consider the following atoms:

1 mol He atoms (6.02×10^{23} atoms He) has a mass of 4.00 g
1 mol Li atoms (6.02×10^{23} atoms Li) has a mass of 6.94 g
1 mol N atoms (6.02×10^{23} atoms N) has a mass of 14.0 g
1 mol Ne atoms (6.02×10^{23} atoms Ne) has a mass of 20.2 g
1 mol Fe atoms (6.02×10^{23} atoms Fe) has a mass of 55.8 g
1 mol U atoms (6.02×10^{23} atoms U) has a mass of 238 g

A mole is the approximate number of grains of sand on all the beaches on earth. One mole of baseballs would cover the entire earth (land and water) to a depth of about 50 miles.

Technically, one mole (1 mol) is defined as the amount of pure substance that contains the same number of particles as there are atoms in exactly 12 grams of ^{12}C (SI definition).

How big is 1 mol? Well, 1 mol is much too colossal for any person to conceive. It is best to consider the fact that the magnitude of a mole is beyond comprehension when considering normal-size objects, but is a convenient number of particles when working with atoms, ions, and molecules. Avogadro's number of atoms can readily be measured on a balance in the lab.

REVIEW PROBLEMS

7.1 What similarities exist between the units dozen and mole?
7.2 Give the accepted SI definition of a mole.
7.3 What is Avogadro's number, and why is it important?
7.4 Calculate how many billions are contained in 1 mol.

7.2 MOLAR MASS OF ATOMS

Each element has a molar mass, the number of grams of that element that contains 6.022×10^{23} atoms (1.000 mol). In all instances, the molar mass is the atomic mass expressed in grams (see Figure 7.1).

One mole of hydrogen atoms weighs 1.0 g, while 1.0 mol of helium atoms weighs 4.0 g. Note that the ratio of molar masses of He to H is 4 to 1, the same

FIGURE 7.1
One mole of like atoms has a mass equal to the atomic mass expressed in *grams.*

Period / Group	IA	IIA	IIIB	IVB	VB	VIB	VIIB	VIIIB			IB	IIB	IIIA	IVA	VA	VIA	VIIA	VIIIA
1	1 H 1.008 g																	2 He 4.003 g
2	3 Li 6.941 g	4 Be 9.012 g											5 B 10.81 g	6 C 12.01 g	7 N 14.01 g	8 O 16.00 g	9 F 19.00 g	10 Ne 20.18 g
3	11 Na 22.99 g	12 Mg 24.31 g											13 Al 26.98 g	14 Si 28.09 g	15 P 30.97 g	16 S 32.06 g	17 Cl 35.45 g	18 Ar 39.95 g
4	19 K 39.10 g	20 Ca 40.08 g	21 Sc 44.96 g	22 Ti 47.90 g	23 V 50.94 g	24 Cr 52.00 g	25 Mn 54.94 g	26 Fe 55.85 g	27 Co 58.93 g	28 Ni 58.70 g	29 Cu 63.55 g	30 Zn 65.38 g	31 Ga 69.72 g	32 Ge 72.59 g	33 As 74.92 g	34 Se 78.96 g	35 Br 79.90 g	36 Kr 83.80 g
5	37 Rb 85.47 g	38 Sr 87.62 g	39 Y 88.91 g	40 Zr 91.22 g	41 Nb 92.91 g	42 Mo 95.94 g	43 Tc 98 g	44 Ru 101.1 g	45 Rh 102.9 g	46 Pd 106.4 g	47 Ag 107.9 g	48 Cd 112.4 g	49 In 114.8 g	50 Sn 118.7 g	51 Sb 121.8 g	52 Te 127.6 g	53 I 126.9 g	54 Xe 131.3 g
6	55 Cs 132.9 g	56 Ba 137.3 g	57* La 138.9 g	72 Hf 178.5 g	73 Ta 180.9 g	74 W 183.8 g	75 Re 186.2 g	76 Os 190.2 g	77 Ir 192.2 g	78 Pt 195.1 g	79 Au 197.0 g	80 Hg 200.6 g	81 Tl 204.4 g	82 Pb 207.2 g	83 Bi 209.0 g	84 Po 209 g	85 At 210 g	86 Rn 222 g
7	87 Fr 223 g	88 Ra 226.0 g	89† Ac 227.0 g	104 Unq 261 g	105 Unp 262 g	106 Unh 263 g												

*	58 Ce 140.1 g	59 Pr 140.9 g	60 Nd 144.2 g	61 Pm 145 g	62 Sm 150.4 g	63 Eu 152.0 g	64 Gd 157.2 g	65 Tb 158.9 g	66 Dy 162.5 g	67 Ho 164.9 g	68 Er 167.3 g	69 Tm 168.9 g	70 Yb 173.0 g	71 Lu 175.0 g
†	90 Th 232.0 g	91 Pa 231.0 g	92 U 238.0 g	93 Np 237.0 g	94 Pu 244 g	95 Am 243 g	96 Cm 247	97 Bk 247 g	98 CF 251 g	99 Es 252 g	100 Fm 257 g	101 Md 258 g	102 No 259 g	103 Lr 260 g

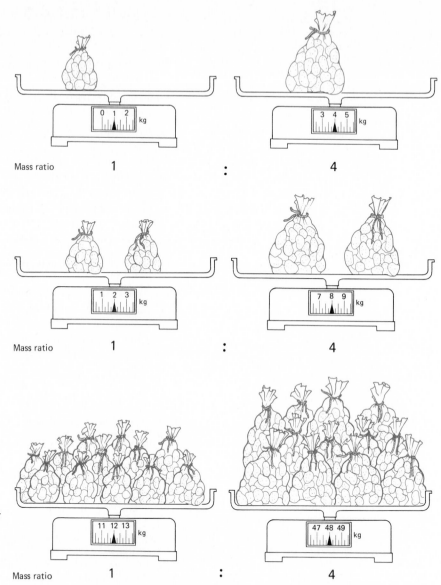

FIGURE 7.2
If the mass ratio of two different bags of potatoes is 1 to 4 (1 kg to 4 kg), then the mass ratio of equal quantities of these two different bags of potatoes is always 1 to 4.

ratio found when comparing an individual He atom with an individual H atom. As long as an equal number of particles is compared, the mass ratio remains fixed.

Let us discuss normal-size objects to help develop a mental picture for what is unobservable. Consider two bags of potatoes, the first a 1-kg bag, and the other a 4-kg bag. What is the mass ratio of the two bags? It is 1 to 4. If two 1-kg bags are compared with two 4-kg bags, the mass ratio is 2 kg to 8 kg, still a 1-to-4 ratio. Comparing a dozen 1-kg bags of potatoes to a dozen 4-kg bags gives a mass ratio of 12 kg to 48 kg: a 1-to-4 ratio of masses still exists. If we compare a thousand

TABLE 7.1   MOLAR MASS OF ATOMS

Element	Atomic mass	Molar mass	Number of atoms
C	12.01	12.01 g	6.022×10^{23}
Na	22.99	22.99 g	6.022×10^{23}
Cl	35.45	35.45 g	6.022×10^{23}
Au	197.0	197.0 g	6.022×10^{23}
Th	232.0	232.0 g	6.022×10^{23}

1-kg bags to a thousand 4-kg bags, the mass ratio is still 1 to 4. The ratio would be unchanged if 1 mol of each is compared! See Figure 7.2.

The same reasoning holds for atoms. For example, an individual He atom has 4 times the mass of a H atom. Therefore, 6.022×10^{23} He atoms (1.000 mol He) have 4 times the mass of 6.022×10^{23} H atoms (1.000 mol H).

The molar mass of any element is obtained by writing the numerical value for the atomic mass (from the periodic table), and placing the unit grams after the number (see Table 7.1). For example, neon's atomic mass is 20, and its molar mass is 20 g. Argon's atomic mass is 40, and its molar mass is 40 g. Both 20 g Ne

Atomic mass of Ne = 20
Molar mass of Ne = 20 g

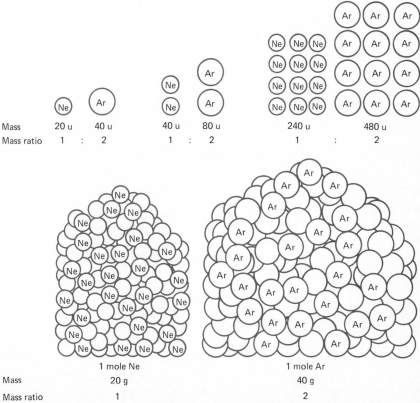

FIGURE 7.3
The atomic mass of neon is 20, and the atomic mass of argon is 40. An argon atom is twice as massive as a neon atom. The mass of 1 dozen argon atoms (480 u) is twice the mass of 1 dozen neon atoms (240 u), and the mass of one mole of argon atoms (40 g) is twice the mass of one mole of neon atoms (20 g).

FIGURE 7.4
To change the number of atoms to moles of atoms, use the conversion factor $\dfrac{1 \text{ mole of atoms}}{6.022 \times 10^{23} \text{ atoms}}$. To change the number of moles of atoms to grams, use the conversion factor $\dfrac{\text{grams of element}}{1 \text{ mole}}$. To go in the opposite direction, mass to moles, and then to atoms, use the reciprocals of these two conversion factors.

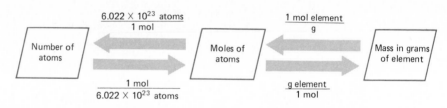

and 40 g Ar contain Avogadro's number of atoms. One mole of Ar is twice as massive as one mole of Ne, because one Ar atom is twice as massive as one Ne atom (see Figure 7.3).

Mole Calculations

The beauty of using moles is that after making a simple mass measurement in a laboratory the investigator can readily calculate the number of moles of atoms or individual atoms contained in the sample. When solving mole problems, it is only necessary to apply the factor-label method, employing a similar procedure to interconverting SI units.

The factor-label method is easily applied to mole calculations.

Two conversion factors are utilized in all mole calculations involving elements:

$$\frac{\text{g}}{\text{mol}} \quad \text{or} \quad \frac{\text{mol}}{\text{g}}$$

and

$$\frac{6.022 \times 10^{23} \text{ atoms}}{1 \text{ mol atoms}} \quad \text{or} \quad \frac{1 \text{ mol atoms}}{6.022 \times 10^{23} \text{ atoms}}$$

For any quantity given—grams, atoms, or moles—the other two amounts can be calculated using either or both conversion factors.

Figure 7.4 illustrates the pathway that is followed in basic mole problems. Refer to this diagram when you are doing the problems.

Example Problem 7.1 illustrates the procedure for converting the mass of a sample to moles.

Example Problem 7.1

How many moles of S atoms are found in a 1.0-g sample of S?

Solution

1. *What is unknown?* Moles of S from a given mass

2. *What is known?* 1.0 g S, and 1 mol S = 32 g or $\dfrac{1 \text{ mol S}}{32 \text{ g S}}$

Molar masses are sometimes called gram atomic masses.

Besides the mass, which is given in the problem, we also know the molar mass of S, 32 g, from the periodic table. In all mole problems, the molar mass is found in the periodic table. Refer to Figure 7.4, which illustrates the pathway taken.

3. Apply the factor-label method.

$$1.0 \text{ g } \cancel{S} \times \frac{1 \text{ mol S}}{32 \text{ g } \cancel{S}} = ? \text{ mol S}$$

Grams are known, and moles are wanted. Thus, the conversion factor indicating the molar mass is written with grams in the denominator and moles (desired unit) in the numerator; grams cancel, leaving moles.

4. Perform the indicated math operations.

$$1.0 \text{ g } \cancel{S} \times \frac{1 \text{ mol S}}{32 \text{ g } \cancel{S}} = 0.031 \text{ mol S}$$

Thirty-two is divided into 1.0, giving 0.031 mol S. Note that in the statement of the problem the mass of S was 1.0 g, which contains two significant figures; the answer, therefore, must also contain two significant figures. Never allow the molar mass to determine the number of significant figures in the answer—always use the same or more significant figures for the molar mass.

Example Problem 7.1 illustrates the general method for performing the majority of mole problems. As in many chemistry problems, careful attention to the units given and those desired normally results in successfully obtaining the correct answer.

If the number of moles of atoms are known, how is the mass obtained? Study Example Problem 7.2, and note that it is similar to Example Problem 7.1 except for proceeding in the opposite direction with respect to the units.

Example Problem 7.2

What is the mass of 2.5 mol Si?

Solution

1. *What is unknown?*   Mass of Si in grams

2. *What is known?*   2.5 mol Si, and $\dfrac{1 \text{ mol Si}}{28 \text{ g Si}}$ (from the periodic table)

A quick glance at Figure 7.4 shows that the mass of 1 mol Si is needed to solve such a problem.

3. Apply the factor-label method.

$$2.5 \text{ mol Si} \times \frac{28 \text{ g Si}}{1 \text{ mol Si}} = ? \text{ g Si}$$

Here, moles are canceled, leaving g, the desired unit.

4. Perform the indicated math operations.

$$2.5 \text{ mol Si} \times \frac{28 \text{ g Si}}{1 \text{ mol Si}} = 70 \text{ g Si} = 7.0 \times 10^1 \text{ g Si}$$

The correct answer is 70 g. We write 7.0×10^1 g to indicate that the final answer contains two significant figures. Leaving the answer as 70 g is ambiguous when considering signficant figures.

Given a specified quantity of moles of atoms, the number of individual atoms contained therein is easily found, somewhat in the same way that the number of eggs is found, given how many dozen. Example Problems 7.3 and 7.4 are models for such conversion problems.

Example Problem 7.3

How many atoms are found in 1.2 mol Ca?

Solution

Calcium metal was discovered in 1808 by Davy and Berzelius, independently of each other. Calcium is the fifth most abundant element in the earth's crust. It is always found as part of a compound, never as the free metal. Limestone, gypsum, apatite, and fluorite are the most common minerals containing calcium.

1. *What is unknown?* Atoms of Ca

2. *What is known?* 1.2 mol Ca, and $\dfrac{1 \text{ mol Ca}}{6.022 \times 10^{23} \text{ atoms Ca}}$

3. Apply the factor-label method.

$$1.2 \text{ mol Ca} \times \frac{6.022 \times 10^{23} \text{ atoms Ca}}{1 \text{ mol Ca}} = ? \text{ atoms Ca}$$

4. Perform the indicated math operations.

$$1.2 \text{ mol Ca} \times \frac{6.022 \times 10^{23} \text{ atoms Ca}}{1 \text{ mol Ca}} = 7.2 \times 10^{23} \text{ atoms Ca}$$

A sample of 1.2 mol Ca contains 7.2×10^{23} atoms Ca.

Calcium's molar mass is not needed in this problem; no mass is specified, just the number of moles. It should be evident that any atom could be substituted for Ca, and the answer would be the same: 1.2 mol of any element contain exactly the same number of atoms (think about dozens if this statement does not register).

Example Problem 7.4

A sample of Au is found to contain 9.7×10^{23} atoms Au. How many moles of Au are in the sample?

Solution

Two forms of gold occur naturally, native gold (1–50% silver alloy) and telluride ore ($AuTe_2$). Native gold is found as veins and dust in quartzite rock or in deposits that result after the rock weathers. Most of the world's gold is obtained from South Africa.

1. *What is unknown?* Moles Au

2. *What is known?* 9.7×10^{23} atoms Au, $\dfrac{6.022 \times 10^{23} \text{ atoms Au}}{1 \text{ mol Au}}$

3. Apply the factor-label method.

$$9.7 \times 10^{23} \text{ atoms Au} \times \frac{1 \text{ mol Au}}{6.022 \times 10^{23} \text{ atoms Au}} = \text{? mol Au}$$

4. Perform the indicated math operations.

$$9.7 \times 10^{23} \text{ atoms Au} \times \frac{1 \text{ mol Au}}{6.022 \times 10^{23} \text{ atoms Au}} = 1.6 \text{ mol Au}$$

When you divide with numbers expressed in scientific notation, remember to divide the coefficients and subtract the exponents.

In Example Problems 7.1 through 7.4 all conversions were accomplished by applying one conversion factor. To convert masses of elements to atoms, or atoms to their masses, more than one conversion factor is required. Given the mass of an element, no simple conversion is known; hence, two conversion factors are utilized: molar mass (g/mol), and Avogadro's number (atoms/mol) (refer to Figure 7.6). Study Example Problems 7.5 and 7.6 as illustrative models for such problems.

Example Problem 7.5

How many Cr atoms are found in a 100.0-g sample of pure Cr?

Solution

Chromium was discovered in 1797 by Vauquelin. Chromium is a steel-hardening metal. It is also plated on metal surfaces to give a bright shiny look.

1. *What is unknown?*   Number of atoms of Cr

2. *What is known?* 100.0 g Cr, $\dfrac{1 \text{ mol Cr}}{52.00 \text{ g Cr}}$, and $\dfrac{6.022 \times 10^{23} \text{ atoms Cr}}{1 \text{ mol Cr}}$

3. Apply the factor-label method.

$$100.0 \text{ g Cr} \times \frac{1 \text{ mol Cr}}{52.00 \text{ g Cr}} \times \frac{6.022 \times 10^{23} \text{ atoms Cr}}{1 \text{ mol Cr}} = ? \text{ atoms Cr}$$

4. Perform the indicated math operations.

$$100.0 \text{ g Cr} \times \frac{1 \text{ mol Cr}}{52.00 \text{ g Cr}} \times \frac{6.022 \times 10^{23} \text{ atoms Cr}}{1 \text{ mol Cr}}$$

$$= 1.158 \times 10^{24} \text{ atoms Cr}$$

Our answer is expressed to four significant figures because the initial mass of Cr is expressed to four significant figures. A good habit to develop is to ask yourself if the answer is reasonable after performing the indicated math operations. In this example, the mass 100.0 g is slightly less than twice the molar mass, which would give an answer around 2 mol or 2 times Avogadro's number.

Example Problem 7.6

What is the mass of 5.64×10^{23} atoms of uranium?

Solution

Uranium was discovered by Klaproth in 1789. He found it in an ore called pitchblende. All of uranium's isotopes are radioactive.

1. *What is unknown?*   Mass of U or grams of U

2. *What is known?* 5.64×10^{23} atoms U, $\dfrac{1 \text{ mol U}}{6.022 \times 10^{23} \text{ atoms U}}$,

$$\frac{238 \text{ g U}}{1 \text{ mol U}}$$

3. Apply the factor-label method.

$$5.64 \times 10^{23} \text{ atoms U} \times \frac{1 \text{ mol U}}{6.022 \times 10^{23} \text{ atoms U}} \times \frac{238 \text{ g U}}{1 \text{ mol U}} = ? \text{ g U}$$

4. Perform the indicated math operations.

$$5.64 \times 10^{23} \; \text{atoms U} \times \frac{1 \; \text{mol U}}{6.022 \times 10^{23} \; \text{atoms U}} \times \frac{238 \; \text{g U}}{1 \; \text{mol U}} = 223 \; \text{g U}$$

Is the answer reasonable? Yes, 5.64×10^{23} atoms U is less than 1 mol; therefore, the answer should be below the molar mass of 238 g.

REVIEW PROBLEMS

7.5 How many moles of atoms are contained in each of the following samples: (a) 8.0 g He, (b) 0.391 g K, (c) 750 g As, (d) 6.35 g I?

7.6 Find the mass of each of the following samples: (a) 3.0 mol Tc, (b) 0.065 mol Y, (c) 1.3×10^2 mol Te, (d) 750 mol Li.

7.7 How many individual atoms exist in: (a) 1.00 g He, (b) 1.00 g Ne, (c) 1.0 g Ar, (d) 1.0 g Kr?

7.8 Determine the number of moles of atoms represented by the following:
(a) 4.0×10^{23} atoms Al
(b) 1.75×10^{22} atoms P
(c) 7.1×10^{25} atoms Zn
(d) 6.02×10^{22} atoms Xe

7.3 MOLAR MASS OF COMPOUNDS

A compound is a chemical combination of atoms. Each compound has a fixed ratio of elements, or on a smaller scale, each molecule is composed of a fixed ratio of atoms. Hence, we can apply moles to molecules, as we did to atoms.

Molecular Mass

Molecular mass (traditionally, molecular weight) is the sum of the atomic masses of atoms within a molecule. Let us use water, H_2O, as an example. Each water molecule contains two H atoms and one O atom. H and O atomic masses are 1.0 and 16.0, respectively. To find the molecular mass of water, multiply 2 times 1.0 to obtain the total mass of H, and add that to 1 times 16.0, the total mass of O, resulting in 18 as the molecular mass.

While water appears to be a simple substance, it is not. Water exhibits "strange" properties for such a "simple" compound.

$$
\begin{array}{lll}
\text{H} & \text{2 atoms} \times 1.0 = & 2.0 \\
\text{O} & \text{1 atom} \times 16.0 = & +16.0 \\
\text{H}_2\text{O} & = & 18.0 \\
\end{array}
$$

If the molecular mass is 18, then the molar mass is 18.0 **grams,** and 18.0 g H_2O contain Avogadro's number of molecules, 6.02×10^{23} H_2O molecules. Example Problem 7.7 illustrates molecular mass determinations.

Example Problem 7.7

Find the molecular mass of (a) CH_4, (b) HNO_3, (c) $B_{10}H_{16}$.

Solution

(a) Molecular mass of CH_4:

$$
\begin{array}{llll}
C & 1 \text{ atom} \times 12.0 = & 12.0 \\
H & 4 \text{ atoms} \times 1.0 = & +4.0 \\
CH_4 & & = & 16.0
\end{array}
$$

(b) Molecular mass of HNO_3:

$$
\begin{array}{llll}
H & 1 \text{ atom} \times 1.0 & = & 1.0 \\
N & 1 \text{ atom} \times 14.0 & = & 14.0 \\
O & 3 \text{ atoms} \times 16.0 & = & +48.0 \\
HNO_3 & & = & 63.0
\end{array}
$$

(c) Molecular mass of $B_{10}H_{16}$:

$$
\begin{array}{llll}
B & 10 \text{ atoms} \times 10.81 & = & 108.1 \\
H & 16 \text{ atoms} \times 1.01 & = & +16.1 \\
B_{10}H_{16} & & = & 124.2
\end{array}
$$

In each of the molecular mass determinations the number of atoms in each molecule is multiplied by its atomic mass, and then the quantities are added to obtain the sum.

Mole Calculations: Compounds

Molar masses of compounds are sometimes called gram molecular masses.

Mole calculations involving molecules are the same as those pertaining to atoms, except that the fundamental particle is a molecule. Therefore, the molar mass of a compound is the molecular mass expressed in grams. Within the molar mass of a compound there are 6.022×10^{23} molecules.

What is the molar mass of carbon dioxide, CO_2, and how many particles are contained in that mass? Calculate the molecular mass of CO_2; one C atom, 12, and two O atoms, each 16, gives a molecular mass of 44. Placing grams as the unit after 44 gives 44 g as the molar mass of CO_2. Within 44 g CO_2 there are 6.0×10^{23} molecules.

$1 \text{ mol } CO_2 = 44 \text{ g}$
$1 \text{ mol } CO_2 = 6.022 \times 10^{23} CO_2$ molecules

In Example Problem 7.7, CH_4, HNO_3, and $B_{10}H_{16}$ have molar masses of 16 g, 63 g, and 124.2 g, respectively. Avogadro's number of CH_4 molecules are found in 16 g CH_4. Likewise, 63 g HNO_3 and 124.2 g $B_{10}H_{16}$ each contain 6.02×10^{23} molecules.

As was the case with atoms, most mole calculations regarding compounds involve interconversions of units—moles, molecules, and grams (see Figure 7.5).

Carefully study Example Problems 7.8 through 7.10, considering Figure 7.5, and note that they are similar to mole calculations involving atoms except that we use molecular masses instead of atomic masses.

FIGURE 7.5
To change the number of molecules to moles of molecules, use the conversion factor $\dfrac{1 \text{ mole of molecules}}{6.022 \times 10^{23} \text{ molecules}}$. To change the number of moles of molecules to grams, use the conversion factor $\dfrac{\text{grams}}{1 \text{ mol}}$. To go in the opposite direction, mass to moles to molecules, use the reciprocals of these two factors.

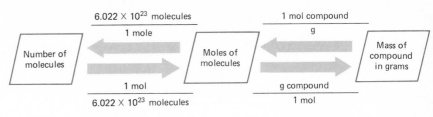

Example Problem 7.8

What is the mass of 4.05 mol N_2O?

Solution

N$_2$O is nitrous oxide (dinitrogen oxide). Its common name "laughing gas" was coined by Sir Humphry Davy, after he discovered its anesthetic effects. It is still used today as an anesthetic.

1. *What is unknown?*   Mass of N_2O or grams of N_2O
2. *What is known?*   4.05 mol N_2O, and the molar mass of N_2O:

$$N \quad 2 \text{ mol N} \times \frac{14.0 \text{ g N}}{\text{mol N}} = \quad 28.0 \text{ g N}$$

$$O \quad 1 \text{ mol O} \times \frac{16.0 \text{ g O}}{\text{mol O}} = +16.0 \text{ g O}$$

$$N_2O \quad\quad\quad\quad\quad\quad = \quad 44.0 \text{ g/mol } N_2O$$

3. Apply the factor-label method.

$$4.05 \text{ mol } N_2O \times \frac{44.0 \text{ g } N_2O}{1 \text{ mol } N_2O} = ? \text{ g } N_2O$$

4. Perform indicated math operations.

$$4.05 \text{ mol } N_2O \times \frac{44.0 \text{ g } N_2O}{1 \text{ mol } N_2O} = 178 \text{ g } N_2O$$

The mass of 4.05 mol N_2O is 178 g.

Example Problem 7.9

How many silane, SiH_4, molecules are contained in a 0.100-g sample of SiH_4?

SiH_4, silane, was first prepared by Wöhler in 1851. He found that SiH_4 reacts spontaneously with the oxygen in the air, producing a white smoke.

Solution

1. *What is unknown?*   Molecules of SiH_4

2. *What is known?* 0.100 g SiH_4, $\dfrac{1 \text{ mole } SiH_4}{6.022 \times 10^{23} \text{ molecules } SiH_4}$, and the molar mass of SiH_4, which is calculated from its atomic masses:

$$\begin{array}{lll} \text{Si} & 1\ \cancel{\text{mol Si}} \times \dfrac{28.1\ \text{g Si}}{\cancel{\text{mol Si}}} & = 28.1\ \text{g Si} \\ \text{H} & 4\ \cancel{\text{mol H}} \times \dfrac{1.0\ \text{g H}}{\cancel{\text{mol H}}} & = +4.0\ \text{g H} \\ \hline SiH_4 & & = 32.1\ \text{g/mol } SiH_4 \end{array}$$

3. Apply the factor-label method.

$$0.100\ \cancel{\text{g } SiH_4} \times \frac{1\ \cancel{\text{mol } SiH_4}}{32.1\ \cancel{\text{g } SiH_4}} \times \frac{6.022 \times 10^{23}\ \text{molecules } SiH_4}{1\ \cancel{\text{mol } SiH_4}} = ?\ \text{molecules } SiH_4$$

4. Perform indicated math operations.

$$0.100\ \cancel{\text{g } SiH_4} \times \frac{1\ \cancel{\text{mol } SiH_4}}{32.1\ \cancel{\text{g } SiH_4}} \times \frac{6.022 \times 10^{23}\ \text{molecules } SiH_4}{1\ \cancel{\text{mol } SiH_4}} = 1.88 \times 10^{21}\ \text{molecules } SiH_4$$

A 0.100-g SiH_4 sample would contain 1.88×10^{21} molecules SiH_4.

Example Problem 7.10

A sample of carbon tetrabromide, CBr_4, contains 8.1×10^{22} molecules CBr_4. What is the mass of the sample?

Solution

1. *What is unknown?* Mass of CBr_4 or grams of CBr_4

2. *What is known?* 8.1×10^{22} molecules CBr_4, $\dfrac{1 \text{ mol } CBr_4}{6.02 \times 10^{23} \text{ molecules } CBr_4}$, and the molar mass of CBr_4:

$$\begin{array}{lll} \text{C} & 1\ \cancel{\text{mol C}} \times \dfrac{12\ \text{g C}}{\cancel{\text{mol C}}} & = 12\ \text{g C} \\ \text{Br} & 4\ \cancel{\text{mol Br}} \times \dfrac{80\ \text{g Br}}{\cancel{\text{mol Br}}} & = +320\ \text{g Br} \\ \hline CBr_4 & & = 332\ \text{g/mol } CBr_4 \end{array}$$

3. Apply the factor-label method and perform indicated math operations.

$$8.1 \times 10^{22} \text{ molecules } \cancel{CBr_4} \times \frac{1 \cancel{\text{ mol } CBr_4}}{6.02 \times 10^{23} \text{ molecules } \cancel{CBr_4}}$$

$$\times \frac{332 \text{ g } CBr_4}{1 \cancel{\text{ mol } CBr_4}} = 45 \text{ g } CBr_4$$

Moles of Elements Within Compounds

In future calculations, it will sometimes be necessary to determine the number of moles of atoms or the number of individual atoms within a specified quantity of a compound. For example, how many moles of H and O are contained in 1 mol of water, H_2O? Water molecules contain two atoms H and one atom O; therefore, 1 mol H_2O contains 2 mol H (2 g H) and 1 mol O (16 g O).

Using another large-scale analogy, consider a person's face. Every person has 1 nose and 2 eyes. One dozen people, among them, possess 1 dozen noses and 2 dozen eyes (see Figure 7.6). It stands to reason that 1 mole of people would have 1 mole of noses and 2 moles of eyes.

1 mol H_2O contains 2 mol H and 1 mol O.

FIGURE 7.6
One person has 2 eyes and 1 nose. One dozen people have 2 dozen eyes and 1 dozen noses. One mole of people have 2 moles of eyes and 1 mole of noses.

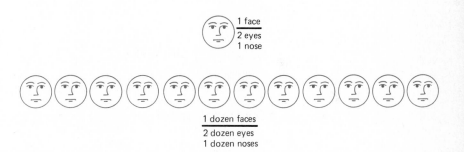

1 face
2 eyes
1 nose

1 dozen faces
2 dozen eyes
1 dozen noses

FIGURE 7.7
To find the number of atoms of a specific type in a given mass of compound, three conversion factors are required. The first is the mass of 1 mole of the compound $\left(\dfrac{1 \text{ mol}}{\text{grams compound}}\right)$; the second factor is the number of moles of the desired element per mole of compound $\left(\dfrac{\text{moles of element}}{1 \text{ mol of compound}}\right)$; and the third factor is the number of atoms per mole of the element $\left(\dfrac{6.022 \times 10^{23} \text{ atoms}}{1 \text{ mol of element}}\right)$.

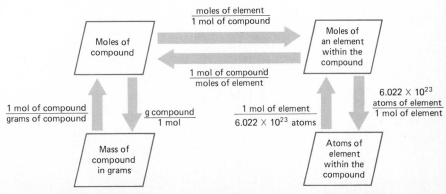

When confronted with such problems, find the number of moles of the compound of interest, and then use the appropriate conversion factor that gives the number of atoms per molecule. Figure 7.7 maps the pathway to be taken. Example Problem 7.11 illustrates such a problem.

Example Problem 7.11

How many moles of C and H atoms are in 9.25 g C_5H_{10}?

Solution

1. *What is unknown?*   Moles of C and moles of H

2. *What is known?*   9.25 g C_5H_{10}, $\dfrac{5 \text{ mol C}}{1 \text{ mol } C_5H_{10}}$, $\dfrac{10 \text{ mol H}}{1 \text{ mol } C_5H_{10}}$, and the molar mass of C_5H_{10}:

$$\text{C} \quad 5 \text{ mol C} \times \frac{12.0 \text{ g C}}{\text{mol C}} = \quad 60.0 \text{ g C}$$

$$\text{H} \quad 10 \text{ mol H} \times \frac{1.0 \text{ g H}}{\text{mol H}} = +10.0 \text{ g H}$$

$$C_5H_{10} \qquad\qquad\qquad = \quad 70.0 \text{ g/mol } C_5H_{10}$$

3. Apply the factor-label method (refer to Figure 7.9).

$$9.25 \text{ g } C_5H_{10} \times \frac{1 \text{ mol } C_5H_{10}}{70.0 \text{ g } C_5H_{10}} \times \frac{5 \text{ mol C}}{1 \text{ mol } C_5H_{10}} = ? \text{ mol C}$$

It is not necessary to write the full setup to find the number of moles of H. The ratio of H to C in C_5H_{10} is 10 to 5 or 2 to 1: $\dfrac{2 \text{ mol H}}{1 \text{ mol C}}$. After finding the number of moles of C, we can use this conversion factor to find the moles of H.

$$\text{moles C} \times \frac{2 \text{ mol H}}{1 \text{ mol C}} = ? \text{ moles H}$$

4. Perform indicated math operations.

$$9.25 \text{ g } C_5H_{10} \times \frac{1 \text{ mol } C_5H_{10}}{70.0 \text{ g } C_5H_{10}} \times \frac{5 \text{ mol C}}{1 \text{ mol } C_5H_{10}} = 0.661 \text{ mol C}$$

$$0.661 \; \cancel{\text{mol C}} \times \frac{2 \; \text{mol H}}{1 \; \cancel{\text{mol C}}} = 1.32 \; \text{mol H}$$

After the number of moles of C in 9.25 g C_5H_{10} is calculated, it is only necessary to double the value to obtain the number of moles of H atoms.

Compounds Not Composed of Molecules

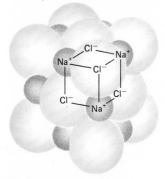

FIGURE 7.8
Sodium chloride, NaCl, is composed of a network of sodium ions, Na^+, surrounded by chloride ions, Cl^- (or chloride ions surrounded by sodium ions). Discrete molecules of sodium chloride do not exist.

Sodium chloride is essential to most living things. Chloride ions are needed for the production of stomach acid, which helps in the digestion of foods. Sodium ions are important in maintaining fluid balance and nerve conduction.

When two nonmetals are combined, the resulting compound is usually composed of individual molecules, e.g., CO, NO, F_2, CH_4, and H_2O. Certain compounds are not composed of separate or discrete molecules. Instead, their structures are a network of ions chemically bonded together. For these compounds, the term molecular mass has no real meaning because there are no molecules present. A different, more accurate term is applied to these compounds: formula mass. **Formula mass** is used for those substances that are not made up of discrete molecules in the same way that molecular mass is used for compounds that do contain molecules.

An example of a compound that is not composed of discrete molecules is sodium chloride, NaCl. Sodium chloride is composed of Na^+ and Cl^- ions in a three-dimensional pattern (see Figure 7.8). Each Na^+ cation is surrounded by Cl^- ions, and each Cl^- anion is surrounded by Na^+ ions. Within NaCl, for every Na^+ there is a Cl^-: the two elements are found in a 1-to-1 ratio. We speak of this as the **formula unit,** or the simplest ratio of atoms within the compound.

In Chapter 8, we shall investigate the structure of these compounds; for now, most compounds that result from combinations of metals and nonmetals (ionic compounds), or metals and charged groups of atoms called polyatomic ions, do not exist as discrete molecules. These substances are called salts. Examples of salts include calcium fluoride, CaF_2; sodium nitrate, $NaNO_3$; copper sulfate, $CuSO_4$; and many others. Table 7.2 lists some commonly encountered ionic substances.

Calculation of the formula mass is the same as finding the molecular mass of a substance composed of molecules. In most situations, the formula mass is treated the same as molecular mass. For instance, NaCl would have a formula mass of 58.5 and a molar mass of 58.5 g, and would contain Avogadro's number of formula units.

TABLE 7.2   EXAMPLES OF IONIC COMPOUNDS

Compound	Formula	Formula mass
Calcium chloride	$CaCl_2$	111.0
Magnesium hydroxide	$Mg(OH)_2$	58.3
Iron(III) oxide	Fe_2O_3	159.7
Titanium(IV) oxide	TiO_2	79.9
Ammonium nitrate	NH_4NO_3	80.0

REVIEW PROBLEMS

7.9 Determine the molar mass of (a) SCl_2, (b) CS_2, (c) H_2SO_4.

7.10 What is the mass in grams of the following substances: (a) 4 mol $AlCl_3$, (b) 3.2×10^{23} molecules PH_3, (c) 0.0015 mol IBr?

7.11 How many molecules are contained in each of the following: (a) 1.00 g NH_3, (b) 1.00 g SF_6, (c) 1.00 g C_3Cl_8?

7.12 Which of the following contains the greatest number of moles of oxygen atoms: (a) 25 g CO_2, (b) 25 g $C_6H_{12}O_6$, or (c) 25 g H_3PO_4?

7.13 Explain why it is incorrect to refer to the "molecular mass" of NaCl.

7.4 MOLES AND CHEMICAL FORMULAS

Armed with the mole concept, chemists can obtain a tremendous amount of information about a compound, solution, or mixture by finding the mass of its components. Moles are used by chemists to determine (1) formulas of compounds, (2) mass relationships in chemical reactions, and (3) concentrations of solutions and mixtures.

In this section, we will look at formula calculations and determinations. Application of moles to equations and solutions shall come in later chapters.

Law of Constant Composition

Each compound is composed of elements in a fixed mass ratio. Any sample of H_2O contains, by mass, 11% H and 89% O. If another compound containing H and O is studied, and a different mass percent is found, it must be concluded that the substance is not pure water. In essence, this is the **law of constant composition:** All samples of a given compound contain the same elements in a fixed mass ratio.

Proust was the son of an apothecary. He was one of the first chemists to study sugars. Proust was an avid balloonist, making one of the first ascensions in 1784.

The law of constant composition was proposed by a French scientist, Joseph-Louis Proust (1754–1826). At the time, it was thought that the composition of a substance could vary, depending on the sample. Proust's "radical" new idea took about 10 years to catch on.

Percent Composition

A direct result of the law of constant composition is that each element within a compound can be expressed as a mass percent. Collectively, all mass percents of elements in a compound are called the **percent composition** of the substance. Water's percent composition is 11% H and 89% O, and that of table sugar (sucrose) is 42.1% C, 6.4% H, and 51.5% O.

Percent (%) is the number of parts per 100 total things. A mass percent is the ratio of the mass of a component to 100 g total mass.

How is the percent composition determined for a compound? Decide on a specific mass of the compound to work with (the total mass), and find the mass of the individual components, the elements. Mass percents are calculated by dividing the mass of each element by the total mass of the compound, and then multiplying by 100.

Conveniently, the molar mass of a compound is utilized in percent composition calculations. In the determination of the molar mass, the mass of each element is calculated. It is then only necessary to convert these masses to percents. In Example Problem 7.12, the percent composition of dinitrogen pentoxide, N_2O_5, is determined.

> **Example Problem 7.12**

What is the percent composition of N_2O_5?

Solution

N_2O_5, dinitrogen pentoxide, is a gas that combines with water to form nitric acid, HNO_3.

1. *What is unknown?* %N and %O in N_2O_5, where %N is $\dfrac{g\ N}{g\ N_2O_5} \times 100$, and %O is $\dfrac{g\ O}{g\ N_2O_5} \times 100$

2. *What is known?* Molecular formula of the compound, N_2O_5, atomic masses of N and O, $\dfrac{14\ g\ N}{mol\ N}$ and $\dfrac{16\ g\ O}{mol\ O}$. From this information, the molar mass is determined:

$$N \quad 2\ \text{mol N} \times \frac{14\ g\ N}{\text{mol N}} = \quad 28\ g\ N$$

$$O \quad 5\ \text{mol O} \times \frac{16\ g\ O}{\text{mol O}} = +80\ g\ O$$

$$N_2O_5 \hspace{4.5cm} = \quad 108\ g/mol\ N_2O_5$$

3. Calculate the percent of N and O in 1 mol N_2O_5:

$$\%\ N = \frac{g\ N}{g\ N_2O_5} \times 100 = \frac{28\ g\ N}{108\ g\ N_2O_5} \times 100 = 26\%\ N$$

$$\%\ O = \frac{g\ O}{g\ N_2O_5 \times 100} = \frac{80\ g\ O}{108\ g\ N_2O_5} \times 100 = 74\%\ O$$

N_2O_5 has a percent composition of 26% N and 74% O. Note that the percents always add up to 100%. You may be tempted to calculate the percent of one element and then subtract that value from 100 to obtain the second percent. This philosophy is fine as long as the first value is correct! A better strategy is to compute both percents and check to see if they add up to 100%.

Example Problem 7.13 illustrates the calculation of the percent composition of a compound containing more than two elements, ethyl alcohol, C_2H_6O, which is called grain or drinking alcohol.

| Example Problem 7.13 |

Find the percent composition of ethyl alcohol, C_2H_6O.

Solution

C$_2$H$_6$O, ethanol, is produced by fermenting various grains or fruits using yeast, a single-cell plant. Higher concentrations of ethanol are obtained through distillation.

1. *What is unknown?*   %C, %H, and %O in C_2H_6O
2. *What is known?*   The molar masses of each element, and the molecular formula. The molar mass of ethanol is:

$$C \quad 2 \text{ mol C} \times \frac{12.0 \text{ g C}}{\text{mol C}} = \quad 24.0 \text{ g C}$$

$$H \quad 6 \text{ mol H} \times \frac{1.0 \text{ g H}}{\text{mol H}} = \quad 6.0 \text{ g H}$$

$$O \quad 1 \text{ mol O} \times \frac{16.0 \text{ g O}}{\text{mol O}} = +16.0 \text{ g O}$$

$$C_2H_6O \qquad = \quad 46.0 \text{ g/mol } C_2H_6O$$

3. Calculate the percent of C, H, and O in 1 mol C_2H_6O.

$$\% \text{ C} = \frac{\text{g C}}{\text{g } C_2H_6O} \times 100 = \frac{24.0 \text{ g C}}{46.0 \text{ g}} \times 100 = 52.2\% \text{ C}$$

$$\% \text{ H} = \frac{\text{g H}}{\text{g } C_2H_6O} \times 100 = \frac{6.0 \text{ g H}}{46.0 \text{ g}} \times 100 = 13\% \text{ H}$$

$$\% \text{ O} = \frac{\text{g O}}{\text{g } C_2H_6O} \times 100 = \frac{16.0 \text{ g O}}{46.0 \text{ g}} \times 100 = 34.8\% \text{ O}$$

Our calculations indicate that the percent composition is 52.2% C, 13% H, and 34.8% O. We can check the answer by seeing that the total equals 100%.

Simplest (Empirical) Formula Calculations

In the laboratory, chemists determine formulas of compounds as one way of characterizing compounds. It is only necessary to obtain the percent composition of the compound to calculate the simplest formula. Simplest formulas express the smallest whole-number ratio of atoms within a molecule.

Finding the simplest formula requires knowledge of the percent composition or the mass ratio of elements within the compound. Converting these masses to moles yields the mole relationship. The simplest formula is then found by obtaining a ratio of whole numbers, such as 2 to 1 or 3 to 5.

TABLE 7.3   SIMPLEST AND MOLECULAR FORMULAS

Compound	Molecular formula	Simplest formula
Benzene	C_6H_6	CH
Glucose	$C_6H_{12}O_6$	CH_2O
Mercury(I) chloride	Hg_2Cl_2	HgCl
A boron hydride	B_4H_{10}	B_2H_5
Propane	C_3H_8	C_3H_8

Consider hydrogen peroxide, H_2O_2, an antiseptic and bleaching agent. Its molecular formula is H_2O_2. A molecular formula expresses the actual number of atoms in the molecule. One hydrogen peroxide molecule contains two H atoms and two O atoms. Its simplest formula is HO (H_1O_1), obtained by dividing the subscripts of the molecular formula by 2. Table 7.3 presents molecular and simplest formulas for selected molecules.

A 3% solution of hydrogen peroxide is used as a germicide. In research and industry, 30% solutions are frequently used. Pure H_2O_2, a pale blue liquid, must be handled with extreme care, for it decomposes violently or readily combines with substances with which it comes in contact.

Many compounds have the same molecular and simplest formula, e.g., propane in Table 7.3. Whenever the subscripts of a compound's molecular formula are not divisible by a common number, the simplest formula is the same as the molecular formula.

Example Problem 7.14 shows how the simplest formula of a compound is obtained, given the percent composition. Each step that is performed is illustrated in Figure 7.9.

FIGURE 7.9
Stepwise procedure to determine the simplest formula of a compound

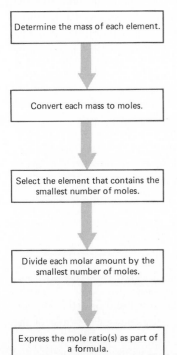

Determine the mass of each element.

Convert each mass to moles.

Select the element that contains the smallest number of moles.

Divide each molar amount by the smallest number of moles.

Express the mole ratio(s) as part of a formula.

Example Problem 7.14

A nitrogen–fluorine compound contains 26.9% N and 73.1% F. What is the simplest formula of the compound?

Solution

1. *What is unknown?*   Simplest formula or the simplest mole ratio of N to F or vice versa
2. *What is known?*   26.9% N and 73.1% F, and the molar masses of N and F, $\dfrac{14.0 \text{ g N}}{\text{mol N}}$, and $\dfrac{19.0 \text{ g F}}{\text{mol F}}$.
3. Determine the number of moles of each element.
   The percentages given are the mass ratios of each element per 100 g of compound; specifically,

$$26.9\% \text{ N} = \frac{26.9 \text{ g N}}{100 \text{ g compound}}$$

$$73.1\% \text{ F} = \frac{73.1 \text{ g F}}{100 \text{ g compound}}$$

Thus, it is convenient to find the number of moles that corresponds to 26.9 g N and 73.1 g F.

$$26.9 \ \cancel{g \ N} \times \frac{1 \ mol \ N}{14.0 \ \cancel{g \ N}} = 1.92 \ mol \ N$$

$$73.1 \ \cancel{g \ F} \times \frac{1 \ mol \ F}{19.0 \ \cancel{g \ F}} = 3.85 \ mol \ F$$

4. Determine the smallest whole-number mole ratio.

Find the element with the smallest number of moles, and then divide this value into the number of moles of each element. In our problem, 1.92 mol N is the smallest quantity of moles; thus, it is divided into itself, giving exactly 1, and then divided into 3.85 mol F, giving 2.01.

$$\frac{1.92 \ \cancel{mol \ N}}{1.92 \ \cancel{mol \ N}} = 1.00$$

$$\frac{3.85 \ mol \ F}{1.92 \ mol \ N} = \frac{2.01 \ mol \ F}{mol \ N}$$

The simplest ratio is 2 mol F to 1 mol N, or translated into a simplest formula, NF_2. If the simplest mole ratio is 1 to 2, this indicates that the atom ratio in the molecule is also 1 to 2. With no other information, the molecular formula cannot be determined. All that is known is the simplest formula; the molecular formula could possibly be NF_2, N_2F_4, N_3F_6, or N_4F_8, all of which have an atom ratio of 1 to 2.

It is not necessary to know the percent composition to determine a compound's empirical formula. All that is required is mass data regarding the compound. As long as the masses of each element are known (refer to Figure 7.9), they can be converted to a mole ratio. Example Problem 7.15 illustrates such a computation.

Example Problem 7.15

Analysis of an 18.169-g sample of an H, S, and O compound finds 11.866 g O and 0.371 g H, and the remainder is S. What is the simplest formula of the compound?

Solution

1. *What is unknown?* Simplest formula of the H, S, and O compound, or the simplest mole ratio of H, S, and O.

2. *What is known?*
In order to perform this calculation, the masses of all substances must be identified. The problem states the total mass of the compound and the masses of H and O. By subtracting, we obtain the mass of S.

$$\begin{aligned} \text{Mass of S} &= \text{total mass} - (\text{mass O} + \text{mass H}) \\ &= 18.169 \text{ g total} - (11.866 \text{ g O} + 0.371 \text{ g H}) \\ &= 5.932 \text{ g S} \end{aligned}$$

From the periodic table the molar masses of H, S, and O are found. They are $\dfrac{1.01 \text{ g H}}{\text{mol H}}$, $\dfrac{32.06 \text{ g S}}{\text{mol S}}$, and $\dfrac{16.00 \text{ g O}}{\text{mol O}}$.

3. Determine the number of moles of each element.

$$0.371 \text{ g H} \times \frac{1 \text{ mol H}}{1.01 \text{ g H}} = 0.367 \text{ mol H}$$

$$5.932 \text{ g S} \times \frac{1 \text{ mol S}}{32.06 \text{ g S}} = 0.1850 \text{ mol S}$$

$$11.866 \text{ g O} \times \frac{1 \text{ mol O}}{15.999 \text{ g O}} = 0.74167 \text{ mol O}$$

4. Find the simplest mole ratio.
Since 0.1850 mol S is the smallest number of moles, divide 0.1850 mol S into each quantity to find the whole-number ratio.

$$\frac{0.367 \text{ mol H}}{0.1850 \text{ mol S}} = \frac{1.98 \text{ mol H}}{\text{mol S}}$$

$$\frac{0.1850 \text{ mol S}}{0.1850 \text{ mol S}} = 1.000$$

$$\frac{0.74167 \text{ mol O}}{0.1850 \text{ mol S}} = \frac{4.009 \text{ mol O}}{\text{mol S}}$$

Final calculated mole ratios will rarely be exact integers, but they should be fairly close, within ±0.05 mol. Our calculations indicate that the simplest formula of the compound is H_2SO_4.

A word of caution: When performing simplest formula calculations, it is imperative to observe all rules regarding significant figures. Failure to observe these rules will result in incorrect ratios.

$1 \times 2 = 2$
$1.5 \times 2 = 3$
$1 \times 3 = 3$
$1.33 \times 3 = 4$
$1 \times 3 = 3$
$1.667 \times 3 = 5$

Sometimes whole numbers are not obtained in the final mole ratio. If this is the case, eliminate the fraction by multiplying by succeedingly larger integers until a ratio of whole numbers is obtained. For example, if a mole ratio of 1 to 1.5 results, this indicates a 2-to-3 ratio; a ratio of 1 to 1.33 indicates a 3-to-4 ratio; and 1 to 1.667 is a 3-to-5 ratio.

Molecular Formula Determination

While knowledge of the simplest formula gives the ratio of atoms within a molecule, the molecular formula expresses the actual quantity of each atom in a molecule.

Once the simplest formula is known, only the molecular mass of a compound is needed to ascertain its molecular formula. Because the molecular formula is a higher multiple of the simplest formula (or the same), if the mass of the simplest formula unit (simplest formula mass) is divided into the molecular mass, a whole number is obtained. This value gives the number of simplest formula units per molecular formula.

For example, if a substance has a simplest formula of CH and a molecular mass of 104, find the simplest formula mass of CH ($12 + 1 = 13$), and divide it into the molecular mass:

$$\frac{\text{Molecular mass}}{\text{Simplest formula unit mass}} = \frac{104}{13} = 8$$

Eight empirical formula units make up this molecule, so the molecular formula of the compound is $8 \times CH$, or C_8H_8. A complete molecular formula calculation is illustrated in Example Problem 7.16.

Example Problem 7.16

The molecular mass of a phosphorus-oxygen compound is 280.4, and a 10.000-g sample contains 4.364 g P. What is the molecular formula of the compound?

Solution

1. What is unknown? Molecular formula of the phosphorus-oxygen compound—the actual number of P and O within each molecule

2. *What is known?*   Molecular mass = 280.4, the total mass of a sample = 10.000 g, and the mass of P in that sample = 4.364 g

The mass of oxygen in the sample is obtained by subtraction:

$$\text{Mass of O} = \text{total mass} - \text{mass of P}$$
$$= 10.000 \text{ g} - 4.364 \text{ g P}$$
$$= 5.636 \text{ g O}$$

3. Find the mole ratio of P and O.

$$4.364 \text{ g P} \times \frac{1 \text{ mol P}}{30.97 \text{ g P}} = 0.1409 \text{ mol P}$$

$$5.636 \text{ g O} \times \frac{1 \text{ mol O}}{16.00 \text{ g O}} = 0.3523 \text{ mol O}$$

4. Find the simplest mole ratio.

$$\frac{0.1409 \text{ mol P}}{0.1409 \text{ mol P}} = 1.000$$

$$\frac{0.3523 \text{ mol O}}{0.1409 \text{ mol P}} = \frac{2.500 \text{ mol O}}{\text{mol P}}$$

Multiplying 2 times $PO_{2.5}$ gives the correct simplest formula, P_2O_5.

5. Determine the molecular formula by dividing the simplest formula unit mass into the molecular mass.

$$\text{Simplest formula unit mass} = (2 \times \text{atomic mass P}) + (5 \times \text{atomic mass O})$$
$$= (2 \times 31.0) + (5 \times 16.0)$$
$$= 142$$

$$\frac{\text{Molecular mass}}{\text{Formula unit mass}} = \frac{280.4}{142} = 2$$

When placed in water, P_4O_{10} produces phosphoric acid, H_3PO_4.

Two formula units make up the total molecular mass; hence, the molecular formula is $2 \times P_2O_5$, or P_4O_{10}. Within this molecule there are four P atoms and ten O atoms.

REVIEW PROBLEMS   **7.14** State the law of constant composition.

7.15 Find the percent composition of (a) BeF_2, (b) KBr, (c) $HClO_4$, (d) $Ca(OH)_2$.

7.16 Find the simplest formula for compounds with the following percent compositions: (a) 75% C and 25% H, (b) 50% S and 50% O.

7.17 A 10.00-g sample contains 5.99 g Ti, and the remaining mass is oxygen. What is the simplest formula of the compound?

7.18 A compound has a molecular mass of 84 and a simplest formula of CH_2. What is the molecular formula of the compound?

7.19 A boron-hydrogen compound is composed of 78.1% B and 21.9% H and has a molecular mass of 27.7. Calculate the molecular formula of the compound.

SUMMARY

A mole is a counting unit that allows chemists to determine the number of individual atoms, molecules, or ions contained in a sample by weighing the sample. One mole contains 6.022×10^{23} particles—Avogadro's number.

When applied to elements, the mass of one mole of atoms is the atomic mass expressed in grams. A mole of molecules has a mass equal to the molecular mass expressed in grams; molecular mass is the sum of the atomic masses of elements contained in a compound. For those compounds whose structures do not contain identifiable molecules, the formula mass is used instead. The formula mass is the mass in grams of 1 mol of formula units of a compound.

Mole calculations involve the application of the factor-label method to change the given quantity to the desired quantity. Two conversion factors are employed: (1) ratio of moles to g, either g/mol or mol/g, and (2) ratio of particles to mole (particles/mol) or mole to particles. For elements, the first ratio (g/mol) is obtained by finding the atomic mass of the element from the periodic table and adding the units "grams"—this is the molar mass. For molecules, grams are the units attached to the molecular mass (molar mass of molecules). The second ratio, particles/mol, is the same for all chemical species, 6.022×10^{23} particles/mol.

Mole calculations are used to determine chemical formulas, both the simplest formulas and the molecular formulas. A compound's molecular formula is the actual number of each atom within the molecular or formula unit. The simplest formula is the smallest ratio of whole numbers of atoms within a molecule. Simplest formulas are determined from the mass data of a compound by converting the masses of each element to moles and determining the smallest ratio of whole numbers. Molecular formulas are determined given the simplest formula and molecular mass.

Formulas of compounds are determined as a direct result of a basic law of chemistry called the law of constant composition. This law states that all samples of a compound contain the same elements in a fixed mass ratio. Hence, all compounds are identified by their percent composition, i.e., the mass percent of each element in a compound.

QUESTIONS AND PROBLEMS

7.20 Define the following: mole, Avogadro's number, molar mass, molecular mass, formula mass, constant composition, percent composition, simplest formula, empirical formula, molecular formula.

Moles

7.21 Why is the mole called a counting unit?

7.22 If Avogadro's number of like atoms is placed on a balance, what mass is observed?

7.23 What is the mass of 1.000 mol of each of the following atoms: (a) N, (b) Cu, (c) Cl, (d) Ga, (e) Ba, (f) Au?

7.24 Perform a rough calculation to determine how many centuries it would take 1 billion people working 24 hours per day, 365 days per year, to produce 1 mol of donuts at a rate of 10 donuts per person each second. (*Hint:* First use conversion factors to determine how many donuts could be produced per year.)

Moles and Atoms

7.25 Complete the following table.

Element	Number of moles	Number of atoms	Mass, g
(a) He	1.00	_____	____
(b) Be	_____	4.21×10^{23}	____
(c) B	_____	_____	5.41
(d) Ne	_____	_____	20.2
(e) V	0.10	_____	____
(f) Rb	_____	6.02×10^{24}	____
(g) ____	1.00	_____	65.4
(h) ____	_____	6.022×10^{23}	112.4
(i) Pd	7.000	_____	____
(j) Sr	_____	1.20×10^{22}	____

7.26 How many moles of Li are in each of the following masses of Li: (a) 0.01 g, (b) 1.0 g, (c) 69.4 g, (d) 100.00 mg?

7.27 Find the number of moles of atoms in each of the following masses of elements: (a) 4.5 g Sc, (b) 2.814 g Si, (c) 7.90 g Se, (d) 8.7 kg Sr.

7.28 How many individual atoms are contained in each mass listed in problem 7.27?

7.29 Determine the mass of each of the following samples of elements: (a) 3.11 mol of helium, (b) 0.500 mol of argon, (c) 1.4 mol of potassium, (d) 9.55 mmol of chromium.

7.30 How many individual atoms are contained in each of the following: (a) 16.5 g C, (b) 0.022 g Al, (c) 7.767 g Zn, (d) 131 mg U?

7.31 How many moles of atoms are found in the following:
(a) 1×10^{23} atoms of neon
(b) 7.2×10^{23} atoms of silver
(c) 1.948×10^{24} atoms of lead
(d) 2.6×10^{21} atoms of mercury?

7.32 What is the mass of the following number of atoms:
(a) 9.17×10^{23} atoms Xe
(b) 8.3×10^{25} atoms Mn
(c) 5.36×10^{20} atoms Ba
(d) 1.423×10^{26} atoms Sn?

7.33 What is the mass of the following number of sodium atoms:
(a) 5.1×10^{23} atoms
(b) 100 billion atoms
(c) 5000 atoms
(d) 1 atom?

7.34 Find the unknown quantity:
(a) 2.7 g of silicon = ? mol of silicon
(b) 4.15×10^{23} atoms of vanadium = ? mol of vanadium
(c) 0.00099 mol of krypton = ? g of krypton
(d) 1.5×10^{-4} g of nickel = ? nickel atoms
(e) 1.05×10^{23} atoms of cesium = ? g cesium.

7.35 Find the unknown quantity:
(a) 41.5 mg Cd = ? mol Cd
(b) 1.000 kg Br = ? atoms Br
(c) 1 atom Os = ? mol Os
(d) 0.00244 mol Zr = ? mg Zr
(e) 5.9×10^{25} atoms Pt = ? kg Pt

7.36 Arrange the following from highest to lowest mass: (a) 0.75 mol H, (b) 1.0 g atoms H, (c) 0.20 mol He, and (d) 1.6×10^{23} atoms He.

7.37 Arrange the following from largest to smallest number of atoms: (a) 15 g Pu, (b) 11 g Ra, (c) 0.50 g H, and (d) 13 g Ne.

Moles and Molecules

7.38 Calculate the molecular mass for (a) CO, (b) NO_2, (c) I_2, (d) CH_4, (e) S_2F_2.

7.39 Calculate the molecular mass of (a) P_4O_6, (b) NI_3, (c) OBr_2, (d) XeF_6, (e) N_2O_5, (f) $H_2C_2O_4$, (g) $POCl_3$, (h) $C_{12}H_{22}O_{11}$.

7.40 Complete the following table:

Compound	Number of moles	Number of molecules	Mass, g
(a) H_2S	1.000	_____	____
(b) BrCl	0.50	_____	____
(c) NH_3	0.175	_____	____
(d) H_2SO_3	____	6.02×10^{22}	____
(e) N_2O_4	____	1.12×10^{23}	____
(f) AlF_3	____	_____	84.0
(g) PBr_5	____	_____	43.1
(h) OCl_2	9.3	_____	____
(i) SO_3	____	3.01×10^{23}	____
(j) C_2H_6	____	1.29×10^{24}	____

7.41 What is the mass of each of the following:
(a) 0.18 mol ClO_2 (b) 5.1 mol HBr
(c) 1.572 mol SeO_2 (d) 8 mol H_2SO_4?

7.42 How many molecules are contained in each of the following: (a) 6.2 g H_2O_2, (b) 0.044 g ClF_3, (c) 12.5 g N_2H_4, (d) 7.8 g H_3PO_3?

7.43 How many moles of oxygen atoms are contained in each of the following?
(a) 9 mol H_2O
(b) 1.1 mol CO_2
(c) 0.66 g P_2O_5
(d) 1.8×10^{23} molecules SO_3

7.44 What mass of H is contained in each of the following: (a) 0.37 mol NH_3, (b) 8.01 g SiH_4, (c) 9.04×10^{23} molecules C_5H_{12}, (d) 10.0 g $B_{10}H_{14}$?

7.45 What are the masses of the following quantities of molecules?
(a) 2×10^{23} molecules PH_3
(b) 5.0×10^{22} molecules H_2Se
(c) 3.99×10^{25} molecules $HClO_3$
(d) 1.000×10^{15} molecules UF_6?

7.46 Find the unknown quantity:
(a) 0.643 g SO_2 = ? mol O
(b) 4.1×10^{23} molecules HBr = ? g HBr
(c) 0.095 mol C_7H_{16} = ? mol H
(d) 1.5×10^{-6} g H_3PO_4 = ? molecules H_3PO_4

7.47 Arrange the following in order, largest to smallest, of total number of phosphorus atoms: (a) 50 mg PCl_3, (b) 5.0×10^{-4} kg P_4O_{10}, (c) 0.050 mol P_4O_6, and (d) 0.050 mol H_3PO_3.

Formula Unit Calculations

7.48 What is the formula mass of each of the following: (a) NaCl, (b) $CaCl_2$, (c) $Cu(NO_3)_2$, (d) Li_3PO_4?

7.49 Find the number of moles of each of the following:
(a) 16 g $NaNO_3$
(b) 100.0 mg $Ca_3(PO_4)_2$
(c) 0.650 kg $KClO_3$

7.50 What is the mass of the following:
(a) 3.9 mol $CuBr_2$
(b) 1.5 mol SnF_2
(c) 0.903 mmol $PbSO_4$?

7.51 How many formula units are in each of the following: (a) 125 g TiO_2, (b) 1.0 kg $NaMnO_4$, (c) 6.7 mol NH_4NO_3, (d) 914.7 mg CaO?

Percent Composition

7.52 Find the percent composition for each of the following: (a) HF, (b) MgO, (c) $HgBr_2$, (d) Si_3N_4, (e) PbO_2.

7.53 Find the percent composition of each compound: (a) $BaCrO_4$, (b) $NaHCO_3$, (c) $Sr(NO_3)_2$, (d) $Ni(CO)_4$, (e) $(NH_4)_2Cr_2O_7$.

7.54 Find the percent of Ag in each compound: (a) AgI, (b) $AgNO_3$, (c) $AgC_2H_3O_2$, (d) $Ag_2S_2O_3$.

7.55 Find the percent of K in each compound: (a) KN_3, (b) $KClO_3$, (c) K_2PdCl_6, (d) KHC_2O_4.

7.56 Hydrated salts are salts that have a definite number of water molecules attached to them. Calculate the percent of water in each of the following hydrated salts:
(a) $CuSO_4 \cdot 5H_2O$
(b) $BaCl_2 \cdot 2H_2O$
(c) $LiClO_4 \cdot 3H_2O$.

7.57 Arrange the following from highest to lowest in percent iron by mass: (a) $FeBr_3$, (b) $Fe(OH)_2$, (c) Fe_3O_4, and (d) $FeSO_4$.

Simplest Formulas

7.58 Given the following percent compositions, find the simplest formulas for each compound:
(a) 46.55% Fe and 53.45% S
(b) 46.67% N and 53.33% O
(c) 80.0% C and 20.0% H
(d) 5.24% Si and 94.76% I
(e) 11.63% N and 88.37% Cl.

7.59 Find the simplest formulas for each of the following compounds:
(a) 60.1% K, 18.4% C, and 21.5% N
(b) 70.2% Pb, 8.1% C, and 21.7% O
(c) 46.53% Cu, 11.72% S, and 41.75% F
(d) 6.90% C, 1.15% H, and 91.95% Br

7.60 A 25.0-g sample of a chromium-oxygen compound contains 13.0 g of chromium, and the remainder is oxygen. What is the simplest formula of the compound?

7.61 A calcium-phosphorus compound is analyzed and is found to contain 0.66 g C and 0.34 g P. Find the simplest formula of the compound.

7.62 After analysis, a 40.0-g sample was found to contain 16.0 g C, 18.7 g N, and 5.3 g H. Determine the simplest formula of the compound.

7.63 A 500.0-mg sample contains 64.1 mg C and 152.1 mg F, and the remainder is Cl. What is the simplest formula of this compound?

7.64 Determine the simplest formula for the compound that contains 28.2% N, 20.8% P, 42.9% O, and 8.1% H.

Molecular Formulas

7.65 The molecular mass of a compound is 140, and it has a percent composition of 85.7% C and 14.3% H. Determine the compound's molecular formula.

7.66 A 1.000-g sample of a compound contains 0.202 g Al and 0.798 g Cl. Its molecular mass is 267. What is the molecular formula?

7.67 Analysis of a compound reveals that it is composed of H, O, and Br. The sample contains 0.64 g H, 10.15 g O, and 50.71 g Br. If its molecular mass is 96.9, calculate its molecular formula.

7.68 Calculate the molecular formula of a compound with molecular mass 90.0 that is composed of 50.00 g C, 66.75 g O, and 8.25 g H.

General Problems

7.69 What quantity of atoms are contained in a pound mole, the atomic mass expressed in pounds?

7.70 An impure sample of $AgNO_3$ contains 59.5% Ag. Calculate the percent of pure $AgNO_3$ in the sample.

7.71 Various minerals are composed of two or more compounds bonded together. Determine the empirical formula of a mineral that contains 60.7% SiO_2, 27.2% MgO, and 12.1% H_2O.

7.72 (a) What mass of 24 carat gold (100% pure) could be obtained from 1 lb of 18 carat (75% pure) gold? (b) How many gold atoms are contained in this sample?

· C H A P T E R ·

8

Compounds, Molecules, and Chemical Bonding

Study Guidelines

After completing Chapter 8, you should be able to:

1. State the properties of ionic and covalent compounds
2. Determine if a compound has more ionic or more covalent character, given the properties of that compound
3. Write the name for simple binary ionic and covalent compounds
4. Discuss the reason that atoms combine to form molecules
5. Illustrate the electron transfer that takes place in the formation of an ionic bond, given a specific metal and nonmetal
6. Write dot formulas for ionic substances
7. Write the formula unit of an ionic compound, given the metal's and nonmetal's group numbers on the periodic table
8. Determine if one chemical species is isoelectronic to another
9. Define electronegativity
10. Rank elements according to their electronegativity, using the periodic table
11. Discuss the reason that some nonmetallic elements exist as diatomic molecules instead of as monatomic atoms
12. List the steps that can be employed to systematically write the dot formula of a simple covalent molecule
13. Identify single, double, and triple covalent bonds
14. Write a dot structure for a molecule, and determine if the bonds are polar or nonpolar
15. Identify and draw dot formulas for simple polyatomic ions
16. Distinguish between bond order, bond length, bond energy, and bond angle

There are many ways to classify compounds, but chemists usually divide compounds into two groups, ionic compounds and covalent compounds. As we shall see, the names given to these groups relate to the forces that bind the atoms together.

Ionic Compounds

Ionic compounds normally result when metallic elements combine with non-metallic elements.

$$\text{Metal} + \text{nonmetal} \longrightarrow \text{ionic compound}$$

Before looking at representative ionic substances, let us see how simple ionic compounds are named. We shall start with **binary** ionic compounds, those that contain two different elements. First name the metal, and then write the nonmetal, replacing its ending with *ide*.

Binary compounds contain two different elements.

1. Write the name of the metal.

Nonmetal – ending + ide

2. Write the name of the nonmetal, replacing its ending with the ending *ide*.

Endings dropped from nonmetals
*ox*ygen
*nitr*ogen
*carb*on
*sulf*ur
*phosph*orus
*fluor*ine
*chlor*ine
*brom*ine
*iod*ine

How is NaCl named? Notice that Na is an alkali metal, and Cl is a halogen, a nonmetal. First, write the name of the metal *sodium*, and then write the name of the nonmetal with an *ide* ending: chlorine − *ine* + *ide* = chloride. NaCl's name, therefore, is sodium chloride.

Other examples of naming binary ionic substances are given in Example Problem 8.1.

Example Problem 8.1

Name the following ionic compounds: (a) KF, (b) CaO, and (c) Mg_3N_2.

Solution

(a) KF
  1. Metal = potassium
  2. Nonmetal − ending + *ide* = fluorine − *ine* + *ide*
  Name = potassium fluoride

CaO is also called lime. The expression "being in the lime-light" stems from the fact that when lime is heated a brilliant white light is released.

(b) CaO
  1. Metal = calcium
  2. Nonmetal − ending + *ide* = oxygen − *ygen* + *ide*
  Name = calcium oxide

(c) Mg_3N_2
  1. Metal = magnesium
  2. Nonmetal − ending + *ide* = nitrogen − *ogen* + *ide*
  Name = magnesium nitride

TABLE 8.1   PROPERTIES OF SIMPLE IONIC COMPOUNDS

Compound	Physical state, 25°C	Color	Melting point, °C	Boiling point, °C	Density, g/mL	Solubility, g/100 mL H_2O
NaCl	Solid	White	801	1413	2.17	35.7 (0°C)
LiF	Solid	White	870	1676	2.60	0.27 (18°C)
$CaCl_2$	Solid	White	782	>1600	2.15	59.5 (0°C)
Fe_2O_3	Solid	Red brown	1565	—	5.24	Insoluble

Ionic compounds share many common properties. At 25°C they are all solids with high melting and boiling points. Most are hard, but brittle. Ionic substances are poor conductors of electricity, except in the molten (liquid) state, when they are good conductors. Dissolved in water, they break up into individual ions that help conduct an electric current. Table 8.1 presents the properties of selected ionic compounds.

Covalent Compounds

Covalent compounds are formed when two or more nonmetals combine.

$$\text{Nonmetal} + \text{nonmetal} \longrightarrow \text{covalent compound}$$

Frequently encountered covalent compounds include water, H_2O; ammonia, NH_3; carbon dioxide, CO_2; and sulfur dioxide, SO_2. In each of these binary examples of covalent compounds two different nonmetallic elements are chemically combined together.

Naming binary covalent compounds is somewhat different from naming ionic compounds. If there is only one atom of the nonmetal that appears first in the formula, write its name with no change. As in naming ionic compounds, replace the ending of the second nonmetal with *ide*. In addition, a prefix indicating the quantity of that element in the compound is also written.

The nonmetal written first in the formula is the one with the most metallic character.

Consider CO_2 as an example. It is named by writing the first nonmetal, carbon, and then modifying the second nonmetal, oxygen, by adding the prefix *di*, meaning "two" and dropping the ending *ygen* and attaching *ide*. Hence, CO_2 is carbon dioxide.

$$CO_2 = \text{carbon } (di + \text{oxygen} - ygen + ide)$$
$$= \text{carbon dioxide}$$

An additional prefix is needed in the names of covalent compounds because nonmetals can combine in a variety of ways. For instance, there is a second oxide of carbon, carbon monoxide, CO. The prefix *mono* or just *mon* is added to indicate that this oxide of carbon only contains one oxygen. Table 8.2 lists the prefixes that are utilized in naming covalent compounds.

Carbon monoxide, CO, is a poisonous gas. It kills by not allowing O_2 to reach the blood's hemoglobin. CO's affinity for hemoglobin is approximately 200 times that of O_2.

If there are two or more of the first nonmetal in a covalent compound, it is necessary to add a prefix indicating how many are in the formula. Example 8.2 illustrates naming different types of covalent compounds.

TABLE 8.2
PREFIXES INDICATING
NUMBER OF ATOMS IN A
COVALENT MOLECULE

Prefix	Number of atoms indicated
mono	1
di	2
tri	3
tetra	4
penta	5
hexa	6
hepta	7
octa	8
ennea (nona)	9
deca	10

The prefix representing 9 was formerly *nona*, and it is still used by many.

Example Problem 8.2

Name the following covalent compounds: (a) SCl_2, (b) N_2O, (c) P_2O_5.

Solution

(a) SCl_2 = sulfur (*di* + chlorine − *ine* + *ide*)
       = sulfur dichloride
(b) N_2O = (*di* + nitrogen) + (oxygen − *ygen* + *ide*)
       = dinitrogen oxide
(c) P_2O_5 = (*di* + phosphorus) + (*pent* + oxygen − *ygen* + *ide*)
       = diphosphorus pentoxide

In the last two examples, a prefix is added to the first nonmetal to indicate how many of this nonmetal are contained within the compound. In the last example, *pent* rather than *penta* is added as the prefix of oxide, in order to generate a word that is easier to pronounce. Normally, if a double vowel is produced, such as *oo* or *ao*, the vowel contributed by the prefix is dropped.

Covalent compounds are markedly different from ionic compounds. Most covalent compounds are either liquids or gases; some are rather soft solids. Compared with average ionic compounds, covalent compounds have lower melting and boiling points, and lower densities. Covalent compounds are poor conductors of both heat and electricity. When dissolved, most do not form ions. Table 8.3 lists some properties of simple covalent compounds.

REVIEW PROBLEMS

8.1 What groups of elements combine together to produce (a) ionic compounds, (b) covalent compounds?

8.2 Write the name of the following ionic compounds: (a) LiF, (b) Na_2S, (c) MgO, (d) K_3P, (e) Ca_3N_2.

8.3 List four properties of (a) ionic and (b) covalent compounds.

8.4 Name the following covalent compounds: (a) SF_2, (b) NBr_3, (c) NO, (d) $BrCl_3$, (e) P_4O_{10}.

8.5 Compare and contrast the properties of NaCl and H_2O.

TABLE 8.3 SELECTED PROPERTIES OF COVALENT COMPOUNDS

Compound	Physical state	Color	Melting point, °C	Boiling point, °C	Density
H_2O, water	Liquid	Colorless	0.0	100	1.0 g/mL
CH_4, methane	Gas	Colorless	−182	−162	0.55 g/L
HF, hydrogen fluoride	Gas	Colorless	−83.1	19.4	0.98 g/L
NH_3, ammonia	Gas	Colorless	−77.7	−33.4	0.77 g/L
CCl_4, carbon tetrachloride	Liquid	Colorless	−23.0	76.8	1.6 g/mL

8.2 CHEMICAL BONDING

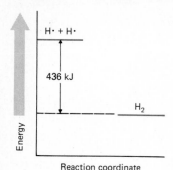

FIGURE 8.1
When 2 moles of hydrogen atoms bond to form 1 mole of H_2, 436 kJ of energy is released. The same quantity of energy, 436 kJ, is required to break the bonds in one mole of H_2.

Diatomic elements are H_2, N_2, O_2, F_2, Cl_2, Br_2, I_2, and At_2.

In addition to the diatomic elements, solid sulfur is found as S_8, and phosphorus exists as P_4.

Chemical bonds are the attractive forces that hold atoms together. In this section we shall attempt to understand what drives atoms to combine and create chemical bonds.

One of the driving forces of nature is the tendency of matter to reach the lowest possible energy state. Generally, a lower energy state implies greater stability. Something that is stable is more resistant to change than something that is less stable.

Elements are ranked according to their degree of stability. Elements like sodium, Na, and chlorine, Cl, are highly reactive (unstable); they tend to undergo chemical changes at the smallest provocation. More stable elements remain unaltered, even under extreme conditions. As a group, the noble gases are quite stable. Helium, for example, does not form any stable compounds.

Certain nonmetallic elements are so unstable that they do not exist as individual atoms, but as diatomic molecules (molecules composed of two atoms). Included in this group are hydrogen (H_2), nitrogen (N_2), oxygen (O_2), plus all of the halogens, fluorine (F_2), chlorine (Cl_2), bromine (Br_2), iodine (I_2), and astatine (At_2). More stable elements, the noble gases, all exist monatomically (as single atoms.)

Let's consider molecular hydrogen, H_2. All samples of hydrogen gas, at room conditions, contain H_2 molecules. How is this explained? One way to answer the question is to consider the stability of two individual H atoms compared with a H_2 molecule. The H_2 molecule is more stable, and the individual H atoms are less stable.

When individual hydrogen atoms are combined, they release energy:

$$\text{H} \cdot + \text{H} \cdot \longrightarrow \text{H}_2 + \frac{436 \text{ kJ}}{\text{mol H}_2}$$

For each mole of H_2 formed, 436 kJ of energy is released. This energy is contained initially in the two H atoms but is released when the atoms combine. Thus, a mole of diatomic hydrogen molecules is 436 kJ/mol more stable than two moles of hydrogen atoms (see Figure 8.1).

FIGURE 8.2
A rock on the ground is in a lower energy state and is more stable than a rock on the top of a cliff. Energy must be added to the rock on the ground to move it to the top of the cliff. Thus, at the top of the cliff it is in a higher energy state and is less stable.

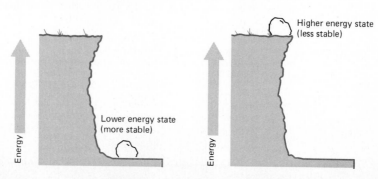

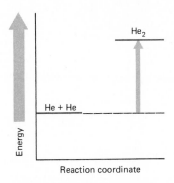

FIGURE 8.3
Energy must be added to He atoms to produce molecules of He_2. Thus, a He_2 molecule is less stable than two He atoms.

As an analogy, think of a boulder on the ground relative to one on top of a hill. Which one has more energy? The boulder on the hill has more energy because kinetic energy was added to the boulder, and stored as potential energy, when it was carried to the top of the hill (see Figure 8.2). A boulder on the top of a hill has the capacity to fall to the ground spontaneously, whereas the opposite is not true. You would be quite amazed to see a boulder jump from the ground to the top of a cliff. Think of the individual H atoms as being at the top of an energy "cliff," and the H_2 molecule as being at the bottom.

Why don't noble gases form diatomic molecules? Individual noble gas atoms exist at a lower energy state (bottom of the hill) compared with diatomic noble gas molecules (top of the hill) (see Figure 8.3). This is verified by the fact that a great deal of energy is required to produce a diatomic noble gas compound, exactly the opposite of what was found for hydrogen gas, where energy is released whenever two H atoms are combined. He_2 does not exist under normal conditions.

Bonding Theory

Throughout the twentieth century, chemists and physicists have attempted to develop and refine a theory that explains why some atoms bond together and others do not. This bonding theory also tries to explain the degree of stability of compounds and to account for the arrangement of atoms within molecules.

G. N. Lewis, in 1916, was one of the first scientists to see that bonding was directly related to the electronic arrangement of atoms. Since that time, a large body of information has been collected to show that chemical bonding is adequately explained in terms of the outer or valence electrons. This bonding theory is now known as the **valence bond** theory.

Bonding involves interactions of electrons among atoms.

As we proceed, do not lose sight of the fact that bonding theories are developed to explain the structure and behavior of molecules and compounds. If we reach an exception to a rule or guideline, it is not a discrepancy in the natural world, but a flaw in the theory. Bonding theories are nothing more than a means for scientists to produce a mental picture of something that they cannot directly see.

Besides the valence bond theory, a second bonding theory, called the molecular orbital (MO) theory, is also used by chemists.

Bond Types

Valence bond theory explains how compounds are held together in terms of electron "transfers" and electron "sharing." When an electron or electrons are transferred from one atom to another in the process of forming a bond, the resultant bond is called an ionic bond. If no electron transfer takes place, but there is "sharing" of electrons between two atoms, the resultant bond is classified as a covalent bond.

Electron transfer = ionic bond
Sharing electrons = covalent bond

While it is convenient to classify bonds in this manner, no bond is purely ionic or covalent. Just as most occurrences in the world are not "black" or "white," bonds have varying degrees of ionic and covalent character. Ionic compounds contain bonds with a higher degree of ionic character, and covalent compounds have a greater degree of covalent character.

NaCl, an "ionic" substance, has 67% ionic character and 33% covalent character.

REVIEW PROBLEMS **8.6** What is a chemical bond?

8.7 List two everyday phenomena that illustrate the tendency of objects to seek their lowest energy states.

8.8 What groups on the periodic table are classified as (a) more stable and (b) less stable elements?

8.9 List elements that normally exist as diatomic molecules.

8.10 Provide an explanation for the fact that helium does not exist diatomically.

8.11 What type of bond results when electrons are shared between two atoms?

8.3 IONIC OR ELECTROVALENT BONDING

When electrons are transferred from a metal to a nonmetal, an **ionic bond** (sometimes called an electrovalent bond) is produced. In this section, we will describe the characteristics and properties of ionic compounds.

Ionic Bonding in Sodium Chloride, NaCl

To illustrate ionic bonding, let us consider sodium chloride, NaCl. When sodium combines with chlorine, the ionic salt sodium chloride is the product. Sodium is an alkali metal with one outer electron:

$$\text{Na} \quad 1s^2 2s^2 2p^6 3s^1$$

The electron dot formula for sodium is

$$\text{Na} \cdot$$

Chlorine's electronic configuration is

$$\text{Cl} \quad 1s^2 2s^2 2p^6 3s^2 3p^5$$

Chlorine's dot formula is

$$:\overset{\cdot\cdot}{\underset{\cdot}{\text{Cl}}}:$$

As you may recall, sodium is an extremely reactive metal, and chlorine is a reactive nonmetal. Sodium has a low ionization energy (little energy is required to remove an electron). Chlorine, in contrast, has a relatively high ionization energy. Sodium's atomic radius is the largest in the third period, while chlorine's is one of the smallest.

Whenever a sodium atom encounters a chlorine atom, sodium's loosely held outer electron ($3s^1$) is pulled away by the more compact chlorine atom, creating two ions. Since sodium loses an electron, it becomes a cation (a positive ion). Chlorine gains sodium's electron, and becomes an anion.

$$\text{Na} \cdot + \; :\overset{\cdot\cdot}{\underset{\cdot\cdot}{\text{Cl}}}: \; \longrightarrow \; \text{Na}^+ \left[:\overset{\cdot\cdot}{\underset{\cdot\cdot}{\text{Cl}}}: \right]^-$$

$$\text{Na} \;\; (1s^2 2s^2 2p^6 3s^1) \longrightarrow e^- + \text{Na}^+ \;\; (1s^2 2s^2 2p^6 3s^0)$$

$$\text{Cl} \;\; (1s^2 2s^2 2p^6 3s^2 3p^5) + e^- \longrightarrow \text{Cl}^- \;\; (1s^2 2s^2 2p^6 3s^2 3p^6)$$

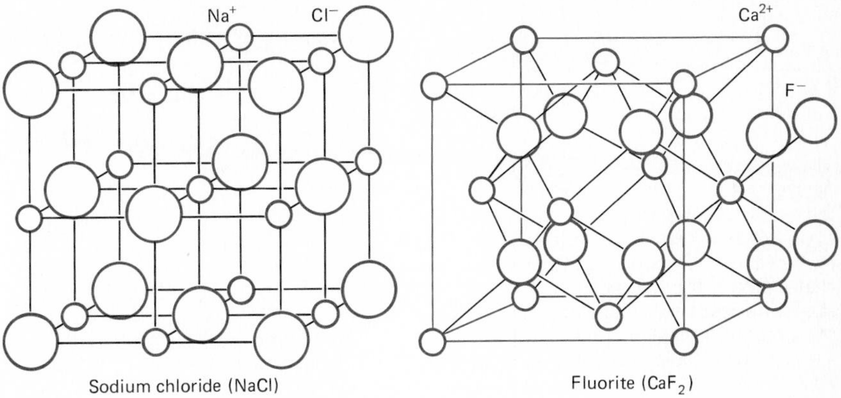

FIGURE 8.4
Ionic solids are composed of regular patterns of anions and cations. This figure shows segments of the crystal lattices of sodium chloride, NaCl, and calcium fluoride, CaF$_2$.

Sodium chloride (NaCl)

Fluorite (CaF$_2$)

Both Na$^+$ and Cl$^-$ have one thing in common: they possess a noble gas electronic configuration. A Na$^+$ ion contains 10 electrons and has the same electronic arrangement as Ne. Chemists say that Na$^+$ is isoelectronic with Ne. **Isoelectronic** is a term that means that two or more chemical species have the same electronic configuration. Therefore, Cl$^-$ is isoelectronic with the noble gas Ar (18 electrons).

Iso is a prefix meaning the "same."

An ionic bond is the force of attraction between unlike charged ions, in this case Na$^+$ and Cl$^-$. Remember that unlike charged particles always attract, but it is not the simple attraction of a pair of Na$^+$ and Cl$^-$ ions that is significant. Ionic compounds, like NaCl, exist in a crystal lattice, a three-dimensional array of Na$^+$ ions surrounded by Cl$^-$ ions, and vice versa. Sodium chloride's crystal lattice is illustrated in Figure 8.4. Except at the surface, each Na$^+$, is surrounded and attracted by 6 Cl$^-$, and each inner Cl$^-$ is surrounded and attracted by 6 Na$^+$.

In Figure 8.4, note that Na$^+$ ions are smaller than Cl$^-$ ions. After a Na atom loses its outer 3s^1 electron, the inner-core electrons are held tightly by the nucleus since there is now a greater number of protons than electrons. In contrast, Cl$^-$ has one more electron than proton; thus, the electrons are not held too tightly to the nucleus. Whereas atomic radii are utilized as a measure of the size of an atom, **ionic radii** are employed to measure the size of ions. Figure 8.5 illustrates the ionic radii of selected ions.

What drives Na and Cl atoms to combine to form NaCl? The combination of these two atoms is accompanied by a loss of heat energy.

$$Na(g) + \tfrac{1}{2} Cl_2(g) \longrightarrow NaCl(g) + \frac{623 \text{ kJ}}{\text{mol NaCl}}$$

NaCl is much more stable than the elements from which it is formed.

When energy is released in chemical reactions, the products are more stable (lower energy) than the reactants. In NaCl, both atoms achieve the stable noble gas configuration—the most stable electronic configuration among atoms.

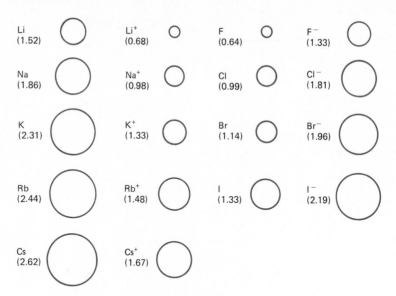

FIGURE 8.5
Ionic radii of alkali metal cations are smaller than the radii of the corresponding alkali metal atoms. The decrease in size is the result of the loss of the outermost electron. Ionic radii of halide anions are larger than the radii of the corresponding halogen atoms. The increase in size is the result of gaining one electron. Atomic radii and ionic radii are measured in angstroms (1 Å = 10⁻¹⁰ m).

The reaction of Na with Cl is a model for the combination of an alkali metal (group IA) with a halogen (group VIIA). All combinations of alkali metals and halogens yield compounds containing ionic bonds with formula units of MX, where M is any alkali metal and X is any halogen. M and X combine in a 1-to-1 ratio because each attains the stable noble gas electronic configuration after transferring one electron.

Other ionic compounds containing an alkali metal and halogen are KI, CsBr, RbCl, and LiF.

A convenient rule to follow is that atoms are most stable when they are isoelectronic with a noble gas. This rule is sometimes called the *octet rule* or *rule of eight* but would be more accurately named the **noble gas rule.** A word of caution when applying the noble gas rule: It is only a generalization that can be applied for many, but not all, cases.

An octet refers to eight things, a group of eight electrons in this instance.

Ionic Bonding in Calcium Fluoride, CaF₂

What happens when calcium atoms bonds with fluorine atoms? Calcium belongs to group IIA, the alkaline earth metals. Each group IIA element has two outer electrons. Fluorine, like chlorine, is a halogen with seven outer electrons. If a fluorine atom removes an electron from a calcium atom, one electron remains in Ca's outer level. Hence, two F atoms accept one electron each from Ca, allowing Ca to attain the noble gas configuration of Ar.

Calcium fluoride, commonly called fluorite, is a high-melting solid (1360°C) with low water solubility.

$$:\ddot{\text{F}}:\quad \text{Ca}:\quad :\ddot{\text{F}}:$$

After the electrons are transferred, Ca is left with a 2+ charge and each F possesses a 1− charge. Both Ca^{2+} and $F^−$ are isoelectronic with noble gases.

$$Ca\ (1s^22s^22p^63s^23p^64s^2) \longrightarrow 2e^- + Ca^{2+}\ (1s^22s^22p^63s^23p^64s^0)$$
$$F\ (1s^22s^22p^5) + e^- \longrightarrow F^-\ (1s^22s^22p^6)$$

Calcium fluoride's electron dot formula is expressed as follows:

$$\text{Ca}^{2+} \; 2\left[:\!\overset{\cdot\cdot}{\underset{\cdot\cdot}{\text{F}}}\!:\right]^{-}$$

Once again we can generalize; alkaline earth metals (except Be whose compounds have little ionic character) and halogens combine to yield ionic compounds with the formula MX_2.

Ionic Bonding in Magnesium Oxide, MgO

Magnesium, Mg, belongs to group IIA. Oxygen, O, is a member of group VIA, the chalcogens. In the formation of magnesium oxide, MgO, magnesium must lose two electrons to become isoelectronic with a noble gas, and the oxygen must gain two electrons to attain a noble gas configuration. Thus, two electrons are transferred from the Mg to the O:

$$\text{Mg}\!: \; + \; \overset{\cdot\cdot}{\underset{\cdot\cdot}{\text{O}}}\!: \; \longrightarrow \; \text{Mg}^{2+}\left[:\!\overset{\cdot\cdot}{\underset{\cdot\cdot}{\text{O}}}\!:\right]^{2-}$$

Both elements achieve the electronic configuration of the noble gas Ne.

$$\text{Mg} \;\; (1s^2 2s^2 2p^6 3s^2) \;\longrightarrow\; 2e^- + \text{Mg}^{2+} \;\; (1s^2 2s^2 2p^6 3s^0)$$

$$\text{O} \;\;\; (1s^2 2s^2 2p^4) + 2e^- \;\longrightarrow\; \text{O}^{2-} \;\; (1s^2 2s^2 2p^6)$$

Magnesium oxide's formula unit is MgO because one pair of electrons is transferred from the Mg to the O.

Ionic Bonding in Potassium Sulfide, K$_2$S

Potassium, K, an alkali metal with one valence electron, and sulfur, a chalcogen with six valence electrons, combine in a 2-to-1 ratio. Potassium achieves the noble gas configuration by losing one electron to S, but the S then has only seven outer electrons. Hence, S removes an electron from another K atom to attain the noble gas configuration:

$$\text{K}\!\cdot \; + \; :\!\overset{\cdot\cdot}{\text{S}}\!: \; + \, \cdot\text{K} \;\longrightarrow\; 2\text{K}^+ + \left[:\!\overset{\cdot\cdot}{\underset{\cdot\cdot}{\text{S}}}\!:\right]^{2-}$$

In potassium sulfide, the K^+ is isoelectronic with Ar, and S^{2-} is isoelectronic with Ar.

Ionic Compound Summary

Table 8.4 lists all possible nontransition metal and nonmetal group combinations that produce ionic substances. Note that in each case both the metal and nonmetal attain a noble gas configuration and each compound is electrically neutral; the sum of positive charges equals the sum of negative charges.

Crystal structures are elucidated through x-ray diffraction analysis.

As mentioned previously, all ionic substances have a three-dimensional crystal lattice structure containing cations and anions. Many different crystal lattice patterns exist. Each has an orderly array of cations surrounded by anions, and vice versa. Figure 8.6 illustrates three crystal lattice patterns.

While the properties of ionic substances are similar, they vary somewhat

TABLE 8.4   IONIC BONDING

Metal group	Nonmetal group	Formula*	Examples
IA	VIIA	MX	NaBr, KI, CsF
IA	VIA	M_2X	Li_2O, K_2O, Rb_2S
IA	VA	M_3X	Na_3N, K_3P
IIA	VIIA	MX_2	$MgCl_2$, $SrBr_2$, CaI_2
IIA	VIA	MX	BaS, SrO, MgS
IIA	VA	M_3X_2	Ca_3N_2, Mg_3P_2
IIIA	VIIA	MX_3	$AlCl_3$, GaF_3
IIIA	VIA	M_2X_3	Al_2O_3, In_2O_3
IIIA	VA	MX	AlN, GaN

*M = metal and X = nonmetal.

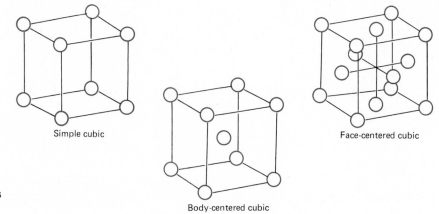

Simple cubic

Body-centered cubic

Face-centered cubic

FIGURE 8.6
Three cubic crystal lattice patterns; simple cubic, body-centered cubic, and face-centered cubic. Many other geometric arrangements of ions are found in crystal lattices.

depending on the (1) charge on the ions, (2) distance between ions, and (3) pattern of ions within the crystal lattice. For example, more highly charged ions in the lattice produce stronger ionic bonds: more energy is needed to break them apart. The ionic bonds of magnesium chloride, $MgCl_2$ (Mg^{2+} $2Cl^-$), are significantly stronger than those of NaCl (Na^+ Cl^-). Greater attraction is found between atoms with higher charges, creating stronger ionic bonds.

REVIEW PROBLEMS

8.12 What is an ionic bond?

8.13 Use electron dot formulas to illustrate the electron transfer that occurs when sodium and chlorine combine to give sodium chloride.

8.14 Write two ions, a cation and anion, that are isoelectronic with (a) Ne, (b) Kr.

8.15 What is the noble gas rule, and how is it applied when considering ionic bonding?

8.16 Write the complete electronic configurations for (a) O^{2-}, (b) Li^+, (c) S^{2-}, (d) Ca^{2+}

8.17 Write the electron dot formula for each of the following ionic substances: (a) SrS, (b) $MgBr_2$, (c) Na_2O.

8.18 What is the formula unit for (a) aluminum phosphide, (b) cesium chloride, (c) calcium nitride, (d) rubidium fluoride?

8.4 INTRODUCTION TO COVALENT BONDING

Many molecules are held together as a result of valence electrons that are "shared" between two atoms. In this case, the properties of the atoms involved are such that one atom is incapable of taking an electron from the other. A chemical bond that forms without the transfer of electrons is classified as a **covalent bond.**

Electronegativity

Linus Pauling (*Linus Pauling Institute of Science and Medicine*)

FIGURE 8.7
Within a period, the electronegativities of atoms increase, excluding the noble gases. Within a group, the electronegativities decrease.

Electronegativity is the capacity of an atom to attract electrons in a chemical bond. An element with a high electronegativity has a greater capacity to attract bonded electrons than one with a smaller electronegativity.

Three or four scales of electronegativity exist, but the original scale developed by Linus Pauling (1901–   ) is the most popular. Pauling collected data on most of the elements, performed various calculations, and produced an electronegativity scale based on the element fluorine, F. Pauling assigned the electronegativity value of 4.0 to fluorine, the most electronegative element.

On the Pauling electronegativity scale two trends are evident: (1) From left to right across the periodic table, electronegativity increases (excluding the noble gases), and (2) with increasing atomic mass within a periodic group, the electronegativity decreases, Francium, Fr, the element at the bottom left corner of the periodic table, is the least electronegative element. The closer an element is to fluorine on the periodic table, the higher its electronegativity. Oxygen is the second most electronegative element, with a value of 3.5. Figure 8.7 shows electronegativity trends of representative elements.

Trends in electronegativity directly parallel trends in ionization energy and are indirectly related to the atomic radii. In other words, elements with high

Increasing electronegativity →

Decreasing electronegativity

Group

Period	IA	IIA										IIIA	IVA	VA	VIA	VIIA	VIIIA
1	H 2.2																He
2	Li 1.0	Be 1.5										B 2.0	C 2.5	N 3.0	O 3.5	F 4.0	Ne
3	Na 0.9	Mg 1.2										Al 1.5	Si 1.8	P 2.1	S 2.5	Cl 3.0	Ar
4	K 0.8	Ca 1.0	Sc 1.3	Ti 1.5	V 1.6	Cr 1.6	Mn 1.5	Fe 1.8	Co 1.8	Ni 1.8	Cu 1.9	Zn 1.6	Ga 1.6	Ge 1.8	As 2.0	Se 2.4	Br 2.8	Kr
5	Rb 0.8	Sr 1.0	Y 1.2	Zr 1.4	Nb 1.6	Mo 1.8	Tc 1.9	Ru 2.2	Rh 2.2	Pd 2.2	Ag 1.9	Cd 1.7	In 1.7	Sn 1.8	Sb 1.9	Te 2.1	I 2.5	Xe
6	Cs 0.7	Ba 0.9	La 1.1	Hf 1.3	Ta 1.5	W 1.7	Re 1.9	Os 2.2	Ir 2.2	Pt 2.2	Au 2.4	Hg 1.9	Tl 1.8	Pb 1.8	Bi 1.9	Po 2.0	At 2.2	Rn
7	Fr 0.7	Ra 0.9	Ac 1.1															

ionization energies, those with small atomic radii, are the most electronegative (excluding the noble gases). Elements with low ionization energies and large atomic radii have lower electronegativities.

Covalent Bonding in Hydrogen, H_2

Hydrogen gas, H_2, is a colorless gas that is highly combustible.

To begin our discussion of covalent bonding it is easiest to start with the simplest molecule, diatomic hydrogen, H_2. Hydrogen gas is composed of H_2 molecules rather than separate H atoms. When one H atom bonds with another H atom, there is no chance of an electron transfer taking place because each H atom has the same electronegativity. Instead H atoms must share their electrons in order to reach the noble gas configuration of two electrons (isoelectronic with He).

$$H\cdot + \cdot H \longrightarrow H:H$$

Sharing electrons implies that 50 percent of the time each H atom within a H_2 molecule has two electrons, or has the stable noble gas configuration (He). At first glance this might not seem to be an effective means for holding atoms together. But in fact the covalent bond between hydrogens is quite strong; 436 kJ/mol are required to break this bond.

Covalent bonds result when an outer-level orbital of one atom overlaps an outer orbital of another atom. A hydrogen atom's $1s$ orbital overlaps the $1s$ orbital of the other H atom. Overlapping orbitals are regions between two nuclei where there is a high probability of finding two electrons. Covalent bonds, like ionic bonds, result from the attraction of positive particles (nuclei) to negative particles (electrons in overlapping orbitals). Figure 8.8 illustrates this important fact.

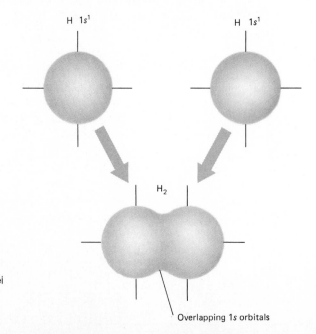

FIGURE 8.8
In the formation of a covalent bond between two hydrogen atoms, the ls orbitals of each hydrogen atom overlap, producing a negative region between the two nuclei. The force of attraction of the nuclei of the two hydrogen atoms for the overlapping orbitals is the covalent bond.

Whenever a pair of electrons is located in a region of space between two nuclei where orbitals overlap, we say that electrons are being shared. In a dot formula, sharing is illustrated by placing the symbols close together and inserting two dots, representing two electrons, between them. Sometimes the electron pair is replaced by a dash (—) which is interpreted as a shared electron pair, a covalent bond.

$$H—H = H:H$$

Covalent Bonding in Fluorine, F_2

Fluorine gas is extremely reactive. Paper, wood, sulfur, and even powdered metals burst into flames when exposed to F_2.

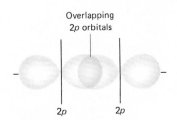

Overlapping
2p orbitals

2p   2p

FIGURE 8.9
In the formation of a covalent bond between two fluorine atoms, the 2p orbitals of each fluorine atom overlap, producing a negative region between the two nuclei.

Fluorine gas is composed of diatomic F_2 molecules. Like H_2, F_2 is composed of two atoms with the same electronegativity. Neither atom can remove an electron from the other; therefore, the bond joining F atoms together is covalent.

Let's consider fluorine's electronic configuration:

$$F \quad 1s^2 2s^2 2p^5$$

If a F atom shares an electron with another F atom, both F atoms would attain the noble gas configuration.

$$:\ddot{F}\cdot + \cdot\ddot{F}: \longrightarrow :\ddot{F}:\ddot{F}:$$

Both fluorine atoms now have eight valence electrons. Fluorine's covalent bond is similar to the one that holds H_2 together, except for the orbitals that are overlapping. In F, the outermost electrons reside in 2p orbitals. Accordingly, it is an electron in a 2p orbital from one F overlapping a 2p orbital from the other F atom that produces the covalent bond. (See Figure 8.9.)

$$:\ddot{F}\overset{2p \quad 2p}{\rule{1.5cm}{0.4pt}}\ddot{F}:$$

In both F_2 and H_2 molecules only one pair of electrons is shared between two nuclei; this is called a single covalent bond. In some other molecules more than one pair of electrons are shared; the bonds are called **multiple covalent bonds,** double and triple covalent bonds.

Multiple Covalent Bonds: Double and Triple Bonds

About 23% of the earth's atmosphere is O_2. Oxygen sustains most life on earth. It supports combustion of objects; without oxygen these objects will not burn.

Oxygen gas exists diatomically as O_2 molecules. Oxygen's electronic configuration is

$$O \quad 1s^2 2s^2 2p^4$$

To attain the noble gas configuration, each O atom shares two electrons. Two orbitals from each O overlap, giving each O atom an electronic configuration that is isoelectronic with Ne.

$$:\ddot{O}::\ddot{O}: \quad :\ddot{O}=\ddot{O}:$$

Now there are a total of four electrons in the region between the two O nuclei. In other words, O molecules contain a double covalent bond.

Double covalent bonds are generally stronger than single bonds: more energy must be added to break them. Having four electrons (four negative charges) between two nuclei produces a stronger attractive force.

An even stronger bond results when six electrons are shared between two nuclei. Six shared electrons form a *triple covalent bond*. Nitrogen gas is a good example of a diatomic molecule with a triple covalent bond. Nitrogen's electronic configuration is

$$N \quad 1s^2 2s^2 2p^3$$

For a N atom to gain the stability of a noble gas configuration, it has to share three of its electrons with another N atom. Nitrogen's original five electrons plus the three shared electrons give it the noble gas configuration of Ne.

$$:N \vdots\vdots N: \qquad :N{\equiv}N:$$

One of the strongest bonds known is the triple bond in N_2.

Not all atoms can form multiple bonds. Some that can are O, N, C, and S. Atoms like H or the halogens only share one electron; consequently, they do not form multiple covalent bonds.

N₂ was discovered by Daniel Rutherford (1749–1819), a Scottish chemist, in 1772. Nitrogen makes up about 77% of the atmosphere. It is quite inert because of the strong triple bond between nitrogens.

O, N, C, and S are capable of forming multiple bonds.

Polar Covalent Bonds: Unequal Sharing of Electrons

In the preceding bonding illustrations, both atoms bonded were the same, resulting in equal sharing. When both atoms are the same, the electronegativities are equal; thus, no atom has a greater control over the shared electron pair. Such a bond is called a **nonpolar covalent bond**—a covalent bond where both atoms have the same electronegativity.

In most covalent bonds, however, the atoms have different electronegativities, and, as a result, one atom has a greater force of attraction for the electrons than the other. Generally, when two different atoms bond together, unequal sharing results. A covalent bond where the electrons are not shared equally is called a **polar covalent bond.**

Polar implies that the bond contains a dipole, or electric charge separation. In other words, one side of the bond is more negative than the other side. To illustrate a molecule that contains a polar covalent bond, let's consider hydrogen chloride gas, HCl(*g*).

Covalent Bonding in Hydrogen Chloride, HCl

A water solution of HCl, HCl(aq), is called hydrochloric acid.

Looking at Figure 8.10, we see that chlorine's electronegativity (3.0) is greater than that of hydrogen (2.2). Whenever H bonds with Cl, a single polar covalent bond results:

$$H{\cdot} + {\cdot}\ddot{Cl}{:} \longrightarrow H{-}\ddot{Cl}{:}$$

Na⁺ Cl⁻ is ionic. H—Cl is polar covalent.

One electron from the H and one from the Cl are shared, giving both H and Cl noble gas configurations. However, since Cl's electronegativity is greater, the

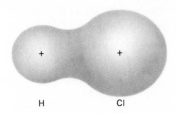

FIGURE 8.10
In HCl, the 3*p* orbital of the chlorine atom overlaps the 1*s* orbital of the hydrogen atom. Since a chlorine atom has a greater electronegativity than a hydrogen atom, the chlorine atom has a greater force of attraction for the shared pair of electrons.

shared pair of electrons spends a larger percentage of time around the Cl nucleus than around the hydrogen nucleus. We show that a dipole exists by writing the symbol delta, δ, followed by a $+$ or $-$ to indicate which atom is more positive and which is more negative:

$$\overset{\delta+}{\text{H}}\!-\!\overset{\delta-}{\text{Cl}}$$

The delta is read as "partial" or "slightly." So δ^- means that an atom has a partial negative charge, and δ^+ indicates a partial positive charge.

Don't confuse a partial charge with the full positive or negative charge that we assigned to ions in ionic compounds. Metals have low electronegativities, and nonmetals have high electronegativities. Consequently, a more complete transfer of electrons occurs in ionic compounds.

In polar covalent bonds a transfer of electrons does not occur; instead a pair (or pairs) of electrons are unequally shared. Ionic bonds can be considered extremely polar covalent compounds, to the point where there is minimal sharing.

Covalent Bonding in Bromine Monochloride, BrCl

Bromine monochloride is another example of a molecule with a polar covalent bond. Chlorine's electronegativity is greater than bromine's (3.0 versus 2.8). Chlorine, therefore, has a slightly greater capacity to attract the shared electron pair. So the electrons spend a slightly larger percentage of their time nearer the chlorine. Whenever bromine monochloride's dot formula is written, a δ^- is placed above the Cl and a δ^+ is placed above the Br.

$$\overset{\delta+}{:\!\overset{\bullet\bullet}{\text{Br}}}\!-\!\overset{\delta-}{\underset{\bullet\bullet}{\overset{\bullet\bullet}{\text{Cl}}}}\!:$$

REVIEW PROBLEMS

8.19 How is a covalent bond different from an ionic bond?

8.20 What is the electronegativity of an atom?

8.21 Arrange the following sets of atoms in decreasing order, highest to lowest, of their electronegativity: (a) P, As, and Sb; (b) Be, Li, and B; (c) Rb, Sr, Cs, and Ba.

8.22 How many electrons are shared in a (a) single, (b) double, and (c) triple covalent bond?

8.23 Draw the electron dot formulas for (a) H_2, (b) O_2, (c) F_2, (d) N_2, (e) Cl_2.

8.24 Give examples and explain the difference between a nonpolar and polar covalent bond.

8.25 Draw the dot formulas for each of the following polar molecules, and indicate the partial charges using δ^+ and δ^-: (a) HI and (b) IF.

8.5 DOT FORMULAS FOR COVALENT MOLECULES

Successfully writing dot formulas for larger and more complex covalent molecules requires a systematic procedure. Most dot formulas for covalent molecules are obtained by following a set of five steps.

Step 1. Determine the total number of outer electrons in all atoms in the molecule. Use the periodic table to find the number of outer electrons in each atom, and then add them together to obtain the total.

Step 2. Identify the central atom (or atoms), and write all other atoms around the central atom. (See Figure 8.11.)

Halogens, except for F, can be central atoms in molecules when they are bonded to more than one oxygen or halogen. In each of the following molecules, Cl is the central atom: $HClO_3$, ClF_3, ClO_2, and Cl_2O_7.

A molecule's central atom is bonded to two or more other atoms and is the atom that determines the overall shape of small molecules. It is not difficult to identify the central atom, and after a little practice it becomes a trivial matter. B, C, Si, N, P, and S are among the most common central atoms encountered.

Hydrogen is never a central atom because it can share only one electron, and thus forms only one bond. For the same reason, halogens are usually not central atoms in the molecules we shall encounter. In binary compounds, oxygen is found as the central atom when bonded to hydrogen or the halogens. In most other binary compounds, and compounds with three or more elements, oxygen is rarely a central atom.

It is fairly standard when writing the name or molecular formula of a compound to write the central atom first, except if the molecule is an acid (HNO_3, H_2SO_4, H_3PO_4, etc.). In each of the following examples, the central atom is written first: CO_2, OF_2, NH_3, and PBr_3.

Step 3. Place a pair of electrons (single bond) between the central atom and each of the other atoms. Then subtract the number of electrons placed in the dot formula from the total number of electrons obtained in Step 1. The resulting number is the quantity of electrons that remain to complete each atom's noble gas configuration.

Step 4. Calculate the number of electrons needed for all atoms to achieve the valence noble gas configuration, and compare the result with the number of electrons that are available (determined in Step 3). If these numbers are equal, place the appropriate number of dots around each symbol so that all atoms have a noble gas configuration. When placing dots into the formula, ask yourself: How many outer electrons does the atom have, and how many does it need to attain a noble gas configuration?

If the number of electrons available is smaller than the number of electrons needed to attain the valence noble gas configuration, then a multiple bond is located in the molecule. Generally, if the deficit is two electrons, a double bond is in the molecule; a shortage of four electrons indicates two double bonds or a triple bond.

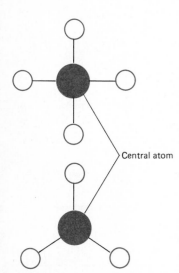

Step 5. Check the final dot formula in two ways: (1) Count the total number of electrons to see that the correct number have been placed in the formula, and (2) see if each atom has a noble gas configuration. An easy way to check if each atom has a noble gas configuration is to circle the electrons surrounding each atom—those belonging to the atom and those that are shared. In all cases, the atom should contain either eight or two electrons. Then and only then are you sure that the dot formula is correct (assuming that it is possible for all atoms to achieve a noble gas configuration).

Central atom

FIGURE 8.11
The central atom in a molecule is bonded to two or more atoms. In the formulas of binary covalent compounds, the central atom is usually written first.

In summary, when writing the dot formula of a covalent compound:

Step 1. Find the total number of outer electrons.

Step 2. Locate the central atom, and write all other atoms around the central atom's symbol.

Step 3. Place a pair of electrons between the central atom and each of the other atoms. Determine the quantity of electrons remaining by subtracting the number of electrons placed in the formula from the total (Step 1).

Step 4. Determine the number of electrons required to achieve the outer configuration of a noble gas for all atoms, and compare with the number of electrons available (Step 3):

1. If the number required equals the number available, place the electrons around each symbol, completing the octets. The molecule contains single bonds only.
2. If the number required is greater than the quantity available, multiple bonds are found in the molecule. A shortage of two electrons indicates a double bond; a deficit of four electrons indicates a triple bond or two double bonds. Locate the atoms that have a multiple bond, and place the correct number of electrons around each symbol.

Step 5. Check the dot formula to see that each atom has a noble gas configuration and that the total number of electrons equals the total found in Step 1.

The five rules for writing dot formulas are general guidelines that can be used for many, but not all, molecules. However, the dot formulas of all molecules encountered in this book can be successfully written using these rules.

Covalent Bonding in Water, H_2O

Water is the most important liquid on earth. A significant portion of the earth's surface is covered with water, and living tissues contain approximately 80% water.

Water's properties, as we shall see in Chapter 13, are somewhat unique among low-molecular-mass liquids. Water's special properties can be understood after learning how it is bonded together.

Water's dot formula is determined by following the five steps that were just outlined.

Step 1. Find the total number of valence electrons:

$$
\begin{aligned}
2 \text{ H atoms} &= 2 \ e^- \\
1 \text{ O atom} &= \underline{+6 \ e^-} \\
H_2O &= 8 \text{ valence } e^-
\end{aligned}
$$

Step 2. Write the central atom, and place the other atoms around the symbol.

Water's central atom, by default, is the oxygen atom. Hydrogen atoms are never central atoms. So write O with the two Hs adjacent to it.

O H
H

It does not matter how the atoms are written around the central atom; dot formulas do not illustrate the spatial arrangement of atoms within a molecule.

Step 3. Place electron pairs between the central and all other atoms. Calculate the number of electrons remaining.

O:H
··
H

Four electrons were placed in the dot formula, leaving four electrons.

$$\text{Remaining } e^- = \text{total } e^- - \text{electrons in formula}$$
$$= 8 - 4 = 4 \ e^-$$

Step 4. Find the number of electrons needed to give each atom a valence noble gas configuration, and compare with the number of electrons available.

Four more electrons are needed to complete the octet around O, since after sharing two electrons from two hydrogens the O now has four electrons. Each hydrogen already has a noble gas configuration of two electrons.

Four electrons are available, and four electrons are needed. Therefore, place the four electrons around the O to complete its octet.

··
:O:H
··
H

Step 5. Check to see if the proper number of electrons are contained in the formula.

1. Check to see that each atom has a noble gas configuration.
Each H has two electrons, and the O has eight.
2. Count the total number of dots in the formula.
Eight electrons are contained in the dot formula, corresponding to the total number of valence electrons in two H atoms and one O atom.

Looking at water's dot formula, two covalent bonds are indicated. Each bond is classified as a polar covalent bond, since the electronegativity of O (3.5) is greater than that of H (2.2). In addition to the two bonds, the central oxygen has two pairs of electrons that are not bonded. They are called either **lone pair electrons** or **nonbonded electron pairs.**

Since all members of group VIA, the chalcogens, have six valence electrons, you might expect that they should bond in a manner similar to oxygen. In simple covalent molecules this is true. For example, consider the dot formula for hydrogen sulfide, H_2S.

Lone pairs

Bonding pairs

$$\overset{\displaystyle \cdot\cdot}{:\overset{}{\text{S}}:\text{H}}$$
$$\text{H}$$

Sulfur is the element directly below oxygen on the periodic table; therefore, the bonding in H_2S is similar to that in H_2O. H_2S is a toxic yellowish gas with the unpleasant odor of rotting eggs. H_2Se and H_2Te have dot formulas similar to those of both water and hydrogen sulfide.

Covalent Bonding in Ammonia, NH_3

Ammonia, NH_3, is a gas at room temperature. Water solutions of ammonia are commonly used as household cleaners. We shall write the electron dot formula of ammonia as an illustration of covalent bonding of atoms in group VA.

Following our rules, we first determine the total number of valence electrons in an ammonia molecule (Step 1):

$$
\begin{aligned}
3 \text{ H with 1 electron each} &= 3\ e^- \\
1 \text{ N with 5 electrons} &= +5\ e^- \\
\text{NH}_3 &= 8 \text{ valence } e^-
\end{aligned}
$$

Ammonia's central atom is N, so in Step 2 we write the symbol for N with three hydrogens placed around it.

$$\text{H N H}$$
$$\text{H}$$

Placing three electron pairs between each H and N takes six electrons from the total of eight electrons, leaving two electrons to complete N's octet (Step 3).

$$\text{H:N:H}$$
$$\overset{}{\underset{\cdot\cdot}{\text{H}}}$$

All of the hydrogens have a noble gas configuration, and the N only needs two more electrons to complete its octet. Two electrons are required and two electrons are remaining, so those electrons are placed next to the N, completing ammonia's dot formula (Step 4).

$$\text{H:}\overset{\cdot\cdot}{\underset{\cdot\cdot}{\text{N}}}\text{:H}$$
$$\text{H}$$

Each atom in a molecule of ammonia has a noble gas configuration, two or eight, and the total number of valence electrons equals eight (Step 5). Ammonia's dot formula reveals a molecule with three polar covalent bonds and one lone electron pair.

Phosphine, PH_3, is bonded in a similar manner to ammonia. Try writing the dot formula of phosphine, following the five steps. Phosphine's correct dot formula is

$$H : \overset{\displaystyle ..}{\underset{\displaystyle H}{P}} : H$$

Like water, the group VA hydrides (binary hydrogen compounds) have analogous dot formulas: NH_3, PH_3, AsH_3, and SbH_3.

Covalent Bonding in Nitrogen Trifluoride, NF_3

Nitrogen trifluoride, NF_3, is a colorless gas. It is the most stable of the nitrogen halides. Let's draw the dot formula of its molecules.

Step 1. Calculate the total number of valence electrons:

$$
\begin{array}{lr}
\text{3 F atoms with 7 outer electrons} & = 21\ e^- \\
\text{1 N atom with 5 outer electrons} & = \underline{5\ e^-} \\
NF_3 & = 26\ e^-
\end{array}
$$

Step 2. Find the central atom, and write the other atoms around it. The central atom in NF_3 is N, since F forms only one bond and cannot be a central atom.

$$\underset{\displaystyle F}{F\ N\ F}$$

Step 3. Write electron pairs between N and F atoms, and calculate the number of electrons remaining.

$$\underset{\displaystyle \overset{\displaystyle ..}{F}}{F : N : F}$$

$$
\begin{aligned}
e^- \text{ remaining} &= \text{total number of } e^- - e^- \text{ written in dot formula} \\
&= 26 - 6 = 20e^-
\end{aligned}
$$

Step 4. Calculate the number of electrons needed to complete all octets, and place the remaining electrons in the dot formula. Each F requires six electrons to complete its octet, and N requires two electrons $(8 - 6 = 2)$. Hence, $6 \times 3 = 18$ are needed for the three F atoms; these plus the two for nitrogen gives 20 electrons—the number available. Place the 20 available electrons in the formula to complete the octets of all atoms.

$$\underset{\displaystyle :\overset{\displaystyle ..}{\underset{\displaystyle ..}{F}}:}{:\overset{\displaystyle ..}{\underset{\displaystyle ..}{F}} : \overset{\displaystyle ..}{N} : \overset{\displaystyle ..}{\underset{\displaystyle ..}{F}}:}$$

Step 5. Check the dot formula for the proper number of electrons. Each F atom has eight electrons, as does the N, and the total number of electrons equals 26; consequently, the dot formula is correct.

Covalent Bonding in Methane, CH_4

Methane, CH_4, commonly called marsh gas, is the main component of natural gas. Methane is produced from the decay of dead plants by certain species of microorganisms.

Following the five steps in writing methane's dot formula results in the following structure:

$$
\begin{array}{c}
\text{H} \\
\text{H} : \overset{..}{\underset{..}{\text{C}}} : \text{H} \\
\text{H}
\end{array}
$$

Verify the dot formula of methane by going through each step. Note that methane's dot formula shows that methane molecules have four single polar covalent bonds, and no lone pair electrons.

Covalent Bonding in Carbon Tetrachloride, CCl_4

Carbon tetrachloride, CCl_4, at one time was found in fire extinguishers as a noncombustible liquid to smother fires and also in cleaning fluids. However, research studies indicated that it was toxic and inhalation of its vapors caused liver damage. CCl_4 was also found to be carcinogenic (cancer producing). Carbon tetrachloride's dot formula is

$$
\begin{array}{c}
: \overset{..}{\underset{..}{\text{Cl}}} : \\
: \overset{..}{\underset{..}{\text{Cl}}} : \overset{..}{\underset{..}{\text{C}}} : \overset{..}{\underset{..}{\text{Cl}}} : \\
: \overset{..}{\underset{..}{\text{Cl}}} :
\end{array}
$$

Again verify CCl_4's dot formula by following each step. Note that the only difference in writing the dot formulas for CF_4, CBr_4, or CI_4 is the symbol for the halogen atom within the molecule.

REVIEW PROBLEMS

8.26 List the five steps that are followed in writing the correct dot formula for a covalent molecule.

8.27 Illustrate the above method for writing dot formulas using (a) OF_2, (b) NI_3, (c) CBr_4, (d) AsH_3, (e) SiH_4, (f) H_2Te.

8.6 OTHER COVALENT BONDING CONSIDERATIONS

If the number of electrons remaining after initially placing electron pairs in the formula is less than the number of electrons required to complete the octets of all atoms; then **multiple bonds** are located in the molecule. We shall use carbon dioxide as an example.

Covalent Bonding in Carbon Dioxide, CO_2

Carbon dioxide, CO_2, is a gas at room temperature. It is an important gas because it is a by-product of cellular respiration—a living organism's means for producing energy. Carbon dioxide is also a product of the combustion of materials containing carbon. Carbonated beverages contain dissolved CO_2; they become "flat" when the CO_2 escapes.

To write CO_2's dot formula, we start by finding the total number of outer electrons, four in C and 12 in the two O atoms, giving a total of 16 electrons.

Peroxides are compounds that contain an O—O single bond. Most peroxides are reactive.

Carbon is the central atom. In molecules that have more than one oxygen, the oxygen is usually not the central atom. Also, oxygen–oxygen single bonds are rather weak; rarely will you come across a compound with two oxygen atoms bonded together. Thus we write:

$$O \quad C \quad O$$

Placing two pairs or four electrons in the formula leaves 12 electrons. But to complete the octets of a carbon and two oxygens, we need 16 electrons—six for each oxygen and four for the carbon. The electron deficit is $16 - 12 = 4$ electrons, indicating that two double covalent bonds are located in CO_2. If we place two more electrons between each C and O, and write the remaining eight electrons around O, four each, the dot formula is then correct.

$$\ddot{\text{O}}::\text{C}::\ddot{\text{O}}: \quad \text{or} \quad :\ddot{\text{O}}=\text{C}=\ddot{\text{O}}:$$

In the second formula, we use two dashes to indicate a double covalent bond (four shared electrons).

Covalent Bonding in Formaldehyde, H₂CO

Formaldehyde, H_2CO, is another molecule with a multiple bond. Formaldehyde's central atom is C with two H and one O bonded to it. By following the five steps for writing dot formulas, we find that two less electrons are available than are required to complete C's octet—an indication of a double bond.

Formaldehyde's correct dot formula is:

A water solution of formaldehyde is called formalin. Formalin is used to preserve dead biological specimens and is sometimes used as an embalming fluid for humans.

$$\begin{matrix} \text{H}:\text{C}::\ddot{\text{O}} \\ \text{H} \end{matrix}$$

Verify this structure by going through each step.

Resonance

Some molecules do not have a unique dot formula. Instead, these molecules have two or more dot formulas that differ only with respect to the placement of the electrons. When this situation is encountered, we say that the molecule exhibits *resonance*.

Each possible dot formula is called a *contributing structure*. The arrangement of atoms in each contributing structure is the same; only the electrons are in different places. The actual arrangement of electrons in molecules exhibiting resonance most closely resembles the average of all contributing structures. To symbolize resonance, the contributing structures are separated by a double-headed arrow ($\longleftrightarrow$).

Only molecules containing multiple bonds can exhibit resonance.

Sulfur dioxide, SO_2, is an example of a molecule that exhibits resonance. Sulfur dioxide is a gas that is released into the atmosphere when sulfur-containing fossil fuels are burned. It contributes to the production of "acid rain." Following the above rules, we find two dot structures:

$$:\!S::\!\ddot{O}: \longleftrightarrow :\!S:\!\ddot{O}:$$
$$:\!\ddot{O}: \qquad\quad :\!\ddot{O}:$$
(I) (II)

If in the formation of a covalent bond one atom donates both electrons, as is the case in SO₂, the bond is classified as a coordinate covalent bond. After their formation, coordinate bonds have the same properties as other covalent bonds.

Contributing structures I and II differ with respect to the placement of the double bond in the dot formula. Both contributing structures suggest that one sulfur-oxygen bond is a single bond and the other is a double bond. However, analysis of this molecule indicates that both sulfur-oxygen bonds are exactly the same, intermediate between a double and single bond.

Sulfur dioxide's formula is best written as follows:

A dotted line indicates a partial bond, the case where electrons are spread out over the entire molecule as opposed to being localized between the nuclei of two atoms.

The actual structure of a molecule that exhibits resonance is considered a hybrid (a blend) of all contributing structures.

Molecules that exhibit resonance are more stable than similar molecules that do not exhibit resonance. Greater stability directly results from spreading electrons over more than two atoms.

Polyatomic Ions

A polyatomic ion is an ion composed of more than one atom. Polyatomic ions are held together by covalent bonds. Examples of common polyatomic ions include (1) hydroxide, OH^-; (2) nitrate, NO_3^-; (3) sulfate, SO_4^{2-}; (4) phosphate, PO_4^{3-}; and (5) ammonium, NH_4^+. Polyatomic ions are ionically bonded to a metal ion or another polyatomic ion.

Poly **is a prefix that means many.**

When writing dot formulas for polyatomic ions, you must add or subtract the number of electrons indicated by the ion's charge in Step 1. For instance, phosphate has a 3− charge or possesses three extra valence electrons in addition to the valence electrons of the phosphorus and oxygen atoms; thus, 32 valence electrons are found in phosphate.

To illustrate writing dot formulas of polyatomic ions, let's write the dot formula for the nitrate ion, NO_3^-.

The "extra" electrons in polyatomic ions come from metals and other polyatomic ions.

Step 1. Total electrons = $5e^-$ from N + $18\ e^-$ from O + $1\ e^-$ extra
$$= 24\ e^-$$

Step 2. N is the central atom and the oxygen atoms are bonded to it.

$$O$$
$$O\ N$$
$$O$$

Steps 3 and 4. After placing six electrons in the dot formula (24 − 6 = 18),

18 electrons remain. To complete the noble gas configurations of all atoms ($3 \times 6 = 18$ O $e^- + 2$ N $e^- = 20$ e^-), 20 electrons are needed. Since there are two less electrons than are necessary, a double bond is indicated. Thus, three contributing structures can be written for the nitrate ion, an ion that exhibits resonance.

$$
\left[\begin{array}{c} :\ddot{O}: \\ :\ddot{O}::N \\ :\ddot{O}: \end{array} \right]^- \longleftrightarrow \left[\begin{array}{c} :\ddot{O}: \\ :\ddot{O}:N \\ :\ddot{O}: \end{array} \right]^- \longleftrightarrow \left[\begin{array}{c} :O: \\ :\ddot{O}:N \\ :\ddot{O}: \end{array} \right]^-
$$

(I)   (II)   (III)

Step 5. Check to see that 24 electrons are found in the structure, including the extra electron, which makes nitrate a $1-$ ion.

    Dot formulas of polyatomic ions are enclosed in square brackets with their charge to indicate that these ions are not molecules: polyatomic ions cannot exist alone.

Molecules with Atoms That Do Not Achieve the Noble Gas Configuration

The set of rules presented are somewhat artificial generalizations that apply to simple molecular systems; there are molecules that don't follow these rules. Remember this is not a fault of the natural world, it is a limitation of using a system of artificial rules!

    Some molecules possess a bonded atom with less than eight electrons. Boron trifluoride, BF_3, is an example of such a molecule.

$$
\begin{array}{c}
:\ddot{F}: \\
\dot{B} \\
:F: \quad :F:
\end{array}
$$

Noble gas compounds were first synthesized by Neil Bartlett in 1962. It was once thought that noble gases could not form stable compounds. Bartlett's first noble gas compound was $XePtF_6$, xenon hexafluoroplatinate.

The dot formula of NO is

$$\cdot \dot{N} = \ddot{O}:$$

The dot formula of NO_2 is

$$:\ddot{O} - \dot{N} = \ddot{O}:$$

Is BF_3 a very stable molecule? No! Boron trifluoride is reactive and combines with almost any molecule that can donate an electron pair.

    Numerous molecules exist with a central atom that has more than eight valence electrons (frequently called an expanded octet). Only certain elements in the third period and beyond have the capacity to accommodate more than eight valence electrons after bonding. Some of these molecules have atoms with 10, 12, and 14 electrons. Noble gas compounds are good examples of molecules with expanded octets; examples include XeF_2, XeF_4, XeF_6, $XeOF_4$, and XeO_3.

    Still other molecules possess an atom with an odd number of electrons, which precludes the possibility of obtaining eight valence electrons. Attempt to draw the dot formulas of nitrogen monoxide, NO, or nitrogen dioxide, NO_2, and you will see that the total number of valence electrons is 11 and 17, respectively.

REVIEW PROBLEMS **8.28** Draw the electron dot formula of carbon monoxide, CO, a molecule with a multiple bond.

8.29 (a) Write a definition for resonance. (b) Illustrate resonance by drawing the electron dot formulas for the three contributing structures of sulfur trioxide, SO_3.

8.30 Draw the dot formulas for the polyatomic ions (a) OH^- and (b) PO_4^{3-}.

8.31 Which of the following molecules does not obey the octet rule? (a) BeF_2, (b) PF_3, (c) IF_3.

8.7 PROPERTIES OF MOLECULES

Molecular Geometry

Bonding electron pairs and lone pairs are negative regions of space.

A molecule's *molecular geometry* is its three-dimensional structure. More simply, molecular geometry refers to the average shape of a molecule. One way of predicting a molecule's shape is to consider the quantity of bonding electrons and lone pair electrons. Molecular shapes can be predicted by knowing that the atoms bonded to the central atom are most stable when they are as far apart as possible from each other, and from the lone pair electrons.

Molecular geometries that are commonly encountered in simple covalent molecules are (1) linear, (2) angular, (3) tetrahedral, (4) pyramidal, and (5) triangular or trigonal planar. Figure 8.12 illustrates these molecular geometries.

Table 8.5 summarizes molecular properties related to molecular geometry and gives examples of each.

Bond Properties

Four properties of bonds are commonly considered when studying molecules: (1) bond length, (2) bond order, (3) bond angle, and (4) bond energy.

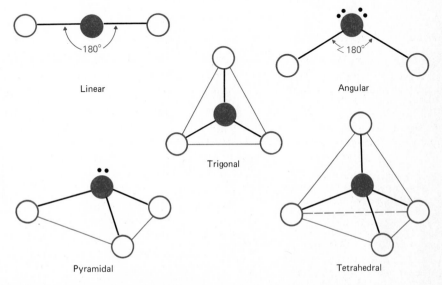

FIGURE 8.12
In linear molecules, all of the atoms are in a straight line (180° bond angle). In angular molecules, three atoms form an angle less than 180°. Trigonal planar molecules contain three atoms around the central atom (all in one plane). The three atoms are separated by 120° in trigonal planar molecules. Pyramidal molecules have three bonded atoms and one lone pair around the central atom. Tetrahedral molecules have four atoms bonded to a central atom separated by an angle of 109.5°.

Linear

Angular

Trigonal

Pyramidal

Tetrahedral

TABLE 8.5   MOLECULAR GEOMETRIES

No. of e^- pairs on central atom	No. of bonds	No. of lone pairs	Geometry	Examples
2	2	0	Linear	BeH_2, $BeCl_2$
3	3	0	Trigonal planar	BF_3, $AlCl_3$
4	4	0	Tetrahedral	CH_4, SiF_4
4	3	1	Pyramidal	NH_3, PF_3
4	2	2	Angular	H_2O, H_2S

C——C
C=C
C≡C

A *bond length* is the average distance between two nuclei of atoms in a covalent bond. Carbon–carbon single bonds are approximately 154 pm (picometers, 10^{-12}m). Carbon–carbon double and triple bonds are even shorter, 133 pm and 120 pm, respectively.

Indirectly related to bond length is bond order. *Bond order* is the number of covalent bonds that link two atoms. A single covalent bond has a bond order of 1, a double bond has a bond order of 2, and a triple bond has a bond order of 3. In resonance structures, fractional bond orders (1.5, 1.33, 1.23, etc.) are found.

Bond order	Type of bond
1	Single
2	Double
3	Triple

A *bond angle* is the angle between two imaginary lines passing through the nucleus of the central atom and the nuclei of two atoms bonded to it. For instance, the bond angle (H—O—H) between hydrogen atoms in water is 104.5 degrees, and the bond angle between hydrogen atoms in methane, CH_4, is 109.5 degrees (see Figure 8.13).

We have already mentioned bond energies as being the amount of energy required to break one mole of a specific bond in the gas phase. Bond energies are related to how stable a bond is. In general, bond energies for triple bonds are higher than for double bonds, which are higher than those for single bonds. Carbon–carbon triple bonds have bond energies in excess of 800 kJ/mol. Whereas most carbon–carbon double-bond energies are near 600 kJ/mol, only 350 kJ/mol is required to break carbon–carbon single bonds.

FIGURE 8.13
(*a*) Water is composed of angular molecules that contain two bonding pairs and two lone pairs on the oxygen atom. The bond angle in a water molecule is 104.5°. (*b*) Methane is composed of tetrahedral molecules. The bond angle between each pair of hydrogen atoms in methane is 109.5°.

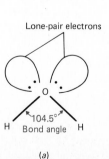

(a)

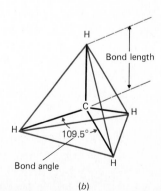

(b)

REVIEW PROBLEMS
   8.32 What are the most common molecular geometries exhibited by simple molecules?

   8.33 Compare single, double, and triple bonds with respect to (a) bond order, (b) bond length, (c) bond energy.

   8.34 What is a bond angle? Give an illustration.

SUMMARY

Compounds are divided into two groups: (1) ionic, and (2) covalent. Ionic compounds form when a metal combines with a nonmetal or polyatomic ion. Covalent compounds result when two or more nonmetals combine. Ionic compounds are named by writing the name of the metal followed by the nonmetal, dropping the nonmetal's ending, and adding *ide*. Covalent compounds are named in a similar fashion, except that a prefix is added to indicate the total number of each atom bonded in the molecule.

Ionic compounds have properties that differ from covalent compounds. Generally, ionic compounds have higher melting points, higher boiling points, and greater densities than covalent compounds. All ionic compounds are solids at room temperature, while covalent compounds are more frequently liquids and gases. Both groups are nonconductors of electricity, although ionic compounds do conduct an electric current after melting.

A chemical bond is the force of attraction that results when atoms either transfer or share electrons. If atoms transfer electrons in the formation of a chemical bond, the bond is classified as an ionic bond. Covalent bonds results when atoms share electrons.

After one or more electrons are transferred from a metal to a nonmetal, two ions result. The metal atom loses electrons, becoming a cation, and the nonmetal atom gains electrons and becomes an anion. Both ions achieve the stable noble gas configuration. The unlike charged ions attract, creating an ionic bond.

Covalent compounds share electrons through overlapping valence orbitals in order to attain a stable noble gas electronic configuration. A covalent bond in which the electrons are equally shared is called a nonpolar bond. Sometimes, electrons are unequally shared; this unequal sharing is called a polar covalent bond.

Electronegativity is a measure of an atom's ability to attract electrons in a covalent bond. When two elements with the same electronegativity bond together, a nonpolar covalent bond is produced. Atoms that combine covalently with atoms of different electronegativities yield polar bonds.

To identify the bonds in a molecule, we write the dot structure. A dot structure accounts for all valence electrons in the molecule. Usually, the correct dot structure is the one whereby all atoms achieve the noble gas configuration. A series of steps are required to successfully write dot formulas for molecules.

Covalent molecules exhibit a variety of covalent bonds: single, double, and triple bonds. A single covalent bond has one shared electron pair. Double bonds have two pairs of shared electrons; triple bonds have three electron pairs shared. Molecules can have any combination of single, double, or triple covalent bonds.

When studying molecules, chemists are concerned with bond lengths, bond orders, bond energies, and bond angles. Bond length is the distance between nuclei in a covalent bond. Bond order indicates if the bond is a single, double or triple bond. Bond energy is the amount of energy required to break one mole of bonds in the gas phase. Lastly, bond angles are the angles between two atoms bonded to the same central atom.

QUESTIONS AND PROBLEMS

8.35 Define the following: ionic compound, covalent compound, chemical bond, stability, valence electron, electrovalent bond, ionic radius, crystal lattice, electronegativity, nonpolar covalent bond, polar covalent bond, multiple bond, resonance, contributing structure, polyatomic ion, bond order, bond length, bond angle, molecular geometry.

Compounds

8.36 How are the rules for naming ionic compounds different from those for naming covalent compounds?

8.37 Name the following ionic compounds: (a) LiBr, (b) MgS, (c) CsBr, (d) Ca_3N_2, (e) $BaCl_2$, (f) PbS, (g) AgF, (h) AlP.

8.38 How do the properties of ionic compounds differ from those of covalent compounds? Be specific.

8.39 What quantity of atoms do the following prefixes refer to: (a) tetra, (b) hexa, (c) di, (d) octa, (e) tri?

8.40 Write the name for the following covalent compounds: (a) N_2O, (b) PF_5, (c) N_2O_4, (d) XeF_2, (e) S_2Cl_2, (f) AsF_5, (g) IF_3, (h) $SeBr_4$.

8.41 Name the following compounds: (a) I_2O_4, (b) K_2S, (c) Cl_2O_7, (d) I_4O_9, (e) Al_2S_3, (f) $CaCl_2$, (g) Te_2F_{10}, (h) SeF_6.

8.42 On an average, predict whether an ionic or a covalent compound would have a higher (a) melting point, (b) boiling point, (c) density.

8.43 Explain why ionic substances conduct an electric current in the molten state but are nonconductors when solids.

Chemical Bonding

8.44 Explain why nitrogen gas is composed of diatomic nitrogen molecules instead of individual nitrogen atoms.

8.45 How do chemists measure the stability of a chemical bond? Give an example.

8.46 What is incorrect about the following statement? The bond that joins Na^+ and Cl^- ions is purely ionic.

8.47 What accounts for the fact that diatomic noble gas compounds, such as Ne_2, Ar_2, or Xe_2, are not found at room conditions?

Ionic Bonding

8.48 What noble gas is isoelectronic with each of the following ions: (a) Ba^{2+}, (b) P^{3-}, (c) Cl^-, (d) Cs^+, (e) Se^{2-}, (f) Li^+, (g) N^{3-}, (h) Te^{2-}?

8.49 Write the complete electronic configuration for each of the following ions: (a) Mg^{2+}, (b) N^{3-}, (c) S^{2-}, (d) K^+, (e) Br^-, (f) Sr^{2+}.

8.50 Select the ion with the largest ionic radius from the following sets:
(a) O^{2-}, S^{2-}, and Se^{2-}
(b) N^{3-}, O^{2-}, and F^-
(c) Na^+, Mg^{2+}, and Al^{3+}
(d) Na^+, Mg^{2+}, K^+, and Ca^{2+}

8.51 Use electron dot formulas to illustrate the electron transfer that occurs when the following ionic substances form: (a) Na_2S, (b) MgF_2, (c) Al_2O_3.

8.52 Draw the electron dot formula for each of the following ionic compounds: (a) KCl, (b) $CaCl_2$, (c) SrS, (d) Rb_2O, (e) Mg_3N_2.

8.53 What is the formula of the ionic compound that results from combinations of the following metals and nonmetals: (a) aluminum-nitrogen, (b) calcium-sulfur, (c) magnesium-iodine, (d) gallium-oxygen, (e) potassium-bromine, (f) lithium-oxygen?

8.54 What is a crystal lattice? Give examples.

Covalent Bonding

8.55 What two trends are observed in electronegativities of atoms?

8.56 Why are electronegativity values generally not assigned to noble gases?

8.57 From each group, select the element with the lowest electronegativity:
(a) F, Cl, and Br
(b) Ge, As, Sn, and Sb
(c) Rb, Cs, and Fr

8.58 Draw a diagram indicating the overlapping orbitals in H_2.

8.59 Explain how the sharing of electrons leads to bonding atoms together.

8.60 In what electron sublevels does sharing occur in each of the following covalent molecules: (a) HBr, (b) IF, (c) BrCl.

8.61 Label each diatomic molecule as polar or nonpolar: (a) O_2, (b) CO, (c) NO, (d) I_2, (e) ICl, (f) F_2.

8.62 Write the dot formula for ClF, and indicate the partial positive and negative charges located in the molecule.

8.63 Follow each step of the suggested procedure for writing dot formulas to write the dot structures of (a) CF_4, (b) OCl_2, (c) H_2Se, (d) PBr_3, (e) SF_2, (f) N_2Cl_4, (g) $C_2H_2Br_2$, (h) Si_2Cl_6, (i) H_2O_2, (j) H_4SiO_4.

8.64 Write the dot formula for the following ions: (a) OH^-, (b) CN^-, (c) NH_4^+, (d) BH_4^-, (e) ClO_4^-, (f) O_2^{2-}.

8.65 Draw the dot formulas for three different compounds containing two carbons: (a) C_2H_6, (b) C_2H_4, and (c) C_2H_2.

8.66 Draw the dot formula for the following acids (H atoms are bonded to O atoms):
(a) Carbonic acid, H_2CO_3
(b) Sulfuric acid, H_2SO_4
(c) Phosphoric acid, H_3PO_4
(d) Nitric acid, HNO_3

8.67 Give an example of a molecule containing: (a) a single and double bond, (b) two double bonds, (c) a triple bond, (d) a triple and single bond, and (e) an atom with fewer than eight valence electrons.

8.68 Which of the following molecules contain atoms with expanded octets: (a) CF_4, (b) PF_5, (c) IF_5, (d) $SnBr_4$?

8.69 Draw all contributing structures for (a) HCO_3^-, (b) CO_3^{2-}, (c) NO_2^-, (d) O_3, and (e) N_2O.

8.70 Use Table 8.5 to predict the shape of the following molecules: (a) NF_3, (b) H_2Se, (c) SiH_4, (d) AlF_3, (e) BeF_2.

8.71 Explain why triple bonds are shorter than double bonds. Consider the number of electrons in the bonds.

8.72 What is the bond order of the bonds contained in the following diatomic molecules: (a) N_2, (b) F_2, (c) O_2, (d) HBr?

8.73 What could account for the fact that the average bond length of a Si—O bond, 166 pm, is longer than a C—O bond, 143 pm.

General Questions

8.74 Correct the following incorrect statements:
(a) In covalent bonds, electrons are always equally shared.
(b) Electronegativity is a measure of the force of attraction between a nucleus and its electrons.
(c) Average bond energies are larger for single covalent than for double covalent bonds.
(d) Ionic compounds result when two nonmetals combine.
(e) The name of the compound SO_3 is sulfur oxide.
(f) Calcium chloride molecules have the formula $CaCl_2$.
(g) Hydrogen's electron is transferred to fluorine in hydrogen fluoride.
(h) All nonpolar covalent bonds result from the same atoms bonding together.

8.75 (a) Acetic acid, $C_2H_4O_2$, contains a carbon-carbon single bond, and both oxygens are bonded to one carbon; determine the correct dot formula for acetic acid.
(b) Draw two more correct dot formulas of compounds with the molecular formula of $C_2H_4O_2$.

8.76 Propane, C_3H_8, is a fuel used to heat houses. Determine the dot formula for propane.

8.77 One form of phosphorus exists as P_4 molecules. Write two different dot formulas for P_4.

8.78 Write dot formulas for molecules described by the following:
(a) Is diatomic with a bond order of 3
(b) Has tetrahedral geometry, but does not contain C or Si
(c) Has a single, a double, and a triple bond
(d) Has two oxygen atoms bonded together
(e) Has both ionic and covalent bonds within the molecule
(f) Is a noble gas compound
(g) Has B, F, N, and H atoms

Chemical Nomenclature

Study Guidelines

After completing Chapter 9, you should be able to:

1. State the basic rules used to assign oxidation numbers for elements
2. Assign oxidation numbers for elements in compounds
3. Write the formulas or names of ionic binary compounds containing metals with either fixed or variable oxidation states
4. Write the formulas or names of covalent binary compounds
5. Identify the principal oxidation states, and name frequently encountered metal ions
6. Distinguish a binary compound from a ternary compound
7. Write the name and formulas for at least 12 polyatomic ions
8. Write names and formulas of ionic ternary compounds
9. Write names and formulas of oxyacids
10. Distinguish between binary and ternary acids
11. Name ternary acids that contain nonmetals that exist in variable oxidation states
12. Write the name of an oxyanion, given the acid that it is derived from, and vice versa

Our chemical nomenclature system must provide names for more substances than there are words in the English language.

Chemical nomenclature is a system used by chemists to name and write formulas of compounds. A method is needed to name the vast quantity of compounds that are known, plus the new ones that are synthesized each day. It is unfortunate that even though a systematic procedure is available to name compounds, many retain common or trivial names. Table 9.6 lists the common and systematic names of a number of compounds (see p. 204).

Our standards for naming compounds are established by the International Union of Pure and Applied Chemistry, IUPAC for short. Because of this international organization, chemists from all countries should, at least in theory, assign the same name for a newly discovered compound—which should diminish the confusion that reigned in the past.

IUPAC adopted the Stock system of nomenclature for naming inorganic compounds. We shall learn a segment of the Stock system along with various common names. Before we can get to the specifics of nomenclature, we need to learn a method for keeping track of the apparent charge, called the oxidation number, that an atom has within a compound.

9.1 OXIDATION NUMBERS (OXIDATION STATES)

Oxidation numbers are used to (1) write formulas, (2) predict properties, and (3) balance oxidation-reduction equations.

Oxidation numbers, sometimes called oxidation states, are signed numbers that are assigned to atoms within molecules. They allow us to keep track of the electrons that are associated with each atom. Oxidation numbers are frequently used to write chemical formulas, to predict properties of compounds, and to help balance equations where electrons are transferred (oxidation-reduction reactions, discussed in Chapter 18).

Knowing the oxidation state of an atom gives us an idea about the atom's positive or negative character. In themselves, oxidation numbers have no physical meaning; they are employed to simplify tasks that are more difficult to accomplish without them.

Oxidation numbers are assigned by following a set of general rules:

Rule 1. All uncombined elements (as they exist naturally) are assigned the oxidation number of zero.

Rule 2. All monatomic ions are assigned oxidation numbers equal to their charges.

Rule 3. Certain elements usually possess a fixed oxidation number in compounds. Those that are most important include:

1. Oxygen's oxidation number is usually -2.
2. Hydrogen's oxidation number is usually $+1$.
3. Halogens normally have a -1 oxidation number in binary compounds.
4. Alkali metals and alkaline earth metals are assigned $+1$ and $+2$, respectively, as their oxidation numbers.

Oxidation numbers are the result of the stated rules, nothing more.

Rule 4. The sum of all oxidation numbers in a compound equals zero, and the sum of oxidation numbers in a polyatomic ion equals its charge.

Let's now look at each rule individually, starting with Rule 1. All uncombined elements are assigned the oxidation number of zero, regardless of how they exist in nature—by themselves, diatomically, or in larger aggregates (P_4 and S_8). It is common practice to write oxidation numbers above the symbols of the elements.

Oxidation numbers are written with the sign preceding the number, i.e., +1, −1, +2, etc. Charges on ions are written with the sign after the number.

$$\overset{0}{\text{Na}} \quad \overset{0}{\text{Fe}} \quad \overset{0}{\text{H}_2} \quad \overset{0}{\text{O}_2} \quad \overset{0}{\text{F}_2} \quad \overset{0}{\text{Ne}} \quad \overset{0}{\text{P}_4} \quad \longleftarrow \text{Oxidation numbers}$$

Rule 2 states that all monatomic ions are assigned an oxidation number equal to their charge:

An ion's charge and oxidation number are exactly the same.

$$\overset{-1}{\text{Cl}^-} \quad \overset{+3}{\text{Al}^{3+}} \quad \overset{+2}{\text{Cu}^{2+}} \quad \overset{-2}{\text{O}^{2-}} \quad \overset{+1}{\text{K}^+} \quad \overset{+2}{\text{Mg}^{2+}}$$

In Rule 3, note that the oxidation numbers correspond to the number of electrons that an atom loses or gains within a binary ionic compound. Halogens gain one electron, chalcogens gain two electrons, alkali metals lose one electron, and alkaline earth metals lose two electrons.

If oxygen is bonded to fluorine, the oxygen is in the +2 oxidation state. This is a result of fluorine's higher electronegativity.

There are some exceptions to Rule 3. For instance, in peroxides, the oxidation number of oxygen is -1, and not -2. Peroxides are compounds that contain an O—O single bond. Two examples of peroxides are hydrogen peroxide, H_2O_2, and sodium peroxide, Na_2O_2. Metallic hydrides, compounds containing hydrogen bonded to a metal with a smaller electronegativity, are another exception to Rule 3. Hydrogen's oxidation number is -1 in these compounds. Examples of metallic hydrides include sodium hydride, NaH, and calcium hydride, CaH_2.

By applying Rule 4, oxidation numbers of other elements are obtained. For example, what is the oxidation number of N in NO? Oxygen's oxidation number is -2, and the sum of NO's oxidation numbers equals zero. Thus N's oxidation number is $+2$:

To keep track of oxidation numbers, write the known oxidation states for individual atoms below the formula and the total oxidation number for all atoms of that type above the symbol.

$$\text{Total oxidation numbers} \longrightarrow \quad \overset{? \; -2 \; = \; 0}{\underset{-2}{\text{NO}}} \qquad \overset{+2-2 \; = \; 0}{\underset{-2}{\text{NO}}}$$
$$\text{Individual oxidation numbers} \longrightarrow$$

What is the oxidation number of S in SO_2? Again following Rule 4, we assign -4 to the two O atoms; hence, the oxidation number of S must be $+4$ in order for the sum to equal zero.

$$\overset{? \; -4 \; = \; 0}{\underset{-2}{\text{SO}_2}} \qquad \overset{+4 \; -4 \; = \; 0}{\underset{-2}{\text{S O}_2}}$$

In a polyatomic ion the reasoning is the same except that the sum of the oxidation numbers is equal to the ion's charge. What is the oxidation number of P in PO_4^{3-}? Four oxygens would have a total of -8, so P's oxidation state is $+5$ to add up to -3.

$$\overset{?-8\,=\,-3}{\underset{-2}{PO_4^{3-}}} \qquad \overset{+5-8\,=\,-3}{\underset{-2}{PO_4^{3-}}}$$

Additional illustrations for assigning oxidation numbers are given in Example Problem 9.1.

Example Problem 9.1

Determine the oxidation state for each element in the following: (a) CuF_2, (b) HNO_3, (c) SO_4^{2-}, (d) $C_{12}H_{22}O_{11}$.

Solution

(a) CuF_2. Each fluorine atom has an oxidation number equal to -1; two fluorines have a total of -2; it follows that Cu's oxidation number is $+2$.

$$\overset{+2\ -2}{\underset{-1}{CuF_2}}$$

(b) HNO_3. H's oxidation number is $+1$, and three O's have a total of -6; therefore, N's oxidation number is $+5$ to yield a total of zero.

$$\overset{+1+5\ -6}{\underset{+1\quad -2}{HNO_3}}$$

(c) SO_4^{2-}. Each oxygen's oxidation number is -2, so four O's have a total of 8. To add up to -2, the charge on SO_4^{2-}, the oxidation number of S is $+6$.

$$\overset{+6\ -8}{\underset{-2}{SO_4^{2-}}}$$

(d) $C_{12}H_{22}O_{11}$. Twenty-two H's give $+22$, and 11 O's give -22, so the 12 C's have a combined oxidation number of zero—each C has an oxidation state of zero.

$$\overset{0\quad +22\ -22}{\underset{+1\quad -2}{C_{12}H_{22}O_{11}}}$$

Group

IA	IIA	IIIB	IVB	VB	VIB	VIIB	←——VIIIB——→			IB	IIB	IIIA	IVA	VA	VIA	VIIA	VIIIA
1 H +1																	2 He
3 Li +1	4 Be +2											5 B +3	6 C ±4,+2	7 N ±3,5,4,2	8 O −2,1	9 F −1	10 Ne
11 Na +1	12 Mg +2											13 Al +3	14 Si +4	15 P ±3,5,4	16 S ±2,4,6	17 Cl ±1,3,5,7	18 Ar
19 K +1	20 Ca +2	21 Sc +3	22 Ti +4,3	23 V +5,4,3,2	24 Cr +6,3,2	25 Mn +7,6,4,3,2	26 Fe +2,3	27 Co +2,3	28 Ni +2,3	29 Cu +1,2	30 Zn +2	31 Ga +3	32 Ge +4	33 As ±3,5	34 Se −2,4,6	35 Br ±1,5	36 Kr
37 Rb +1	38 Sr +2	39 Y +3	40 Zr +4	41 Nb +5,3	42 Mo +6,5,4,3,2	43 Tc +7	44 Ru +8,6,4,3,2	45 Rh +4,3,2	46 Pd +2,4	47 Ag +1	48 Cd +2	49 In +3	50 Sn +4,2	51 Sb ±3,5	52 Te −2,4,6	53 I ±1,5,7	54 Xe
55 Cs +1	56 Ba +2	57 La +3	72 Hf +4	73 Ta +5	74 W +6,5,4,3,2	75 Re +7,6,4,2	76 Os +8,6,4,3,2	77 Ir +6,4,3,2	78 Pt +2,4	79 Au +1,3	80 Hg +2,1	81 Tl +1,3	82 Pb +4,2	83 Bi +3,5	84 Po +2,4	85 At ±1,3,5,7	86 Rn
87 Fr +1	88 Ra +2	89 Ac +3	104 Unq	105 Unp	106	107	108										

58 Ce +3,4	59 Pr +3,4	60 Nd +3	61 Pm +3	62 Sm +2,3	63 Eu +2,3	64 Gd +3	65 Tb +3,4	66 Dy +3	67 Ho +3	68 Er +3	69 Tm +2,3	70 Yb +2,3	71 Lu +3

90 Th +4	91 Pa +4,5	92 U +6,5,4,3	93 Np +6,5,4,3	94 Pu +6,5,4,3

FIGURE 9.1

The most common oxidation states of each element are listed under its symbol.

A large percentage of elements exist in a variety of oxidation states.

With the exception of the metals in groups IA, IIA, and IIIB, metals generally exist in more than one oxidation state. Chromium, Cr, for example, is found in the +6, +3, and +2 oxidation states. Gold is found in both +3 and +1 oxidation states. Nonmetals, except F, also exhibit a range of oxidation states. Sulfur's oxidation states include +6, +4, +2, and −2. Figure 9.1 shows the common oxidation states of elements.

REVIEW PROBLEMS

9.1 What is chemical nomenclature?

9.2 (a) What does the oxidation number of an element within a compound represent? (b) Is the oxidation number for an element exactly the same as its charge in all cases? Explain.

9.3 List the four rules that are employed to assign oxidation numbers to elements in compounds.

9.4 Determine the oxidation numbers for all elements in each compound: (a) NH_3, (b) MnO_2, (c) $AlCl_3$, (d) N_2O, (e) P_2O_5.

9.5 Find the oxidation state for all nonoxygen elements in each of the following polyatomic ions: (a) $C_2O_4^{2-}$, (b) NO_2^-, (c) ClO_3^-, (d) CO_3^{2-}.

9.2 NAMING AND WRITING FORMULAS OF BINARY COMPOUNDS

Binary compounds are those that contain two different elements. Ternary compounds contain three different elements.

Binary compounds are those that contain two different elements. **Ternary compounds,** discussed in the next section, contain three different elements. Within each of these two groups there are various categories of compounds that have

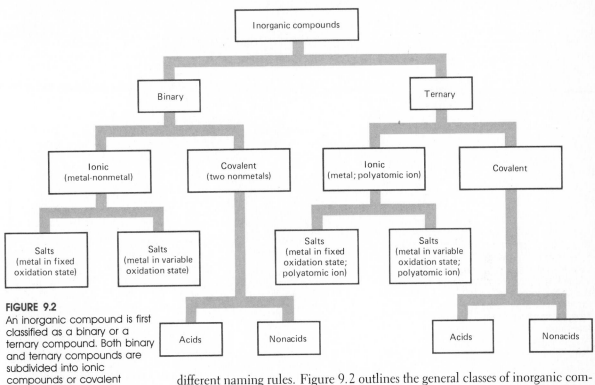

FIGURE 9.2
An inorganic compound is first classified as a binary or a ternary compound. Both binary and ternary compounds are subdivided into ionic compounds or covalent compounds. Ionic compounds contain either a metal that exists in a fixed oxidation state or a metal that exists in a variable oxidation state. Covalent substances are divided into acids and nonacids.

different naming rules. Figure 9.2 outlines the general classes of inorganic compounds.

When naming a compound, first identify what class of compound is being named. Using the flowchart in Figure 9.3, decide if the compound is binary or ternary. Then decide if the compound is ionic or covalent. Ionic compounds are those that contain a metal and a nonmetal or a metal and a polyatomic ion bonded together. Covalent compounds are those that have two nonmetals bonded together or a nonmetal bonded to a polyatomic ion.

Binary Ionic Compounds Containing Metals with Fixed Oxidation Numbers

In Chapter 8, we learned the rules for naming simple binary ionic compounds, i.e., those that have a metal ion with a fixed oxidation number and a nonmetal ion. To review, first state the metal's name, and then write the nonmetal's name, removing its ending and adding the suffix *ide*.

$$\text{Name} = \text{metal} + (\text{nonmetal} - \text{ending} + ide)$$

If you have forgotten how to name these compounds, refer back to Section 8.1.

Binary Ionic Compounds Containing Metals with Variable Oxidation Numbers

What if the metal ion component of a binary ionic substance has a variable oxidation state? We mentioned that metals can exist in more than one oxidation state; consequently, more than one binary compound can result when such a metal combines with a nonmetal. For example, two chlorides of copper are known, $CuCl$ and $CuCl_2$. Our naming system must provide a means for distinguishing between the two. In the Stock system, we modify the above rules for

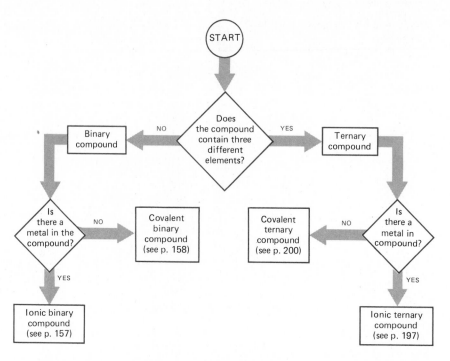

FIGURE 9.3
To use this nomenclature flowchart, begin at the top and answer each question concerning the compound of interest. When you answer either Yes or No, the chart poses another question or gives the type of compound with a page reference.

naming ionic substances to include the oxidation state of the metal ion.

Specifically, we write the full name of the metal followed by its oxidation number expressed in roman numerals in parentheses:

$$\text{Metal name(oxidation number)}$$

The left parenthesis is written next to the last letter in the metal's name, and the oxidation state is written as a roman numeral, followed by the right parenthesis. Accordingly, the two oxidation states of copper are expressed as follows:

$$Cu^+ = \text{copper(I)}$$
$$Cu^{2+} = \text{copper(II)}$$

To write the complete names of $CuCl$ and $CuCl_2$, it is only necessary to append the modified name of the nonmetal (nonmetal − ending + *ide*), chloride. Therefore, $CuCl$ is copper(I) chloride, and $CuCl_2$ is copper(II) chloride. Note that a space is placed between the right parenthesis and the nonmetal name, but no space is written between the left parenthesis and the end of the metal name. When reading names expressed this way, state the metal name followed by the number—"copper-one" and "copper-two."

The old system of nomenclature handles this situation in a different manner. To name the chlorides of copper, two endings, *ic* and *ous*, are added to the classic name of copper, *cuprum*. The ending *ic* designates the higher oxidation

Modern and old names for the elements:

Modern	Old
Antimony	Stibium
Copper	Cuprum
Gold	Aurum
Iron	Ferrum
Lead	Plumbum
Silver	Argentum
Tin	Stannum

state, while *ous* identifies the lower oxidation state of the metal. For CuCl, where Cu is in the +1 state, we attach *ous* to cuprum − *um*, giving cuprous:

$$Cu^+ = cuprum - um + ous = cuprous$$

In a similar manner, *ic* is affixed to cuprum − *um*, yielding cupric:

$$Cu^{2+} = cuprum - um + ic = cupric$$

In the old system, the names of the chlorides of Cu are cuprous chloride, CuCl, and cupric chloride, $CuCl_2$.

Naming binary compounds using the Stock system only requires knowledge of the oxidation states of the elements in the compound. Additional knowledge of the old names for metals is required when using the old system. Table 9.1 lists both the old and new names of frequently encountered metal ions.

Refer to Table 9.1 initially when writing the names of metals with more than one oxidation state. After some practice, you will be able to remember them with the aid of a periodic table. Example Problem 9.2 illustrates the naming of compounds with metals that have variable oxidation states.

Example Problem 9.2

Write names for the following compounds using the stock system and the old system: (a) HgI_2, (b) PbO_2, (c) SnF_2, (d) FeO.

Mercury(I) is a diatomic cation, i.e., it exists in pairs—Hg_2^{2+}. Whenever mercury(I) is written, it indicates a pair of mercury ions.

TABLE 9.1 **NAMES OF METALS WITH MULTIPLE OXIDATION STATES**

Metal	Oxidation state	Stock name	Old name
Copper	+1	Copper(I)	Cuprous
	+2	Copper(II)	Cupric
Iron	+2	Iron(II)	Ferrous
	+3	Iron(III)	Ferric
Mercury	+1	Mercury(I)	Mercurous
	+2	Mercury(II)	Mercuric
Lead	+2	Lead(II)	Plumbous
	+4	Lead(IV)	Plumbic
Chromium	+2	Chromium(II)	Chromous
	+3	Chromium(III)	Chromic
Manganese	+2	Manganese(II)	Manganous
	+3	Manganese(III)	Manganic
Cobalt	+2	Cobalt(II)	Cobaltous
	+3	Cobalt(III)	Cobaltic
Tin	+2	Tin(II)	Stannous
	+4	Tin(IV)	Stannic
Titanium	+3	Titanium(III)	Titanous
	+4	Titanium(IV)	Titanic

Solution

(a) HgI_2. Determine the oxidation state of Hg in HgI_2, using the rules for assigning oxidation numbers. Iodine is a halogen with an oxidation number of -1 in binary compounds; two I's have a total of -2. Consequently, Hg's oxidation number is $+2$ (the higher oxidation state of Hg), which is mercury(II) in the new system and mercuric in the old system. Iodine's name is changed by removing the *ine* and adding *ide*.

$$\text{Stock name} = \text{mercury(II) iodide}$$

$$\text{Old name} = \text{mercuric iodide}$$

(b) PbO_2. The total oxidation number for two oxygens is -4. In order to add up to zero, Pb's oxidation state is $+4$. The $+4$ oxidation state of lead is the higher one, necessitating that *ic* is placed at the end of its old name, plumbum.

$$\text{Stock name} = \text{lead(IV) oxide}$$

$$\text{Old name} = \text{plumbic oxide}$$

Various compounds can contain an element that exists in two different oxidation states. Fe_3O_4 is a combination of FeO and Fe_2O_3; hence, it would be best named iron(II, III) oxide.

(c) SnF_2. Following the general rules, we find that Sn is in the $+2$ oxidation state, the lower oxidation state of tin. Consequently, the names for SnF_2 are:

$$\text{Stock name} = \text{tin(II) fluoride}$$

$$\text{Old name} = \text{stannous fluoride}$$

(d) FeO. Oxygen's oxidation number is -2; therefore, the oxidation number for iron is $+2$—iron(II) or ferrous.

$$\text{Stock name} = \text{iron(II) oxide}$$

$$\text{Old name} = \text{ferrous oxide}$$

Naming Covalent Binary Compounds

In Section 8.1, we discussed the rules for naming covalent binary compounds. To review, the nonmetal in the positive oxidation state (more metallic) is written first unchanged, followed by the nonmetal in a negative oxidation state (more nonmetallic) with its ending dropped and the *ide* ending attached. An added prefix is attached to indicate the number of atoms of each nonmetal in the formula.

Binary Acids

Various covalent binary hydrogen compounds are classified as acids. When dissolved, **acids** ionize and add hydrogen ions, H^+, and anions to water. In the following equation, hypothetical acid HA breaks up into H^+ and A^- (anion):

An acid donates H$^+$ to water when it dissolves.

$$HA \xrightarrow{\text{water}} H^+(aq) + A^-(aq)$$

To name binary acids, the prefix *hydro* is added to the nonmetal name, the ending is dropped, and *ic acid* is attached in its place.

When a substance is dissolved in water, (aq) is written next to its name to indicate that it is in aqueous solution.

$$\text{Acid name} = hydro + \text{nonmetal} - \text{ending} + ic\ acid$$

Examples of binary acids include hydrochloric acid, $HCl(aq)$; hydrosulfuric acid, $H_2S(aq)$; hydrofluoric acid, $HF(aq)$; and hydrobromic acid, $HBr(aq)$

Writing Formulas of Binary Compounds

Writing the formula of a binary compound, given its name, first requires you to determine if the compound is ionic or covalent. If it is covalent, the actual number of atoms is specified in the formula. Therefore, you write the formula with the subscripts indicated by the prefixes, for example, carbon disulfide, CS_2; dinitrogen tetroxide, N_2O_4; and carbon tetrabromide, CBr_4.

To write the formula of an ionic compound, write each ion with its correct oxidation state and then determine the quantity of each ion that produces a zero oxidation state for the compound.

The sum of all oxidation numbers in a compound is zero.

What is the formula of chromium(III) oxide? Given the Stock formula, the oxidation state of the metal is readily obtained. Oxygen belongs to group VIA and has a -2 oxidation number in binary compounds. Write both elements with their oxidation numbers:

$$\overset{+3}{Cr} \quad \overset{-2}{O}$$

Since the sum of all oxidation numbers adds up to zero, ask yourself what number of Cr and O atoms would give the same total oxidation numbers. This is easily accomplished by finding the lowest common multiple for the two oxidation numbers; for Cr and O the lowest common multiple is 6. Then, divide each individual oxidation number into the lowest common multiple to obtain the correct formula subscript.

$$\text{Lowest common multiple} = 6$$

$$\overset{+3}{Cr} \quad \overset{-2}{O}$$
$$\tfrac{6}{3} = 2 \qquad \tfrac{6}{2} = 3$$

Thus two Cr atoms and three O atoms are found in chromium(III) oxide.

$$Cr_2O_3$$

Example Problem 9.3 gives additional examples of writing formulas of compounds given their names.

Example Problem 9.3

Write the formulas for each of the following: (a) cobaltous oxide, (b) lead(II) chloride, (c) titanium(III) sulfide.

Solution

(a) Cobaltous oxide.   Referring to Table 9.1, we find that cobaltous is Co^{2+}, and oxygen has an oxidation number of -2; thus, the formula of cobaltous oxide is CoO. Co and O are found in a 1 to 1 ratio in cobaltous oxide.

$$CoO$$

(b) Lead(II) chloride.   Lead(II) is Pb^{2+}, and chlorine is a halogen with an oxidation number of -1. Two chloride ions are needed to balance the Pb^{2+}; hence, lead(II) chloride's formula is $PbCl_2$.

$$PbCl_2$$

(c) Titanium(III) sulfide.   The oxidation number for titanium(III) is $+3$, and that for sulfur is -2; the lowest common multiple is 6. Thus, two titanium atoms and three sulfur atoms are needed to give a zero oxidation state for titanium(III) sulfide.

$$Ti_2S_3$$

REVIEW PROBLEMS

9.6 Write the old name that corresponds to the following Stock names for metal ions:
(a) Copper(II)
(b) Manganese(III)
(c) Lead(II)
(d) Iron(III)
(e) Tin(IV)

9.7 Write the Stock names for each of the following: (a) FeS, (b) Cu_3N, (c) PbO_2, (d) Hg_2Br_2, (e) MnF_3.

9.8 Write the formulas for each of the following: (a) stannous bromide, (b) mercuric oxide, (c) cobaltic nitride, (d) plumbous fluoride, (e) copper(I) sulfide, (f) sulfur dichloride.

9.3 NAMING TERNARY COMPOUNDS

A **ternary compound** is one containing three elements. From Figures 9.2 and 9.3 it should be apparent that ternary compounds are classified in a manner similar to binary compounds. First decide if the ternary compound is ionic or

TABLE 9.2   IMPORTANT POLYATOMIC IONS

Name of polyatomic ion	Formula
Hydroxide	OH^-
Sulfate	$SO_4{}^{2-}$
Nitrate	$NO_3{}^-$
Phosphate	$PO_4{}^{3-}$
Chlorate	$ClO_3{}^-$
Cyanide	CN^-
Permanganate	$MnO_4{}^-$
Carbonate	$CO_3{}^{2-}$
Chromate	$CrO_4{}^{2-}$
Borate	$BO_3{}^{3-}$
Acetate	$C_2H_3O_2{}^-$
Ammonium	$NH_4{}^+$

Two polyatomic ions have the *ide* ending: (1) hydroxide, OH^- and (2) cyanide, CN^-. Don't confuse these ions with nonmetal ions.

covalent. If it is ionic, check to see if the metal has a fixed or variable oxidation number. If it is covalent, determine if it is an acid or not.

We shall concentrate on naming ternary compounds composed of a metal and a polyatomic ion (ionic), and of hydrogen bonded to a polyatomic ion (acid). (Polyatomic ions were introduced in Chapter 8. They are ions that are composed of more than one atom, e.g., hydroxide, OH^-; nitrate, $NO_3{}^-$; and sulfate, $SO_4{}^{2-}$.)

Study hint: Write the formula of the polyatomic ion on one side of an index card, and the name on the other side. Flip through the cards, learning names of formulas; then, turn the deck over and proceed in the opposite direction, learning the formulas of names.

Ionic ternary compounds are named like binary compounds, except that a polyatomic ion name is found in place of either the metal ion or nonmetal ion name. Study Table 9.2 and memorize the 12 polyatomic ions listed. Once these 12 polyatomic ions are learned, naming ternary compounds is simplified. When you learn ions, don't forget about their charges. It is of no value to learn the ions without their charges! (A more complete listing of ions, including polyatomic ions, is found in the Appendix.)

Ionic Ternary Compounds

When naming an ionic ternary compound, first determine if the metal in the ternary compound has a fixed or variable oxidation number. Depending on what metal is found, the compound is named in a similar manner to an ionic binary compound except there is a polyatomic ion name in the formula.

Let's illustrate the naming of ionic ternary compounds by considering $Co(NO_3)_2$. Cobalt is a transition metal with a variable oxidation state. Thus, determine the oxidation state of Co and write its oxidation state in parentheses—cobalt(II). Then identify the polyatomic ion, $NO_3{}^-$, nitrate, and add it to the metal's name—cobalt(II) nitrate.

$$Co(NO_3)_2 = \text{cobalt(II) nitrate}$$

If more than one polyatomic ion is found in a compound, it is placed in parentheses with the appropriate subscript.

Cobalt's oxidation state is obtained by knowing the charge on the nitrate ion, -1. Since there are two nitrate ions, the cobalt exists in the $+2$ oxidation state.

Example Problem 9.4 presents other examples of naming ternary compounds.

Example Problem 9.4

Name the following compounds using the Stock system: (a) K_2CrO_4, (b) $(NH_4)_2S$, (c) $FeCO_3$, (d) $Mn_3(PO_4)_2$.

Solution

(a) K_2CrO_4. The metal is potassium, an alkali metal with a fixed oxidation state, and the polyatomic ion is chromate. Thus:

$$K_2CrO_4 = \text{potassium chromate}$$

A number is not written after potassium because it only exists in one oxidation state, $+1$.

(b) $(NH_4)_2S$. In this compound, we find the ammonium ion in place of a metal, and the nonmetal is sulfur. Remove sulfur's ending and attach *ide*:

$$(NH_4)_2S = \text{ammonium sulfide}$$

(c) $FeCO_3$. Iron is a metal with a variable oxidation state; consequently, the name must indicate which oxidation state iron is in. Given that carbonate, CO_3^{2-}, has a $2-$ charge, then iron's oxidation number is $+2$ in order for the sum to equal zero. Accordingly:

$$FeCO_3 = \text{iron(II) carbonate}$$

(d) $Mn_3(PO_4)_2$. Once again, the oxidation state of the metal has to be expressed because manganese exists in a variety of states. Phosphate has a charge of $3-$; therefore two phosphates have a total charge of $6-$. So the total oxidation number for the three Mn atoms is $+6$, or each Mn has an oxidation number of $+2$.

$$Mn_3(PO_4)_2 = \text{manganese(II) phosphate}$$

To write the formula of a ternary compound, given its name, follow the same procedures used to write formulas of binary compounds, as shown in Example Problem 9.5.

Example Problem 9.5

Write the formulas for: (a) zinc(II) hydroxide, (b) silver(I) cyanide, and (c) cobaltic sulfate.

Solution

(a) Zinc(II) hydroxide. Zinc(II) is Zn^{2+}, and hydroxide is OH^-. To produce a compound with a zero oxidation state, two hydroxides are required per zinc ion.

$$Zn^{2+} \quad OH^-$$
$$\text{Zinc(II) hydroxide} = Zn(OH)_2$$

(b) Silver(I) cyanide. Silver(I) is Ag^+, and cyanide is CN^-. Since they have equal but opposite charges, they are combined in a 1-to-1 ratio.

$$Ag^+ \quad CN^-$$
$$\text{Silver(I) cyanide} = AgCN$$

(c) Cobaltic sulfate. Cobaltic represents the higher of cobalt's oxidation states, Co^{3+}. Sulfate's formula is SO_4^{2-}. Here we have to find the lowest common denominator between 3 and 2: it is 6. Cobaltic sulfate becomes $Co_2(SO_4)_3$.

$$Co^{3+} \quad SO_4^{2-}$$
$$\text{Cobaltic sulfate} = Co_2(SO_4)_3$$

A group of polyatomic ions exists with one less oxygen than some of those listed in Table 9.3. For example, in addition to sulfate, SO_4^{2-}, there is an ion with the formula SO_3^{2-}. How is this ion named? If a polyatomic ion has one less oxygen than the ion ending in *ate*, it is given an *ite* ending. Consequently, SO_3^{2-} is the sul*ite* ion since it has one less oxygen than sul*ate*, SO_4^{2-}. Table

TABLE 9.3   **SELECTED OXYANIONS: *ate* VERSUS *ite* ENDINGS**

Oxyanion ion	Name
SO_4^{2-}	Sulfate
SO_3^{2-}	Sulfite
NO_3^-	Nitrate
NO_2^-	Nitrite
PO_4^{3-}	Phosphate
PO_3^{3-}	Phosphite
AsO_4^{3-}	Arsenate
AsO_3^{3-}	Arsenite
SeO_4^{2-}	Selenate
SeO_3^{2-}	Selenite

An oxyanion is a polyatomic ion that contains oxygen.

9.3 lists other examples of *ite* **oxyanions,** i.e., polyatomic ions containing oxygen.

Once the formula for the *ate* ion is learned, it is only necessary to subtract an oxygen from the formula and replace the ending of the name with *ite* to obtain the correct formula and name of the *ite* ion.

Example Problem 9.6

Write the Stock name for: (a) $Fe_3(PO_3)_2$ and (b) $Ba(NO_2)_2$.

Solution

(a) $Fe_3(PO_3)_2$. Looking at the oxyanion, we notice that it has one less oxygen than phosphate, PO_4^{3-}; thus, it is the phosphite ion, PO_3^{3-}. Since there are two phosphite ions in the formula, each with a charge of 3−, their total charge is 6−. Each iron ion has a charge of 2+ ($3 \times 2+ = 6+$) to balance the charge of the phosphite ions. Accordingly, the name of $Fe_3(PO_3)_2$ is iron(II) phosphite.

$$Fe_3(PO_3)_2 = \text{iron(II) phosphite}$$

(b) $Ba(NO_2)_2$. Barium is an element in group IIA; thus it has a fixed charge of 2+. NO_2^- has one less oxygen than nitrate, NO_3^-, nitrate; so its name is nitrite. Thus:

$$Ba(NO_2)_2 = \text{barium nitrite}$$

REVIEW PROBLEMS

9.9 Identify each of the following polyatomic ions: (a) ClO_3^-, (b) MnO_4^-, (c) CrO_4^{2-}, (d) PO_4^{3-}.

9.10 Write the formula for: (a) ammonium ion, (b) carbonate ion, (c) nitrate ion, (d) cyanide ion, (e) phosphite ion, (f) selenite ion.

9.11 Name the compound: (a) Li_2CO_3, (b) $Ba(ClO_3)_2$, (c) $CuCrO_4$, (d) $Hg_2(NO_3)_2$, (e) Li_3PO_4, (f) $NiNO_2$.

9.12 Write the formulas for: (a) ammonium chromate, (b) iron(II) sulfate, (c) aluminum hydroxide, (d) nickel(II) nitrate, (e) manganous phosphate, and (f) sodium sulfite.

Covalent Ternary Compounds

Since we shall encounter few covalent ternary compounds that are not acids, only acids are discussed in this section. Examples of ternary acids include sulfuric acid, $H_2SO_4(aq)$; nitric acid, $HNO_3(aq)$; phosphoric acid, $H_3PO_4(aq)$; chloric acid, $HClO_3(aq)$; and boric acid, $H_3BO_3(aq)$.

Oxyacids

Oxyacids are ternary acids that contain oxygen. Such compounds have hydrogens bonded to a oxyanion. For instance, if two hydrogens are attached to sulfate, SO_4^{2-}, the resulting acid is called sulfuric acid. This name was obtained by adding *ic acid* to the end of the name of the nonmetal contained in sulfate.

$$\text{Sulfuric acid} = \text{sulfur} + \textit{ic acid}$$

Oxyanions that end in *ate*, when bonded to hydrogen(s), form oxyacids whose names are derived from the name of the nonmetal with *ic acid* as the ending. Table 9.4 presents examples of other *ic acids*.

If the nonmetal in the oxyacid exists in more than one oxidation state, then there is more than one oxyacid, each with a varying number of oxygen atoms. For example, consider two oxyacids containing H, O, and S:

$$\underset{+1 \quad \;\; -2}{\overset{+2 \; +6 \; -8}{H_2 S O_4}} \qquad \underset{+1 \quad \;\; -2}{\overset{+2 \; +4 \; -6}{H_2 S O_3}}$$

$$\underset{+6}{H_2SO_4} = \text{sulfuric acid}$$

$$\underset{+4}{H_2SO_3} = \text{sulfurous acid}$$

Sulfuric acid's sulfur is in the +6 oxidation state, whereas, the sulfur in sulfurous acid, H_2SO_3, is in the lower +4 state.

If two oxyacids exist with the same nonmetal, the acid containing the nonmetal in the higher oxidation state is given the *ic acid* ending (corresponding to the oxyanion sulfate, SO_4^{2-}); *ous acid* is the ending for the oxyacid possessing the nonmetal in the lower oxidation state (corresponding to the oxyanion sulfite, SO_3^{2-}).

Two oxyacids containing nitrogen exist: nitric acid, HNO_3, and nitrous acid, HNO_2. Nitric acid's nitrogen is found in the +5 state, whereas nitrous acid's nitrogen is in the +3 oxidation state.

$$\underset{+5}{HNO_3} = \text{nitric acid}$$

$$\underset{+3}{HNO_2} = \text{nitrous acid}$$

The names are derived from the oxyanions contained within the molecule. A nitrate, NO_3^-, is found in HNO_3; accordingly, *ic acid* is added to the stem *nitr* from nitrogen, giving nitric acid. A nitrite ion, NO_2^-, is found in nitrous acid; thus, *ous acid* is added to the stem *nitr*, giving nitrous acid.

TABLE 9.4   OXYACID NAMES

Oxyacid (formula)	Oxyanion	Name
Nitric acid (HNO_3)	NO_3^-	Nitrate
Carbonic acid (H_2CO_3)	CO_3^{2-}	Carbonate
Sulfuric acid (H_2SO_4)	SO_4^{2-}	Sulfate
Boric acid (H_3BO_3)	BO_3^{3-}	Borate
Chloric acid ($HClO_3$)	ClO_3^-	Chlorate
Selenic acid (H_2SeO_4)	SeO_4^{2-}	Selenate

A few nonmetals form three or four oxyacids. Let's consider the four oxyacids of Cl:

$$HClO_4, \text{ perchloric acid} \quad \overset{+7}{Cl}$$

$$HClO_3, \text{ chloric acid} \quad \overset{+5}{Cl}$$

$$HClO_2, \text{ chlorous acid} \quad \overset{+3}{Cl}$$

$$HClO, \text{ hypochlorous acid} \quad \overset{+1}{Cl}$$

Here, in addition to the acids with the *ic* and *ous* endings, two other acids exist, one with a Cl in a higher oxidation state, and another in a lower state. When this situation arises, the acid containing the nonmetal in the highest oxidation state (contains one more O than the *ic acid*) is named by placing *per* as a prefix, and attaching *ic acid* at the end. Similarly, the acid with the nonmetal in the lowest oxidation state (contains one less O than the *ous acid*) is given a prefix of *hypo* with the *ous acid* ending.

Hypo is a prefix meaning "below." *Hyper* is a prefix meaning "above."

When perchloric acid, $HClO_4$, gives up a H^+, the ClO_4^- or *perchlorate* ion results. Acids with *per* as the prefix and *ic acid* as an ending contain polyatomic ions that are named by dropping the *ic acid* and adding *ate*. Similarly, acids with *hypo* and *ous acid* contain the *hypo . . . ite* ion. Thus when HOCl loses a H^+, a *hypochlorite* ion, OCl^-, results.

Example Problem 9.7

Write the names for the following series of oxyacids, and the polyatomic ions contained within the molecules: (a) $HBrO_4$, (b) $HBrO_3$, (c) $HBrO_2$, (d) HBrO.

Solution

Per . . . ic oxyacids contain one more O than *ic* acids, and *hypo . . . ous* acids contain one less O than *ous* acids.

(a) $HBrO_4$'s Br is found in the +7 oxidation state, the highest of the series. It is named perbromic acid, and it contains the perbromate ion, BrO_4^-.

(b) $HBrO_3$'s Br is found in the +5 state, and it is named bromic acid; the BrO_3^- is the bromate ion.

(c) $HBrO_2$'s Br is in the +3 oxidation state, and $HBrO_2$ is called bromous acid; it contains a bromite ion, BrO_2^-.

(d) HBrO's Br oxidation number is +1, and HBrO is hypobromous acid. BrO^- is the hypobromite ion.

Unfortunately, there is no easy way of knowing which oxyacids exist for a particular nonmetal. It is a matter of sitting down and learning the most impor-

TABLE 9.5   SELECTED OXYACIDS

Nonmetal	Oxyacid	Formula
P	Phosphoric acid	H_3PO_4
	Phosphorous acid	H_3PO_3
	Hypophosphorous acid	H_3PO_2
Se	Selenic acid	H_2SeO_4
	Selenous acid	H_2SeO_3
B	Boric acid	H_3BO_3
C	Carbonic acid	H_2CO_3
Si	Silicic acid	H_4SiO_4
As	Arsenic acid	H_3AsO_4
	Arsenous acid	H_3AsO_3

tant ones, and referring to a table of oxyacids for the others. Table 9.5 gives a partial listing of additional common oxyacids.

Acids containing more than one hydrogen (H_2SO_4, H_3PO_4, . . .) can lose some or all of their hydrogen ions in chemical reactions. If H_2SO_4 loses one H^+, the resulting ion is HSO_4^-:

Both hydrogen sulfate and bisulfate are names for HSO_4^-.

$$H_2SO_4 \longrightarrow H^+(aq) + HSO_4^-(aq)$$

HSO_4^- is named hydrogen sulfate. It is only necessary to affix the word *hydrogen* to the name of the polyatomic ion, SO_4^{2-}.

Phosphoric acid is commercially one of the most important chemicals. Pure phosphoric acid is a white solid that melts at 42°C. It is generally sold as an 80–85% water solution.

Phosphoric acid, H_3PO_4, can give up one, two, or three H^+ ions. If one H^+ is lost, then $H_2PO_4^-$, dihydrogen phosphate, is produced. If two H^+ ions are lost, then HPO_4^{2-}, monohydrogen phosphate results:

$$H_3PO_4 \longrightarrow H^+(aq) + H_2PO_4^-(aq) \quad \text{(dihydrogen phosphate)}$$
$$H_3PO_4 \longrightarrow 2H^+(aq) + HPO_4^{2-}(aq) \quad \text{(monohydrogen phosphate)}$$
$$H_3PO_4 \longrightarrow 3H^+(aq) + PO_4^{3-}(aq) \quad \text{(phosphate)}$$

Sodium hydrogen carbonate's common names are baking soda and bicarbonate of soda. When acid combines with $NaHCO_3$, CO_2 (g) is liberated. Consequently, it is used to help bread rise.

Polyatomic ions containing hydrogen are named exactly the same way as other polyatomic ions. To name $NaHCO_3$, it is only necessary to write sodium hydrogen carbonate. $CaHPO_4$ is named calcium monohydrogen phosphate.

REVIEW PROBLEMS

9.13 Name the binary acid: (a) HI(aq), (b) H_2S(aq), (c) HF(aq).

9.14 What is the name of each of the following oxyacids: (a) HNO_3, (b) H_3PO_3, (c) H_2SO_3, (d) HClO?

9.15 Write the formula for: (a) boric acid, (b) selenic acid, (c) chlorous acid, (d) carbonic acid.

9.16 Write the formulas for: (a) aluminum arsenate, (b) cesium hydrogen sulfite, (c) strontium borate, (d) iron(II) dihydrogen phosphate.

9.17 Write the name for each of the following salts: (a) K_2SO_3, (b) $Ba(NO_2)_2$, (c) $LiHSO_3$, (d) $NaBrO_2$, (e) AgClO.

9.18 Write formulas for: (a) rubidium chlorate, (b) nickel(II) hydrogen carbonate, (c) mercuric selenite, (d) magnesium bromate, (e) lead(II) dihydrogen phosphite.

TABLE 9.6   COMMON AND SYSTEMATIC NAMES FOR SELECTED COMPOUNDS

Common Name	Formula	Systematic name
Alum	$NaAl(SO_4)_2 \cdot 12H_2O$	Sodium aluminum sulfate dodecahydrate
Alumina	Al_2O_3	Aluminum oxide
Baking soda	$NaHCO_3$	Sodium hydrogen carbonate
Barite	$BaSO_4$	Barium sulfate
Blue vitriol	$CuSO_4 \cdot 5H_2O$	Copper(II) sulfate pentahydrate
Borax	$Na_2B_4O_7 \cdot 10H_2O$	Sodium tetraborate decahydrate
Calcite	$CaCO_3$	Calcium carbonate
Calomel	Hg_2Cl_2	Mercury(I) chloride
Cinnabar	$HgCl_2$	Mercury(II) chloride
Dolomite	$CaMg(CO_3)_2$	Calcium magnesium carbonate
Epsom salts	$MgSO_4 \cdot 7H_2O$	Magnesium sulfate heptahydrate
Fluorite	CaF_2	Calcium fluoride
Galena	PbS	Lead(II) sulfide
Germane	GeH_4	Germanium hydride
Glauber's salt	$Na_2SO_4 \cdot 10H_2O$	Sodium sulfate decahydrate
Gypsum	$CaSO_4 \cdot 2H_2O$	Calcium sulfate dihydrate
Halite	$NaCl$	Sodium chloride
Hematite	Fe_2O_3	Iron(III) oxide
Lime	CaO	Calcium oxide
Magnesia	$MgCO_3$	Magnesium carbonate
Muriatic acid	HCl	Hydrochloric acid
Oil of vitriol	H_2SO_4	Sulfuric acid
Plaster of paris	$CaSO_4 \cdot 1/2H_2O$	Calcium sulfate hemihydrate
Pyrite	FeS_2	Iron(IV) sulfide
Quartz	SiO_2	Silicon dioxide

Summary

A convenient system of numbers, called oxidation numbers, is used to help keep track of the electrons associated with elements in compounds. Four rules are applied to identify the oxidation numbers of elements. Oxidation numbers are essential in writing formulas and equations but are not an actual measurable property of elements.

Binary compounds are those containing two different elements. A name is assigned, depending on what two elements are found in the compound. Ionic binary compounds are named by writing the name of the metal and then modifying the nonmetal name by dropping its ending and adding *ide*. If the compound contains a metal that can exist in more than one oxidation state, then the name includes a roman numeral in parentheses, indicating the exact oxidation number of the metal. This number is placed immediately after the metal's name.

An old system of naming binary compounds is still employed by some chemists; this system requires knowledge of the old Latin names for metals. Using the old names as the stem, either *ic* or *ous* is appended as a suffix, denoting the metal with higher and lower oxidation numbers, respectively.

Ternary compounds contain three different elements. Most are composed of a metal and a polyatomic ion. They are named in a similar manner to binary compounds, except the polyatomic ion is written in place of the nonmetal. Various polyatomic ions must be learned in order to write formulas of ternary compounds. Examples of polyatomic ions include chromate, CrO_4^{2-}; nitrate, NO_3^-; cyanide, CN^-; and hydroxide, OH^-.

Acids are substances that yield H^+ when they dissolve in water. Binary acids are named by adding the prefix *hydro* to the nonmetal name and the ending *ic acid* to the stem of the nonmetal. Many ternary acids are oxyacids because they contain an oxygen in addition to the nonmetal and hydrogen. Various prefixes and suffixes are added to the nonmetal stem to indicate the exact oxyacid. Oxyacids have the endings *ic* or *ous*, depending on the oxidation state of the nonmetal. Oxyanions derived from acids are named in a systematic fashion. If the resulting ion is from an *ic* acid, the ending *ate* is added to the oxyanion; *ite* is the suffix given to oxyanions produced from *ous* acids.

QUESTIONS AND PROBLEMS

9.19 Define the following: chemical nomenclature, IUPAC, oxidation number, oxidation-reduction reaction, binary compound, ternary compound, Stock system, polyatomic ion, binary acid, ternary acid, oxyacid, oxyanion.

Oxidation Numbers

9.20 Identify the oxidation numbers of all elements in each of the following compounds: (a) Br_2, (b) H_2S, (c) CaO, (d) NF_3, (e) N_2O_5, (f) PCl_5, (g) Al_2S_3, (h) CF_4.

9.21 Identify the oxidation state of S in each of the following compounds: (a) SO_2, (b) SO_3, (c) H_2SO_3, (d) H_2SO_4, (e) S_8, (f) S_2O_7, (g) S_2O_3, (h) S_2F_{10}.

9.22 Identify the oxidation state of N in each of the following compounds: (a) N_2, (b) N_2O, (c) N_2O_3, (d) N_2O_4, (e) $NOBr$, (f) HNO_2, (g) $H_2N_2O_2$, (h) N_2H_4.

9.23 Find the oxidation number of each element in the following compounds: (a) $K_2Cr_2O_7$, (b) Na_2GeO_3, (c) Na_2UO_4, (d) $RbHSO_4$, (e) $Ca(HS)_2$, (f) $K_4V_2O_7$, (g) $Na_2C_2O_4$, (h) $Cu(CN)_2$, (i) Na_2O_2 (peroxide), (j) MgH_2 (hydride).

9.24 Find the metal's oxidation state in the following: (a) $CePO_4$, (b) Sb_2O_5, (c) $Cd(OH)_2$, (d) $Cr_2(SO_4)_3$, (e) $CoSeO_4$, (f) Cu_2SO_4, (g) $MnHPO_4$, (h) Hg_2CO_3.

Naming and Writing Formulas of Binary Compounds

9.25 Write the name for each of the following covalent substances: (a) CO_2, (b) N_2O, (c) CCl_4, (d) PBr_3, (e) OF_2, (f) SiO_2.

9.26 Write the name for each of the following: (a) $SnBr_2$, (b) CoN, (c) PbS, (d) Cu_3P, (e) HgO, (f) Li_2Se, (g) $SrCl_2$, (h) Sc_2O_3.

9.27 Write the formula for:
(a) Iron(II) iodide
(b) Manganic oxide
(c) Cupric bromide
(d) Chromium(III) fluoride
(e) Magnesium sulfide
(f) Calcium carbide
(g) Mercurous chloride
(h) Stannic nitride

9.28 Write the formulas for each of the following oxides:
(a) Thallium(I) oxide
(b) Uranium(IV) oxide
(c) Gold(III) oxide
(d) Molybdenum(VI) oxide
(e) Manganese(III) oxide
(f) Ruthenium(VIII) oxide
(g) Tungsten(VII) oxide
(h) Vanadium(V) oxide.

Naming and Writing Formulas of Ternary Compounds

9.29 Complete the following table by writing the name of the compound that is a combination of the anion listed horizontally and the cation listed vertically:

	$C_2H_3O_2^-$	PO_4^{3-}	MnO_4^-	CN^-	CO_3^{2-}	CrO_4^{2-}
NH_4^+	____	____	____	____	____	____
Ca^{2+}	____	____	____	____	____	____
Fe^{3+}	____	____	____	____	____	____
Hg^{2+}	____	____	____	____	____	____
Al^{3+}	____	____	____	____	____	____
Sn^{2+}	____	____	____	____	____	____
Cs^+	____	____	____	____	____	____

9.30 Complete the following table by writing the formula of the compound that is a combination of the anion listed horizontally and the cation listed vertically:

	Hydroxide	Phosphite	Sulfate	Chlorate	Selenate
Strontium	____	____	____	____	____
Potassium	____	____	____	____	____
Lead(II)	____	____	____	____	____
Cobaltic	____	____	____	____	____
Ammonium	____	____	____	____	____
Gallium(III)	____	____	____	____	____

9.31 In the following table, the trivial name for a compound is given with either its formula or modern name; complete the table by writing the formula or modern name of the compound:

Old name	Formula	Modern name
Alunogenite		Aluminum sulfate
Aragonite	$CaCO_3$	
Baking soda		Sodium hydrogen carbonate
Blue vitriol	$CuSO_4$	

Old name	Formula	Modern name
Celestite		Strontium sulfate
Chrome yellow	$PbCrO_4$	
Cyanouric acid		Gold(III) cyanide
Glauber's salt	Na_2SO_4	
Hemimorphite		Zinc(II) silicate
Nitrobarite	$Ba(NO_3)_2$	

9.32 Write the formula of the compound that results when the ammonium ion is combined with each of the following ions: (a) selenide, (b) sulfite, (c) nitrate, (d) iodate, (e) perchlorate, (f) hydrogen carbonate, (g) bromate, (h) hypochlorite.

9.33 Write the formulas of the compounds that result when acetate ions are combined with the following ions: (a) barium, (b) cadmium(II), (c) ferric, (d) cuprous, (e) gallium(III), (f) indium(III), (g) plumbous, (h) platinum(II).

Acids

9.34 Write the names of the binary acids: (a) HBr, (b) HI, (c) H_2Se.

9.35 Write the names of the oxyacids: (a) H_3BO_3, (b) HClO, (c) HIO_3, (d) H_3AsO_4, (e) H_2CO_3, (f) $HBrO_4$.

9.36 Lactic acid, $HC_3H_5O_3$, a weak organic acid, ionizes to a small extent in water, producing the $C_3H_5O_3^-$ ion. What is the ion's name?

9.37 Write the names for the following acids: (a) HIO_4, (b) $HC_2H_3O_2$, (c) H_3BO_3, (d) HIO, and (e) HCN.

9.38 Write the name of the polyatomic ion found in each acid in Question 9.37.

9.39 Write the name for each of the following polyatomic ions: (a) HSO_3^-, (b) HPO_3^{2-}, (c) HSO_4^-, and (d) HCO_3^-.

Naming and Writing Formulas in General

9.40 What do the following endings indicate about a compound: (a) ite, (b) ate, (c) ide, (d) ic, and (e) ous?

9.41 Name the following compounds:
(1) $(NH_4)_2S$
(2) SbI_3
(3) H_3AsO_4
(4) As_2O_3
(5) $BaCrO_4$
(6) $BeSeO_3$
(7) $BiCl_4$
(8) BN
(9) BrO_2
(10) $Cd(BrO_3)_2$
(11) $Ca(ClO)_2$
(12) S_2F_{10}
(13) $Ce(OH)_3$
(14) $CsHCO_3$
(15) Cl_2O_7
(16) $Cr_2(SO_3)_3$
(17) CoF_3
(18) $CuSeO_4$
(19) $GaCl_3$
(20) GeS_2
(21) HCN
(22) HCN(aq)
(23) HIO_3(aq)
(24) IF_5

(25) Ir_2S_3
(26) $Fe(H_2PO_2)_3$
(27) $Pb(AsO_2)_2$
(28) $LiClO_3$
(29) $Mg(NO_3)_2$
(30) $MnAs$
(31) $Hg(BrO_3)_2$
(32) $Mo(SO_3)_3$
(33) Si_3Cl_8

(34) OsO
(35) $Pd(NO_3)_2$
(36) PBr_5
(37) $Pt(OH)_2$
(38) KH_2AsO_4
(39) $RaCO_3$
(40) ReF_6
(41) $Rh_2(SO_4)_3$
(42) Rb_2SeO_4

(43) Sc_2O_3
(44) SeF_4
(45) Ag_2TeO_3
(46) NaH_2PO_3
(47) $Sr(MnO_4)_2$
(48) S_2F_2
(49) $TlCN$
(50) $Sn(SO_4)_2$

9.42 Write the formula for each of the following:

(1) Titanium(IV) fluoride
(2) Tungsten(V) bromide
(3) Zinc(II) chromate
(4) Zirconium(IV) oxide
(5) Ammonium sulfate
(6) Cesium hydroxide
(7) Calcium sulfite
(8) Barium bromate
(9) Sodium monohydrogen phosphate
(10) Bismuth(III) iodide
(11) Cupric hypochlorite
(12) Aluminum nitrite
(13) Ferric arsenide
(14) Lead(II) chlorite
(15) Lithium monohydrogen phosphite
(16) Manganese(II) sulfate
(17) Mercurous sulfide
(18) Molybdenum(IV) sulfide
(19) Nickel(II) bicarbonate
(20) Dinitrogen trioxide
(21) Palladium(IV) silicide
(22) Osmium(II) sulfite
(23) Sodium perchlorate
(24) Hypophosphorous acid

(25) Phosphorus nitride
(26) Potassium iodite
(27) Rhenium(VI) chloride
(28) Tin(II) phosphide
(29) Rubidium chlorite
(30) Scandium(III) sulfate
(31) Silicon nitride
(32) Silicic acid
(33) Silver phosphate
(34) Sodium hydrogen sulfite
(35) Strontium iodate
(36) Tantalum(III) nitride
(37) Thallous sulfate
(38) Stannic iodide
(39) Zinc(II) cyanide
(40) Cobaltous phosphate
(41) Ammonium bromate
(42) Aluminum carbide
(43) Calcium telluride
(44) Chloric acid
(45) Ferric boride
(46) Manganese(II) carbonate
(47) Mercury(II) sulfide
(48) Hydrocyanic acid
(49) Potassium permanganate
(50) Cerium(III) carbonate

· C H A P T E R ·

10

Chemical Equations

Study Guidelines

After completing Chapter 10, you should be able to:

 1. Identify the two parts, reactants and products, of a chemical equation
 2. Identify and write all common symbols found in chemical equations
 3. Explain the meaning of (*g*), (*l*), (*s*), and (*aq*)
 4. State the effect of a catalyst on a chemical reaction
 5. List the steps followed when balancing a chemical equation
 6. Explain why subscripts of compounds cannot be changed when balancing a chemical equation
 7. Balance chemical equations, using the inspection method
 8. Translate a word equation into a balanced chemical equation
 9. List the four types of inorganic reactions
10. Write a general equation that represents the format for each general type of inorganic reaction
11. Write a balanced chemical equation that is illustrative of each type of inorganic reaction
12. Identify different classes of (a) combination, (b) decomposition, (c) single displacement, and (d) metathesis (double displacement) reactions
13. Discuss the effects of reaction conditions on specific reactions
14. Use the activity series to decide if a displacement reaction occurs
15. Predict the products and write equations for reactions, given the reactants
16. Predict what substance precipitates or is released as a gas when two aqueous salt solutions undergo a metathesis reaction
17. Explain the difference between endothermic and exothermic reactions
18. Identify endothermic and exothermic reactions from their chemical equations
19. Explain, in terms of bond breaking and bond formation, what happens in endothermic and exothermic reactions

Chemical changes occur when substances undergo changes in their composition. When a substance undergoes a chemical change, we say "a chemical reaction has taken place." It is inconvenient and time-consuming to express what happens during chemical reactions by writing the complete names of all substances involved (word equation); instead, chemists write a concise statement, called a **chemical equation,** using the symbols of the elements and the formulas of compounds. Other special symbols are added to the equation to express exactly what is happening during chemical changes.

A chemical equation has two parts: (1) reactants and (2) products. *Reactants*, sometimes called starting materials, are all substances present prior to the chemical change. Each reactant is listed by writing its symbol or formula separated by a plus (+) sign. All reactants are written to the left of an arrow ($\rightarrow$) that separates the reactants from the products.

Reactants = starting materials
Products = final materials

All *products*, substances produced after the chemical change occurs, are written to the right of the arrow, and are separated by plus signs. Hence, chemical equations have the following format:

$$\text{Reactants} \longrightarrow \text{Products}$$

or

$$A + B \longrightarrow C + D$$

In our example, hypothetical substances A and B are the reactants, and C and D are the products of the reaction. The above equation is translated as "reactant A combines with reactant B to yield product C and product D." Note that the arrow is read as "to yield" or "yields" or "gives."

Frequently, other information is added to the equation. For instance, it is important to know what physical state the reactants and products exist in. Four symbols are commonly employed to indicate physical states: solid (*s*), liquid (*l*), gas (*g*), and water or aqueous solution (aq). Enclosed in parentheses, these symbols are written next to the formula within the equation. Consider the following example:

$$A(g) + B(l) \longrightarrow C(aq) + D(s)$$

Translated to a word equation, this states "reactant A, in the gas phase, combines with reactant B, in the liquid phase, yielding product C, which is dissolved in water, and product D, in the solid phase."

Conditions required for the reaction to take place are also written into the equation. They are placed either above or below the arrow. If heat is needed for the chemical change, the word "heat" or more commonly the Greek letter delta, Δ, is written above or below the arrow. Sometimes the actual temperature is expressed.

TABLE 10.1 SUMMARY OF SYMBOLS FOUND IN CHEMICAL EQUATIONS

Symbol	Meaning
$\rightarrow$	Yields, produces, gives
+	Separates compounds and elements
(s)	Solid state
(l)	Liquid state
(g)	Gaseous state
(aq)	Aqueous solution
$\xrightarrow{\text{cat}}$	Reaction requiring a catalyst
$\xrightarrow{\Delta}$	Reaction requiring heat
$\xrightarrow{\text{uv}}$	Reaction requiring ultraviolet light

Besides heat effects, various chemical reactions take place only in the presence of electromagnetic radiation; e.g., infrared (ir) and ultraviolet light (uv).

$$A(s) \xrightarrow{\Delta} B(g) + C(s)$$
$$A(s) \xrightarrow{\text{heat}} B(g) + C(s)$$

Both of the above equations are read as "solid reactant A is heated to yield gaseous product B and solid product C."

Most chemical reactions in living tissues are catalyzed by special protein structures called enzymes.

Various reactions require **catalysts,** which are substances that increase the rates of reactions and are recovered basically unchanged after the reaction. The word "catalyst," the abbreviation "cat," or the actual name of the catalyst is written above or below the arrow. Any special conditions (high or low pressure, presence or absence of light, etc.) are also written near the arrow.

Table 10.1 summarizes the symbols used in writing chemical equations.

REVIEW PROBLEMS

10.1 (a) What is a chemical change? (b) How is it different from a physical change?

10.2 Collectively, what are the substances called that are located: (a) to the right of the arrow and (b) to the left of the arrow in a chemical equation?

10.3 What is the exact meaning of the following symbols used in chemical equations: (a) +, (b) Δ, (c) (aq), (d) $\rightarrow$?

10.4 Translate the following chemical equation into a word equation:

$$C(s) + O_2(g) \xrightarrow{\Delta} CO_2(g)$$

10.2 BALANCING CHEMICAL EQUATIONS

Equations illustrate the law of conservation of mass: atoms cannot be gained or lost in reactions.

When an equation is written, it must obey the **law of conservation of mass,** i.e., matter cannot be created or destroyed. Specifically, the number of atoms of each different element in an equation must be the same on the left and right sides of the arrow. After a chemical change, the same type and number of atoms are present; they are merely rearranged.

To obey the law of conservation of mass, an equation must be **balanced.** Balancing an equation involves placing coefficients in front of all reactants and products so that the same number of atoms of each element appears on either side of the equation.

The inspection method is utilized to balance equations.

Simple equations are balanced by using the inspection method, which involves equalizing the number of elements of each type by placing **coefficients** in front of all elements and compounds in chemical equations.

Let's illustrate the balancing of equations by the inspection method. We shall use the combination reaction in which hydrogen gas, $H_2(g)$, combines with oxygen gas, $O_2(g)$, to yield water vapor, $H_2O(g)$. First write the unbalanced equation, including all reactants and products.

$$H_2(g) + O_2(g) \longrightarrow H_2O(g) \quad \text{(unbalanced)}$$

After writing the unbalanced equation, we readily see that the number of oxygen atoms is not the same on both sides of the equation. Two oxygen atoms appear on the left side, and only one oxygen atom is found on the right. To balance the oxygen atoms, we place a 2 in front of the formula of water. Two water molecules contain two oxygen atoms.

Don't change subscripts when you are balancing equations.

We cannot place a 2 as a subscript next to the oxygen in water because this will change water's composition. **Never change subscripts when you are balancing a chemical equation.** Incorrectly placing the 2 as a subscript next to oxygen gives a new product of the reaction, H_2O_2—hydrogen peroxide. Hydrogen peroxide's properties are vastly different from those of water!

Changing formula subscripts produces a new compound and totally changes the equation.

After balancing the oxygen atoms, we are left with the following partially balanced equation:

$$H_2(g) + \underline{O_2}(g) \longrightarrow 2H_2\underline{O}(g)$$

In balancing the oxygen atoms, we now know the number of hydrogen atoms that are needed. Two water molecules contain four hydrogen atoms $(2 \times H_2O = 4\ H$ and $2\ O)$. Therefore, we place a 2 in front of $H_2(g)$ on the left side of the equation, giving four hydrogen atoms.

$$2\underline{H_2}(g) + \underline{O_2}(g) \longrightarrow 2\underline{H_2O}(g)$$

After balancing a particular element in an equation, underline that element on either side of the arrow to indicate that it has been balanced. When all elements are underlined, the equation is balanced.

The last step, an important one, is to check to see that you have correctly balanced all atoms. If the same number of atoms of each type appear on both sides, the equation is properly balanced. If you are off by even one atom, then the equation is not balanced. In our example equation, we find that 4 H atoms and 2 O atoms are on either side of the arrow, indicating a correctly balanced equation. It is convenient to check off or underline each atom as you are verifying that the equation is balanced:

$$2\underline{H_2}(g) + \underline{O_2}(g) \longrightarrow 2\underline{H_2O}(g) \quad \text{(balanced)}$$

When the equation is balanced, the coefficients indicate the ratio in which the reactants and products combine and form. Our equation now reads "two

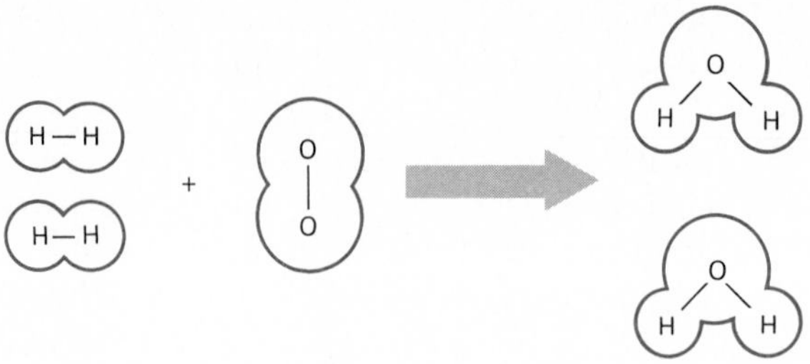

FIGURE 10.1
In the reaction of hydrogen and oxygen, two hydrogen molecules react with one oxygen molecule, producing two water molecules.

H_2 combines with O_2 explosively; this reaction should not be demonstrated on a large scale unless it is carefully controlled.

molecules of hydrogen gas combine with one molecule of oxygen gas to yield two molecules of water vapor." (See Figure 10.1.) As we shall see in Chapter 11, the coefficients also indicate mole relationships in chemical reactions.

Let's tackle a slightly more complex equation to further illustrate balancing techniques. Balance the equation for the reaction in which methane gas, $CH_4(g)$, combines with oxygen gas, $O_2(g)$, to yield carbon dioxide gas, $CO_2(g)$, and water vapor, $H_2O(g)$. As before, write the unbalanced equation:

$$CH_4(g) + O_2(g) \longrightarrow CO_2(g) + H_2O(g) \qquad \text{(unbalanced)}$$

When you are balancing equations where one element appears in many of the reactants and products, it is best to balance that atom last. Here, and in many equations, this atom is oxygen. Oxygen frequently is the last atom balanced; hydrogen is normally balanced next to last. Thus, we should start by balancing carbon atoms. There is 1 C on each side of the arrow; therefore, carbons are balanced as written:

$$\underline{C}H_4(g) + O_2(g) \longrightarrow \underline{C}O_2(g) + H_2O(g)$$

Moving to H, we see that 4 H atoms are located on the left side, and 2 H atoms are on the right. A 2 is placed in front of H_2O to produce four H's on the right:

$$CH_4(g) + O_2(g) \longrightarrow CO_2(g) + 2H_2O(g)$$

Four oxygen atoms

All that remains is to balance the oxygen atoms. Since we have balanced all of the products, we now know the total number of oxygen atoms. One CO_2 molecule contains 2 O atoms, and 2 H_2O molecules contain 2 O atoms, giving a total of 4 O atoms; therefore, there must be 4 O atoms in the reactants. Looking at the left side of the equation, we find only two oxygen atoms. Ask yourself, what number times 2 gives 4? The answer is 2, so place a 2 in front of the O_2 to complete the balancing of the equation.

$$\underline{C}H_4(g) + 2\underline{O}_2(g) \longrightarrow \underline{C}O_2(g) + 2H_2O(g) \qquad \text{(balanced)}$$

Finally, check to see that all atoms are balanced: 1 C atom, 4 H atoms,

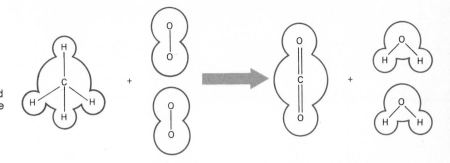

FIGURE 10.2
In the reaction of methane and oxygen, one methane molecule reacts with two oxygen molecules, producing one carbon dioxide molecule and two water molecules.

and 4 O atoms are located on each side of the equation. Our equation states "one molecule (or mole) of methane gas combines with two molecules (moles) of oxygen gas to produce one molecule (mole) of carbon dioxide gas and two molecules (moles) of water vapor." (See Figure 10.2.)

A summary of the rules that can be followed to successfully balance simple chemical equations follows:

Rule 1. Write an unbalanced equation including the correct formulas of all reactants and products.

This first step is most significant, since one incorrect formula alters the way the equation is balanced or in some instances makes it impossible to balance. Always double-check to see that you have written the unbalanced equation properly.

Instead of balancing equations haphazardly, think first of a "plan of attack" and then execute it.

Rule 2. Determine a logical sequence for balancing each atom in the equation, leaving those atoms that appear in more than one compound on each side for last.

It is often best to start with metals, especially those that appear only in one molecule on each side of the equation. Next proceed to nonmetals that also occur in one molecule on each side; then balance the remaining nonmetals that are found in a number of molecules. Step 2 is a planning step; it simplifies balancing equations and gives an orderly approach to the problem.

Rule 3. Balance each atom, one at a time, by placing appropriate coefficients in front of the atoms and molecules in the equation. If possible, proceed in the predetermined order arrived at in Step 2.

After balancing an atom, underline it on both sides of the equation to indicate to yourself that it has been balanced. When all atoms are underlined, the equation should be balanced. If after balancing a couple of atoms you see that your predetermined order for balancing the atoms is not the most efficient, drop it, and proceed in the most efficient way to successfully balance the equation.

Rule 4. Check to see that all atoms are balanced in the equation. If an equal number of atoms are on each side of the equation, for each different atom, then the equation is balanced; if not, repeat Step 3. These steps for balancing equations are summarized in Figure 10.3.

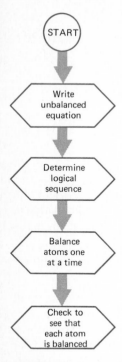

FIGURE 10.3
Steps to follow when you
balance an equation

Some other helpful hints to consider when balancing equations are:

1. Although whole numbers are generally preferred when balancing equations, it is not incorrect to use fractional coefficients occasionally. Certain equations are more easily balanced with one fraction. Consider the following correctly balanced equation:

$$Na(s) + H_2O(l) \longrightarrow NaOH(aq) + \tfrac{1}{2}H_2(g)$$

To eliminate the fraction, the coefficient of each substance in the equation is multiplied by 2.

$$2Na(s) + 2H_2O(l) \longrightarrow 2NaOH(aq) + H_2(g)$$

2. If you find that all of the coefficients are divisible by a small whole number, divide by that number to reduce to the lowest set of coefficients. Balanced equations are correct when they are expressed in the lowest possible multiple of coefficients. For example:

$$4C(s) + 2O_2(g) \longrightarrow 4CO(g)$$

is incorrect. Dividing by 2 gives the correctly balanced equation:

$$2C(s) + O_2(g) \longrightarrow 2CO(g)$$

3. If polyatomic ions are found as reactants and are unchanged after the reaction, they can be balanced as a unit. For instance, if two nitrate ions, NO_3^-, are found within a reactant, you can place a 2 in front of the product containing NO_3^- to balance the nitrates.
4. If an odd number of atoms appear on one side and an even number of atoms are found on the other side, multiply the odd number by 2 in order to give an even number.

Example Problem 10.1 gives additional examples of balancing equations.

Example Problem 10.1

Balance the following equations:
(a) $NH_3(g) + O_2(g) \longrightarrow NO(g) + H_2O(g)$
(b) $C_2H_6(g) + O_2(g) \longrightarrow CO_2(g) + H_2O(g)$

Solution

(a) $NH_3(g) + O_2(g) \longrightarrow NO(g) + H_2O(g)$ (unbalanced)

Oxidizing ammonia to produce NO and H_2O is the first step in the Ostwald process for the production of nitric acid from ammonia. To most efficiently conduct this reaction, the ammonia is combined with an excess of air, heated to about 650°C, and then passed over a special metal catalyst.

Observe that O atoms are found in all but one compound; balance them last. If we balance the H atoms first, then we will know the total number of N atoms in the equation. Therefore, balance H atoms and N atoms before O atoms.

Balance the H atoms by finding the lowest multiple of 3 and 2; this is 6. Place a 2 as the coefficient of NH_3, and a 3 in front of H_2O.

$$2N\underline{H}_3(g) + O_2(g) \longrightarrow NO(g) + 3\underline{H}_2O(g)$$

Now that we know that 2 NH_3 molecules are required, we know the number of N atoms needed—2. However, if we place a 2 in front of the NO, that will leave us with 5 O atoms—an odd number. Thus we cannot balance the O atoms without using a fraction, 2.5. The equation can be balanced by placing 2.5 as the coefficient of O_2. Fractions can be avoided by changing the coefficients of NH_3 and H_2O so that an odd number of O atoms does not result. This is accomplished by multiplying the coefficients by 2, yielding 4 and 6.

$$4N\underline{H}_3(g) + O_2(g) \longrightarrow NO(g) + 6\underline{H}_2O(g)$$

Now we have 12 H atoms on either side, and we know that we need 4 N atoms. Place 4 as the coefficient in front of NO to balance the N atoms.

$$4\underline{N}H_3(g) + O_2(g) \longrightarrow 4\underline{N}O(g) + 6H_2O(g)$$

O atoms are the only ones remaining to be balanced. All of the products have their correct coefficients; thus, we need 10 O atoms, 4 O from the 4 NO and 6 O from the 6 H_2O. Only 2 O atoms are on the left side; so, we can multiply by 5 to give 10 O.

$$4\underline{N}H_3(g) + 5\underline{O}_2(g) \longrightarrow 4\underline{N}O(g) + 6\underline{H}_2O(g) \quad \text{(balanced)}$$

Check to see that there are 4 N atoms, 12 H atoms, and 10 O atoms on either side of the equation.

(b) $C_2H_6(g) + O_2(g) \longrightarrow CO_2(g) + H_2O(g)$ (unbalanced)
In this equation it is convenient to start with C atoms, proceed to H atoms, and balance the O atoms last.

Two C atoms are on the left; they are balanced by placing 2 in front of CO_2.

$$\underline{C}_2H_6(g) + O_2(g) \longrightarrow 2\underline{C}O_2(g) + H_2O(g)$$

Six H atoms in C_2H_6 are balanced by placing a 3 in front of the H_2O.

$$\underline{C}_2H_6(g) + O_2(g) \longrightarrow 2\underline{C}O_2(g) + 3\underline{H}_2O(g)$$

It might be helpful to think in terms of moles when using fractional coefficients. In this equation, 1 mol C_2H_6 combines with 3.5 mol O_2.

A total of 7 O atoms are found in the products. This time let's balance the equation using fractional coefficients. Since there are 2 O atoms on the left, ask yourself "what number multiplied by 2 gives 7?" The answer is $\frac{7}{2}$ or 3.5; thus, the correct coefficient is 3.5.

$$\underline{C_2H_6}(g) + 3.5\underline{O_2}(g) \longrightarrow 2\underline{CO_2}(g) + 3\underline{H_2O}(g) \qquad \text{(balanced)}$$

A check reveals that 2 C atoms, 6 H atoms, and 7 O atoms are in both the reactants and products.

REVIEW PROBLEMS

10.5 Balance the following equations:
(a) $N_2O_4 \longrightarrow NO_2$ (b) $Al + F_2 \longrightarrow AlF_3$
(c) $P_4 + Br_2 \longrightarrow PBr_3$ (d) $ZnS + O_2 \longrightarrow ZnO + SO_2$
(e) $C + SO_2 \longrightarrow CS_2 + CO$

10.6 Change the following word equation to a balanced chemical equation: Calcium phosphide, Ca_3P_2, combines with water, H_2O, to yield calcium hydroxide, $Ca(OH)_2$, and phosphine, PH_3.

10.7 Translate the following word equation into a balanced chemical equation: Aqueous sulfuric acid combines with aqueous aluminum hydroxide, producing liquid water and aqueous aluminum sulfate.

10.3 CLASSIFICATION OF INORGANIC CHEMICAL REACTIONS

Inorganic chemical reactions can be classified in many different ways. A common grouping of inorganic reactions is as follows:

1. Combination reactions
2. Decomposition reactions
3. Single displacement reactions
4. Metathesis reactions

You should realize that the above classification is an artificial grouping of reactions that helps chemists to better understand chemical reactions. Not all chemical reactions fit into this simple grouping, and there is some overlapping among groups.

Don't lose sight of the fact that all of the information presented in this section is the result of studying each reaction experimentally (in the laboratory). Products obtained in chemical reactions depend mainly on the specific conditions: different conditions give different products.

It is helpful to organize and study the various classes of inorganic reactions in a systematic fashion. Many students find that writing general and specific equations on small index cards greatly assists in learning the information. Write an equation on one side of the card, and the reaction type on the other side. (See

FIGURE 10.4
One way to study chemical reactions is to write the equation on one side of a small index card and the name of the reaction type on the other side of the card. Prepare one card for each reaction to be learned.

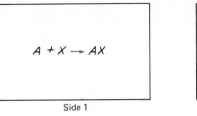

A + X → AX

Side 1

Combination reaction

Side 2

Figure 10.4.) This is an extremely effective technique for learning chemical equations.

Combination Reactions

Combination reactions (sometimes called addition reactions) have the following form:

$$A + X \longrightarrow AX$$

In a combination reaction, two reactants combine to produce one product.

Three possibilities exist for combination reactions: (1) two elements, (2) an element and a compound, or (3) two compounds are united. Let us look at each case individually.

Following are equations of combination reactions in which two elements combine:

$$\text{(1)} \qquad 4Na(g) + O_2(g) \xrightarrow{\Delta} 2Na_2O(s)$$

$$\text{(2)} \qquad 2H_2(g) + O_2(g) \xrightarrow{\text{spark}} 2H_2O(g)$$

$$\text{(3)} \qquad P_4(s) + 5O_2(g) \xrightarrow{\Delta} P_4O_{10}(s)$$

$$\text{(4)} \qquad H_2(g) + Cl_2(g) \xrightarrow{\text{light}} 2HCl(g)$$

Combining the metal U with F_2 produces the gas UF_6. Uranium hexafluoride is used to separate U isotopes through a diffusion process, in which the heavier isotopes diffuse more slowly than the lighter ones.

$$\text{(5)} \qquad U(s) + 3F_2(g) \longrightarrow UF_6(g)$$

In each of these equations we see either a metal or nonmetal combining with a reactive nonmetal, oxygen or a halogen. When metals combine with oxygen (Equation 1), the product is called a **metal oxide:**

$$\text{Metal} + O_2 \longrightarrow \text{metal oxide}$$

Nonmetal oxides result when a nonmetal combines with oxygen (Equations 2 and 3):

$$\text{Nonmetal} + O_2 \longrightarrow \text{nonmetal oxide}$$

Whenever O_2 combines with another substance, the reaction is referred to as an oxidation reaction. Oxidations occur when substances lose electrons.

Formation of oxides is sensitive to the reaction conditions and the availability of oxygen. Carbon, a nonmetal, combines with oxygen to form two oxides of carbon, carbon dioxide and carbon monoxide. If an excess amount of O_2 is present and a limited quantity of C, the main product is CO_2; however, if the O_2 is limited and the C is in excess, CO is the main product.

(6) $$C(s) + O_2(excess) \xrightarrow{\Delta} CO_2(g) \quad \text{(major product)}$$

(7) $$2C(s) + O_2(limited) \xrightarrow{\Delta} 2CO(g) \quad \text{(major product)}$$

Metal hydrides, when placed into water, give off H_2 gas. Thus, metal hydrides are used to store hydrogen that is released at a later time.

Hydrogen combines with both metals and nonmetals to produce **metal** and **nonmetal hydrides**:

(8) $$2Na(l) + H_2(g) \xrightarrow{\Delta} 2NaH(s) \quad \text{(sodium hydride)}$$

(9) $$Ca(s) + H_2(g) \xrightarrow{\Delta} CaH_2(s) \quad \text{(calcium hydride)}$$

(10) $$N_2(g) + 3H_2(g) \xrightarrow{\Delta} 2NH_3(g)$$

(11) $$F_2(g) + H_2(g) \longrightarrow 2HF(g)$$

When hydrogen combines with various metals, metal hydrides result. In these ionic compounds, the H is in a negative oxidation state (-1) as a result of hydrogen's higher electronegativity, i.e., its greater capacity to hold onto electrons.

In the second category of combination reactions, an element combines with a compound. The following are examples:

KO_2 is potassium superoxide. KO_2 is used in special survival masks for miners. Water from a miner's breath reacts with KO_2, giving off life-sustaining O_2 and producing KOH, which combines with exhaled CO_2 to form $KHCO_3$.

(12) $$CO(g) + \tfrac{1}{2}O_2(g) \xrightarrow{\Delta} CO_2(g)$$

(13) $$2K_2O(s) + 3O_2(g) \xrightarrow{\Delta} 4KO_2(s)$$

(14) $$PbCl_2(s) + Cl_2(g) \longrightarrow PbCl_4(l)$$

(15) $$PF_3(g) + F_2(g) \longrightarrow PF_5(g)$$

In Equations 12 and 13, oxygen combines with a nonmetal and metal oxide, respectively, giving products containing a larger number of oxygens. These two equations are also examples of **oxidation reactions** (addition of oxygen). Equations 14 and 15 are examples of **halogenation** reactions (addition of halogens).

In the third group of combination reactions, two compounds join together.

(16) $$SO_3(g) + H_2O(l) \longrightarrow H_2SO_4(aq)$$

(17) $$BaO(s) + H_2O(l) \longrightarrow Ba(OH)_2(aq)$$

(18) $$CO_2(g) + MgO(s) \longrightarrow MgCO_3(s)$$

Metal oxides are sometimes called basic anhydrides—bases without water. Placing a metal oxide into water gives a basic solution. Similarly, nonmetal oxides are called acidic anhydrides.

In Equation 16, a nonmetal oxide, SO_3, combines with water to produce sulfuric acid, H_2SO_4. Many nonmetal oxides, when added to water, produce acids. In contrast, metal oxides combine with water to form bases (metal hydroxides). The product of Equation 17 is aqueous barium hydroxide, a base. In equation 18, a nonmetal oxide, CO_2, combines with a metal oxide, MgO, to produce a salt.

Decomposition Reactions

Decomposition reactions have the general form:

$$AX \longrightarrow A + X$$

Only one reactant is found in a **decomposition reaction,** and under the proper conditions it breaks down to two or more products. Decomposition reactions always have a compound as the reactant; elements cannot be chemically decomposed.

Many decomposition reactions are initiated by heat, which is required to break the bonds in the starting materials. Certain less stable compounds will spontaneously decompose without heat. Other compounds require electricity, light, or special catalysts to decompose.

Examples of decomposition reactions are:

(19) $\qquad 2H_2O(l) \xrightarrow{\text{elec}} 2H_2(g) + O_2(g)$

(20) $\qquad 2N_2O(g) \xrightarrow{\Delta} 2N_2(g) + O_2(g)$

Equation 21 is of historical significance. After hearing about the reaction from Priestley, Lavoisier showed that O_2, not phlogiston, is the substance that supports combustion.

(21) $\qquad 2HgO(s) \xrightarrow{\Delta} 2Hg(l) + O_2(g)$

(22) $\qquad CaCO_3(s) \xrightarrow{\Delta} CaO(s) + CO_2(g)$

(23) $\qquad 2NaHCO_3(s) \xrightarrow{\Delta} Na_2CO_3(s) + CO_2(g) + H_2O(g)$

(24) $\qquad 2KClO_3(s) \xrightarrow[\text{cat}]{\Delta} 2KCl(s) + 3O_2(g)$

In Equations 19 to 21, compounds are decomposed to elements. With the addition of energy (heat or electricity), oxygen gas is driven off. Equation 24 illustrates the case in which a compound is not fully decomposed to elements. Potassium chlorate, $KClO_3$, when heated with MnO_2 (catalyst) liberates $O_2(g)$, leaving a salt, KCl, behind.

Equations 22 and 23 are decomposition reactions in which the products include both compounds and elements. Equation 22 shows a carbonate, $CaCO_3$, decomposing to two oxides, CaO and CO_2, a characteristic reaction of carbonate salts. Since hydrogen carbonates are similar to carbonates, they also liberate $CO_2(g)$ when heated (Equation 23).

A group of compounds called hydrated salts, or hydrates, release water when heated.

(25) $\qquad \underset{\substack{\text{Hydrated}\\\text{salt}}}{CaSO_4 \cdot 2H_2O(s)} \longrightarrow \underset{\substack{\text{Anhydrous}\\\text{salt}}}{CaSO_4(s)} + 2H_2O(g)$

Placement of a dot between the $CaSO_4$ and $2\,H_2O$ in the formula of the hydrated salt indicates that the water molecules are loosely bonded to the salt. On heating, the hydrated salt is dehydrated, resulting in the anhydrous salt, i.e., the salt without water.

Single Displacement Reactions

A **displacement reaction** (also called replacement reaction) occurs when an element displaces another element that is a part of a compound:

$$A + BX \longrightarrow AX + B$$

$$A + BX \longrightarrow AX + B$$

In this general equation, element A replaces element B in compound BX. Typically in displacement reactions, a metal displaces another metal or hydrogen in the compound with which it is combining. Following are some examples of displacement reactions:

(26) $\quad Zn(s) + Pb(NO_3)_2(aq) \longrightarrow Zn(NO_3)_2(aq) + Pb(s)$

(27) $\quad Mg(s) + H_2SO_4(aq) \longrightarrow MgSO_4(aq) + H_2(g)$

(28) $\quad 2K(s) + 2HOH(l) \longrightarrow 2KOH(aq) + H_2$

When (aq) is written next to an ionic substance, MX, it indicates that the substance is broken up into ions, each ion surrounded by many water molecules.

In Equation 26, a metal, Zn, replaces lead, Pb, in aqueous solution. After the reaction, solid lead, Pb, is produced; the zinc is dissolved in water, where it becomes $Zn^{2+}(aq)$. In Equation 27, the metal Mg replaces hydrogen in sulfuric acid, H_2SO_4, yielding the dissolved salt $MgSO_4(aq)$ and hydrogen gas, H_2. More reactive metals can displace hydrogen in water, as shown in Equation 28, where potassium metal replaces hydrogen and produces aqueous potassium hydroxide, $KOH(aq)$, and hydrogen gas, H_2.

Metals are ranked according to their ability to displace other metals and hydrogen from compounds.

This ranking of metals and hydrogen is called the **activity series of metals.** A partial activity series is shown in Table 10.2. Metals on the left side of the list displace metals farther to the right. For example, Ni can displace Pb from compounds; Pb is unable to displace Ni from its compounds.

(29) $\quad Ni(s) + Pb(NO_3)_2(aq) \longrightarrow Ni(NO_3)_2(aq) + Pb(s)$

(30) $\quad Pb(s) + Ni(NO_3)_2(aq) \longrightarrow$ no reaction (NR)

Halogens (not included in Table 10.2) can displace halides, which are halogens in the -1 oxidation state.

(31) $\quad Cl_2(g) + 2NaI(aq) \longrightarrow 2NaCl(aq) + I_2(s)$

(32) $\quad Br_2(l) + 2NaI(aq) \longrightarrow 2NaBr(aq) + I_2(s)$

TABLE 10.2 ACTIVITY SERIES OF SELECTED METALS

Most active ————————————————————————————————————— Least active

Li	K	Ba	Sr	Ca	Na	Mg	Al	Mn	Zn	Fe	Ni	Sn	Pb	H	Cu	Hg	Ag	Pt	Au
Liberates H_2 in cold H_2O, steam, or acid						Liberates H_2 in steam or acid					Liberates H_2 only in acid solutions								

Halogen's order of reactivity is $F_2 > Cl_2 > Br_2 > I_2$. Consequently, Cl_2 displaces both Br^- and I^-; Br_2 is capable of displacing only I^-.

Metathesis Reactions   Two substances are displaced in **metathesis reactions** (sometimes called double displacement reactions):

$$AY + BX \longrightarrow AX + BY$$

$$AY + BX \longrightarrow AX + BY$$

In the general form of the equation, we see that A is initially combined with Y, and B is combined with X. A and B are the elements that exist in positive oxidation states; X and Y have negative oxidation numbers. When AY and BX combine, A displaces B from BX and B then attaches to Y, yielding AX and BY.

Neutralization reactions, in which acids combine with bases (metallic hydroxides) to form salts and water, are good examples of metathesis reactions. Consider the following neutralization reactions:

The net equation for the neutralization reaction of a strong acid by a strong base is $H^+(aq) + OH^-(aq) \rightarrow H_2O(l)$.

$$(33) \qquad HCl(aq) + NaOH(aq) \longrightarrow NaCl(aq) + H_2O(l)$$
$$(34) \qquad HNO_3(aq) + KOH(aq) \longrightarrow KNO_3(aq) + H_2O(l)$$
$$(35) \qquad H_2SO_4(aq) + Mg(OH)_2(aq) \longrightarrow MgSO_4(aq) + 2H_2O(l)$$

In Equations 33 to 35 hydrogen ions, $H^+(aq)$, from each acid combine with the $OH^-(aq)$ from each base, giving water. Metals in the bases then become associated with the anions originally attached to H^+.

Many reactions that occur in aqueous solution are classified as metathesis reactions. Aqueous reactions are those that take place when two or more solutions are poured together, yielding either a solid insoluble substance (a substance that does not dissolve in water), a precipitate, or a gaseous product that bubbles out of the water. Examples of aqueous reactions are as follows:

$$(36) \qquad NaBr(aq) + AgNO_3(aq) \longrightarrow AgBr(s) + NaNO_3(aq)$$
$$(37) \qquad BaCl_2(aq) + K_2SO_4(aq) \longrightarrow BaSO_4(s) + 2KCl(aq)$$
$$(38) \qquad (NH_4)_2S(aq) + Cu(C_2H_3O_2)_2(aq) \longrightarrow CuS(s) + 2NH_4C_2H_3O_2(aq)$$
$$(39) \qquad 2HCl(aq) + Na_2CO_3(s) \longrightarrow 2NaCl(aq) + CO_2(g) + H_2O(l)$$
$$(40) \qquad KCN(aq) + HNO_3(aq) \longrightarrow KNO_3(aq) + HCN(g)$$

Equations 36 to 38 are examples of metathesis reactions in which precipitates form; Equations 39 and 40 show reactions in which a gas is released when solutions are mixed. Equation 39, at first glance, might not look like a metathesis reaction. Initially, $H_2CO_3(aq)$, carbonic acid, is formed and immediately decomposes to produce CO_2 and H_2O.

$$(41) \qquad H_2CO_3(aq) \longrightarrow CO_2(g) + H_2O(l)$$

A summary of the reaction types is presented in Table 10.3.

TABLE 10.3   SUMMARY OF SELECTED INORGANIC REACTIONS

I. Combination reactions (A + X $\longrightarrow$ AX)
   A. Metal + nonmetal $\longrightarrow$ salt
   B. Metal + oxygen $\longrightarrow$ metal oxide (basic oxide)
   C. Nonmetal + oxygen $\longrightarrow$ nonmetal oxide (acidic oxide)
   D. Metal + hydrogen $\longrightarrow$ metal hydride
   E. Nonmetal + hydrogen $\longrightarrow$ nonmetal hydride
   F. Metal oxide + water $\longrightarrow$ base (metal hydroxide)
   G. Nonmetal oxide + water $\longrightarrow$ acid (oxyacid)
   H. Metal oxide + nonmetal oxide $\longrightarrow$ salt
II. Decomposition reactions (AX $\longrightarrow$ A + X)
   A. Oxide $\longrightarrow$ element + oxygen gas
   B. Carbonate $\longrightarrow$ oxide + carbon dioxide
   C. Hydrogen carbonate $\longrightarrow$ carbonate + carbon dioxide + water
   D. Hydrate $\longrightarrow$ anhydrous salt + water
III. Single displacement reactions (A + BX $\longrightarrow$ AX + B)
   A. Active metal + water $\longrightarrow$ hydroxide (or oxide) + hydrogen gas
   B. Metal + acid $\longrightarrow$ salt solution + hydrogen gas
   C. Metal + salt solution $\longrightarrow$ displaced metal + new salt solution
   D. Metal + salt solution $\longrightarrow$ gas + new salt solution
   E. Halogen + halide solution $\longrightarrow$ displaced halogen + new halide solution
IV. Metathesis reactions (AY + BX $\longrightarrow$ AX + BY)
   A. Acid + base $\longrightarrow$ salt + water
   B. Two aqueous solutions $\longrightarrow$ precipitate + salt solution
   C. Two aqueous solutions $\longrightarrow$ gas + salt solution
   D. Acid + carbonate solution $\longrightarrow$ salt solution + carbon dioxide + water
   E. Metal oxide + acid $\longrightarrow$ salt + water

REVIEW PROBLEMS

10.8 Write the general form for each of the following reactions: (a) single displacement, (b) decomposition, (c) metathesis, (d) combination.

10.9 Write a balanced equation that illustrates the following: (a) combination of two compounds, (b) decomposition of a metal oxide, (c) displacement of hydrogen by a metal, and (d) decomposition of a hydrate salt.

10.10 Explain how a metathesis reaction is different from a single displacement reaction. Give two examples.

10.11 Write three equations that illustrate different types of metathesis reactions.

10.12 Identify the class of equations to which each of the following belongs:
   (a) $H_2O_2(l) \longrightarrow H_2O(l) + \frac{1}{2}O_2(g)$
   (b) $2Na(s) + 2H_2O(l) \longrightarrow 2NaOH(aq) + H_2(g)$
   (c) $P_4O_{10}(s) + 6H_2O(l) \longrightarrow 4H_3PO_4(aq)$
   (d) $2Al(s) + 3Zn(NO_3)_2(aq) \longrightarrow 2Al(NO_3)_3(aq) + 3Zn(s)$
   (e) $CaCl_2(aq) + Na_2CO_3(aq) \longrightarrow CaCO_3(s) + 2NaCl(aq)$

10.4 WRITING CHEMICAL EQUATIONS

Writing equations is an important skill that is developed through experience and practice. In this section we shall concentrate on predicting the products of a reaction, given the reactants, and then balancing the equation. Use the following guidelines to help write and balance chemical equations.

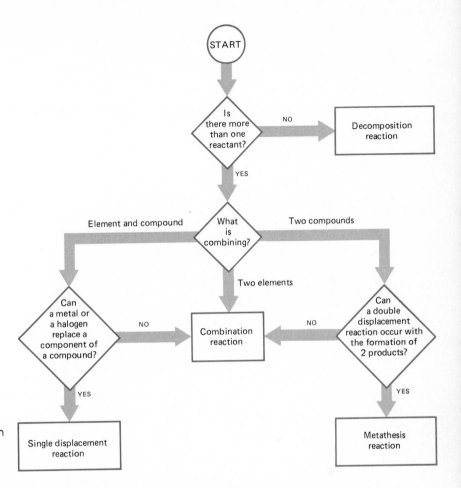

FIGURE 10.5
To identify the reaction type, begin at the top of the flowchart, correctly answer each question, and then proceed in the direction indicated by the arrow with the correct answer.

Guidelines for Writing Chemical Equations

There are three steps in writing chemical equations, given the reactants.

Step 1. Classify the reaction being considered. To assist in the task of identifying the reaction class, refer to Figure 10.5, a flowchart for finding the class of inorganic reaction. Start at the top of the chart and answer the questions, following the arrows until the reaction class is identified.

Step 2. After you decide on the reaction type, combine the reactants in a manner consistent with the reaction type. If the reaction is a single displacement, use the activity series to determine if the metal is capable of displacing the metal ion in the solution. Completion of Step 2 yields a reasonable unbalanced equation.

Summary of guidelines:

1. Classify reaction.
2. Combine or decompose in a manner consistent with reaction class.
3. Balance and check equation.

Step 3. Balance the equation using the rules learned in Section 10.2. If the equation does not balance after following all of the rules, it is possible that you have written the wrong equation or the wrong formula for one of the reactants or products. Go back and double-check your reasoning and the formulas that you have written. Maybe you have omitted a substance or made a careless mistake. Figure 10.6 summarizes the three steps required to write chemical equations.

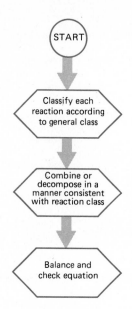

FIGURE 10.6
Steps to follow when you write
equations

Now let's use the guidelines to write equations for the reactions in the
following examples.

Example 1 Predict the products, and write a balanced equation for the reaction when magnesium combines with sulfur.
First, write the correct symbols of the reactants left of an arrow:

$$Mg + S \longrightarrow ?$$

We can eliminate three of the four general types of reactions. Decomposition, displacement, and metathesis reactions are eliminated because they all
contain at least one compound as a reactant. Thus, the reaction is a combination
reaction of a metal (Mg) combining with a nonmetal (S) to produce a salt.
What salt forms when Mg and S are combined? Magnesium is an alkaline
earth metal (group IIA), and when bonded it exists in the $+2$ oxidation state;
sulfur, a chalcogen (group VIA), in binary compounds exists in the -2 oxidation
state. Consequently, we would expect the product to be magnesium sulfide,
MgS, giving the following balanced equation:

*Sulfur does not normally exist as
individual S atoms. Sulfur is
found mainly in rings of 6, 8, 10,
or 12 sulfur atoms. The most
commonly found form of sulfur
is S_8, which exists in three
different crystalline forms.*

$$Mg + S \longrightarrow MgS \quad \text{(balanced)}$$

Example 2 Write and balance the equation for the reaction in which an aqueous solution of
sodium phosphate is combined with an aqueous solution of zinc(II) chloride.
Write the reactants.

$$Na_3PO_4(aq) + ZnCl_2(aq) \longrightarrow ?$$

TABLE 10.4   SOLUBILITY RULES FOR COMMON COMPOUNDS IN WATER

Water-Soluble Compounds

1. Most salts containing alkali metal ions (IA) or the ammonium ion (NH_4^+) are soluble.
2. Most salts containing acetate, $C_2H_3O_2^-$, and nitrate, NO_3^-, are soluble.
3. Most salts containing sulfate, SO_4^{2-}, are soluble except those in which the SO_4^{2-} is combined with Pb^{2+}, Sr^{2+}, and Ba^{2+}. Sulfate salts of Ca^{2+} and Ag^+ are only slightly soluble.
4. Chlorides, Cl^-, are soluble except for chlorides of Ag^+, Hg_2^{2+}, and Pb^{2+} ($PbCl_2$ is soluble in boiling water).

Water-Insoluble Compounds

1. Virtually all oxides, O^{2-}, and hydroxides, OH^-, are insoluble, except for those salts with alkali metal (IA) and alkaline earth metal (IIA) ions.
2. Carbonates, CO_3^{2-}, and phosphates, PO_4^{3-}, are insoluble except those salts with alkali metal (IA) and ammonium, NH_4^+, ions.
3. Sulfides, S^{2-}, are insoluble except those combined with alkali metals (IA), alkaline earth (IIA) metals, and ammonium, NH_4^+, ions.

Learning solubility rules is also simplified by using index cards. Place the various groups on one side, and their solubilities on the other side.

Here we have two salt solutions combining together. Decomposition reactions are ruled out because more than one reactant is present. When two compounds combine together, there are two possibilities of reaction types: metathesis or combination. Combination reactions are eliminated, since it is quite unlikely that four ions (Na^+, PO_4^{3-}, Zn^{2+}, and Cl^-) would combine together to produce a single compound. A more reasonable outcome is a double displacement reaction; therefore, our reaction belongs to the metathesis class.

We now have a seemingly formidable problem. If this is a metathesis reaction, which of the two products precipitates out of solution? Such a problem is solved by using a water-solubility table. **Solubility** is the degree to which a substance dissolves in a solvent. Table 10.4 presents a series of general rules that are followed to determine the expected water solubility of a compound.

Because we are dealing with a metathesis reaction, the Na^+ replaces the Zn^{2+}, producing NaCl(aq), which is soluble in water. The Zn^{2+} ion then attaches to the PO_4^{3-}, giving the insoluble $Zn_3(PO_4)_2$(s). (See Table 10.4.) Thus, the unbalanced equation is:

$$Na_3PO_4(aq) + ZnCl_2(aq) \longrightarrow Zn_3(PO_4)_2(s) + NaCl(aq) \quad \text{(unbalanced)}$$

Once the equation is balanced, we obtain

$$2Na_3PO_4(aq) + 3ZnCl_2(aq) \longrightarrow Zn_3(PO_4)_2(s) + 6NaCl(aq) \quad \text{(balanced)}$$

Example 3   Write and balance the equation for the reaction that occurs when a pure zinc bar is placed in a mercury(II) nitrate solution.

$$Zn(s) + Hg(NO_3)_2(aq) \longrightarrow ?$$

In the equation, Zn, a metal, is immersed in a $Hg(NO_3)_2$ solution. Decomposition and metathesis reactions are easily eliminated because an element is combining with a compound. Checking the activity series in Table 10.2, we find that zinc is a more active metal than Hg; hence, Zn can displace the Hg in a single displacement reaction. Accordingly, the equation becomes

$$Zn(s) + Hg(NO_3)_2(aq) \longrightarrow Zn(NO_3)_2(aq) + Hg(l) \quad \text{(balanced)}$$

Check the equation to see that it is balanced as written.

REVIEW PROBLEMS

10.13 Classify the expected type of reaction that the following reactants would undergo under appropriate conditions: (a) two different nonmetals, (b) metal oxide plus heat, (c) metal oxide and a nonmetal oxide, (d) acid and a base, (e) nonmetal oxide and oxygen, (f) halogen and halide salt, (g) acid and cyanide salt, and (h) carbonate plus heat.

10.14 Predict the products of the following reactions; write NR (no reaction) if they do not combine:
(a) $MgSO_4 \cdot 7H_2O \xrightarrow{\Delta}$ (b) $AgNO_3(aq) + Na_2S(aq) \longrightarrow$
(c) $P_4(s) + O_2(g) \longrightarrow$ (d) $Ag(s) + ZnSO_4(aq) \longrightarrow$
(e) $KOH(aq) + HClO_3(aq) \longrightarrow$

10.15 Predict the products of the following reactions, and write a complete balanced equation:
(a) $Ca(s) + O_2(g) \longrightarrow$ (b) $KClO_3(s) \xrightarrow{\Delta}$
(c) $SO_2(g) + H_2O(l) \longrightarrow$ (d) $Cs(l) + H_2O(l) \longrightarrow$
(e) $Pb(NO_3)_2(aq) + K_2SO_4(aq) \longrightarrow$ (f) $H_2(g) + S(s) \longrightarrow$

10.5 ENERGY CONSIDERATIONS IN CHEMICAL REACTIONS

So far, we have concentrated on the changes that matter undergoes during chemical reactions. Even though the matter appears to be the primary concern, chemical reactions take place only when the proper amount of energy is present.

Heat energy, and other sources of energy, are extremely important to chemical reactions. In all reactions, heat is either absorbed or liberated during the course of the reaction. When heat from the surroundings is absorbed, the reaction is called **endothermic,** which literally means "to take in heat." The opposite of an endothermic reaction is an **exothermic** reaction, i.e., one that releases heat to its surroundings. (See Figure 10.7.)

To understand endothermic and exothermic reactions, think of the reactants as having a fixed quantity of chemical potential energy. If an endothermic

An exothermic reaction releases heat energy, while an endothermic reaction absorbs heat energy.

FIGURE 10.7
If heat flows from a chemical reaction to the surroundings, the reaction is classified as exothermic. If heat flows from the surroundings into the reaction, it is classified as an endothermic reaction.

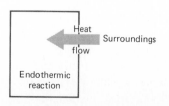

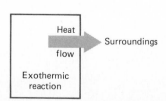

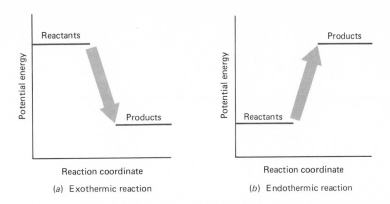

FIGURE 10.8
If the enthalpy of the products is less than the enthalpy of the reactants, an exothermic reaction has taken place. If the enthalpy of the products is greater than the enthalpy of the reactants, an endothermic reaction has taken place.

In endothermic reactions, the enthalpy of the reactants is less than the enthalpy of the products. In exothermic reactions, the enthalpy of the reactants is greater than the enthalpy of the products.

reaction occurs, energy is absorbed by the reactants and is ultimately stored in the products. Therefore, the products have more potential energy than the reactants. In contrast, if an exothermic reaction occurs, the reactants lose energy to the surroundings, producing products with a smaller amount of potential energy. (See Figure 10.8.) The measure of chemical potential energy of reactants and products is called **enthalpy.**

Endothermic reactions are sometimes expressed as follows:

$$A + B + heat \longrightarrow C + D$$

In this equation, heat is added to the reactants, A and B, and stored in the products, C and D. In other words, the total enthalpy of the products is greater than that of the reactants.

In equations illustrating exothermic reactions, heat is written as a product:

The enthalpy change in a chemical reaction is called the heat of reaction, symbolized by writing ΔH.

$$\Delta H = H_{products} - H_{reactants}$$

$$A + B \longrightarrow C + D + heat$$

Heat is liberated in this hypothetical reaction, indicating that the enthalpy of the products is less than the enthalpy of the reactants (see Figure 10.8). In actual equations the word "heat" is not written; instead, the number of kilocalories (kcal) or kilojoules (kJ) is written for a specified amount of substance reacting—usually kilocalories or kilojoules per mole of reactant. For example:

Some believe that CO_2 is becoming a major pollutant. If the level of CO_2 is increased significantly, it would trap energy normally radiated from the earth to outer space ("greenhouse" effect). As a result, the temperature of the earth would rise, melting the polar ice caps and flooding numerous populated areas on earth.

$$CH_4(g) + 2O_2(g) \longrightarrow CO_2(g) + 2H_2O(g) + 213 \text{ kcal } (891 \text{ kJ})$$
$$2Ag_2O(s) + 14.8 \text{ kcal } (61.9 \text{ kJ}) \longrightarrow 4Ag(s) + O_2(g)$$

In the first equation, methane, CH_4, is combined with oxygen gas, O_2, producing carbon dioxide, CO_2, and water vapor, H_2O, but more importantly, 213 kcal/(mol CH_4) of heat energy is produced. This is one of the more significant reactions that occur when natural gas is burned. Natural gas is used to heat houses and as a fuel for industry. Heat energy is by far the most valuable product of this reaction.

TABLE 10.5   HEATS OF COMBUSTION FOR SELECTED SUBSTANCES

Substance	Formula	State	Heat of combustion, −kJ/mol
Octane	C_8H_{18}	Liquid	5450
Propane	C_3H_8	Gas	2200
Carbon	C	Solid	400
Hydrogen	H_2	Gas	240
Benzene	C_6H_6	Liquid	3270

Reactions that are similar to the burning of methane are called **combustion reactions**. A combustion reaction is one in which substances rapidly combine with oxygen, giving off heat and light. Combustible substances are those that undergo combustion reactions. The amount of heat liberated per mole of combustible substance is called the **heat of combustion**. Heats of combustion for various substances are listed in Table 10.5.

Many metals used by people are locked in ores as metal ions. After the ores are extracted from the earth, they are transported to a place where they are heated under the proper conditions to liberate the metal.

In the second of the above equations, 14.8 kcal of heat is required per 2 mol of silver(I) oxide, Ag_2O, to produce silver, Ag, and oxygen gas. A constant heat source is necessary to apply heat to Ag_2O in order to obtain the silver metal, Ag, locked within.

Generally, exothermic reactions are more spontaneous and self-sustaining than endothermic reactions. Heat produced by an exothermic reaction provides a constant source of energy. Endothermic reactions, in contrast, proceed only when heat energy is applied. Once the heat is removed, the reaction usually ceases.

Breaking a bond is an endothermic process. In contrast, bond formation is exothermic.

Heat energy is absorbed or released by chemical reactions as a result of breaking and producing chemical bonds. The breaking of chemical bonds is an endothermic process, while bond formation is an exothermic process. Therefore, exothermic reactions release heat because more energy is released in the formation of bonds than is required to break bonds. In other words, stronger bonds are found in the products than in the reactants. Endothermic reactions are the opposite: more energy is required to break bonds in the reactants than is given off during the reaction—bonds are stronger in the reactants than in the products.

REVIEW PROBLEMS   **10.16** (a) What is an exothermic reaction? (b) What is an endothermic reaction?

10.17 On what side of the arrow is the heat written in an (a) exothermic and (b) endothermic reaction?

10.18 Classify each as either exothermic or endothermic:
   (a) $H_2 + Cl_2 \longrightarrow 2HCl + 44$ kcal
   (b) $Cu_2O + 40$ kcal $\longrightarrow 2Cu + \frac{1}{2}O_2$
   (c) $I_2 + Br_2 + 20$ kcal $\longrightarrow 2IBr$
   (d) $C + O_2 \longrightarrow CO_2 + 94$ kcal

10.19 Explain what happens, in terms of bond breaking and formation, when an endothermic reaction occurs.

SUMMARY

Chemical equations are a shorthand notation that chemists use to indicate what happens during chemical reactions. Each equation has substances separated by an arrow (meaning "yields"). All substances written to the left of the arrow are called reactants, and the substances written to the right of the arrow are called products. Each chemical equation is initially determined by performing laboratory studies to determine the products that result from a given set of reactants and conditions.

Chemical equations obey the law of conservation of mass, which states that matter cannot be created or destroyed during normal chemical changes. To obey this law, an equation must be balanced. A correctly balanced equation indicates the ratio in which the reactants combine to yield the products and the ratio in which the products are formed. Equations are balanced by placing coefficients in front of each substance in the equation so that the same number of atoms of each element appears on both sides of the equation. Usually, the correct coefficients are determined by comparing the number of each different element on either side of the equation in what is called the inspection method.

To better understand reactions, chemists group similar equations into a general class. Elementary inorganic reactions are normally classified into four groups: (1) combination, (2) decomposition, (3) single displacement, and (4) metathesis reactions. Combination reactions are those in which two or more reactants unite to produce a single product. Decomposition reactions are the opposite of combination reactions: a single reactant is broken up into two or more products. Single displacement reactions are those in which an element replaces another element within a compound. Finally, in metathesis reactions (double displacements) two compounds react and the more positive part of one combines with the negative part of the other.

Besides the elements and compounds in a reaction, energy effects must be considered. Reactions are classified according to their ability to either release or take in heat energy. Those reactions in which heat is absorbed are called endothermic reactions. Reactions in which there is a net release of heat energy are called exothermic reactions.

QUESTIONS AND PROBLEMS

10.20 Define the following terms: chemical equation, reactant, product, aqueous solution, catalyst, balancing, combination reaction, decomposition reaction, single displacement reaction, metathesis reaction, oxide, hydride, hydrate, activity series, precipitation, exothermic, endothermic, enthalpy.

Format of Chemical Equations

10.21 What is the meaning of each of the following symbols that are used in chemical equations: (a) (aq), (b) (g), and (c) $\xrightarrow{\text{cat}}$?

10.22 Write the names for all reactants and products in

$BaBr_2 + (NH_4)_2CO_3 \longrightarrow BaCO_3 + 2NH_4Br.$

10.23 For each of the following, write a word equation that expresses exactly what is indicated by the following chemical equations:

(a) $2SO_3(g) \xrightarrow{\Delta} 2SO_2(g) + O_2$

(b) $Hg(l) + Cl_2(g) \longrightarrow HgCl_2(s)$

(c) $N_2(g) + 3H_2(g) \xrightarrow{cat} 2NH_3(g)$

(d) $Al_2S_3(s) + 6H_2O(l) \longrightarrow 2Al(OH)_3(s) + 3H_2S(aq)$

Balancing Equations

10.24 Balance the following equations:

(a) $P_4 + O_2 \longrightarrow P_2O_5$

(b) $Mg + N_2 \longrightarrow Mg_3N_2$

(c) $Li_2O + H_2O \longrightarrow LiOH$

(d) $Cl_2 + KI \longrightarrow I_2 + KCl$

(e) $Cu + O_2 \longrightarrow CuO$

(f) $FeO + SiO_2 \longrightarrow FeSiO_3$

(g) $Na_2CO_3 + C \longrightarrow Na + CO$

(h) $WO_3 + H_2 \longrightarrow W + H_2O$

(i) $B_2O_3 + H_2O \longrightarrow H_3BO_3$

(j) $H_2S + O_2 \longrightarrow SO_2 + H_2O$

10.25 Balance the following slightly more complex equations:

(a) $C_4H_{10} + O_2 \longrightarrow CO_2 + H_2O$

(b) $POF_3 + H_2O \longrightarrow H_3PO_4 + HF$

(c) $Cu(NO_3)_2 \longrightarrow CuO + NO_2 + O_2$

(d) $CaCO_3 + H_3PO_4 \longrightarrow Ca_3(PO_4)_2 + H_2O$

(e) $FeS_2 + O_2 \longrightarrow FeO + SO_2$

(f) $Al + CuSO_4 \longrightarrow Al_2(SO_4)_3 + Cu$

(g) $LiH + AlCl_3 \longrightarrow LiAlH_4 + LiCl$

(h) $Cr_2O_3 + C \longrightarrow Cr + CO$

(i) $(NH_4)_2Cr_2O_7 \longrightarrow Cr_2O_3 + N_2 + H_2O$

(j) $C_{10}H_{22} + O_2 \longrightarrow CO_2 + H_2O$

10.26 Balance the following more difficult equations:

(a) $B_3N_3H_6 + O_2 \longrightarrow N_2O_5 + B_2O_3 + H_2O$

(b) $IBr + NH_3 \longrightarrow NH_4Br + NI_3$

(c) $KAlSi_3O_8 + H_2O + CO_2 \longrightarrow$
$\qquad K_2CO_3 + Al_2Si_2O_5(OH)_4 + SiO_2$

(d) $Co_3O_4 + Al \longrightarrow Co + Al_2O_3$

(e) $Al(OH)_3 + NaOH \longrightarrow NaAlO_2 + H_2O$

(f) $C_7H_6O_2 + O_2 \longrightarrow CO_2 + H_2O$

(g) $HClO_4 + P_4O_{10} \longrightarrow H_3PO_4 + Cl_2O_7$

(h) $XeF_2 + H_2O \longrightarrow Xe + O_2 + HF$

(i) $Na_2H_3IO_6 + AgNO_3 \longrightarrow$
$\qquad Ag_5IO_6 + NaNO_3 + HNO_3$

(j) $XeF_4 + SF_4 \longrightarrow Xe + SF_6$

10.27 Translate the following word equations into correctly balanced chemical equations, and classify each as combination, decomposition, single displacement, or metathesis reactions, or none of these:

(a) Sodium chloride + silver nitrate solutions yields sodium nitrate + silver(I) chloride

(b) Aluminum hydroxide + nitric acid solutions yields aluminum nitrate + water

(c) Chlorine gas + rubidium bromide solution yields rubidium chloride + bromine liquid

(d) Iron(III) acetate + sodium sulfide solutions yields sodium acetate + iron(III) sulfide

(e) Silicon tetrafluoride gas + water yields silicon dioxide solid + hydrofluoric acid solution

(f) Manganese (IV) oxide + hydrochloric acid yields manganese(II) chloride + chlorine gas + water

(g) Dinitrogen tetroxide gas when heated yields nitrogen dioxide

(h) Calcium phosphide + water yields calcium hydroxide + phosphine, PH_3

(i) Silver(I) nitrate on heating yields silver + nitrogen dioxide gas + oxygen gas

(j) Aluminum metal + copper(II) sulfate yields copper metal + aluminum sulfate

(k) Mercury(I) nitrate + potassium chloride yields mercury(I) chloride + potassium nitrate

(l) Calcium sulfate dihydrate yields calcium sulfate + water

(m) Ammonium sulfide + cadmium(II) chloride yields cadmium sulfide + ammonium chloride

(n) Phosphorous acid + sodium hydroxide yields sodium phosphite + water

(o) Ammonia + sulfuric acid yields ammonium sulfate

(p) Carbon disulfide + chlorine gas yields disulfur dichloride + carbon tetrachloride

(q) Calcium phosphate + sulfuric acid yields phosphoric acid + calcium sulfate

(r) Ammonia + copper(I) oxide yields nitrogen gas + water + copper metal

(s) Ammonium nitrate yields water + dinitrogen oxide gas

(t) Magnesium hydroxide + zinc(II) nitrate solutions yields zinc hydroxide + magnesium nitrate

Classes of Inorganic Reactions

10.28 Write an equation to illustrate the following types of combination reactions: (a) a compound combines with an element, (b) two compounds combine together, (c) two elements combine.

10.29 Write an equation for a combination reaction in which each of the following products results: (a) CO, (b) SO_2, (c) NO_2, (d) Cs_2O, (e) PH_3, (f) MgH_2.

10.30 What metal oxide (basic oxide), when added to water, produces each of the following bases: (a) NaOH, (b) $Sr(OH)_2$, (c) $Mg(OH)_2$, (d) $Al(OH)_3$?

10.31 What nonmetal oxide (acidic oxide), when added to water, produces the acid: (a) H_2CO_3, (b) H_2SO_3, (c) H_3PO_4, (d) H_2SO_4?

10.32 Write the formula for a compound that decomposes to the following products:
(a) $MgO + CO_2$ (b) $KCl + O_2$
(c) $K_2CO_3 + CO_2 + H_2O$ (d) $Ba(NO_2)_2 + 1H_2O$
(e) $H_2 + O_2$ (f) $NaNO_2 + O_2$
(g) $SO_2 + O_2$

10.33 Write a balanced single displacement equation illustrating each of the following:
(a) A metal replacing $H_2(g)$ from water
(b) A metal replacing $H_2(g)$ from an acid
(c) A metal replacing copper in a copper(II) nitrate solution
(d) Bromine liquid replacing iodine in a solution of rubidium iodide

10.34 What substances combine in a neutralization reaction? Give three examples.

10.35 Write an equation illustrating a metathesis reaction in which:
(a) An insoluble hydroxide forms
(b) Carbon dioxide gas is one of the products
(c) An insoluble carbonate is produced
(d) The soluble salt NH_4I results.

10.36 Using Table 10.3, predict whether each of the following is water soluble or water insoluble: (a) Na_2SO_4, (b) $Fe(OH)_2$, (c) $MgCO_3$, (d) Hg_2Cl_2, (e) NH_4Br, (f) VO_2, (g) $CsC_2H_3O_2$, (h) Al_2S_3.

Writing Chemical Equations

10.37 Complete and balance each of the following combination equations:
(a) $Zn(s) + S_8 \longrightarrow$ (b) $Al(s) + O_2(g) \longrightarrow$
(c) $Na(s) + F_2(g) \longrightarrow$ (d) $H_2(g) + N_2(g) \longrightarrow$
(e) $Br_2(g) + I_2(g) \longrightarrow$ (f) $CO(g) + O_2(g) \longrightarrow$
(g) $Mg(s) + N_2(g) \longrightarrow$ (h) $Ag(s) + Cl_2(g) \longrightarrow$

10.38 Complete and balance the equations for the following decomposition equations:
(a) $PtO_2 \longrightarrow$ (b) $CuSO_4 \cdot 5H_2O \longrightarrow$
(c) $NaNO_3 \longrightarrow$ (d) $SrCO_3 \longrightarrow$
(e) $LiHCO_3 \longrightarrow$ (f) $H_2O \longrightarrow$

(g) $MgSO_3 \cdot 6H_2O \longrightarrow$ (h) $H_2O_2 \longrightarrow$

10.39 Complete and balance the equation for the following single displacement reactions:
(a) $Zn + Pb(NO_3)_2 \longrightarrow$
(b) $Ba + H_2O \longrightarrow$
(c) $Ni + SnBr_2 \longrightarrow$
(d) $Hg + Fe(NO_3)_2 \longrightarrow$
(e) $KCl + I_2 \longrightarrow$
(f) $Cu(ClO_3)_2 + Mn \longrightarrow$
(g) $Pb + H_2SO_4 \longrightarrow$
(h) $HC_2H_3O_2 + Zn \longrightarrow$

10.40 Complete and balance the equation for the following aqueous metathesis equations:
(a) $NiCl_2 + Ca(OH)_2 \longrightarrow$
(b) $Hg(C_2H_3O_2)_2 + K_2CO_3 \longrightarrow$
(c) $H_3PO_4 + AgNO_3 \longrightarrow$
(d) $H_2SO_3 + Al(OH)_3 \longrightarrow$
(e) $(NH_4)_2S + BaI_2 \longrightarrow$
(f) $Cs_2CO_3 + HC_2H_3O_2 \longrightarrow$
(g) $Li_2SO_4 + Co(NO_3)_2 \longrightarrow$
(h) $NH_4CN + HBr \longrightarrow$

10.41 Translate the following to symbols and formulas, and then complete and balance the equation:
(a) Silver(I) nitrate + copper(II) chloride $\longrightarrow$
(b) Iron + water $\longrightarrow$
(c) Potassium hydroxide + sulfuric acid $\longrightarrow$
(d) Ammonia + oxygen gas $\longrightarrow$
(e) Sulfur trioxide + water $\longrightarrow$
(f) Sulfuric acid + zinc $\longrightarrow$
(g) Nitrogen monoxide + oxygen gas $\longrightarrow$
(h) Aluminum + oxygen gas $\longrightarrow$
(i) Hydrogen gas + tin(II) nitrate $\longrightarrow$
(j) Sodium sulfate decahydrate $\overset{\Delta}{\longrightarrow}$
(k) Ammonia + hydrochloric acid $\longrightarrow$
(l) Sodium nitrite + hydrochloric acid $\longrightarrow$
(m) Sodium acetate + lead(II) acetate $\longrightarrow$
(n) Gold(III) sulfide + iron $\longrightarrow$
(o) Arsenic + chlorine (g) $\longrightarrow$
(p) Calcium nitrate $\overset{\Delta}{\longrightarrow}$
(q) Lithium oxide + water $\longrightarrow$
(r) Tin(IV) chloride + magnesium $\longrightarrow$
(s) Nitric acid + calcium oxide $\longrightarrow$
(t) Boron + fluorine gas $\longrightarrow$

Energy in Chemical Reactions

10.42 Give an example of (a) an exothermic reaction and (b) an endothermic reaction.

10.43 Using the data contained in Table 10.5:
(a) Calculate the amount of heat that is liberated when 1.0 g of octane is combusted.
(b) If all this heat could be transfered to 1.00 kg of water at 25°C, what is the maximum temperature that the water would reach? The specific heat of water is 4.184 J/(g°C).

10.44 Classify the following as endothermic or exothermic:
(a) $PCl_5 + 90 \text{ kcal} \longrightarrow P + \frac{5}{2}Cl_2$
(b) $CaCO_3 + 43 \text{ kcal} \longrightarrow CaO + CO_2$
(c) $FeS + 2H^+ \longrightarrow Fe^{2+} + H_2S + 13 \text{ kJ}$
(d) $C + 2F_2 \longrightarrow CF_4 + 220 \text{ kcal}$

10.45 (a) What is enthalpy?
(b) What enthalpy change occurs in an endothermic reaction?
(c) What enthalpy change occurs in an exothermic reaction?

10.46 Most decomposition reactions are endothermic. Write an explanation to account for this observation.

General Questions

10.47 Correct the following incorrect statements:
(a) Most metal oxides produce acidic solutions when placed into water.
(b) All inorganic reactions belong to four different classes.
(c) Fractional coefficients can never be used to balance chemical equations.
(d) All phosphate salts are insoluble.
(e) Zinc metal, when placed in liquid water, replaces hydrogen and dissolves immediately.
(f) The product of a metathesis reaction is a precipitate.
(g) Substances that undergo endothermic reactions are used as fuels.
(h) When bonds are broken heat energy is released.

10.48 Write and balance the following equations.
(1) The Haber reaction occurs when nitrogen and hydrogen gases are combined at high pressure, at 550°C, and with a metal catalyst, producing ammonia gas.
(2) Ammonia when combusted gives nitrogen monoxide and water vapor (Ostwald process).
(3) Nitrogen monoxide is further oxidized to nitrogen dioxide.
(4) The nitrogen dioxide is pumped through water to yield both nitric acid and nitrous acid.

10.49 Write and balance equations for reactions 1 through 7.
(1) Calcium carbonate, when heated to about 850°C, decomposes to calcium oxide and carbon dioxide.
(2) The carbon dioxide is used to manufacture sodium bicarbonate and sodium carbonate. Carbon dioxide is combined with ammonia, water, and sodium chloride to produce sodium bicarbonate plus ammonium chloride (the Solvay process).
(3) Sodium carbonate is formed along with carbon dioxide and water when sodium bicarbonate is decomposed.
(4) The calcium oxide produced in Equation 1 is combined with carbon at high temperatures to produce calcium carbide and carbon monoxide.
(5) Calcium carbide is then heated with nitrogen gas at 1100°C, producing calcium cyanamide, $CaCN_2$, and carbon.
(6) This calcium cyanamide is combined with the sodium carbonate produced in Equation 3 and carbon to give sodium cyanide, NaCN, and calcium carbonate.
(7) Sodium cyanide is the main source of hydrogen cyanide gas. Sodium cyanide is combined with sulfuric acid, producing hydrogen cyanide and sodium sulfate.

Stoichiometry: Application of the Mole Concept

Study Guidelines

After completing Chapter 11, you should be able to:
1. Explain the meaning of a stoichiometric relationship
2. List reasons for experimentally obtaining quantities different from those predicted by stoichiometric calculations
3. Determine molecular and mole relationships, given a balanced chemical equation
4. Show that in a correctly balanced equation the sum of the masses of the reactants equals the sum of the masses of the products
5. Calculate the number of moles or mass of product, given either the number of moles or mass of a reactant, or vice versa
6. Calculate the amount of heat liberated from a reaction, given the mass or number of moles of starting material
7. Explain how heat energy transfers are measured through calorimetry
8. Determine if a reactant is a limiting reagent
9. Find the maximum mass of product formed, given the masses of all reactants combined
10. Calculate the percent yield of a reaction, given the actual yield and masses of reactants.

The term "stoichiometry" comes from the Greek terms *stoicheion,* meaning "element," and *metron,* "to measure."

What is stoichiometry (stoy-key-ahm-uh-tree)? **Stoichiometry** is the study of mole, mass, energy, and volume relationships in chemical reactions. In stoichiometry, we usually look at the quantities of reactants that combine together to produce various amounts of products.

Many industries make use of stoichiometric relationships to predict the masses of raw materials required to produce the desired amounts of final products. Knowledge of stoichiometry is applied to solve production problems in the recovery of metals from ores, synthesis of medicines and drugs, and manufacture of explosives.

Jeremias Benjamin Richter (1762–1807) was the scientist to coin the term "stoichiometry." Richter applied basic math principles to what was known about combining masses of compounds to explain how substances reacted quantitatively.

Stoichiometric relationships are initially found by investigating a chemical reaction in the laboratory, and then determining the ratios in which the reactants combine and the ratios in which the products form. When these ratios are determined, the correct balanced equation can be written. As we saw in Chapter 10, the coefficients placed in front of substances in chemical equations indicate the theoretical ratio in which reactants combine and products form.

Stoichiometry predicts what should happen, not what will happen, in chemical changes.

A word of caution is in order before we get to the specifics of stoichiometry. Predictions made from stoichiometric considerations are theoretical; this means our answers are not what *will* happen, but what *should* happen, under ideal conditions. If we predict the quantity of a product that forms from specified amounts of reactants, our prediction is the theoretical maximum that could result, but rarely will the maximum quantity be obtained in the laboratory.

Even though a chemical equation appears as a simple relationship, it generally is not. Reactants normally have more than one pathway to follow as they react. In other words, one or more "side reactions" may occur at the same time as the principal one, thus decreasing the yield of the product of interest. Small quantities of impurities can alter the pathway of the reaction—sometimes significantly. Additionally, meticulous attention must be given to the energy requirements of a reaction. If these requirements are not met, different products may result, or the reaction may not occur at all!

REVIEW PROBLEMS
11.1 (a) What is stoichiometry? (b) How is stoichiometry applied?
11.2 What factors tend to decrease the observed yield of a reaction compared with the computed theoretical yield?

11.2 CHEMICAL EQUATION CALCULATIONS

To begin our study of stoichiometry, let us consider a relatively simple chemical reaction, the combination of hydrogen and chlorine gases to produce hydrogen chloride gas:

$$H_2(g) + Cl_2(g) \longrightarrow 2HCl(g)$$

FIGURE 11.1
One molecule of diatomic hydrogen, H_2, collides with one molecule of diatomic chlorine, Cl_2, producing 2 molecules of hydrogen chloride, HCl.

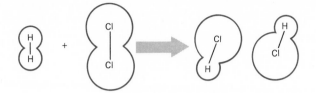

Formation of HCl is the result of a series of intermediate steps that the reactants follow; these steps are called the reaction mechanism.

In words, the equation states that a hydrogen molecule combines with a chlorine molecule to produce two molecules of hydrogen chloride gas. A diagram of this reaction is shown in Figure 11.1.

Knowing the correct coefficients in the equation provides us with a wealth of information. As already stated, the coefficients give an indication of what is happening quantitatively at the molecular level—one H_2 molecule combining with one Cl_2 molecule, producing two HCl molecules.

Hydrogen chloride is a colorless, poisonous gas. HCl melts at $-112°C$ and boils at $-84°C$. When HCl is dissolved in water, it almost totally breaks up into ions—this solution is called hydrochloric acid.

Normal laboratory conditions do not allow us to work with individual molecules; instead we are more concerned with the number of moles of substances. We can move from the molecular level to the mole level by multiplying all components of the equation by Avogadro's number, 6.022×10^{23}. Consequently, 6.022×10^{23} H_2 molecules combine with 6.022×10^{23} Cl_2 molecules to produce 1.204×10^{24} ($2 \times 6.022 \times 10^{23}$) HCl molecules. More simply, 1 mol H_2 combines with 1 mol Cl_2 yielding 2 mol HCl.

$$1 \text{ mol } H_2 + 1 \text{ mol } Cl_2 \longrightarrow 2 \text{ mol HCl}$$

The correct coefficients in a chemical equation give the mole ratios in which the substances combine.

Mass relationships follow directly from the mole relationships. Thus 1 mol H_2, 2.0 g H_2, combines with 1 mol Cl_2, 70.9 g Cl_2, to produce 2 mol HCl, 72.9 g HCl. Notice that the law of conservation of mass is upheld; the sum of the masses of the reactants, 72.9 g (2.0 g H_2 + 70.9 g Cl_2), equals the mass of the product, 72.9 g.

Table 11.1 summarizes molecule, mole, and mass relationships in the formation of HCl from the elements.

Given the balanced equation and quantity of starting materials, we can determine the products' masses, the number of moles, or the number of molecules (or vice versa). To show this, let's determine how many moles of Cl_2 are needed to combine with 2 mol H_2 and how many moles HCl result.

Since H_2 and Cl_2 combine in a 1-to-1 mole ratio, exactly the same number

TABLE 11.1   REACTION STOICHIOMETRY FOR THE FORMATION OF HYDROGEN CHLORIDE, HCl

	H_2	$+$	Cl_2	$\longrightarrow$	$2HCl$
Molecules	1 molecule		1 molecule		2 molecules
Molecules	6.02×10^{23} molecules		6.02×10^{23} molecules		$2 \times 6.02 \times 10^{23}$ molecules
Moles	1 mol		1 mol		2 mol
Mass	2.0 g		70.9 g		72.9 g

of moles of Cl_2 as moles of H_2 are required. Thus, 2 mol Cl_2 combines with 2 mol H_2. For each mole of reactant, two moles of product, HCl, are formed. Thus, 4 mol HCl is produced. In terms of mass, 2 mol H_2 (2 × 2.0 g H_2/mol), 4.0 g H_2, combines with 2 mol Cl_2 (2 × 70.9 g Cl_2/mol), 141.8 g Cl_2, to produce 4 mol HCl (4 × 36.45 g/mol), 145.8 g HCl.

	H_2	+	Cl_2	$\longrightarrow$	2HCl
Moles	2 mol		2 mol		4 mol
Mass	4.0 g		141.8 g		145.8 g

It should be apparent that the coefficients of the reactants and products in a chemical equation give the mole ratios in which substances combine. Thus, we use the coefficients to generate conversion factors. For instance, if we are interested in the number of moles of HCl produced per mole of H_2, it is only necessary to write:

$$\frac{2 \text{ mol HCl}}{1 \text{ mol } H_2}$$

A similar conversion factor relates moles of Cl_2 to formation of HCl:

$$\frac{2 \text{ mol HCl}}{1 \text{ mol } Cl_2}$$

The mole relationship between the two reactants is expressed as follows:

$$\frac{1 \text{ mol } Cl_2}{1 \text{ mol } H_2}$$

Mole-Mole Calculations

Turning our attention to the equation that represents the formation of ammonia, NH_3, from its elements (Haber reaction), we shall illustrate stoichiometry problems that involve mole relationships exclusively.

$$N_2(g) + 3H_2(g) \xrightarrow[\Delta]{cat} 2NH_3(g)$$

In the Haber reaction, 1 mol N_2 gas combines with 3 mol H_2 yielding 2 mol NH_3.

How many moles of H_2 are required to combine exactly with 10 mol N_2? Utilizing conversion factors from the equation, we find that 3 mol H_2 is required per mole N_2:

$$\frac{3 \text{ mol } H_2}{1 \text{ mol } N_2}$$

Fritz Haber (1868–1934), a German chemist, developed a method of mixing N_2 and H_2 under pressure with an iron catalyst to produce ammonia. Unfortunately this discovery prolonged World War I. Before Haber's discovery, Germany's capacity to produce explosives was limited because most explosives were made from nitrate-rich deposits found in Chile, a region inaccessible to Germany because of a blockade by the British navy.

FIGURE 11.2
To determine the number of moles of product formed, given the number of moles of a reactant, multiply the number of moles of reactant by the conversion factor that gives the number of moles of product per mole of reactant. This conversion factor is obtained from the coefficients of the reactants and products in the balanced equation.

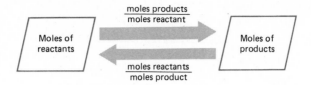

Just multiply this conversion factor by the number of moles of nitrogen given, 10 mol.

$$10 \; \text{mol N}_2 \times \frac{3 \; \text{mol H}_2}{1 \; \text{mol N}_2} = 30 \; \text{mol H}_2$$

If 30 mol H_2 is needed to combine with 10 mol N_2, how many moles of NH_3 result? Again, look at the equation and extract out the mole relationship between either moles N_2 and moles NH_3 produced or moles H_2 and moles NH_3 produced.

$$\frac{2 \; \text{mol NH}_3}{1 \; \text{mol N}_2} \qquad \text{or} \qquad \frac{2 \; \text{mol NH}_3}{3 \; \text{mol H}_2}$$

Using either of these conversion factors gives us the correct answer:

or

$$10 \; \text{mol N}_2 \times \frac{2 \; \text{mol NH}_3}{1 \; \text{mol N}_2} = 20 \; \text{mol NH}_3$$

$$30 \; \text{mol H}_2 \times \frac{2 \; \text{mol NH}_3}{3 \; \text{mol H}_2} = 20 \; \text{mol NH}_3$$

Thus 20 mol NH_3 theoretically results when 10 mol N_2 combines with 30 mol H_2. Chemists call this calculated amount of product the **theoretical yield.**

Theoretical yields are the amounts of products predicted by applying the principles of stoichiometry.

Example Problems 11.1 and 11.2 illustrate the application of the factor-label method and our problem-solving techniques to solve mole-mole stoichiometry problems. Refer to Figure 11.2, which graphically illustrates how mole-mole problems are solved.

Example Problem 11.1

Calculate the theoretical maximum number of moles of NH_3 that results when 0.55 mol H_2 combines with excess N_2.

$$N_2 + 3H_2 \longrightarrow 2NH_3$$

Solution

1. *What is unknown?*   Moles NH_3

2. *What is known?* 0.55 mol H_2, $\dfrac{2 \text{ mol } NH_3}{3 \text{ mol } H_2}$

3. Apply the factor-label method.

$$0.55 \text{ mol } H_2 \times \frac{2 \text{ mol } NH_3}{3 \text{ mol } H_2} = ? \text{ mol } NH_3$$

4. Perform the indicated math operations.

$$0.55 \text{ mol } H_2 \times \frac{2 \text{ mol } NH_3}{3 \text{ mol } H_2} = 0.37 \text{ mol } NH_3$$

Theoretically, 0.37 mol NH_3 results when 0.55 mol H_2 combines with excess N_2 in the above reaction. An assumption is made that an excess of N_2 is present, or at least the exact amount needed to combine with the 0.55 mol H_2; if not, the maximum product yield is not obtained. If 0.0 g N_2 is present, the reaction does not take place! A common practice is to assume that sufficient quantities of all the other reactants are present unless told otherwise.

If not stated otherwise, assume in stoichiometry calculations that sufficient quantities of all substances are present for the reaction to take place.

Example Problem 11.2

How many moles of butane, C_4H_{10}, are required to combine with excess oxygen, O_2, to produce 6.44 mol of carbon dioxide, CO_2, in the following reaction:

$$C_4H_{10} + 6.5O_2 \longrightarrow 4CO_2 + 5H_2O$$

Butane is used as a fuel in cigarette lighters. It is a colorless gas that boils at $-0.3°C$. It is a liquid in lighters because it is under pressure.

Solution

1. *What is unknown?*   Moles of C_4H_{10}

2. *What is known?* 6.44 mol CO_2, $\dfrac{1 \text{ mol } C_4H_{10}}{4 \text{ mol } CO_2}$

3. Apply the factor-label method.

$$6.44 \text{ mol } CO_2 \times \frac{1 \text{ mol } C_4H_{10}}{4 \text{ mol } CO_2} = ? \text{ mol } C_4H_{10}$$

4. Perform the indicated math operations.

$$6.44 \text{ mol } CO_2 \times \frac{1 \text{ mol } C_4H_{10}}{4 \text{ mol } CO_2} = 1.61 \text{ mol } C_4H_{10}$$

Initially, 1.61 mol C_4H_{10} must have been present to combine with excess oxygen to produce 6.44 mol CO_2.

REVIEW PROBLEMS

11.3 Using the Haber equation, calculate the number of moles of NH_3 produced when 0.29 mol N_2 combines with excess H_2.

11.4 Calculate the number of moles of O_2 that combine with 5.19 mol C_4H_{10} to produce CO_2 and H_2O.

11.5 How many moles of H_2O are given off when 1.2×10^3 mol C_4H_{10} is combined with excess oxygen?

Mole-Mass Calculations

Usually, chemists are concerned with finding masses of reactants and products. One additional conversion factor is needed to calculate masses, the molar mass of the reactant or product (grams per mole, g/mol). See Figure 11.3, which illustrates the steps required to solve mole-mass problems.

An important industrial reaction is the water gas formation reaction. Steam, $H_2O(g)$, is passed over red-hot coke, $C(s)$, yielding a mixture of carbon monoxide, $CO(g)$, and hydrogen gas, $H_2(g)$. This mixture of CO and H_2, water gas, can be burned, giving off a great deal of energy.

In addition to being used as a fuel, water gas is combined with steam on a special catalyst to produce CO_2, and more H_2.

$$H_2O(g) + C(s) \longrightarrow CO(g) + H_2(g)$$
$$\text{Water gas}$$

Let's calculate the mass of CO produced when 125 mol of steam is combined with excess coke.

From the equation, we obtain the mole ratio of CO formed per mole of H_2O reacted, a 1-to-1 ratio. Accordingly, we solve the problem as we did previous example problems, adding one more conversion factor that converts moles of CO to grams of CO.

$$\text{Mass CO} = \text{moles of } H_2O \times \frac{\text{moles of CO}}{\text{mole of } H_2O} \times \frac{\text{grams of CO}}{\text{mole of CO}}$$

$$= 125 \text{ mol } H_2O \times \frac{1 \text{ mol CO}}{1 \text{ mol } H_2O} \times \frac{28.0 \text{ g CO}}{1 \text{ mol CO}}$$

$$= 3500 \text{ g CO} = 3.5 \times 10^3 \text{ g CO} = 3.5 \text{ kg CO}$$

FIGURE 11.3
Two conversion factors are required to determine the mass of product formed from a given number of moles of reactant. They are $\dfrac{\text{moles of product}}{\text{moles of reactant}}$, obtained from the balanced equation, and $\dfrac{\text{mass of product}}{1 \text{ mole product}}$, the molar mass of the product.

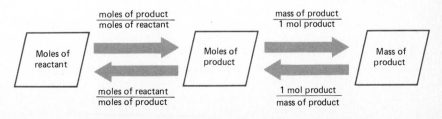

Study Example Problems 11.3 and 11.4 as examples of mole-mass calculations, i.e., stoichiometry problems involving moles and mass.

Example Problem 11.3

What mass of coke, C(s), must be present to combine with 125 mol of steam in the water gas reaction?

$$H_2O(g) + C(s) \longrightarrow CO(g) + H_2(g)$$

Solution

1. *What is unknown?* Grams of C(s)

2. *What is known?* 125 mol $H_2O(g)$, $\dfrac{1 \text{ mol C}}{1 \text{ mol } H_2O}$, and $\dfrac{12.0 \text{ g C}}{1 \text{ mol C}}$

3. Apply the factor-label method.

$$125 \text{ mol } H_2O \times \frac{1 \text{ mol C}}{1 \text{ mol } H_2O} \times \frac{12.0 \text{ g C}}{1 \text{ mol C}} = ? \text{ g C}$$

4. Perform the indicated math operations.

$$125 \text{ mol } H_2O \times \frac{1 \text{ mol C}}{1 \text{ mol } H_2O} \times \frac{12.0 \text{ g C}}{1 \text{ mol C}} = 1500 \text{ g C}$$
$$= 1.50 \times 10^3 \text{ g C, or } 1.50 \text{ kg C}$$

Whenever 125 mol of steam is combined with coke, 1.50 kg of coke is needed to react.

Example Problem 11.4

Another name for MgO is magnesia. It is sometimes used to prepare $Mg(OH)_2(aq)$, called milk of magnesia. This solution is used to decrease excess stomach acid.

How many moles of magnesium oxide, MgO, form when 65 g Fe_2O_3 combines with excess magnesium in the following reaction?

$$3Mg + Fe_2O_3 \longrightarrow 3MgO + 2Fe$$

Solution

1. *What is unknown?* Moles of MgO

2. *What is known?*   65 g Fe_2O_3, $\dfrac{1 \text{ mol } Fe_2O_3}{159.6 \text{ g } Fe_2O_3}$, and $\dfrac{3 \text{ mol MgO}}{1 \text{ mol } Fe_2O_3}$

3. Apply the factor-label method.

$$65 \text{ g } \cancel{Fe_2O_3} \times \frac{1 \text{ mol } \cancel{Fe_2O_3}}{159.6 \text{ g } \cancel{Fe_2O_3}} \times \frac{3 \text{ mol MgO}}{1 \text{ mol } \cancel{Fe_2O_3}} = ? \text{ mol MgO}$$

4. Perform the indicated math operations.

$$65 \text{ g } \cancel{Fe_2O_3} \times \frac{1 \text{ mol } \cancel{Fe_2O_3}}{159.6 \text{ g } \cancel{Fe_2O_3}} \times \frac{3 \text{ mol MgO}}{1 \text{ mol } \cancel{Fe_2O_3}} = 1.2 \text{ mol MgO}$$

When 65 g Fe_2O_3 is combined with excess Mg, a maximum of 1.2 mol MgO could be expected to form.

Mass-Mass Calculations

Phosphorus exists in two different forms. White phosphorus is composed of P_4 molecules; red phosphorus is made up of long chains of P_4 molecules bonded together. White P is very reactive and spontaneously bursts into flames when exposed to the air. Red phosphorus is much less reactive and less toxic.

Mass-mass calculations are those in which a mass is initially given and the final answer is also a mass. This type of stoichiometric calculation is frequently encountered. Let's look at some examples of such problems.

When phosphorus, P_4, is combined with chlorine gas, Cl_2, a colorless fuming liquid, phosphorus trichloride, PCl_3, is produced.

$$P_4(s) + 6Cl_2(g) \longrightarrow 4PCl_3(l)$$

What mass of PCl_3 forms when 100.0 g P_4 is combined with excess chlorine gas?

We should first convert the mass of P_4 to moles of P_4. We do this because the balanced equation directly shows the mole ratio, not the mass ratio. Once the number of moles of P_4 is known, the problem is exactly the same as a mole to mass conversion.

Mass of PCl_3 formed

$$= 100.0 \text{ g } \cancel{P_4} \times \frac{1 \text{ mol } \cancel{P_4}}{123.9 \text{ g } \cancel{P_4}} \times \frac{4 \text{ mol } \cancel{PCl_3}}{1 \text{ mol } \cancel{P_4}} \times \frac{137.3 \text{ g } PCl_3}{1 \text{ mol } \cancel{PCl_3}}$$

$$= 443.3 \text{ g } PCl_3$$

The first conversion factor,

$$\frac{1 \text{ mol } P_4}{123.9 \text{ g } P_4}$$

converts the mass of P_4 to moles. From the equation, it is evident that 4 mol PCl_3 is formed per mole of P_4; thus, the second conversion factor yields the number of moles of PCl_3 produced. We complete the problem by changing the number of moles of PCl_3 to grams of PCl_3, using the molecular mass of PCl_3, 137.3 g.

Carefully go through Example Problems 11.5 and 11.6, which illustrate mass-mass stoichiometry problems.

Example Problem 11.5

What mass of Cl_2 is needed to combine with excess P_4 to yield 0.927 g PCl_3?

$$P_4(s) + 6Cl_2(g) \longrightarrow 4PCl_3(l)$$

Solution

1. *What is unknown?* Grams of Cl_2

2. *What is known?*   0.927 g PCl_3, $\dfrac{1 \text{ mol } PCl_3}{137.3 \text{ g } PCl_3}$, $\dfrac{4 \text{ mol } PCl_3}{6 \text{ mol } Cl_2}$

3. Apply the factor-label method.

$$0.927 \text{ g } PCl_3 \times \frac{1 \text{ mol } PCl_3}{137.3 \text{ g } PCl_3} \times \frac{6 \text{ mol } Cl_2}{4 \text{ mol } PCl_3} \times \frac{70.9 \text{ g } Cl_2}{1 \text{ mol } Cl_2} = ? \text{ g } Cl_2$$

4. Perform the indicated math operations.

$$0.927 \text{ g } PCl_3 \times \frac{1 \text{ mol } PCl_3}{137.3 \text{ g } PCl_3} \times \frac{6 \text{ mol } Cl_2}{4 \text{ mol } PCl_3} \times \frac{70.9 \text{ g } Cl_2}{1 \text{ mol } Cl_2} = 0.718 \text{ g } Cl_2$$

A sample of 0.718 g Cl_2, when combined with an excess of P_4, yields 0.927 g PCl_3.

Example Problem 11.6

The common name for $NaNO_3$ is Chile saltpeter. Both $NaNO_3$ and $NaNO_2$ are used to cure meat, to prevent dangerous bacterial growth. They are frequently added to hot dogs, bacon, and hams.

What mass of oxygen gas, O_2, is liberated when a 2.5-g sample of sodium nitrate is heated?

$$2NaNO_3(s) \longrightarrow 2NaNO_2(s) + O_2(g)$$

Solution

1. *What is unknown?*   Grams of O_2

2. *What is known?*   2.5 g NaNO$_3$, $\dfrac{85 \text{ g NaNO}_3}{1 \text{ mol NaNO}_3}$, $\dfrac{1 \text{ mol O}_2}{2 \text{ mol NaNO}_3}$, and

$\dfrac{32 \text{ g O}_2}{1 \text{ mol O}_2}$.

3. Apply the factor-label method.

$$2.5 \text{ g NaNO}_3 \times \frac{1 \text{ mol NaNO}_3}{85 \text{ g NaNO}_3} \times \frac{1 \text{ mol O}_2}{2 \text{ mol NaNO}_3} \times \frac{32 \text{ g O}_2}{1 \text{ mol O}_2} = ? \text{ g O}_2$$

4. Perform the indicated math operations.

$$2.5 \text{ g NaNO}_3 \times \frac{1 \text{ mol NaNO}_3}{85 \text{ g NaNO}_3} \times \frac{1 \text{ mol O}_2}{2 \text{ mol NaNO}_3} \times \frac{32 \text{ g O}_2}{1 \text{ mol O}_2}$$
$$= 0.47 \text{ g O}_2$$

When a 2.5-g sample of NaNO$_3$ decomposes, it liberates 0.47 g O$_2$.

Figure 11.4 illustrates the steps required to solve a mass-mass stoichiometry problem.

REVIEW PROBLEMS   **11.6** Calculate the number of moles of SO$_2$ that is produced when 844 g S$_8$ is combined with excess O$_2$.

$$S_8(s) + 8O_2(g) \longrightarrow 8SO_2(g)$$

11.7 Consider the following equation:

$$P_4(s) + 6Cl_2(g) \longrightarrow 4PCl_3(l)$$

(a) What mass of P$_4$ is required to combine with 51.2 g Cl$_2$ to produce PCl$_3$?
(b) What mass of PCl$_3$ results when 51.2 g Cl$_2$ is combined with the calculated mass of P$_4$?

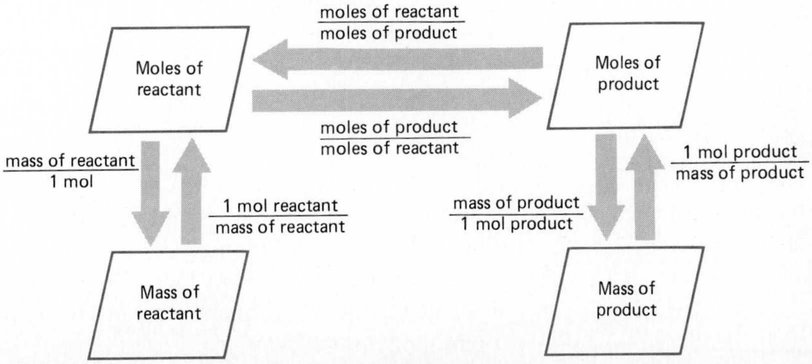

FIGURE 11.4
Mass-mass problems are solved in a similar manner to mole-mass problems except that a third conversion factor is needed to change the mass of the reactant to moles. This conversion factor is the molar mass of the reactant, $\dfrac{\text{mass of reactant}}{1 \text{ mole of reactant}}$.

11.8 Determine the mass of F_2 required to combine with N_2 to produce 712 g NF_3 in the following reaction:

$$N_2(g) + 3F_2(g) \longrightarrow 2NF_3(g)$$

11.3 QUANTITATIVE ENERGY EFFECTS

Exothermic reactions are those in which heat energy is released to the surroundings. Endothermic reactions are those in which heat is absorbed from the surroundings.

As mentioned in Chapter 10, heat energy is either released or absorbed in chemical reactions. Whenever heat is liberated, the reaction is called an **exothermic** reaction; when heat is absorbed, the reaction is called an **endothermic** reaction. The amount of heat transferred in chemical reactions is related to the number of moles of reactants that undergo chemical change. A fixed quantity of heat is transferred per mole of reactant consumed.

Through the application of stoichiometric principles, we can calculate the energy change in a chemical reaction. Let us consider the reaction in which HCl is synthesized from its elements. This time the heat evolved by the reaction is also shown.

$$H_2(g) + Cl_2(g) \longrightarrow 2HCl(g) + 44 \text{ kcal}$$

In addition to the 2 mol HCl produced, 44 kcal (184 kJ) is released (an exothermic reaction). In other words, for each mole of H_2 and Cl_2 combined, 2 mol HCl and 44 kcal of heat are produced.

If we have the balanced equation and the quantity of heat transferred per mole of reactant, we can calculate the amount of energy released for any mass of reactant. What quantity of heat is liberated when 1.0 g H_2 is combined with excess Cl_2 to produce HCl? To solve this problem, we need a conversion factor that relates moles of H_2 to amount of heat liberated; it is

$$\frac{44 \text{ kcal}}{1 \text{ mol } H_2}$$

Therefore, it is necessary to calculate the number of moles of H_2 and apply the above conversion factor.

$$1.0 \text{ g } H_2 \times \frac{1 \text{ mol } H_2}{2.0 \text{ g } H_2} \times \frac{44 \text{ kcal}}{1 \text{ mol } H_2} = 22 \text{ kcal}$$

For each gram of H_2 that combines with Cl_2, 22 kcal (92 kJ) of heat energy is liberated to the surroundings.

See Example Problems 11.7 and 11.8 for more examples of problems dealing with energy effects in chemical reactions. Refer to Figure 11.5, which illustrates the required steps.

FIGURE 11.5
To calculate the energy consumed or liberated by a chemical reaction, the conversion factor relating the number of moles of reactant and the amount of energy in kilocalories or kilojoules transferred is required. Therefore, if the mass of reactant is given, convert the mass to moles and then multiply by the energy conversion factor.

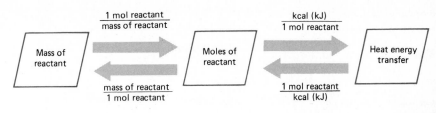

Acetylene is an industrially important gas; more than a billion pounds are produced per year in the United States. It is used in oxyacetylene torches and miners' lamps and is the starting material for the synthesis of many other carbon compounds.

Example Problem 11.7

Formation of acetylene, C_2H_2, from its elements requires the addition of 227 kJ of energy.

$$2C(s) + H_2(g) + 227 \text{ kJ} \longrightarrow C_2H_2(g)$$

Calculate the amount of heat energy that should be available to produce 545 g C_2H_2.

Solution

1. *What is unknown?*   Kilojoules of energy

2. *What is known?*   545 g C_2H_2, $\dfrac{227 \text{ kJ}}{1 \text{ mol } C_2H_2}$, $\dfrac{26.0 \text{ g } C_2H_2}{1 \text{ mol } C_2H_2}$

3. Apply the factor-label method.

$$545 \text{ g } C_2H_2 \times \frac{1 \text{ mol } C_2H_2}{26.0 \text{ g } C_2H_2} \times \frac{227 \text{ kJ}}{1 \text{ mol } C_2H_2} = ? \text{ kJ}$$

4. Perform the indicated math operations.

$$545 \text{ g } C_2H_2 \times \frac{1 \text{ mol } C_2H_2}{26.0 \text{ g } C_2H_2} \times \frac{227 \text{ kJ}}{1 \text{ mol } C_2H_2} = 4.76 \times 10^3 \text{ kJ}$$

To produce 545 g C_2H_2, 4.76×10^3 kJ of energy is required in addition to the proper amount of reactants. Acetylene is used as a fuel; it is commonly used in torches for welding metals. As is the case with other fuels, acetylene's formation from its elements is endothermic.

Example Problem 11.8

Blood sugar, or glucose, $C_6H_{12}O_6$, is one of the main sources of energy for living systems. It is broken down in cells to CO_2 and H_2O, releasing 2816 kJ per mole of $C_6H_{12}O_6$:

Glucose is the most abundant sugar structure in nature. Cellulose, a major component of plant cells, is a high-molecular-mass molecule containing thousands of glucose molecules bonded together. Starch, another component of plants, is also a molecule composed of bonded glucose molecules.

$$C_6H_{12}O_6(s) + 6O_2(g) \longrightarrow 6CO_2(g) + 6H_2O(g) + 2816 \text{ kJ}$$

Calculate the number of kilocalories of heat released per gram of glucose.

Solution

1. *What is unknown?* $\dfrac{\text{kcal}}{1 \text{ g } C_6H_{12}O_6}$

2. *What is known?* $1.00 \text{ g } C_6H_{12}O_6$, $\dfrac{2816 \text{ kJ}}{1 \text{ mol } C_6H_{12}O_6}$,

$\dfrac{180 \text{ g } C_6H_{12}O_6}{1 \text{ mol } C_6H_{12}O_6}$, and $4.184 \text{ J} = 1 \text{ cal}$ or $4.184 \text{ kJ} = 1 \text{ kcal}$

3. Apply the factor-label method.

$$1.00 \text{ g } C_6H_{12}O_6 \times \frac{1 \text{ mol } C_6H_{12}O_6}{180 \text{ g } C_6H_{12}O_6} \times \frac{2816 \text{ kJ}}{1 \text{ mol } C_6H_{12}O_6} \times \frac{1 \text{ kcal}}{4.184 \text{ kJ}} = ? \text{ kcal}$$

4. Perform the indicated math operations.

$$1.00 \text{ g } C_6H_{12}O_6 \times \frac{1 \text{ mol } C_6H_{12}O_6}{180 \text{ g } C_6H_{12}O_6} \times \frac{2816 \text{ kJ}}{1 \text{ mol } C_6H_{12}O_6} \times \frac{1 \text{ kcal}}{4.184 \text{ kJ}}$$

$$= 3.74 \text{ kcal}$$

For each gram of glucose "burned" by your cells, 3.74 kcal of energy is released.

Calorimetry   Heat energy transfers in chemical reactions are measured by instruments called calorimeters. A **calorimeter** is nothing more than the reaction vessel equipped in such a manner that the energy either evolved or absorbed is detected.

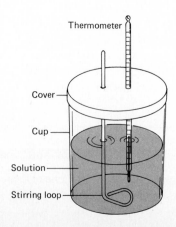

Thermometer

Cover

Cup

Solution

Stirring loop

FIGURE 11.6
A simple calorimeter can be constructed from a styrofoam cup because styrofoam is a poor heat conductor. Thus, the energy released or taken in is calculated from the temperature change of the solution in the calorimeter.

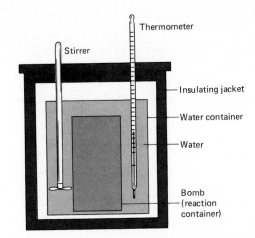

FIGURE 11.7
A bomb calorimeter is used to obtain the quantity of heat transferred in a chemical reaction at constant volume.

The heat capacity of a calorimeter is measured by finding the heat absorbed by the calorimeter and then dividing that quantity by the temperature increase.

Crude estimates of the heat transferred in chemical reactions can be made with the simple calorimeter made from Styrofoam cups that is pictured in Figure 11.6. Two solutions are poured together inside of the cup. Temperature readings are taken before and after the solutions react. The only other quantity that must be determined is the heat capacity of the calorimeter, i.e., the quantity of heat absorbed by the calorimeter and its contents.

By constructing the calorimeter out of Styrofoam, which does not absorb too much heat, only the temperature change of the resulting solution is considered. Rough estimates of heat transfers are obtained using Styrofoam cups. Calorimeters used in research laboratories are more sophisticated. One such calorimeter is pictured in Figure 11.7.

REVIEW PROBLEMS

11.9 What mass of Cl_2 would combine with H_2 to produce 8.5 kcal of energy in the following reaction?

$$H_2 + Cl_2 \longrightarrow 2HCl + 44 \text{ kcal}$$

11.10 How much energy, in kilojoules, is released when 454 g $C_6H_{12}O_6$, glucose, is combined with O_2 in human cells?

$$C_6H_{12}O_6 + 6O_2 \longrightarrow 6CO_2 + 6H_2O + 2816 \text{ kJ}$$

11.11 (a) What is the function of a calorimeter?
(b) What is meant by the heat capacity of a calorimeter?

11.4 LIMITING REAGENT PROBLEMS

So far we have assumed in our calculations that sufficient quantities of all reactants were present to combine with the specified mass of one reactant of interest. However, there are instances when specific masses for all the reactants are given

TABLE 11.2   LIMITING REAGENT IN THE OXIDATION OF CARBON

$$C(s) + O_2(g) \longrightarrow CO_2(g)$$

Moles C	Moles O_2	Moles CO_2 produced	
5	10	0	(initial amounts)
4	9	1	(after 1 mol reacts)
3	8	2	(after 2 mol reacts)
2	7	3	(after 3 mol reacts)
1	6	4	(after 4 mol reacts)
0*	5	5	(after 5 mol reacts)

*Once the carbon is consumed, the reaction cannot continue. At that time, 5 mol of unreacted O_2 and 5 mol CO_2 remain. Carbon is the limiting reagent, and oxygen is in excess.

A limiting reagent or limiting reactant is the reactant that is consumed first, and thus it determines the maximum amount of products formed.

and we must decide which reactant limits the reaction and determines the maximum yield of product; such problems are called **limiting reagent** problems.

To illustrate limiting reagent problems, let us consider the oxidation of carbon to carbon dioxide:

$$C(s) + O_2(g) \longrightarrow CO_2(g)$$

One mole of carbon combines with one mole of oxygen gas to produce one mole of carbon dioxide. As long as equal molar quantities of reactants are combined, a limiting reagent situation is not observed. If 5 mol C and 5 mol O_2 are combined, 5 mol CO_2 is formed. However, if 5 mol C is combined with 10 mol O_2, only 5 mol CO_2 results. After the 5 mol C is reacted, the reaction stops. Carbon is therefore the limiting reagent. Table 11.2 shows stepwise what happens when 5 mol C combines with 10 mol O_2.

A number of everyday experiences provide good analogies to the situation that is found when reactants combine. Take for instance the case where pages are being stapled together to produce a pamphlet. If 25 first pages, 50 second pages, and 100 third pages are being stapled to produce a three-page pamphlet, how many complete pamphlets could be produced? Here, the first page is the "limiting reagent." Once 25 pamphlets are stapled, no more complete pamphlets can be produced because of the lack of first pages. It doesn't matter than 25 second pages and 75 third pages are unstapled; there is no way to produce a complete pamphlet without adding more first pages. (See Figure 11.8.)

Our initial limiting reagent example was straightforward because the reactants combined in a 1 to 1 ratio. What if the reactants combine in some other ratio? Again we return to the Haber reaction:

$$N_2(g) + 3H_2(g) \longrightarrow 2NH_3(g)$$

What is the limiting reagent if 6 mol N_2 combines with 6 mol H_2? In the Haber reaction, H_2 is consumed 3 times faster than the N_2. Initially, 1 mol N_2 combines with 3 mol H_2, leaving 5 mol N_2 and 3 mol H_2. When the next 1 mol N_2

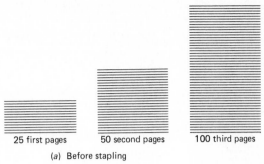

25 first pages   50 second pages   100 third pages

(a) Before stapling

FIGURE 11.8

A pamphlet is produced by stapling three pages together. If there are 25 first pages, 50 second pages, and 100 third pages, the maximum number of complete pamphlets that can be produced is 25. After the first page runs out, no more completed pamphlets can be produced.

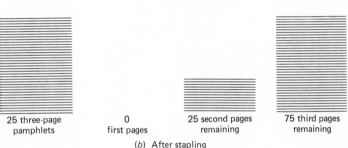

25 three-page pamphlets   0 first pages   25 second pages remaining   75 third pages remaining

(b) After stapling

reacts, it consumes all of the remaining H_2—the reaction then stops for lack of H_2. Hydrogen is therefore the limiting reagent. Thus 4 mol of unreacted N_2 remains in the reaction vessel, with the 4 mol NH_3 that was produced. See Table 11.3.

As shown in the above illustrations, the limiting reagent is found by comparing the number of moles of reactants, taking into account the mole ratio in which they combine. In most cases, masses of the reactants are given, and it is necessary to convert the masses to moles and then decide which reactant is limiting (consumed first).

What is the limiting reagent when 10.0 g H_2 and 50.0 g O_2 are combined to produce water vapor?

$$2H_2(g) + O_2(g) \longrightarrow 2H_2O(g)$$

First calculate the number of moles of each reactant:

TABLE 11.3   **LIMITING REAGENT IN THE HABER REACTION**

$$N_2 + 3H_2 \longrightarrow 2NH_3$$

Moles of N_2	Moles of H_2	Moles of NH_3 produced	
6	6	0	(initial amounts)
5	3	2	(after 1 mol N_2 reacts)
4	0	4	(after 2 mol N_2 reacts)

$$10.0 \text{ g } H_2 \times \frac{1 \text{ mol } H_2}{2.00 \text{ g } H_2} = 5.00 \text{ mol } H_2$$

$$50.0 \text{ g } O_2 \times \frac{1 \text{ mol } O_2}{32.0 \text{ g } O_2} = 1.56 \text{ mol } O_2$$

After calculating the number of moles of each reactant, select one of the reactants and determine how many moles of the other are needed to combine with it completely. Since it doesn't matter which reactant we select, let's determine how many moles H_2 are required to combine with the O_2 that is present.

$$1.56 \text{ mol } O_2 \times \frac{2 \text{ mol } H_2}{1 \text{ mol } O_2} = 3.13 \text{ mol } H_2 \text{ required}$$

We take the number of moles of O_2 and multiply it by 2 because, in the equation, 2 mol H_2 is required per 1 mol O_2. Looking at the results, we see that 3.13 mol H_2 is required to combine with the 1.56 mol O_2. Is there enough H_2 to combine with the O_2? Yes, only 3.13 mol H_2 is needed, and 5.00 mol is available. Hence, O_2 is the limiting reagent. After the 1.56 mol O_2 combines with the excess H_2, the reaction ceases.

Once the limiting reagent is identified, all other calculations are completed using the number of moles of limiting reagent. What total mass of H_2O is produced in the above problem?

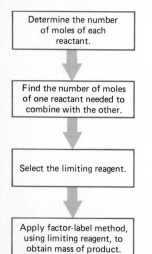

FIGURE 11.9
Steps required to solve limiting reagent problems

Determine the number of moles of each reactant.

Find the number of moles of one reactant needed to combine with the other.

Select the limiting reagent.

Apply factor-label method, using limiting reagent, to obtain mass of product.

$$1.56 \text{ mol } O_2 \times \frac{2 \text{ mol } H_2O}{1 \text{ mol } O_2} \times \frac{18.0 \text{ g } H_2O}{1 \text{ mol } H_2O} = 56.2 \text{ g } H_2O$$

When 10.0 g H_2 and 50.0 g O_2 are combined together, the theoretical yield of water is 56.2 g.

Note that if you had incorrectly decided on the limiting reagent, the calculated mass of water obtained would have been larger than the theoretical yield.

$$5.00 \text{ mol } H_2 \times \frac{2 \text{ mol } H_2O}{2 \text{ mol } H_2} \times \frac{18.0 \text{ g } H_2O}{1 \text{ mol } H_2O} = 90.0 \text{ g } H_2O \quad \text{(incorrect answer)}$$

This answer, 90.0 g H_2O, could only result if a sufficient quantity of O_2 is available (2.5 mol O_2). It is impossible for 90.0 g H_2O to form with the stated quantities of starting materials!

Example Problems 11.9 and 11.10 provide additional illustrations of limiting reagent problems. The steps required to solve limiting reagent problems are listed in Figure 11.9.

Example Problem 11.9

What mass of SO_3 is produced when 0.600 g SO_2 is combined with 0.400 g O_2?

$$2SO_2(g) + O_2(g) \longrightarrow 2SO_3(g)$$

Solution

1. Identify the limiting reagent.

First, calculate which reactant is the limiting reagent by comparing the initial number of moles.

$$0.600 \text{ g } SO_2 \times \frac{1 \text{ mol } SO_2}{64.0 \text{ g } SO_2} = 0.00938 \text{ mol } SO_2$$

$$0.400 \text{ g } O_2 \times \frac{1 \text{ mol } O_2}{32.0 \text{ g } O_2} = 0.0125 \text{ mol } O_2$$

Find the number of moles of SO_2 needed to combine with the number of moles of O_2 actually present, or vice versa.

$$0.0125 \text{ mol } O_2 \times \frac{2 \text{ mol } SO_2}{1 \text{ mol } O_2} =$$

$$0.0250 \text{ mol } SO_2 \text{ required to combine with } 0.0125 \text{ mol } O_2$$

In order for 0.0125 mol O_2 to combine completely with SO_2, 0.0250 mol SO_2 is required. However, only 0.00938 mol SO_2 is present; consequently, SO_2 is the limiting reagent.

2. Apply the factor-label method. Complete the problem by using conversion factors to find the mass of SO_3 formed (mole-mass problem).

$$0.00938 \text{ mol } SO_2 \times \frac{2 \text{ mol } SO_3}{2 \text{ mol } SO_2} \times \frac{80.0 \text{ g } SO_3}{1 \text{ mol } SO_3} = 0.750 \text{ g } SO_3$$

If 0.600 g SO_2 is combined with 0.400 g O_2, 0.750 g SO_3 is the maximum quantity of product that could be produced.

Example Problem 11.10

Hydrogen gas combines with nitrogen dioxide to produce ammonia and water. What mass of ammonia results when 25.0 g of hydrogen combines with 185 g of nitrogen dioxide?

Solution

1. Start by writing and balancing the equation.

$$H_2 + NO_2 \longrightarrow NH_3 + H_2O \qquad \text{(unbalanced)}$$
$$7H_2 + 2NO_2 \longrightarrow 2NH_3 + 4H_2O \qquad \text{(balanced)}$$

2. Identify the limiting reagent.

$$25.0 \text{ g } H_2 \times \frac{1 \text{ mol } H_2}{2.00 \text{ g } H_2} = 12.5 \text{ mol } H_2$$

$$185 \text{ g } NO_2 \times \frac{1 \text{ mol } NO_2}{46.0 \text{ g } NO_2} = 4.02 \text{ mol } NO_2$$

Find the number of moles of NO_2 needed to combine with 12.5 mol H_2.

$$12.5 \text{ mol } H_2 \times \frac{2 \text{ mol } NO_2}{7 \text{ mol } H_2} = 3.57 \text{ mol } NO_2$$

Since 3.57 mol NO_2 (4.02 mol are actually present) is needed to react with 12.5 mol H_2, the H_2 is the limiting reagent. When all 12.5 mol of H_2 is combined, an excess of NO_2 remains.

$$12.5 \text{ mol } H_2 \times \frac{2 \text{ mol } NH_3}{7 \text{ mol } H_2} \times \frac{17.0 \text{ g } NH_3}{1 \text{ mol } NH_3} = 60.7 \text{ g } NH_3$$

When 25.0 g H_2 is combined with 185 g NO_2, the theoretical yield of NH_3 is 60.7 g.

REVIEW PROBLEMS

11.12 What is the limiting reagent in a chemical reaction?

11.13 For the combination reaction of carbon and oxygen gas to produce carbon dioxide, determine if carbon or oxygen is the limiting reagent, given the following amounts:
(a) 3 mol C + 3 mol O_2 (b) 1 mol C + 2 mol O_2
(c) 1.0 g C + 1.0 g O_2 (d) 11 g C + 35 g O_2
(e) 0.0750 g C + 0.175 g O_2

11.14 For each part of Review Problem 11.13, calculate the amount of excess reactant after the reaction is allowed to go to completion.

11.15 What is the maximum mass of sodium bromide that results when 73.6 g of sodium metal combines with 250.5 g of bromine in the reaction $2Na + Br_2 \rightarrow 2NaBr$?

11.5 THEORETICAL AND ACTUAL YIELDS

Frequently the mass of product obtained in a reaction is less than the predicted theoretical yield. The amount of product isolated is called the **actual yield**.

Chemists commonly report the percent yield for products, where the *percent yield* is

$$\text{Percent yield} = \frac{\text{actual yield}}{\text{theoretical yield}} \times 100$$

If the theoretical yield of a product is 5 g, and only 4 g is isolated, the percent yield is reported as 80% ($\frac{4}{5} \times 100$). Example Problem 11.11 gives an example of a percent yield problem.

Example Problem 11.11

$Ca_3(PO_4)_2$ is found in phosphate rock. When it is heated to a high temperature, free phosphorus, P_4 is produced. However, most phosphate rock is used to produce phosphate fertilizers—$Ca(H_2PO_4)_2$, for example.

Calcium phosphate, $Ca_3(PO_4)_2$, combines with silicon dioxide, SiO_2, to produce calcium silicate, $CaSiO_3$, and P_4O_{10}. Initially, 50.0 g $Ca_3(PO_4)_2$ is combined with excess SiO_2. After the reaction, 16.9 g P_4O_{10} is isolated; calculate the percent yield of P_4O_{10}.

$$2Ca_3(PO_4)_2 + 6SiO_2 \longrightarrow 6CaSiO_3 + P_4O_{10}$$

Solution

1. Find the theoretical yield of P_4O_{10}.

$$50.0 \text{ g } Ca_3(PO_4)_2 \times \frac{1 \text{ mol } Ca_3(PO_4)_2}{3.10 \times 10^2 \text{ g } Ca_3(PO_4)_2} \times \frac{1 \text{ mol } P_4O_{10}}{2 \text{ mol } Ca_3(PO_4)_2}$$
$$\times \frac{284 \text{ g } P_4O_{10}}{1 \text{ mol } P_4O_{10}} = 22.9 \text{ g } P_4O_{10}$$

Without going through each step, the above factor-label setup determines the mass of P_4O_{10} that forms when 50.0 g of calcium phosphate combines with excess silicon dioxide. Thus, the theoretical yield is 22.9 g P_4O_{10}.

2. Determine the percent yield.

$$\text{Percent yield } P_4O_{10} = \frac{\text{actual yield}}{\text{theoretical yield}} \times 100$$
$$= \frac{16.9 \text{ g}}{22.9 \text{ g}} \times 100$$
$$= 73.8\%$$

A yield of 16.9 g P_4O_{10} represents a 73.8% yield. Somehow during the course of the reaction and isolation of the product, 26.2% P_4O_{10} was lost.

SUMMARY

Stoichiometry is the study of quantitative relationships in chemical reactions. With a correctly balanced equation, mole and mass relationships are easily obtained. Predictions that result from stoichiometric calculations are the theoretical maximum amounts that could be obtained in a reaction.

Commonly, masses of starting materials are given, and quantities of products are sought. Initially, the masses of reactants are converted to moles because the balanced equation indicates the mole ratio between each reactant and product. After the number of moles of desired product is calculated, it is only necessary to convert the moles of product to grams, using the molecular or atomic mass. Stoichiometry problems are ideally suited to be solved by the factor-label method.

Energy transfers can also be predicted from chemical equations. Given the quantity of heat evolved or absorbed per mole of reactant, a conversion factor relating moles and energy is determined. Energy transfers in chemical reactions are experimentally measured through calorimetry, i.e., the measurement of heat transferred in chemical reactions using an instrument called a calorimeter.

When the initial quantities of all reactants are given and no assumption is made that one or more reactants is in excess, it is necessary to find the reactant (limiting reagent) that is consumed first. This is called a limiting reagent problem. After the limiting reagent is consumed, the reaction ceases.

After chemical reactions are conducted, the amounts of products isolated differ from the amounts predicted. Chemists regularly calculate the percent yield of the reaction, which is the ratio of the actual yield to the theoretical yield times 100.

QUESTIONS AND PROBLEMS

11.16 Define the following terms: stoichiometry, theoretical yield, calorimetry, heat capacity, limiting reagent, actual yield, percent yield.

Quantitative Equation Relationships

11.17 Complete the following table for the reaction in which N_2 and O_2 combine to yield NO:

	N_2	+	O_2	$\longrightarrow$	2NO
Molecules	1 molecule		___		___
Molecules	6.0×10^{23} molecules		___		___
Moles	1 mol		___		___

	N_2	+	O_2	$\longrightarrow$	2NO
Mass, g	___		32 g		___
Mass, g	___		___		6.0 g

11.18 Complete the following table for the reaction in which pentane, C_5H_{12}, combines with O_2 to produce CO_2 and H_2O:

	C_5H_{12}	+	$8O_2$	$\longrightarrow$	$5CO_2$	+	$6H_2O$
Molecules	1 molecule		___		___		___
Molecules	6.02×10^{23} molecules		___		___		___

$$C_5H_{12} \;+\; 8O_2 \;\longrightarrow\; 5CO_2 \;+\; 6H_2O$$

Mole	1 mol	____	____	____
Mole	____	1 mol	____	____
Mass, g	72 g	____	____	____

11.19 When Ca combines with F_2, CaF_2 is produced: $Ca + F_2 \rightarrow CaF_2$. If 1 mol Ca and 1 mol F_2 are combined:
(a) How many moles of CaF_2 result?
(b) What mass of CaF_2 forms?
(c) Show that the balanced equation illustrates the law of conservation of mass—the sum of masses of reactants equals the mass of product.

11.20 Consider the reaction in which bromine combines with chlorine: $Br_2 + Cl_2 \rightarrow 2BrCl$. If 1 mol Br_2 combines with 1 mol Cl_2 to produce 2 mol BrCl, show that the sum of masses of reactants equals the mass of product.

Mole-Mole Relationships

11.21 For the reaction $2Al + 3Cl_2 \rightarrow 2AlCl_3$, determine the number of moles of $AlCl_3$ that is produced from the stated quantity of reactant (assume the other reactant is always in excess): (a) 2 mol Al, (b) 3 mol Cl_2, (c) 10 mol Al, (d) 12 mol Cl_2, (e) 5 mol Cl_2.

11.22 Write all possible conversion factors that relate the number of moles of each reactant to the number of moles of each product:

$$C_3H_8 + 5O_2 \longrightarrow 3CO_2 + 4H_2O$$

11.23 For the following equation, calculate the requested quantities:

$$3KCl + 4HNO_3 \longrightarrow Cl_2 + NOCl + 2H_2O + 3KNO_3$$

(a) Number of moles of Cl_2 produced for each mole of KCl reacted
(b) Number of moles of H_2O produced for 3 mol HNO_3 reacted
(c) Number of moles of HNO_3 reacted to produce 6 mol NOCl
(d) Number of moles of Cl_2 and KNO_3 produced when 11 mol KCl reacts
(e) Number of moles of each product produced when 0.1 mol HNO_3 reacts.

11.24 Consider the equation:

$$4FeS_2 + 11O_2 \longrightarrow 2Fe_2O_3 + 8SO_2$$

For each of the given molar quantities of reactant, determine the number of moles of each product: (a) 2 mol FeS_2, (b) 6 mol FeS_2, (c) 1.7 mol O_2, (d) 22 mol O_2, (e) 9 mol O_2.

Mole-Mass Calculations

11.25 A simple laboratory preparation of O_2 gas is to decompose $KClO_3$ by heating it in the presence of MnO_2, a catalyst:

$$2KClO_3 \xrightarrow[\Delta]{MnO_2} 2KCl + 3O_2$$

(a) Determine the mass of O_2 produced when 2 mol $KClO_3$ is decomposed.
(b) Calculate the number of moles of O_2 produced when 1.00 g $KClO_3$ is heated.
(c) Find the number of moles of KCl and O_2 produced from 125 g $KClO_3$.
(d) Find the number of moles needed to produce 3.0 kg O_2.

11.26 Sodium bicarbonate, $NaHCO_3$, combines with HCl to produce a salt, NaCl, carbon dioxide, CO_2, and water, H_2O:

$$NaHCO_3 + HCl \longrightarrow NaCl + CO_2 + H_2O$$

(a) What mass of HCl should be present to totally combine with 1.00 mol $NaHCO_3$?
(b) How many moles of CO_2 are produced when 72 g $NaHCO_3$ reacts?
(c) Determine the mass of NaCl that is formed when 3.5 mol HCl combines with excess $NaHCO_3$.
(d) What mass of $NaHCO_3$ is required to produce 0.68 mol H_2O?

11.27 Copper metal is isolated from its oxide, Cu_2O, by heating the oxide in the presence of copper(I) sulfide, Cu_2S:

$$2Cu_2O + Cu_2S \longrightarrow 6Cu + SO_2$$

Calculate the mass of Cu produced for each of the following quantities of reactants (assume excess amount of the other reactant):
(a) 1 mol Cu_2O (b) 12 mol Cu_2O
(c) 0.055 mol Cu_2O (d) 5.905 mol Cu_2S
(e) 217 mol Cu_2S (f) 17.15 mmol Cu_2O

Mass-Mass Calculations

11.28 When silver(I) oxide, Ag_2O, is heated, it readily liberates oxygen, O_2, and leaves free silver, Ag:

$$2Ag_2O \longrightarrow 4Ag + O_2$$

What mass of silver results when the following quantities of silver(I) oxide are completely heated?
(a) 1.00 g Ag_2O (b) 0.232 g Ag_2O
(c) 9.13 g Ag_2O (d) 2.6 kg Ag_2O

11.29 Lead nitrate, $Pb(NO_3)_2$, on heating, decomposes to lead(II) oxide, PbO, oxygen, O_2, and nitrogen dioxide, NO_2:

$$2Pb(NO_3)_2 \longrightarrow 2PbO + O_2 + 4NO_2$$

What mass of $Pb(NO_3)_2$ must be heated to produce the following masses of products?
(a) 37 g PbO   (b) 812 g O_2
(c) 0.792 kg NO_2 (d) 584.1 mg PbO

11.30 Carbon tetrachloride, CCl_4, once used as a cleaning fluid and as a fire extinguisher, is produced by heating methane, CH_4, and chlorine, Cl_2:

$$CH_4 + 4Cl_2 \longrightarrow CCl_4 + 4HCl$$

(a) What mass of CH_4 is needed to combine with exactly 10.0 g Cl_2?
(b) How many grams of Cl_2 are required to produce 295 g CCl_4, assuming excess CH_4?
(c) What mass of CH_4 must have reacted if 64 g HCl is liberated?
(d) Determine the mass of both CH_4 and Cl_2 that would be required to produce exactly 100.0 kg CCl_4.

11.31 Pure boron can be prepared by combining boron trichloride, BCl_3, with hydrogen gas, H_2, at high temperatures:

$$2BCl_3 + 3H_2 \longrightarrow 2B + 6HCl$$

(a) Calculate the mass of boron produced when 93.2 g BCl_3 is combined with excess hydrogen gas.
(b) What mass of hydrogen gas is needed to completely combine with 465 g BCl_3?
(c) How many grams of BCl_3 must be present to produce 12.0 g B?
(d) Determine the mass of H_2 that combines exactly with 3.5 kg BCl_3, and how much B and HCl is produced.

11.32 Xenon tetrafluoride, XeF_4, a noble gas compound, is highly reactive. When placed in water, XeF_4 undergoes the following reaction:

$$6XeF_4 + 8H_2O \longrightarrow 2XeOF_4 + 4Xe + 16HF + 3O_2$$

(a) Calculate the mass of XeF_4 that is needed to produce 1.00 g $XeOF_4$.
(b) After a measured mass of XeF_4 is placed in water, 3.26 g HF is given off. What mass of XeF_4 was combined with water?
(c) How much water should be combined with exactly 0.00255 g XeF_4 in the above reaction?
(d) Determine the mass of each product produced when 11.9 g XeF_4 is combined with water.

11.33 Nickel chloride hexahydrate, $NiCl_2 \cdot 6H_2O$, on heating in a vacuum yields nickel chloride, $NiCl_2$, and water, H_2O:

$$NiCl_2 \cdot 6H_2O \longrightarrow NiCl_2 + 6H_2O$$

(a) How many grams of the hydrate should be heated to produce 3.42 g H_2O?
(b) When 58.1 g of the hydrate is heated, what mass of the anhydrous compound, $NiCl_2$, results?
(c) What mass of hydrate is heated to yield 533 g $NiCl_2$?
(d) Find the mass of each product when 745 mg of nickel chloride hexahydrate is heated.

11.34 Potassium nitrate decomposes to potassium nitrite and oxygen gas.
(a) Write the balanced equation for this reaction.
(b) What mole relationships exist in this equation?
(c) How many grams of potassium nitrite form when 300.0 g of potassium nitrate is heated?
(d) What mass of oxygen is liberated when 1.445 g of potassium nitrate is decomposed?

11.35 Silicon tetrachloride and carbon monoxide are produced when silicon dioxide, carbon, and chlorine gas are heated.
(a) Write the balanced equation for this chemical change.
(b) What mass of silicon tetrachloride is produced when 159 g of silicon dioxide is combined with excess carbon and chlorine?
(c) How many molecules of carbon monoxide are produced when 0.00476 g chlorine combines with excess reactants?

11.36 Stibnite, antimony(III) sulfide, when heated with iron

undergoes a single replacement reaction, liberating free antimony metal and iron(II) sulfide.
(a) Write the balanced equation for the reaction.
(b) What mass of iron is required to combine with 77.2 g of stibnite?
(c) How many grams of antimony are produced when 4.56 kg of stibnite is reacted?
(d) What masses of stibnite and iron should be combined to yield 810.5 g of antimony?

Energy Effects

11.37 When hydrogen, H_2, is combined with oxygen, O_2, water vapor is formed plus 136 kcal of heat energy:

$$2H_2 + O_2 \longrightarrow 2H_2O + 136 \text{ kcal}$$

(a) Rewrite the equation with the energy expressed as kilojoules instead of kilocalories.
(b) What masses of H_2 and O_2 liberate 136 kcal?
(c) If 1.00 mol H_2 is combined with excess O_2, what quantity of heat is given off?
(d) If 1.00 g O_2 is combined with excess hydrogen, what quantity of heat is liberated?

11.38 Consider the following equation, showing the oxidation of nitrogen monoxide:

$$2NO + O_2 \longrightarrow 2NO_2 + 116.7 \text{ kJ}$$

(a) What mass of NO produces 50.0 kJ of heat energy?
(b) If 11.7 g NO is reacted, what quantity of heat is liberated?
(c) How many moles of oxygen are needed to combine with excess NO to produce 500 J?
(d) What mass of NO is required to produce 27.9 kcal of energy?

11.39 When combusted, ethylene, C_2H_4, gives off CO_2 and H_2O:

$$C_2H_4 + 3O_2 \longrightarrow 2CO_2 + 2H_2O + 337 \text{ kcal}$$

(a) How much heat is liberated per gram of C_2H_4?
(b) What mass of C_2H_4 must be combusted to liberate 1000 kcal?
(c) If 35.7 g O_2 is available, with excess C_2H_4, what quantity of heat is released?
(d) How much C_2H_4 must be combusted to liberate 1 kJ of energy?

11.40 The decomposition of aluminum oxide in the following reaction requires 798 kcal:

$$2Al_2O_3 + 798 \text{ kcal} \longrightarrow 4Al + 3O_2$$

(a) How much energy is required to decompose 8.0 mol Al_2O_3?
(b) What quantity of heat is needed to decompose 504 g Al_2O_3?
(c) If 66 g Al metal is obtained, what amount of heat energy was required to decompose the Al_2O_3?
(d) How many kilojoules of energy are required to decompose a gram of Al_2O_3?

11.41 When 1.001 g of calcium carbonate is decomposed to calcium oxide and carbon dioxide, 1.78 kJ of energy is needed.
(a) Write a balanced equation for the reaction, including the energy (kJ) required per mole of calcium carbonate.
(b) Rewrite the equation with the energy expressed in kilocalories.

11.42 In a simple calorimeter, two 50.0-g solutions of reacting chemicals are poured together, each initially at 25.0°C. After they are allowed to react, the final temperature of the resulting mixture is found to be 38.0°C. Assuming that all of the heat released by the reaction is absorbed by the water, and knowing that the heat capacity of the solution is 4.18 J/(g·°C), calculate the amount of heat released by the reaction.

Limiting Reagent Problems

11.43 Consider the equation $H_2 + I_2 \rightarrow 2HI$.
(a) Calculate the mass of HI that forms when 1.05 g H_2 combines with 122 g I_2.
(b) What is the maximum yield of HI when 9.60 g H_2 combines with 1225 g I_2?

11.44 Phosgene, $COCl_2$, combines with water to produce carbon dioxide, CO_2, and hydrochloric acid, HCl: $COCl_2 + H_2O \rightarrow CO_2 + 2HCl$.
(a) What mass of CO_2 is produced when 4.36 g $COCl_2$ is combined with 0.859 g H_2O?
(b) Determine the mass of HCl formed when 15 g of each reactant is combined?

11.45 Consider the equation: $2NaCl + H_2SO_4 \rightarrow Na_2SO_4 + 2HCl$.
(a) How many grams of Na_2SO_4 are produced when 100.0 g of each reactant are combined?
(b) Find the mass of Na_2SO_4 that forms when 31.5 g NaCl and 22.7 g H_2SO_4 are combined.

11.46 Lead(II) nitrate solution combines with a solution of potassium iodide yielding a precipitate, lead(II) iodide, and aqueous potassium nitrate.

(a) Write the balanced equation for this aqueous reaction.
(b) What mass of lead iodide precipitates from solution if 5.00 g of lead(II) nitrate combines with 2.50 g of potassium iodide.
(c) Find the amount of uncombined reactant.

11.47 If sodium carbonate is combined with carbon and nitrogen gas, carbon monoxide and sodium cyanide result.
(a) What mass of sodium cyanide results when 30.0 g of carbon, 75.0 g of sodium carbonate, and excess nitrogen gas are combined?
(b) What mass of nitrogen gas would have been consumed in the reaction?

Percent Yield

11.48 When benzene, C_6H_6, is combined with chlorine, Cl_2, chlorobenzene, C_6H_5Cl, and hydrogen chloride gas, HCl, are produced.
(a) Exactly 13.0 g of benzene is combined with excess chlorine. After the reaction, 11.3 g of chlorobenzene is isolated. Calculate the theoretical yield of chlorobenzene.
(b) If 11.3 g of chlorobenzene is isolated after the reaction, calculate the percent yield of chlorobenzene.

11.49 Rare germanium metal is isolated by heating GeO_2 in the presence of pure carbon: $GeO_2 + 2C \rightarrow Ge + 2CO$. After 692 g GeO_2 is reacted with carbon, 415 g Ge is obtained. Calculate the percent yield of germanium.

General Problems

11.50 Calcium hydroxide (slaked lime), $Ca(OH)_2$, is prepared by the following reactions:

$$CaCO_3 \longrightarrow CaO + CO_2$$
$$CaO + H_2O \longrightarrow Ca(OH)_2$$

(a) What mass of $Ca(OH)_2$ results from 795 g $CaCO_3$?
(b) If 89% is the maximum yield of the above reactions, what mass of $Ca(OH)_2$ forms when 397 kg $CaCO_3$ decomposes?

11.51 Zinc blende or zinc(II) sulfide ore is one source of metallic zinc. Initially, the zinc sulfide is combined with oxygen yielding zinc(II) oxide and sulfur dioxide. The resulting zinc(II) oxide is then heated with carbon to produce free zinc metal and carbon monoxide.
(a) Write and balance both equations.
(b) Calculate the mass of zinc obtained from 1.00 metric ton (1000 kg = 1 metric ton) of zinc(II) sulfide, assuming a percent yield of 79.0%.
(c) What mass of zinc forms when 20.0 g of zinc(II) sulfide combines with 10.0 g of oxygen (assume 100% yield)?

11.52 Consider the following aqueous reaction:

$$2KMnO_4 + 10KI + 8H_2SO_4 \longrightarrow$$
$$6K_2SO_4 + 2MnSO_4 + 5I_2 + 8H_2O$$

Calculate the mass of I_2 that is produced when 8.00 g $KMnO_4$, 41.0 g KI, and 18.0 g H_2SO_4 are combined.

· C H A P T E R ·

12

Gases

Chapter Guidelines

After completing Chapter 12, you should be able to:
1. List and discuss the principal assumptions of the kinetic molecular theory
2. Give the properties of ideal gases
3. State a definition for pressure, and give three examples of units that are commonly used to measure pressure
4. Convert any given pressure unit to another
5. State in words and as a mathematical expression: (a) Boyle's law, (b) Charles' law, (c) Avogadro's law, (d) Gay-Lussac's law, and (e) Dalton's law
6. Correctly interpret graphs of Boyle's law and Charles' law
7. Calculate the final pressure, volume, or temperature of an ideal gas, given the initial and final set of conditions
8. Provide an explanation for the fact that equal volumes of ideal gases at the same pressure and temperature contain the same number of molecules
9. State the molar volume of a gas at standard temperature and pressure (STP), and use the molar volume to find the mass or number of moles of an ideal gas
10. Calculate the density or molecular mass of a gas, given necessary information and the molar gas constant
11. Write and apply the ideal gas equation to find unknown properties of gases
12. Given the volume of a gas at a specified set of conditions and a chemical equation, determine the volume of the other gases that combine with this gas, and the volume of gases that are produced
13. Perform mass-volume and mole-volume stoichiometry problems
14. Explain why the total pressure of a mixture of gases is equal to the sum of the individual pressures of each gas in the mixture
15. Determine the partial pressure exerted by a gas in a mixture, given the total pressure and the partial pressures of all other gases
16. List the five gases found in greatest amount in the atmosphere
17. Discuss the discovery, properties, and uses of N_2, O_2, and the noble gases

The behavior and properties of gases can be theoretically explained using the **kinetic molecular theory,** or literally, the moving molecule theory. This theory is a model that explains the behavior of gases using generalizations about the randomly moving molecules or atoms within a gas. Some of the assumptions of the kinetic molecular theory are:

1. All gases contain atoms or molecules moving rapidly and randomly in straight lines.
2. Individual atoms or molecules are widely separated from each other and do not exert forces on other atoms or molecules, except when colliding. Nearly all of the volume of a gas is empty space.
3. Collisions of atoms or molecules with the walls of the container and with each other are perfectly elastic. This means there is no net loss of kinetic energy on colliding.
4. The average kinetic energy $(\text{KE} = \frac{1}{2}mv^2)$ of atoms or molecules is proportional to the Kelvin temperature of the gas. The average energy of gas molecules does not change unless the temperature changes.

Kinetic energy is the energy possessed by something that is moving. Faster-moving objects have more kinetic energy than slower-moving ones with the same mass.

A gas that behaves exactly according to the above assumptions is an **ideal gas,** one that exhibits perfect behavior. In actuality, there is no such thing as an ideal gas. However, real gases under conditions of low pressure and high temperature approach the behavior of ideal gases. Relationships are more involved and complicated for real gases; so, we will concentrate on the perfect behavior of gases, realizing that our predictions are not totally correct for real gases.

An ideal gas is one that behaves exactly as predicted by the ideal gas equations.

12.2 GAS LAWS

Gas laws are empirical relationships regarding the volume (V), pressure (P), temperature (T), and amount of gas in moles (n).

Pressure   One property of gases, pressure, has not yet been discussed. *Gas pressure* is defined as the force exerted by a gas on a unit area.

$$\text{Pressure} = \frac{\text{force}}{\text{area}}$$

Torricelli was educated in Rome as a mathematician; however, his primary interest was mechanics. He wrote a book on mechanics, which was noticed by Galileo and led to their meeting. From his association with Galileo, Torricelli moved to the forefront of the Italian scientific community.

Molecules moving throughout a gas sample hit the container's walls, producing this force.

Two instruments are used to measure gas pressure: barometers and manometers. A barometer is the instrument used to measure the pressure exerted by the atmosphere, whereas a manometer measures the pressure of isolated gas samples.

Evangelista Torricelli (1608–1647), a student of Galileo, designed the first mercury barometer. A barometer is constructed by filling a glass tube with Hg and then placing the open end of the tube vertically into a container of Hg.

FIGURE 12.1

(a) A barometer is an instrument for measuring atmospheric pressure. One atmosphere supports 760 mm Hg in a vertical closed-end tube. (b) A manometer is an instrument that is used to measure the pressure of isolated gas samples. This figure is a diagram of an open-end manometer, a U-shaped tube that contains mercury. The pressure of a gas sample is calculated from the difference in height of the mercury columns in each arm of the tube, and the atmospheric pressure.

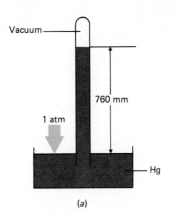

(a)

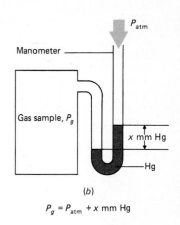

(b)

$$P_g = P_{atm} + x \text{ mm Hg}$$

Some of the Hg initially spills out of the tube into the container of Hg. When the force exerted by the Hg column equals the force exerted by the atmosphere on the surface of the Hg in the container, no more Hg spills out (see Figure 12.1). If the atmospheric pressure increases, more Hg is pushed up into the tube, and if the pressure decreases, Hg spills out of the tube.

Torricelli measured atmospheric pressure in terms of the total height of Hg supported in the glass tube by the atmosphere. At sea level, the atmosphere supports a column of Hg about 760 mm high. Today, gas pressures are still measured in terms of the height of mercury that a gas supports. Units of pressure most frequently employed are atmospheres (atm), millimeters of mercury (mm Hg), torr, and pascals (Pa). One atmosphere is defined as 760 mm Hg.

$$1.00 \text{ atm} = 760 \text{ mm Hg} = 760 \text{ torr}$$

Numerically, 1 mm Hg is equal to 1 torr (named for Torricelli).

Since these units are not derived from the base SI units, they are not SI units. The SI unit for pressure is the pascal (Pa), a small unit of pressure. A pascal is derived from the SI unit of force, newton (1 kg·m/s^2) and the unit for area, meter squared (m^2).

$$1 \text{ atm} = 101,325 \text{ Pa} = 101.3 \text{ kPa}$$

Because of the pascal's small size, the unit kilopascal (kPa) is normally used.

Example Problem 12.1 provides an example of interconverting pressure units.

Blaise Pascal (1623–1662), at the age of 16, published a book on geometry and conic sections. Pascal repeated Torricelli's famous experiment, and carried it further by getting someone to take the barometer to the top of a mountain to see if the Hg level dropped.

Example Problem 12.1

Convert 796 mm Hg to torr, atmospheres, and kilopascals.

Solution

1. *What is unknown?*   Torr, atm, and kPa

2. *What is known?*   796 mm Hg, $\dfrac{1 \text{ torr}}{1 \text{ mm Hg}}$, $\dfrac{760 \text{ mm Hg}}{1 \text{ atm}}$, and

$\dfrac{101{,}300 \text{ Pa}}{1 \text{ atm}}$

3. Apply the factor-label method.

$$796 \text{ mm Hg} \times \frac{1 \text{ torr}}{1 \text{ mm Hg}} = 796 \text{ torr}$$

$$796 \text{ mm Hg} \times \frac{1 \text{ atm}}{760 \text{ mm Hg}} = 1.05 \text{ atm}$$

$$796 \text{ mm Hg} \times \frac{1 \text{ atm}}{760 \text{ mm Hg}} \times \frac{101{,}300 \text{ Pa}}{1 \text{ atm}} \times \frac{1 \text{ kPa}}{1000 \text{ Pa}} = 106 \text{ kPa}$$

Given the number of millimeters of mercury, one immediately knows the number of torr since they are numerically equal. Converting to atmospheres is accomplished by multiplying by $\dfrac{1 \text{ atm}}{760 \text{ mm Hg}}$. Finally, changing to kPa requires that the pressure in mm Hg be first converted to atmospheres, and then converted to pascals and ultimately to kilopascals.

REVIEW PROBLEMS

12.1 What are the main assumptions of the kinetic molecular theory?
12.2 (a) What is an ideal gas? (b) Why do chemists develop relationships for an ideal gas when no real gas exactly behaves in this manner?
12.3 How is gas pressure measured?
12.4 Perform the following pressure unit conversions:
(a) 133 Pa = ? torr   (b) 390 mm Hg = ? atm
(c) 1.4 atm = ? kPa   (d) 1250 torr = ? atm

Boyle's law

Boyle's law is the gas law that relates the volume and pressure of an ideal gas at constant temperature and constant number of moles. Intuitively, it is easy to understand Boyle's law. As shown in Figure 12.2, if we increase the external pressure on a gas sample, the volume decreases. Diminishing the external pressure allows the gas to expand and increase in volume. Therefore, a statement of **Boyle's law** is that the volume of a gas is inversely proportional to its pressure when the temperature and number of moles of gas are held constant.

Mathematically, a statement of Boyle's law is as follows:

$$PV = k \qquad \text{(at constant } T \text{ and } n\text{)}$$

Boyle was the first scientist to collect gases and systematically study their properties. He discovered the P-V relationship using a 17-ft tube that contained a trapped gas and Hg.

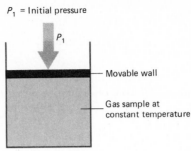

P_1 = Initial pressure

P_1

Movable wall

Gas sample at
constant temperature

Increase pressure, $P_2 > P_1$

Volume decreases

Decrease pressure, $P_3 < P_1$

Volume increases

P_3

FIGURE 12.2
At constant temperature and for a constant number of moles, if the pressure on a gas sample is increased, the volume of the gas decreases, and if the pressure is decreased, the volume increases. This relationship is known as Boyle's law.

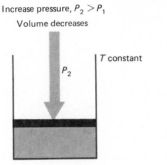

T constant

P_2

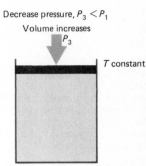

T constant

where P is the pressure of the gas, V is its volume, and k is a constant of proportionality.

To illustrate the inverse proportionality between the volume and pressure of an ideal gas, consider the data presented in Table 12.1. Initially, the pressure on the gas is 20 atm and the volume is 1 L. When the pressure is decreased to 10 atm (halved), the volume increases to 2 L (doubles). After decreasing the pressure to 5 atm, the volume increases to 4 L (doubles again). Notice that in each case the product of the pressure times the volume remains 20 L·atm; $P \times V$ remains constant as long as no other factors (T, n, or other experimental factors) are changed. The P versus V inverse relationship is plotted in Figure 12.3 on page 264.

A more useful expression of Boyle's law is:

$$P_1V_1 = P_2V_2$$

where P_1 is the initial pressure, V_1 is the initial volume, and V_2 is the final volume after the pressure has been changed to P_2. A mathematical expression of Boyle's law allows us to calculate a new volume or pressure for an ideal gas after the pressure has changed.

What is the new volume of a gas that initially is in a 5.0-L cylinder under a pressure of 15 atm when the pressure is decreased to 1.0 atm? Using Boyle's law, first rearrange the equation to isolate V_2 on one side:

TABLE 12.1
BOYLE's LAW RELATIONSHIP

P, atm	V, L	k, L·atm
20	1	20
10	2	20
5	4	20
4	5	20
2	10	20
1	20	20

$P_1V_1 = k$ and $P_2V_2 = k$, and therefore $P_1V_1 = P_2V_2$, a mathematical expression of Boyle's law.

FIGURE 12.3
The initial gas pressure is 1.0 atm, and the initial volume is 20 L. When the pressure is increased to 2.0 atm, the volume decreases to 10 L. When the pressure is increased to 5.0 atm, the volume decreases to 4.0 L, one-fifth the initial volume.

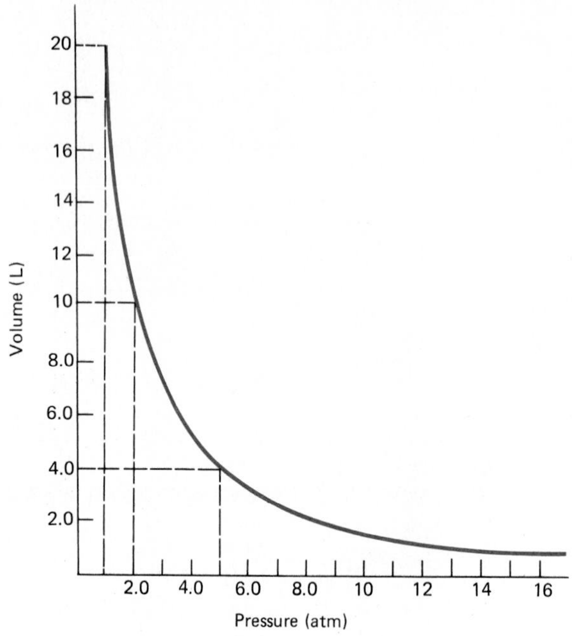

$$\frac{P_1V_1}{P_2} = \frac{\cancel{P_2}V_2}{\cancel{P_2}}$$

After dividing by P_2 and changing the equation around, we get

$$V_2 = V_1 \times \frac{P_1}{P_2}$$

It is convenient to express the equation as above, where the pressure terms are isolated from the initial volume. The ratio of pressures is sometimes called the *pressure factor*, and is another conversion factor. Therefore, to solve the problem, we multiply the initial volume (V_1), 5.0 L, times the ratio of initial pressure to final pressure (P_1/P_2), 15 atm/1.0 atm.

$$V_2 = 5.0 \text{ L} \frac{15 \cancel{\text{atm}}}{1.0 \cancel{\text{atm}}}$$

$$= 75 \text{ L}$$

With the decrease in pressure the volumes expands to 75 L, exactly what we would expect for an inverse relationship: decreasing the pressure increases the volume.

Even if you forget the mathematical equation for Boyle's law, you can calculate the final volume by realizing that P and V are inversely related. Ac-

FIGURE 12.4
To find the final volume V_2 of a gas after the pressure is changed at constant temperature and for a constant number of moles, the initial volume V_1 is multiplied by the pressure factor. The pressure factor P_1/P_2 is the conversion factor that relates the initial pressure P_1 to the final pressure P_2.

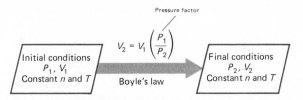

cordingly, the pressure factor must increase the magnitude of the initial volume if the pressure is decreased. In other words, the larger pressure must be written in the numerator and the smaller pressure written in the denominator.

The procedure for determining the final volume of an ideal gas after a pressure change is diagrammed in Figure 12.4. Example Problem 12.2 is another example of a problem involving Boyle's law.

Example Problem 12.2

Find the final volume of a gas when the pressure on 375 mL He is increased from 428 mm Hg to 1657 mm Hg.

Solution

1. *What is unknown?* mL He (final volume, V_2)
2. *What is known?* 375 mL (initial volume, V_1), 428 mm Hg (initial pressure, P_1), and 1657 mm Hg (final pressure, P_2)
3. Apply Boyle's law expression.

Instead of blindly plugging numbers into the expression, ask yourself what is happening. In this problem, the pressure is increased; thus, the volume decreases. The pressure factor should be written with the smaller pressure divided by the larger pressure (P_1/P_2).

$$V_2 = V_1 \times \frac{P_1}{P_2}$$

$$= 375 \text{ mL} \times \frac{428 \text{ mm Hg}}{1657 \text{ mm Hg}}$$

$$= 96.9 \text{ mL}$$

After completing the problem, look at the answer and see if it is reasonable. A final volume of 96.9 mL is significantly less than the initial volume, exactly what we expect when the pressure is increased.

Charles' Law   Charles' law is the relationship between the volume and Kelvin temperature of a gas at constant pressure and constant number of moles. If a gas sample is heated,

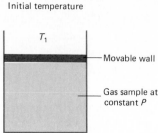

Initial temperature

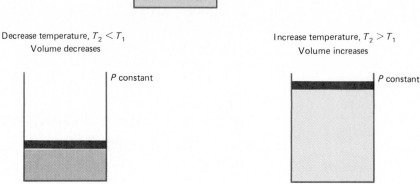

Decrease temperature, $T_2 < T_1$
Volume decreases

Increase temperature, $T_2 > T_1$
Volume increases

FIGURE 12.5
If the pressure and number of moles of gas are constant, the volume of a gas is directly proportional to its Kelvin temperature. Thus, if the temperature of a gas sample is decreased, the volume decreases; and if the temperature is increased, the volume increases.

Jacques Alexandre Cesar Charles (1746–1823) was the first to develop H_2 balloons, and on a few occasions he ascended to over a mile in one of his balloons. Charles repeated an experiment performed by Guillaume Amontons (1663–1705) regarding the expansion of gases on heating. Charles is mainly credited with the accurate determination that at 0°C the volume of a gas increases by 1/273 for each degree change.

the average kinetic energy of the molecules increases—they move faster. Because they move faster, they hit the walls of the container more frequently and with a greater force of impact. Hence, if the pressure is to remain constant, the volume must increase (see Figure 12.5). Likewise, if the temperature is decreased, the volume decreases because the molecules slow down, hitting the walls less frequently and with less force.

Charles' law states that the volume of a gas is directly proportional to its Kelvin temperature when the pressure and number of moles are held constant. Mathematically, a statement of Charles' law is as follows:

$$\frac{V}{T} = k \quad \text{(at constant } P \text{ and } n)$$

where V is volume, T is the Kelvin temperature of the gas, and k is a proportionality constant. Now we are dealing with a direct relationship, one in which both variables (V and T) change in the same direction. Increase the temperature and the volume increases; decrease the temperature and the volume decreases.

Figure 12.6 presents a representative graph of T versus V for an ideal gas. Looking at Figure 12.6, we see a linear relationship in which the volume increases with increasing temperature. Of special interest is the point where the line touches the temperature axis (point of zero volume), $-273.15°C$, or absolute zero. Don't rashly jump to any false conclusions! It's fun to extrapolate data to see what might be expected above and below the range of collected data. Remember, we are describing ideal gases (they exist mainly in the minds of chemists). Gases, on cooling, liquefy before reaching 0 K.

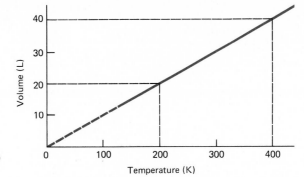

FIGURE 12.6
The volume of a gas sample is directly proportional to its Kelvin temperature. The graph shows that if the Kelvin temperature is doubled from 200 K to 400 K, the volume doubles from 20 L to 40 L.

Changing the Charles' law expression to a workable equation, we get:

$$\frac{V_1}{T_1} = \frac{V_2}{T_2}$$

Rearranging to solve for the final volume, V_2, we obtain:

$$V_2 = V_1 \times \frac{T_2}{T_1}$$

Once again, the final volume of the gas is calculated by multiplying the initial volume of the gas times a conversion factor, the temperature factor. In contrast to the Boyle's law pressure factor, where the initial pressure is divided by the final pressure, in the Charles' law temperature factor the final Kelvin temperature is divided by the initial Kelvin temperature.

If an ideal gas initially occupied 25.4 L at 25.0°C, what is the final volume when the temperature is changed to 78.2°C? In this problem, the temperature is increased; consequently, the volume increases—a direct relationship. Before applying the Charles' law relationship, we must change the Celsius temperatures to Kelvin temperatures. Remember that the Kelvin temperature is calculated by adding 273.15 to the Celsius temperature:

$$T_1 = 25.0°C + 273.2 = 298.2 \text{ K}$$
$$T_2 = 78.2°C + 273.2 = 351.4 \text{ K}$$

Applying the Charles' law relationship:

$$V_2 = V_1 \times \frac{T_2}{T_1}$$

$$= 25.4 \text{ L} \times \frac{351.4 \cancel{K}}{298.2 \cancel{K}}$$

$$= 29.9 \text{ L}$$

FIGURE 12.7
To find the final volume V_2 of a constant number of moles of a gas after the temperature is changed at constant pressure, the initial volume V_1 is multiplied by the temperature factor. The temperature factor T_2/T_1 is the conversion factor that relates the final temperature T_2 to the initial temperature T_1.

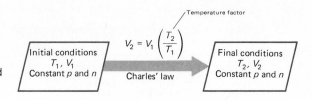

we find that the final volume of the gas sample is 29.9 L, an increase of 4.50 L from the initial volume. Figure 12.7 illustrates the procedure for finding the final volume of an ideal gas after a temperature change.

Example Problem 12.3 illustrates another problem involving Charles' law.

Example Problem 12.3

Determine the final volume of a 178-mL Xe gas sample when the temperature is decreased from 63.5°C to −155.3°C.

Solution

1. *What is unknown?* Final volume, V_2, in milliliters
2. *What is known?*   178 mL (initial volume, V_1), 63.5°C (initial temperature, T_1), −155.3°C (final temperature, T_2)
3. Convert Celsius temperatures to Kelvin temperatures.

$$T_1 = 63.5°C + 273.2 = 336.7 \text{ K}$$
$$T_2 = -155.3°C + 273.2 = 117.9 \text{ K}$$

4. Apply Charles' law.

$$V_2 = V_1 \times \frac{T_2}{T_1}$$
$$= 178 \text{ mL} \times \frac{117.9\ \cancel{K}}{336.7\ \cancel{K}}$$
$$= 62.3 \text{ mL}$$

After the gas is cooled from 336.7 K to 117.9 K, the volume decreases to 62.3 mL, if the pressure and number of moles of gas are constant.

Combined Gas Law Boyle's and Charles' laws are mathematically combined together to yield what is called the **combined gas law.**

$$V_2 = V_1 \times \frac{P_1}{P_2} \times \frac{T_2}{T_1}$$

In the combined gas law equation, there are two conversions factors to adjust the initial volume to the final volume:

$$V_2 = V_1 \times \underbrace{\frac{P_1}{P_2}}_{\substack{\text{Pressure} \\ \text{factor}}} \times \underbrace{\frac{T_2}{T_1}}_{\substack{\text{Temperature} \\ \text{factor}}} \quad (\text{constant } n)$$

Multiplying each side of this equation by $\dfrac{P_2}{T_2}$ gives another expression of the combined gas law equation:

$$\frac{P_1 V_1}{T_1} = \frac{P_2 V_2}{T_2}$$

Combined gas law

With the combined gas law, there is no need to deal with the individual gas laws. If temperature is held constant ($T_1 = T_2$), the temperature terms drop out of the equation, giving Boyle's law:

$$V_2 = V_1 \times \frac{P_1}{P_2} \times \frac{\cancel{T_2}}{\cancel{T_1}} = V_1 \times \frac{P_1}{P_2}$$

Boyle's law

Gay-Lussac is usually given credit for identifying the direct relationship between pressure and the absolute temperature of a gas when the volume and number of moles are constant.

In a similar manner, if the pressures are held constant ($P_1 = P_2$), the pressure terms cancel, leaving the Charles' law expression. (Prove this to yourself.)

If the volume is constant ($V_1 = V_2$), the volume terms divide out of the equation, giving us the third gas law, called **Gay-Lussac's law**. This states that pressure is directly proportional to the Kelvin temperature at constant volume and constant number of moles.

When both pressure and temperature changes are made on a gas, the final volume depends on the direction and magnitude of the two changes. For in-

FIGURE 12.8
To find the final volume V_2 of a constant number of moles of a gas after both a temperature and a pressure change, the initial volume V_1 is multiplied by both the pressure factor P_1/P_2 and the temperature factor T_2/T_1.

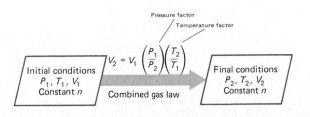

stance, if the pressure is increased and the temperature is decreased, the final volume of the gas is decreased. Whenever the pressure is increased, the volume is decreased, and in a similar manner, when the temperature is decreased the volume also decreases; both changes decrease the gas volume.

Example Problems 12.4 and 12.5 provide illustrations of the application of the combined gas law. Use Figure 12.8 as a guide for solving combined gas law problems with pressure and temperature changes.

Example Problem 12.4

STP means a standard temperature of 273 K and a standard pressure of 1 atm.

A gas initially occupies a volume of 5.2 L at 44°C and 718 mm Hg. What volume does the gas occupy at standard temperature and pressure (STP), 273 K (0°C) and 1.00 atm?

Solution

1. *What is unknown?* Final volume, V_2, in liters
2. *What is known?* 5.2 L, V_1; 44°C, T_1; 718 mm Hg, P_1; 273 K, T_2; and 1.00 atm, P_2
3. Change given units to kelvins and atmospheres.

$$T_1 = 44°C + 273 = 317 \text{ K}$$

$$P_1 = 718 \text{ mm Hg} \times \frac{1.00 \text{ atm}}{760 \text{ mm Hg}} = 0.945 \text{ atm}$$

T_1 is changed to kelvins, and P_1 is changed to atmospheres so that the units will cancel in the combined gas law equation.
4. Apply the combined gas law.

$$V_2 = V_1 \times \frac{P_1}{P_2} \times \frac{T_2}{T_1}$$

$$= 5.2 \text{ L} \times \frac{0.945 \text{ atm}}{1.00 \text{ atm}} \times \frac{273 \text{ K}}{317 \text{ K}}$$

$$= 4.2 \text{ L}$$

Is our answer reasonable? Pressure is increased and temperature is decreased. An increase in P decreases the volume, and a decrease in T also decreases the volume; accordingly, the final volume is less than the initial volume.

Example Problem 12.5

A gas occupies a volume of 444 mL at 0°C and 593 mm Hg. What must be the final temperature of the gas if, after changing the pressure to 290 mm Hg, the gas occupies 1.88 L?

Solution

In this problem, instead of solving for a new volume, the final temperature is requested, given the pressure and volume change. Rearranging the combined gas law, we obtain the following equation to solve for the final temperature.

$$T_2 = T_1 \times \frac{P_2}{P_1} \times \frac{V_2}{V_1}$$

Verify this for yourself.

1. *What is unknown?* T_2 (final temperature), in kelvins
2. *What is known?*   0°C, T_1; 444 mL, V_1; 1.88 L, V_2; 593 mm Hg, P_1; and 290 mm Hg, P_2
3. Change T_1 to kelvins, and V_1 to liters.

$$T_1 = 0°C + 273 = 273 \text{ K}$$

$$V_1 = 444 \text{ mL} \times \frac{1 \text{ L}}{1000 \text{ mL}} = 0.444 \text{ L}$$

4. Apply the combined gas law.

$$T_2 = 273 \text{ K} \times \frac{290 \text{ mm Hg}}{593 \text{ mm Hg}} \times \frac{1.88 \text{ L}}{0.444 \text{ L}}$$

$$= 565 \text{ K} = 292°C$$

When the original gas sample is heated and the pressure is decreased to 290 mm Hg, the gas occupies 1.88 L at 565 K (292°C).

REVIEW PROBLEMS   **12.5** Write a mathematical expression for: (a) Boyle's law, (b) Charles' law, and (c) the combined gas law.
12.6 Distinguish between an inverse and direct proportion by sketching a graph for each.
12.7 Calculate the final volume of an ideal gas if initially it occupies 1.0 L at a pressure of 1.0 atm and the pressure is changed to 3.5 atm.

12.8 What is the final volume of a gas that initially occupies 55.2 mL at 13°C when the temperature is increased to 37°C?

12.9 A gas occupies 75 L at STP. What is its volume at 294 torr and 3.00×10^2 K?

Avogadro's Law

So far in our discussion of gases, the number of moles of particles within the gas sample has been held constant. We did not want any of the gas to escape, or other gases to be added, during the time the pressure and temperature were varied. Now let's look at what happens when the number of moles of gas is varied, keeping P and T constant.

Amedeo Avogadro was the first to learn that if pressure and temperature are kept constant, the volume of a gas is directly proportional to the number of moles of gas molecules:

$$V = kn$$

Avogadro's hypothesis went mainly unnoticed. About 50 years later, Stanislao Cannizzaro (1826–1910), Avogadro's countryman, applied the hypothesis and determined the molecular masses of gases.

where V is volume, n is number of moles, and k is a proportionality constant. A balloon is a good illustration of Avogadro's law. A balloon expands when air is added to it (moles of air increase), and it contracts when air escapes (moles of air decrease).

An important relationship develops out of Avogadro's law. If equal volumes of different gases are compared, they must contain the same number of moles of particles. Let's compare equal volumes of He(g) and Ne(g). Using the equation of Avogadro's law, we would obtain the following:

$$V_{He} = kn_{He} \quad \text{and} \quad V_{Ne} = kn_{Ne}$$

Since the volumes are equal ($V_{He} = V_{Ne}$), then

$$kn_{He} = kn_{Ne}$$

At the same temperature and pressure the proportionality constants k in the equations are equal; therefore, the constants can be divided out of the equation, leaving:

$$n_{He} = n_{Ne}$$

Thus, the number of moles of He gas equals the number of moles of Ne gas if equal volumes are compared at the same temperature and pressure.

Avogadro's law may be stated as follows: Equal volumes of different gases contain the same number of molecules if their pressure and temperature are the same. If we select standard conditions (STP) of 1.00 atm and 273 K, 1.00 mol of an ideal gas occupies 22.4 L. This volume is called the **molar volume** of an ideal gas.

The volume of a basketball is approximately 22.4 L.

$$\text{Molar volume}_{STP} = \frac{22.4 \text{ L}}{1.00 \text{ mol}}$$

FIGURE 12.9
Avogadro's law calculations require the use of the molar gas volume at STP conditions, 22.4 L/mol. The molar volume of a gas is the conversion factor that relates the volume and number of moles of an ideal gas.

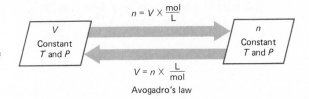

$$n = V \times \frac{mol}{L}$$

V
Constant
T and P

n
Constant
T and P

$$V = n \times \frac{L}{mol}$$

Avogadro's law

The molar volume of a gas is a conversion factor that allows us to calculate volume and mole relationships for gases (see Figure 12.9). To illustrate this, let's calculate the number of moles of He gas contained in a 1.00-L container at 273 K and 1.00 atm (STP).

$$V_{He} = 1.00 \, \cancel{L} \times \frac{1.00 \, mol}{22.4 \, \cancel{L}}$$

$$= 0.0446 \, mol$$

At STP, 0.0446 mol He is contained in a volume of 1.00 L. Example Problem 12.6 is another illustration of Avogadro's law.

Example Problem 12.6

What volume would 8.4 g Ne gas occupy at STP?

Solution

1. *What is unknown?* Liters Ne

2. *What is known?* 8.4 g Ne, $\dfrac{1.00 \, mol \, Ne}{20.0 \, g \, Ne}$, and $\dfrac{22.4 \, L \, Ne}{1.00 \, mol \, Ne}$

3. Apply Avogadro's law.

$$V = 8.4 \, \cancel{g \, Ne} \times \frac{1.00 \, \cancel{mol \, Ne}}{20.0 \, \cancel{g \, Ne}} \times \frac{22.4 \, L \, Ne}{1.00 \, \cancel{mol \, Ne}}$$

$$= 9.4 \, L \, Ne$$

First, we convert the mass of Ne to moles, using its molar mass, and then we compute the number of liters by applying the molar volume relationship. We determine that 8.4 g Ne occupies 9.4 L when the temperature is 273 K and the pressure is 1.00 atm.

What if the gas is not at standard conditions? We first determine the volume the gas occupies at standard conditions, and then change to the desired set of conditions by applying the combined gas law. Example Problem 12.7 illustrates mole-volume relationships at nonstandard conditions.

Example Problem 12.7

What volume would 0.25 g Ar gas occupy at 298 K and 0.54 atm?

Solution

1. *What is unknown?* Liters Ar at 298 K and 0.54 atm

2. *What is known?* 0.25 g Ar, $\dfrac{1.0 \text{ mol Ar}}{4.0 \times 10^1 \text{ g Ar}}$, 298 K, and 0.54 atm

3. Determine the volume of Ar at STP. It is not necessary to solve the problem in two steps, but for illustrative purposes we shall find the volume at STP and then correct the volume to the stated conditions.

$$V_{STP} = 0.25 \text{ g Ar} \times \frac{1.0 \text{ mol Ar}}{4.0 \times 10^1 \text{ g Ar}} \times \frac{22.4 \text{ L Ar}}{1.00 \text{ mol Ar}}$$

$$= 0.14 \text{ L Ar at STP}$$

4. Apply the combined gas law.

$$V_2 = V_{STP} \times \frac{P_1}{P_2} \times \frac{T_2}{T_1}$$

$$= 0.14 \text{ L} \times \frac{1.0 \text{ atm}}{0.54 \text{ atm}} \times \frac{298 \text{ K}}{273 \text{ K}}$$

$$= 0.28 \text{ L Ar}$$

REVIEW PROBLEMS

12.10 State Avogadro's law in two different ways.

12.11 Find the number of moles of gas molecules in each volume at STP: (a) 0.224 L, (b) 103 L, (c) 75 mL.

12.12 At STP, what volume would each of the following gases occupy? (a) 4.0 g H_2, (b) 8.0 g He, (c) 56 g CO.

12.13 Determine the mass of 0.0259 L of radon gas, Rn, at STP.

12.14 What volume would 0.921 g F_2 gas occupy at 2.50 atm and 25.0°C?

12.3 IDEAL GAS EQUATION

Boyle's law, Charles' Law, and Avogadro's law are mathematically combined to give a relationship for P, V, T, and n of an ideal gas. This relationship is called the **ideal gas equation**.

$$PV = nRT$$

where P is pressure, V is volume, T is temperature, n is moles, and R is the ideal gas constant. Rearranging the equation and solving for R, we find the numerical value of R:

$$\frac{PV}{nT} = \frac{\not{n}R\not{T}}{\not{n}\not{T}}$$

Therefore

$$R = \frac{PV}{nT}$$

If we select a gas sample of 1.00 mol, we know that at STP the volume of an ideal gas is 22.4 L. Substituting these numbers into the equation, we get

$$R = \frac{1 \text{ atm} \times 22.4 \text{ L}}{1 \text{ mol} \times 273 \text{ K}}$$

$$= 0.0821 \frac{\text{L·atm}}{\text{mol·K}}$$

Thus, the value of the ideal gas constant R is 0.0821 L·atm/(mol·K) for all ideal gases.

To use the ideal gas equation, we must adjust the given units so that they cancel the units in the ideal gas constant R or change the units of the ideal gas constant to match those given.

If we are given three variable properties of a gas and the ideal gas equation, the fourth variable can be calculated. When solving such a problem, rearrange the ideal gas equation algebraically, isolating the unknown variable on one side. For instance, if P, V, and T are known, rearrange the equation to solve for n. Thus we have

$$PV = nRT$$

and we divide both sides by RT

$$\frac{PV}{RT} = \frac{n\not{R}\not{T}}{\not{R}\not{T}}$$

giving
$$n = \frac{PV}{RT}$$

When the equation is successfully rearranged, the units cancel, leaving only the desired units, moles.

$$n = \frac{\cancel{atm} \times \cancel{L}}{\frac{\cancel{L} \cdot \cancel{atm} \times \cancel{K}}{mol \cdot \cancel{K}}} = \frac{1}{\frac{1}{mol}} = mol$$

Example Problems 12.8 and 12.9 apply the ideal gas equation to find an unknown gas property.

Example Problem 12.8

What is the volume of 2.2 mol of a gas at 440 K and 3.6 atm?

Solution

1. *What is unknown?*   Volume, L

2. *What is known?*   2.2 mol, 440 K, 3.6 atm, and $R = 0.0821 \dfrac{L \cdot atm}{mol \cdot K}$

3. Rearrange the ideal gas equation.

$$\frac{P\cancel{V}}{\cancel{P}} = \frac{nRT}{P}$$

$$V = \frac{nRT}{P}$$

4. Substitute known values into the equation. All units given in this problem correspond to those in the ideal gas constant, so it is only necessary to substitute the numbers into the equation, and solve for V.

$$V = \frac{2.2 \text{ mol} \times 0.0821 \dfrac{L \cdot atm}{K \cdot mol} \times 440 \text{ K}}{3.6 \text{ atm}}$$

$$= 22 \text{ L}$$

When 2.2 mol of an ideal gas is found at 440 K and 3.6 atm, it will occupy 22 L.

Example Problem 12.9

Find the temperature of 0.39 mol of gas that occupies 9850 mL under a pressure of 631 torr.

Solution

1. *What is unknown?* T in kelvins

2. *What is known?* $n = 0.39$ mol, $V = 9850$ mL, and $P = 631$ torr

3. Change given units to those that correspond to the ideal gas constant. Since the ideal gas constant is expressed in L·atm/(mol·K), the volume must be converted to liters and pressure to atmospheres.

$$V = 9850 \ \cancel{mL} \times \frac{1 \ L}{1000 \ \cancel{mL}} = 9.850 \ L$$

$$P = 631 \ \cancel{torr} \times \frac{1.00 \ atm}{760 \ \cancel{torr}} = 0.830 \ atm$$

4. Rearrange and substitute into the ideal gas equation.

$$\frac{PV}{nR} = \frac{\cancel{n}\cancel{R}T}{\cancel{n}\cancel{R}}$$

$$T = \frac{PV}{nR}$$

$$T = \frac{0.830 \ \cancel{atm} \times 9.850 \ \cancel{L}}{0.39 \ \cancel{mol} \times 0.0821 \ \dfrac{\cancel{L} \cdot \cancel{atm}}{\cancel{mol} \cdot K}}$$

$$= 255 \ K = 2.6 \times 10^2 \ K$$

At the stated conditions, the temperature of the gas is 2.6×10^2 K.

Modification of the ideal gas equation allows us to find molecular masses of gases. Remember that the number of moles of a substance is calculated by dividing the mass by the molar mass (MM):

$$Moles = \frac{mass}{mass/mol}$$

$$= \frac{g}{g/mol}$$

Therefore, we can substitute g/MM into the ideal gas equation for n, giving:

$$PV = nRT$$

Substituting g/MM for n

$$PV = \frac{g}{MM} RT$$

Rearranging, and isolating molar mass, MM, we have

$$MM = \frac{gRT}{PV}$$

Molar masses of gases are found by determining their masses at a known set of conditions and applying the above equation. See Example Problem 12.10.

Example Problem 12.10

The mass of a gas sample is 61.5 g, and it occupies 37.2 L at 313 K and 0.924 atm. What is the molar mass of the gas?

Solution

1. *What is unknown?* g/mol of gas, MM
2. *What is known?* 61.5 g, 37.2 L, 313 K, 0.924 atm, and

$$R = 0.0821 \frac{L \cdot atm}{mol \cdot K}$$

3. Apply the modified form of the ideal gas equation.

$$MM = \frac{gRT}{PV}$$

$$= \frac{61.5 \text{ g} \times 0.0821 \frac{\cancel{L} \cdot \cancel{atm}}{\cancel{K} \cdot mol} \times 313 \cancel{K}}{0.924 \cancel{atm} \times 37.2 \cancel{L}}$$

$$= 46.0 \text{ g/mol}$$

Calculations indicate that the molecular mass of the gas is 46.0 g/mol.

REVIEW PROBLEMS

12.15 Determine the numerical value for the ideal gas constant R in L·torr/(mol·K).

12.16 What volume would 6.3 mol of gas occupy at 32 atm and 273 K?

12.17 Determine the temperature, in Celsius, of 0.0532 mol of a gas that occupies 0.280 L under a pressure of 8.50 atm.

12.18 How many moles of gas are contained in a gas sample that has a measured volume of 65 mL under a pressure of 12.6 atm and a temperature of −37°C?

12.4 VOLUME RELATIONSHIPS IN CHEMICAL REACTIONS

Volume-Volume Relationships

Gay-Lussac was the first to recognize that the volumes of combining gases can be expressed as a ratio of whole numbers at constant temperature and pressure. This relationship is now known as the **law of combining volumes.** Let's illustrate the law of combining volumes by considering the gas phase reaction of $H_2(g)$ and $Cl_2(g)$ to produce $HCl(g)$:

$$H_2(g) + Cl_2(g) \longrightarrow 2HCl(g)$$

As we have already seen, the coefficients within an equation indicate the mole ratio in which the reactants combine and the products form. According to the law of combining volumes, the coefficients also represent the volume ratios. Thus, the equation indicates that at a specified temperature and pressure, one volume of H_2 combines with one volume of Cl_2 to produce two volumes of HCl. See Figure 12.10.

$$H_2(g) + Cl_2(g) \longrightarrow 2HCl(g)$$
$$\text{1 vol} \quad \text{1 vol} \quad\quad \text{2 vol}$$

Solutions to problems involving gas volumes are solved in a similar manner to the stoichiometry problems in Chapter 11, except that the conversion factors obtained from the equation's coefficients contain volume units. Some conversion factors derived from the above equation are as follows:

$$\frac{1\ L\ H_2}{1\ L\ Cl_2} \quad \frac{1\ L\ Cl_2}{1\ L\ H_2} \quad \frac{2\ L\ HCl}{1\ L\ H_2}$$

Study Example Problem 12.11 as a sample problem involving volume relationships.

FIGURE 12.10
If 1 L $H_2(g)$ is combined with 1 L $Cl_2(g)$ at the same set of conditions, 2 L HCl is produced. This is an example of the law of combining volumes.

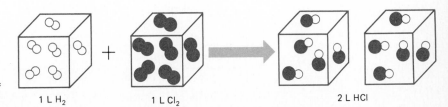

1 L H_2 1 L Cl_2 2 L HCl

Example Problem 12.11

Consider the Haber reaction:

$$N_2(g) + 3H_2(g) \longrightarrow 2NH_3(g)$$

What volume of ammonia gas, NH_3, forms when 35 L H_2 combines with excess N_2 at fixed temperature and pressure?

Solution

1. *What is unknown?*   Liters NH_3

2. *What is known?*   35 L H_2, $\dfrac{2 \text{ L } NH_3}{3 \text{ L } H_2}$

3. Apply the factor-label method.

$$35 \, \cancel{\text{L } H_2} \times \frac{2 \text{ L } NH_3}{3 \, \cancel{\text{L } H_2}} = ? \text{ L } NH_3$$

4. Perform the indicated math operations.

$$35 \, \cancel{\text{L } H_2} \times \frac{2 \text{ L } NH_3}{3 \, \cancel{\text{L } H_2}} = 23 \text{ L } NH_3$$

A yield of 23 L NH_3 is expected when 35 L H_2 combines with excess N_2.

Mass-Volume Relationships

On many occasions, solids, liquids, and gases are encountered in chemical reactions. For instance, if $KClO_3(s)$ is heated in the presence of small amounts of MnO_2, $KCl(s)$ and $O_2(g)$ are produced:

$$2KClO_3(s) \xrightarrow[\Delta]{MnO_2} 2KCl(s) + 3O_2(g)$$

From the coefficients in the equation, we see that 2 mol $KClO_3$ decomposes to produce 2 mol KCl and 3 mol O_2. Stated another way, 2 mol $KClO_3$ decomposes to produce 2 mol KCl and 67.2 L O_2 at STP (3 mol × 22.4 L/mol). Example Problem 12.12 is an illustration of a mass-volume problem.

Example Problem 12.12

What volume of hydrogen gas, at STP, is liberated when 5.0 g of iron, Fe, is combined with excess hydrochloric acid, HCl, in the following reaction:

$$Fe(s) + 2HCl(aq) \longrightarrow FeCl_2(aq) + H_2(g)$$

Solution

1. *What is unknown?* Liters H_2 gas at STP

2. *What is known?* 5.0 g Fe, $\dfrac{1 \text{ mol Fe}}{55.8 \text{ g Fe}}$, $\dfrac{1 \text{ mol Fe}}{1 \text{ mol } H_2}$, and $\dfrac{22.4 \text{ L } H_2}{1 \text{ mol } H_2}$

3. Apply the factor-label method.

$$5.0 \text{ g Fe} \times \frac{1 \text{ mol Fe}}{55.8 \text{ g Fe}} \times \frac{1 \text{ mol } H_2}{1 \text{ mol Fe}} \times \frac{22.4 \text{ L } H_2}{1 \text{ mol } H_2} = ? \text{ L } H_2$$

4. Perform the indicated math operations.

$$5.0 \text{ g Fe} \times \frac{1 \text{ mol Fe}}{55.8 \text{ g Fe}} \times \frac{1 \text{ mol } H_2}{1 \text{ mol Fe}} \times \frac{22.4 \text{ L } H_2}{1 \text{ mol } H_2} = 2.0 \text{ L } H_2$$

When 5.0 g Fe is placed in a beaker of hydrochloric acid, 2.0 L H_2 forms if the conditions are adjusted to STP.

While many gas stoichiometry problems specify STP conditions, it should be noted that few reactions actually occur at STP conditions. Problems involving reactions that are not at standard conditions are solved in one of two ways: (1) Assume STP conditions, solve the stoichiometry problem, and then apply the combined gas law, or (2) from the equation, find the number of moles of gaseous products, and then use the ideal gas equation. Both of these methods are demonstrated in Example Problem 12.13.

Example Problem 12.13

Barium carbonate, $BaCO_3$, is heated to produce barium oxide, BaO, and carbon dioxide, CO_2:

$$BaCO_3(s) \xrightarrow{\Delta} BaO(s) + CO_2(g)$$

What volume of CO_2 is produced, at 710 mm Hg and 425 K, when 3.0 g $BaCO_3$ is decomposed?

Solution

1. *What is unknown?* Liters CO_2 at 710 mm Hg and 425 K

2. *What is known?*   3.0 g $BaCO_3$, $\dfrac{1 \text{ mol } BaCO_3}{197 \text{ g } BaCO_3}$, $\dfrac{1 \text{ mol } CO_2}{1 \text{ mol } BaCO_3}$,

$\dfrac{22.4 \text{ L } CO_2}{1 \text{ mol } CO_2}$

Method 1. Assume STP and correct to desired conditions.

$$3.0 \text{ g } BaCO_3 \times \frac{1 \text{ mol } BaCO_3}{197 \text{ g } BaCO_3} \times \frac{1 \text{ mol } CO_2}{1 \text{ mol } BaCO_3} \times \frac{22.4 \text{ L } CO_2}{1 \text{ mol } CO_2}$$

$$\times \frac{760 \text{ mm Hg}}{710 \text{ mm Hg}} \times \frac{425 \text{ K}}{273 \text{ K}} = 0.57 \text{ L } CO_2$$

Here, we have solved the problem assuming STP conditions. The molar volume is used to find the number of liters of CO_2, and then pressure and temperature factors are employed to change to the desired conditions. Standard temperature and pressure are T_1 and P_1, respectively, and 425 K and 710 mm Hg are T_2 and P_2.

Method 2. Find moles of CO_2, and substitute into the ideal gas equation.

$$3.0 \text{ g } BaCO_3 \times \frac{1 \text{ mol } BaCO_3}{197 \text{ g } BaCO_3} \times \frac{1 \text{ mol } CO_2}{1 \text{ mol } BaCO_3} = 0.15 \text{ mol } CO_2$$

$$710 \text{ mm Hg} \times \frac{1 \text{ atm}}{760 \text{ mm Hg}} = 0.934 \text{ atm}$$

Substitute into the ideal gas equation.

$$PV = nRT \quad \text{or} \quad V = \frac{nRT}{P}$$

$$V = \frac{0.015 \text{ mol} \times 0.0821 \dfrac{\text{L·atm}}{\text{K·mol}} \times 425 \text{ K}}{0.934 \text{ atm}}$$

$$= 0.57 \text{ L}$$

REVIEW PROBLEMS   **12.19** Consider the following gas phase reaction:

$$2NO(g) + O_2(g) \longrightarrow 2NO_2(g)$$

(a) What volume of NO_2 forms at STP when 26 L NO combines with excess O_2?
(b) At STP, how many liters of NO are required to combine with exactly 159 L O_2?
(c) At 25°C and 2.3 atm, what volume of O_2 is required to produce 88 L NO_2?

12.20 What volume of O_2 gas is released at STP when 35 g H_2O_2 decomposes to H_2O and O_2?

$$2H_2O_2(aq) \longrightarrow 2H_2O(l) + O_2(g)$$

12.21 What volume of O_2 gas is produced at 799 mm Hg and 321 K when 0.759 g $KClO_3$ is decomposed to KCl and O_2?

12.5 MIXTURE OF GASES

One of the properties of gases is that any number of nonreacting gases can be mixed together in any proportion to produce a gaseous solution (homogeneous mixture). How do we deal with gas mixtures?

From Avogadro's law, we know that an ideal gas occupies 22.4 L at STP conditions and 2 mol occupies 2 × 22.4 L, or 44.8 L. What if we initially placed 1 mol of a gas in a container and then placed 1 mol of another gas in the same container? Since there are now 2 mol of gas in the container, the expected volume of the gas mixture is also 2 × 22.4 L, or 44.8 L (see Figure 12.11). In other words, if the pressure and temperature are constant, the total volume of the gas mixture depends on the total number of moles of gas particles, which is equal to the sum of the moles of each gas found in the mixture:

$$n_{total} = n_1 + n_2 + n_3 + \cdots$$

where n_1, n_2, n_3 are the number of moles of gases 1, 2, and 3.

Looking at gas mixtures from a different perspective, if the volume and temperature are constant, what happens when gases are mixed? If a 22.4-L container at 273 K contains 1 mol of gas, the pressure gauge on the container reads 1 atm. When 1 mol of another gas is pumped into the container, 2 mol total, the pressure gauge increases to 2 atm. Addition of 1 mol of a third gas increases the pressure to 3 atm, and so on. Hence, the total pressure of a mixture of gases is equal to the sum of the individual pressures (partial pressures) of each different gas in the mixture.

FIGURE 12.11
One mole of an ideal gas occupies 22.4 L at 1.00 atm and 273 K. If 1.00 mol of another gas is added to the container, with no change in volume or temperature, the pressure increases to 2.00 atm. The total pressure of a mixture of gases is equal to the sum of the partial pressures of each gas in the mixture.

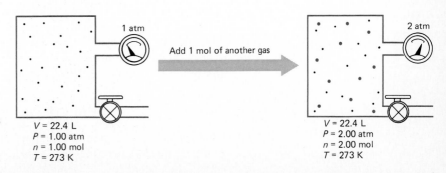

1 atm

Add 1 mol of another gas

2 atm

$V = 22.4$ L
$P = 1.00$ atm
$n = 1.00$ mol
$T = 273$ K

$V = 22.4$ L
$P = 2.00$ atm
$n = 2.00$ mol
$T = 273$ K

Even though Dalton is most famous for his proposals regarding the atomic nature of matter, he was keenly interested in meteorology. Dalton's drive to understand the weather led him to study air, a mixture of gases.

John Dalton was the first scientist to explain pressures of gas mixtures; accordingly, this relationship is now known as **Dalton's law of partial pressures.** Mathematically, the law of partial pressures is as follows:

$$P_{total} = P_1 + P_2 + P_3 + \cdots$$

where P_1, P_2, P_3 are partial pressures of gases in the mixture, and P_{total} is the total pressure of all gases. Dalton's law is illustrated in Example Problem 12.14.

Example Problem 12.14

What is the pressure exerted by Ar gas in a mixture of Ar, He, and Ne that has a total pressure of 753 torr? The partial pressures of He and Ne are 321 torr and 254 torr, respectively.

Solution

1. *What is unknown?* P_{Ar} in torr
2. *What is known?* $P_{He} = 321$ torr, $P_{Ne} = 254$ torr, and $P_{total} = 753$ torr
3. Apply Dalton's law of partial pressures.

$$P_{total} = P_{He} + P_{Ne} + P_{Ar}$$
$$753 \text{ torr} = 321 \text{ torr} + 254 \text{ torr} + P_{Ar}$$
$$P_{Ar} = 753 \text{ torr} - (321 + 254) \text{ torr}$$
$$= 178 \text{ torr}$$

In the three-component gas mixture, Ar's partial pressure is 178 torr. Adding the three partial pressures together gives the total pressure, 753 torr.

Dalton's law can be used to find the dry volume of insoluble gases that are collected over water. One method used to collect a gas over water is as follows: Water is added to a gas bottle (or other container), which is then inserted over a trough. The gas is allowed to displace the water in the gas bottle (see Figure 12.12). Water vapor is mixed with the collected gas when this procedure is used.

Where did the water vapor come from? It came from the evaporation of water in the gas bottle and trough. The total pressure exerted by the collected gas is the sum of the pressures of the gas and water vapor.

$$P_{total} = P_{gas} + P_{H_2O}$$

To find the pressure of the pure gas, subtract the pressure of the water vapor (vapor pressure of water) from the total pressure.

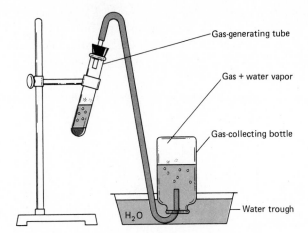

FIGURE 12.12
Gas produced in a gas-generating tube is collected by water displacement in a gas-collecting bottle. Water evaporates and mixes with the gas produced. Hence, to find the pressure of the dry gas, the vapor pressure of water is subtracted from the total pressure of the gas mixture.

Gas-generating tube

Gas + water vapor

Gas-collecting bottle

Water trough

H_2O

$$P_{gas} = P_{total} - P_{H_2O}$$

The dry volume of a gas is the volume a pure gas occupies if no water vapor is present.

Table 12.2 presents the vapor pressures of water at various temperatures. Vapor pressure is discussed in greater detail in the next chapter; at this time, just keep in mind that water's vapor pressure depends primarily on the water's temperature.

Example Problem 12.15 illustrates Dalton's law and the collection of gases over water.

Example Problem 12.15

Hydrogen gas, H_2, was generated by placing magnesium metal, Mg, in an acid solution:

Water's vapor pressure is directly proportional to temperature: the higher the temperature, the higher the vapor pressure.

TABLE 12.2   VAPOR PRESSURES OF WATER FROM 20°C TO 30°C

Temperature, °C (K)	Vapor pressure, mm Hg or torr	Vapor pressure, atm
20 (293)	17.5	0.0230
21 (294)	18.7	0.0246
22 (295)	19.8	0.0261
23 (296)	21.1	0.0278
24 (297)	22.4	0.0295
25 (298)	23.8	0.0313
26 (299)	25.2	0.0332
27 (300)	26.7	0.0351
28 (301)	28.3	0.0372
29 (302)	30.0	0.0395
30 (303)	31.8	0.0418

$$Mg(s) + 2HCl(aq) \longrightarrow MgCl_2(aq) + H_2(g)$$

The H_2 generated was collected over water in a trough; 63.4 mL H_2 was formed at 27.0°C and 752.1 mm Hg. What mass of magnesium metal was necessary to produce the H_2?

Solution

1. *What is unknown?*   Grams Mg

2. *What is known?*   63.4 mL H_2,   752.1 mm Hg,   27.0°C, $\dfrac{1 \text{ mol } H_2}{1 \text{ mol } Mg}$,

$\dfrac{1 \text{ mol } Mg}{24.3 \text{ g } Mg}$, $P_{H_2O} = 26.7$ mm Hg at 27°C

3. Apply Dalton's law and find the pressure due to H_2 only.

From Dalton's law, we know that the total pressure, 752 mm Hg, is equal to the pressure of H_2 and H_2O:

$$P_{total} = P_{H_2} + P_{H_2O}$$
$$752.1 \text{ mm Hg} = P_{H_2} + 26.7 \text{ mm Hg}$$

Hence, if 26.7 mm Hg is subtracted from both sides of the equation, the pressure of pure H_2 is obtained:

$$P_{H_2} = P_{total} - P_{H_2O}$$
$$= 752.1 \text{ mm Hg} - 26.7 \text{ mm Hg}$$
$$= 725.4 \text{ mm Hg}$$

Of the total pressure measured for the gas mixture, 725.4 mm Hg is exerted by H_2; therefore, our remaining calculations should only use the dry pressure.

4. Find the number of moles of H_2. Finding the moles of H_2 can be accomplished in two ways; let's determine the quantity of moles by applying the ideal gas equation.

$$P = 725.4 \text{ mm Hg} \times \frac{1 \text{ atm}}{760 \text{ mm Hg}} = 0.9545 \text{ atm}$$

$$T = 27.0°C + 273.2 = 300.2 \text{ K}$$

$$V = 63.4 \text{ mL} \times \frac{1 \text{ L}}{1000 \text{ mL}} = 0.0634 \text{ L}$$

Rearranging and substituting values into the ideal gas equation, we get:

$$n = \frac{PV}{RT}$$

$$= \frac{0.9545 \text{ atm} \times 0.0634 \text{ L}}{0.08205 \frac{\text{L·atm}}{\text{mol·K}} \times 300.2 \text{ K}}$$

$$= 0.00246 \text{ mol H}_2$$

5. Apply the factor-label method to find the mass of Mg.

$$0.00246 \text{ mol H}_2 \times \frac{1 \text{ mol Mg}}{1 \text{ mol H}_2} \times \frac{24.3 \text{ g Mg}}{1 \text{ mol Mg}} = 0.0598 \text{ g Mg}$$

Initially, 0.0598 g Mg was present to produce 63.4 mL H_2 over water at the stated conditions.

REVIEW PROBLEMS

12.22 State Dalton's law of partial pressures.
12.23 Using the kinetic molecular theory, how can you explain Dalton's law?
12.24 Find the total pressure of a gaseous mixture containing the following pressures of gases: 1.23 atm H_2, 541 torr He, and 99.1 kPa Ne.
12.25 A sample of N_2 gas is collected over water. The pressure of the wet N_2 gas at 28°C is 749 mm Hg. What is the dry pressure of N_2?

12.6 ATMOSPHERE

The atmosphere is the mixture of gases that surrounds the earth. Earth's atmosphere acts as a protective "blanket" that traps energy from the sun and blocks out various types of radiation.

The gas mixture most important to living things is the earth's atmosphere. Nitrogen, N_2, and oxygen, O_2, are the primary gases in the atmosphere, making up about 99% of its volume. All other gases are found in relatively small amounts. Table 12.3 lists the principal atmospheric gases and their percent by volume.

TABLE 12.3 COMPOSITION OF THE ATMOSPHERE AT SEA LEVEL (PERCENT BY VOLUME)

Gas	Formula	Percent
Nitrogen	N_2	78
Oxygen	O_2	21
Argon	Ar	0.9
Carbon dioxide	CO_2	0.03–0.04
Neon	Ne	0.0012
Helium	He	0.0005
Krypton	Kr	0.0001
Ozone	O_3	0.00006
Hydrogen	H_2	0.00005
Xenon	Xe	0.000009

FIGURE 12.13
Besides the normal components of the atmosphere, human activities contribute gases such as sulfur oxides, nitrogen oxides, hydrocarbons, and carbon monoxide. These gases are responsible for air pollution. (*Documerica*)

Some of the pollutant gases in the atmosphere are SO_2, SO_3, CO, NO, NO_2, and various hydrocarbons (carbon-hydrogen compounds).

In addition to the gases listed in Table 12.3, variable amounts of water vapor, sulfur oxides, nitrogen oxides, hydrocarbons, carbon monoxide, and other trace gases are in the atmosphere.

Let's take a closer look at the principal gases in the atmosphere: nitrogen, oxygen, and the noble gases.

Nitrogen

Nitrogen, N_2, was discovered by Daniel Rutherford in 1772. It is a colorless, odorless, and tasteless gas. N_2 boils at $-195.8°C$ at 1 atm and freezes at $-210°C$. Nitrogen's density, 1.261 g/L, is slightly less than air's density.

Rutherford was remotely related to Sir Walter Scott, and was a firm believer in Stahl's phlogiston theory. When Rutherford discovered nitrogen, he called it "dephlogisticated air" because it did not support combustion or sustain life.

A nitrogen molecule possesses a strong triple bond that requires a large amount of energy to break (226 kcal/mol or 945 kJ/mol). Thus, N_2 is quite unreactive. Since it is stable, nitrogen gas is used as an "inert atmosphere." Canned foods remain fresher if they are sealed with pure nitrogen gas instead of air. When chemists handle substances that react with air (as a result of air's O_2 content), they place the substances in a container filled with N_2 or another unreactive gas.

Because nitrogen is inert, it is difficult to tap its huge reserve in the atmosphere. The Haber process is an important procedure for **fixing** N_2, i.e., converting atmospheric N_2 to nitrogen compounds.

$$N_2 + 3H_2 \longrightarrow 2NH_3$$

Haber reaction

In nature, N_2 is fixed by microorganisms that have the capacity to convert N_2 to NH_3. Synthesis of NH_3 from N_2 is the primary step in what is referred to

Proteins are one of the four principal groups of compounds found in living systems; the others are lipids, carbohydrates, and nucleic acids.

as the **nitrogen cycle.** Figure 12.14 illustrates the primary components of the nitrogen cycle. After NH_3 is formed, it is converted by other microorganisms to nitrites, NO_2^-, nitrates, NO_3^-, and ultimately into proteins and other biologically important nitrogen compounds.

Oxygen

Karl Wilhelm Scheele (1742–1786) was self-educated in chemistry and was initially an apothecary. Scheele's accomplishments in chemistry include the discoveries of citric acid, hydrogen cyanide, lactic acid, oxalic acid, and hydrogen fluoride gas. He was also part of a research group that discovered the elements W, Mo, Ba, and Mg.

Credit for the discovery of oxygen, O_2, is usually given to Joseph Priestley, who in 1774 heated mercury(II) oxide, HgO, liberating O_2. Karl Wilhelm Scheele had discovered O_2 about 3 years prior to Priestley, but he neglected to have his findings published until 1777.

Like nitrogen, O_2 gas is colorless, tasteless, and odorless. Oxygen is slightly more dense than air; its density is 1.429 g/L, compared with air's density of 1.292 g/L. The water solubility of oxygen is relatively low—only 31.6 mL O_2 gas dissolves per liter of water. Nevertheless, this small quantity of O_2 sustains the lives of animals that live in water. Oxygen melts at −218°C and boils at −183°C at 1 atm.

Oxygen reacts with most other elements. We have already seen how O_2 reacts with both metals and nonmetals to form a diversity of oxides. Oxygen molecules contain oxygen-oxygen double bonds (O=O) that are quite stable. Most reactions that O_2 undergoes require temperatures and pressures above room conditions. But in some cases, as in rusting, O_2 combines slowly at room temperature.

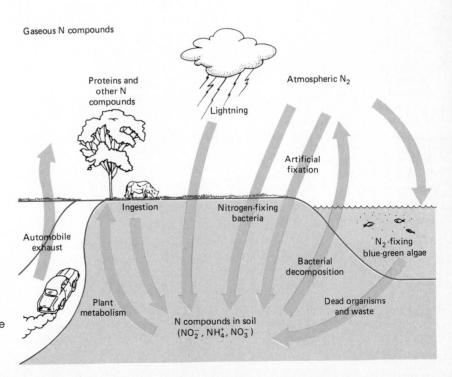

Gaseous N compounds

Proteins and other N compounds

Atmospheric N_2

Lightning

Artificial fixation

Ingestion

Nitrogen-fixing bacteria

N_2-fixing blue-green algae

Automobile exhaust

Bacterial decomposition

Plant metabolism

N compounds in soil (NO_2^-, NH_4^+, NO_3^-)

Dead organisms and waste

FIGURE 12.14
All of the nitrogen on earth is part of the nitrogen cycle. Nitrogen in many forms moves continually through the atmosphere, lithosphere, and biosphere. Nitrogen gas is the main component of the atmosphere. In the lithosphere, nitrogen-containing salts are the most abundant forms. In the biosphere, proteins are the primary nitrogen compounds.

Gases are liquefied by compressing, cooling, and then allowing the gases to expand through a small opening, which further cools them. The cycle is then repeated until the gases liquefy.

Industrially, most O_2 is prepared by isolating it from air that has been cooled to very low temperatures; a similar method is employed to produce N_2 gas. O_2 is used in the steel industry to accelerate combustion and increase reaction temperatures. Smaller quantities of O_2 are used in the medical industry as part of life-support systems. The aerospace industry uses O_2 as a component of liquid rocket fuels. Each year, larger quantities of O_2 are being used to treat sewage and more effectively reduce the concentration of impurities in dirty water.

Cellular respiration occurs in areas inside the cell called mitochondria. A complex series of reactions take place in mitochrondia, producing energy for the cell.

Oxygen is the sustainer of life. Animals need a constant supply of O_2. Even a brief interruption of O_2 can cause brain damage or death. Oxygen is needed by cells to produce energy in the respiratory chain. To complete the respiratory chain, O_2 must be present to combine with hydrogen atoms to produce water. O_2 is transported from the lungs to the cells by a large molecule called hemoglobin. Each hemoglobin molecule contains four iron ions that bond to four O_2.

Noble Gases

While they are not major components of the atmosphere, the noble gases are quite interesting. When Mendeleev proposed his famous periodic table, the noble gases had not been discovered; thus, in Mendeleev's periodic table the alkali metals directly followed the halogens.

Janssen traveled throughout the world to observe astronomical phenomena. He made his famous discovery of He in the sun while in India.

A French astronomer, Pierre Janssen (1824–1907), discovered the first noble gas, helium, in 1868. Helium was not discovered on earth; Janssen detected helium in the sun. Using a technique called spectroscopic analysis, Janssen identified He from the light it gives off at high temperatures. Many scientists were skeptical of his discovery, and it was not until 1895 when Sir William Ramsay (1852–1916) found He on earth (trapped in rocks) that He was generally accepted. Helium's name is derived from the Greek word *helios* meaning "the sun."

Rayleigh was a physicist who became interested in a controversy regarding the scale of atomic weights. To gather information about atomic weights, he needed the densities of gases. During the process of measuring nitrogen's density, Rayleigh found the apparent discrepancy in mass.

Helium's discovery sparked a great deal of interest once chemists realized that He did not fit into their periodic table. In 1894, Lord Rayleigh (John William Strutt Rayleigh, 1842–1919) showed that nitrogen gas isolated from the air was heavier than nitrogen gas obtained from decomposition reactions of nitrogen-containing salts. Rayleigh tried to solve this puzzle but became frustrated and finally wrote a letter to *Nature* (one of Britain's most respected scientific journals) asking for help or suggestions. Sir William Ramsay answered Rayleigh's plea for help, and later that year, using spectroscopy techniques, he discovered that argon (meaning the "lazy one" or "inert one") was mixed with the nitrogen, producing the higher density.

Spectroscopy involves passing light through a prism, which breaks the light up into its characteristic colors (wavelengths). When substances are heated, they give off a light that is used to identify the substance.

Ramsay continued the search for other noble gases; in 1898, he identified and isolated neon ("new one"), Ne; krypton ("hidden one"), Kr; and xenon ("stranger"), Xe. Radon, the remaining noble gas, was discovered by Friedrich Dorn as a radioactive decay product of radium; however, Rn was named and further studied by Marie and Pierre Curie.

Physical properties of the noble gases are listed in Table 12.4. The boiling points of noble gases are among the lowest in the periodic table. Helium remains in the gas phase to within a few degrees of absolute zero. He's low boiling point is

TABLE 12.4   PHYSICAL PROPERTIES OF NOBLE GASES

Noble gas	Melting point, °C	Boiling point, °C	Density (STP), g/L
Helium	−272.2	−268.9	0.179
Neon	−248.7	−246.1	0.900
Argon	−189.2	−185.9	1.78
Krypton	−156.6	−153.4	3.75
Xenon	−111.9	−108.1	5.90

accounted for in terms of weak forces among the atoms, making He one of the most "ideal gases." As a result of its low boiling point, liquid He is used in cryogenic research, i.e., in the study of very low temperatures. Helium is also placed in "lighter than air" balloons, and it is mixed with O_2 in diver's tanks to prevent the narcotic effect of N_2 breathed at high pressures.

Neon is used in lamps and signs; when a current is passed through the neon-filled tube, it causes neon to give off its characteristic red glow. Argon, the most abundant noble gas on earth, is placed in incandescent light bulbs to decrease the vaporization and oxidation of the bulb's tungsten (W) filament, resulting in a longer life for the light bulb.

As we have seen, the noble gases are very unreactive and only combine with highly electronegative nonmetals like F and O. Serious studies of noble gas compounds began in the early 1960s, when Neil Bartlett synthesized xenon tetrafluoride, XeF_4, which exists as colorless crystals that melt near 90°C and can be stored at room temperature in a glass container for relatively long periods without decomposition.

REVIEW PROBLEMS

12.26 What are the five most abundant gases in the atmosphere?

12.27 What is the nitrogen cycle, and why is it important to living things?

12.28 Write a common use for the following gases: (a) N_2, (b) O_2, and (c) He.

12.29 Which noble gas would be expected to exhibit the most ideal gas properties? Explain.

SUMMARY

The gaseous state is the least dense and most fluid of the physical states. The properties of gases have been studied for hundreds of years, and a theory has been developed to explain these properties. This theory, called the kinetic molecular theory, is a set of assumptions concerning the behavior of atoms and molecules that make up an ideal gas.

Gas laws explain relationships among volume, pressure, temperature, and number of moles of gas particles in ideal gas samples. Boyle's law states that the volume of an ideal gas is inversely proportional to the applied pressure at constant temperature and number of moles. Gas pressure is the force that gas particles exert on their containers. Pressure is measured in atmospheres, kilopascals, millimeters of mercury, and torr, where 1 atm = 760 mm Hg = 760 torr = 101 kPa.

Charles' law states that the volume of a gas is directly proportional to its absolute temperature at constant pressure and constant number of moles. Avogadro's law states that the volume of a gas is directly proportional to the number of moles of gas at constant pressure and temperature, or stated another way, at a constant set of conditions equal volumes of ideal gases contain the same number of molecules. At standard temperature and pressure (273 K and 1 atm), the volume of 1.00 mol of an ideal gas is 22.4 L.

Various mathematical expressions are derived from the gas laws. Two of the most important ones are the combined gas law

$$\frac{P_1V_1}{T_1} = \frac{P_2V_2}{T_2}$$

which is used to determine the final set of conditions, given the initial conditions, and the ideal gas equation:

$$PV = nRT$$

where one of the four gas variables can be calculated, given the other three. R in the equation is the ideal gas constant, 0.0821 L·atm/(mol·K), which describes the relationship between gas pressure and volume, and moles and kelvin temperature.

In chemical reactions that involve gases, Gay-Lussac was the first scientist to show that the volumes of combining gases can be expressed as ratios of whole numbers; in other words, given a correctly balanced equation, the coefficients indicate the volume ratios in which gases combine. Therefore, volume relationships are determined using procedures similar to those used to find mass and mole relationships in chemical reactions.

When gases are mixed, the sum of the pressures of each individual gas, called partial pressures, equals the total pressure exerted by the gaseous mixture. This relationship is called Dalton's law of partial pressures.

Earth's atmosphere is composed of a mixture of gases, primarily N_2 and O_2 (99%). Argon, CO_2, H_2O vapor, and approximately 10 other gases make up the remaining 1% of the atmosphere. Nitrogen gas is made up of stable diatomic nitrogen molecules, which are converted (fixed) by special microorganisms, in nature, to nitrogen compounds that ultimately enter living systems as proteins and other biologically important compounds.

The physical properties of oxygen are similar to, but different than, those of nitrogen. Oxygen is stable but more reactive than N_2. When heated, O_2 combines with most other elements. Oxygen plays a central role in living systems. Argon is the third most abundant gas in the atmosphere. Argon and the other noble gases are the most inert elements. Historically, their rarity and inertness made them elusive. Noble gases were not identified until the end of the nineteenth century.

QUESTIONS AND PROBLEMS

12.30 Define the following terms: kinetic molecular theory, kinetic energy, ideal gas, pressure, pascal, mm Hg, torr, atmosphere, barometer, manometer, inverse proportion, direct proportion, absolute zero, STP, partial pressure, inert atmosphere, nitrogen cycle, cryogenics.

Kinetic Molecular Theory

12.31 Using the kinetic molecular theory, explain each of the following:
(a) Gases are the most compressible state of matter.
(b) Two nonreacting gases can be mixed in any proportions.
(c) When heated, gas volumes increase.
(d) Gases always take the shape of their container.
(e) Gas particles do not settle to the bottom of their containers on standing.

12.32 Why do gases exhibit more ideal properties as their pressures are decreased and their temperatures are increased?

Pressure

12.33 (a) How is gas pressure defined?
(b) What is the SI unit for pressure?

12.34 Explain how a mercury barometer measures atmospheric pressure.

12.35 Why could a barometer also be used as an altimeter, a device for measuring height above the ground?

12.36 Account for the fact that, even at sea level, atmospheric pressure is constantly changing.

12.37 Convert 1.35×10^5 Pa to: (a) kPa, (b) torr, (c) atm, and (d) mm Hg.

12.38 Change 0.809 atm to: (a) torr, (b) Pa, and (c) kPa.

12.39 Convert 1214 mm Hg to: (a) torr, (b) atm, (c) Pa, and (d) kPa.

12.40 Change 0.25 torr to: (a) atm, (b) Pa, and (c) kPa.

12.41 Perform the following pressure conversions:
(a) 64.2 atm = ? mm Hg
(b) 0.011 mm Hg = ? Pa
(c) 4.1×10^3 mm Hg = ? atm
(d) 0.0043 kPa = ? torr
(e) 971 mm Hg = ? torr

12.42 A unit of pressure used in the United States is pounds per square inch, abbreviated $lb/in.^2$ or psi. Standard atmospheric pressure is 14.7 $lb/in.^2$. Calculate the pressure, in torr, inside a tire with an internal pressure of 29 psi.

Boyle's Law

12.43 State Boyle's law, specifying the variables that are held constant.

12.44 Using the mathematical statement of Boyle's law, $PV = k$, determine the volume of a gas for the following pressures, if it is known that $k = 0.33$ L·atm: (a) 1 atm, (b) 5 atm, (c) 0.5 atm, (d) 150 atm, (e) 88 kPa.

12.45 Consider the data in the graph of P versus V in Figure 12.15 to answer the following questions:
(a) What is the volume of the gas at 5.0 atm?

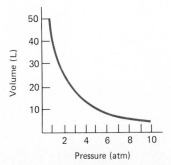

FIGURE 12.15

(b) At what pressure will the gas occupy a volume of 20.0 L?

(c) At an infinite pressure (if that were possible), what should happen to the gas volume?

12.46 What is the final volume of an ideal gas if initially it occupies 6.4 L at a pressure of 1.5 atm and the following pressure changes are made: (a) 2.0 atm, (b) 1.0 atm, (c) 25 atm, (d) 0.19 atm, (e) 815 mm Hg?

12.47 A sample of N_2 gas occupies 67.5 mL at 345 atm. What is the volume of the nitrogen gas at standard atmospheric pressure?

12.48 If a sample of Ne gas occupies 9.11 L at 543 mm Hg, what volume will the Ne occupy at 63.5 kPa?

12.49 Initially, a Ne gas sample occupies 22 L at 795 mm Hg. Determine the volume the Ne gas occupies under each of the following pressures: (a) 0.25 atm, (b) 0.25 torr, (c) 0.25 Pa, (d) 0.25 kPa.

12.50 What is the final pressure that would have to be exerted on 583 mL H_2 at 1 atm to change its volume to: (a) 1000 mL, (b) 340 mL, (c) 10 mL, (d) 5.13 L?

12.51 A 55-L helium tank contains compressed helium at 62 atm. How many 1.3-L balloons can be filled with the He in the cylinder, assuming that atmospheric pressure is exactly 1 atm?

Charles' Law

12.52 State Charles' law, and indicate which variables are held constant.

12.53 Using the mathematical statement of Charles' law, $\frac{V}{T} = k$, determine the volume of an ideal gas at the following temperatures if it is known that $k = 0.067$ L/K: (a) 251 K, (b) 25°C, (c) 0.0°C, (d) −109°C, (e) 607°F.

12.54 Consider the data in the graph of T versus V in Figure 12.16 to answer the following questions:

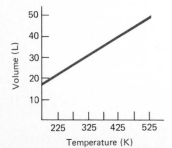

FIGURE 12.16

(a) At what temperature will the gas occupy 35.0 L?

(b) What is the temperature when the gas occupies 20.0 L?

(c) What is the volume of the gas at −10.0°C?

12.55 Why is there no simple relationship between the volume of an ideal gas and its temperature expressed in degrees Celsius?

12.56 What is the final volume of a gas that initially occupies 4.9 L at 298 K when the temperature is changed to: (a) 149 K, (b) 596 K, (c) 298°C, (d) −65°C?

12.57 Calculate the volume of O_2 gas that initially occupies 4.398 L at 100.0°C after the temperature is changed to 450.1 K.

12.58 An ideal gas, initially occupying 8.33 L at 0°C, would have to be at what temperature for it to occupy: (a) 10.0 L, (b) 6.00 L, (c) 15.2 L?

12.59 (a) Theoretically, what volume would a 25-L sample of He gas at 298 K occupy if its temperature is decreased to 0.10 K?

(b) What pressure is required to decrease the volume of the 25-L He sample to the same volume that results when the He is cooled to 0.10 K if initially the gas sample is at 1.0 atm?

Combined Gas Law

12.60 What happens to the volume of a gas sample when the following pressure and temperature changes are made?

(a) P increased, T decreased

(b) P decreased, T increased

(c) Small P increase, large T increase

(d) Large P increase, small T decrease.

12.61 What are the volumes of the following gas samples when the conditions are changed to STP?

(a) $V_1 = 5.5$ L, $T_1 = 85°C$, and $P_1 = 1.2$ atm

(b) $V_1 = 468$ mL, $T_1 = -72°C$, and $P_1 = 345$ torr

(c) $V_1 = 0.911$ L, $T_1 = 125$ K, and $P_1 = 121$ kPa

(d) $V_1 = 742$ L, $T_1 = 273°C$, and $P_1 = 92.1$ kPa

12.62 For each of the following, determine the final volume that the gas occupies:

(a) $V_1 = 0.54$ L $V_2 = ?$ L

     $T_1 = 300$ K $T_2 = 400$ K

     $P_1 = 8.2$ atm $P_2 = 5.4$ atm

(b) $V_1 = 444$ mL $V_2 = ?$ L

     $T_1 = -15°C$ $T_2 = -144°C$

     $P_1 = 215$ kPa $P_2 = 0.432$ atm

(c) $V_1 = 3.12$ L $V_2 = ?$ mL
  $T_1 = 0°C$ $T_2 = 310$ K
  $P_1 = 532$ torr $P_2 = 320$ kPa

(d) $V_1 = 6.32$ L $V_2 = ?$ mL
  $T_1 = -75°F$ $T_2 = -110°C$
  $P_1 = 403$ torr $P_2 = 0.27$ atm

12.63 Determine the final pressure of the following gases when the stated volume and temperature changes are made:

(a) $V_1 = 54.3$ L $V_2 = 47.8$ L
  $T_1 = 37°C$ $T_2 = 73°C$
  $P_1 = 0.923$ atm $P_2 = ?$ atm

(b) $V_1 = 3746$ mL $V_2 = 4.53$ L
  $T_1 = 155$ K $T_2 = -82.2°C$
  $P_1 = 78.4$ kPa $P_2 = ?$ Pa

(c) $V_1 = 62.8$ L $V_2 = 77.4$ L
  $T_1 = -23°F$ $T_2 = 35°F$
  $P_1 = 912$ torr $P_2 = ?$ mm Hg

(d) $V_1 = 0.236$ mL $V_2 = 1.72$ L
  $T_1 = 823$ K $T_2 = 723$ K
  $P_1 = 104$ kPa $P_2 = ?$ atm

12.64 If 235 L of a gas is prepared at 721 mm Hg and 316°C and then pumped into a 3.70-L steel tank at 25.0°C, what pressure will the tank have to withstand?

12.65 Calculate the Celsius temperature that is required to maintain a gas that is housed in a 300.0-L container under a pressure of 3.00 atm; the gas initially occupies 450.0 L at STP.

Avogadro's Law

12.66 Calculate the number of moles in each of the following gas samples at STP: (a) 202 L, (b) 493 L, (c) 0.064 L, (d) 0.33 mL, (e) 1.0 qt.

12.67 Find the volumes occupied by the following gases at STP:
(a) 4.55 g Ne   (b) 0.034 g SO_2
(c) 597 mg CH_4 (d) 2.9 g Rn
(e) 57.3 g O_3

12.68 Determine the masses of the following gases, given their volumes at STP:
(a) 475 L N_2O (b) 0.24 L HBr
(c) 12.9 L C_2H_2 (d) 4.1 L UF_6
(e) 76 mL Ar.

12.69 What volumes would the following gases occupy at 792 mm Hg and 310 K:
(a) 6.11 g H_2 (b) 214 g ClF
(c) 0.0261 g SO_3?

12.70 Calculate the volumes of the following at the stated conditions:
(a) 49.5 g He at 348 K and 0.924 atm
(b) 0.830 mol O_2 at 288 K and 838 mm Hg
(c) 1.97 g Cl_2 at 80°C and 333 kPa
(d) 0.512 g H_2S at 712°C and 0.936 atm

12.71 What are the densities, in grams per liter, of the following gases at standard conditions: (a) N_2, (b) CO, (c) Kr, (d) C_2H_6, and (e) $C_2F_2Cl_2$.

12.72 Determine the densities, in grams per liter, of the following gases at the stated conditions:
(a) O_2 at 0.607 atm and 94.5°C
(b) C_2H_6O at 0.550 atm and 460 K
(c) CF_4 at 803 mm Hg and 124°C.

Ideal Gas Equation

12.73 Solve the ideal gas equation, $PV = nRT$, for each of the following variables: (a) P, (b) n, (c) T, and (d) V, and show that the units cancel to give the correct units of the variable.

12.74 Use the factor-label method to determine the value of the ideal gas constant, R, with each of the following sets of units:

(a) $\dfrac{mL \cdot atm}{mol \cdot K}$ (b) $\dfrac{mL \cdot mm\ Hg}{mol \cdot K}$ (c) $\dfrac{kPa \cdot L}{mol \cdot K}$.

12.75 What are the volumes, in liters, of the following gases at the stated conditions?
(a) 5.26 mol at 436 K and 0.259 atm
(b) 0.988 mol at 745 mm Hg and 92.4°C
(c) 0.411 mol at 526 torr and 802 K

12.76 Find the number of moles of gas molecules in the following gas samples:
(a) 1.68 L at 0.992 atm and 77.7°C
(b) 75.1 L at 147 kPa and 49.8°C
(c) 0.0627 L at 158 K and 899 torr

12.77 Calculate the pressure, in atmospheres, exerted by the following gases:
(a) 0.771 mol occupying 83.8 L at 287 K
(b) 4.68 mol occupying 9762 mL at −43.9°C

(c) 0.0236 mol occupying 86.6 mL at 77.8 K

12.78 Determine the density of SF_6 at 176°C and 0.499 atm, using the ideal gas equation.

12.79 If 21.7 g of a gas sample occupies 18.9 L at 0.86 atm and 310 K, what is the molecular mass of the gas?

12.80 Calculate the molecular mass of 285 g of a gas that occupies 38.2 L at 125 kPa and −8.0°C.

Volume Relationships in Chemical Reactions

12.81 For the gas-phase reaction of the formation of HCl, $H_2(g) + Cl_2(g) \rightarrow 2HCl(g)$, determine the volume of H_2 required to combine with excess Cl_2 to produce the following volumes of HCl at STP conditions:
(a) 73.8 L HCl   (b) 4.1 L HCl
(c) 0.27 L HCl   (d) 360.0 L HCl.

12.82 Consider the equation:

$$2SO_2(g) + O_2(g) \longrightarrow 2SO_3(g)$$

What volume of SO_3 is produced when each of the following volumes of O_2 gas is combined with excess amounts of SO_2 at 1.5 atm and 298 K?
(a) 1.84 L   (b) 0.327 L O_2
(c) 24.7 mL O_2 (d) 525 L O_2

12.83 Oxygen, O_2, is prepared by placing sodium peroxide, Na_2O_2, in water:

$$2Na_2O_2(s) + 2H_2O(l) \longrightarrow O_2(g) + 4NaOH(aq)$$

Calculate the volume of O_2 gas liberated, at STP, when the following masses of Na_2O_2 are combined with excess water:
(a) 9.3 g Na_2O_2 (b) 34.2 g Na_2O_2
(c) 0.793 kg Na_2O_2 (d) 585 mg Na_2O_2.

12.84 Lead(II) sulfide, PbS, combines with ozone, O_3, to produce $PbSO_4$ and O_2 gas:

$$PbS(s) + 4O_3(g) \longrightarrow PbSO_4(s) + 4O_2(g)$$

Determine the mass of PbS(s) needed to combine exactly with the following volumes of O_3 at standard conditions: (a) 2.9 L, (b) 0.623 mL, (c) 909 L, (d) 0.827 L.

12.85 Nitrogen trichloride, NCl_3, decomposes to nitrogen gas, N_2, and chlorine gas, Cl_2:

$$2NCl_3(l) \longrightarrow N_2(g) + 3Cl_2(g)$$

Find the volume of $Cl_2(g)$ produced when the following quantities of NCl_3 are decomposed at the stated conditions:
(a) 3.69 g NCl_3 at 887 mm Hg and 270°C
(b) 29.4 g NCl_3 at 0.175 atm and 375 K
(c) 0.602 mol NCl_3 at 592 K and 316 kPa
(d) 0.0979 g NCl_3 at 818 mm Hg and 29.3°C

12.86 Ammonia gas is combined with oxygen gas on a catalyst to produce nitrogen monoxide and steam at 975°C and 2.3 atm. Calculate the volume of nitrogen monoxide produced when the following are combined with an excess of the other reactant:
(a) 17.1 L ammonia   (b) 0.329 g ammonia
(c) 44 L oxygen   (d) 553 g oxygen

12.87 When a sugar, $C_6H_{12}O_6$, is fermented with yeast, ethanol, C_2H_6O, and carbon dioxide gas, CO_2, are produced. If 47.6 g of sugar is completely fermented, what volume of carbon dioxide is given off at 31°C and 765 mm Hg?

Mixture of Gases

12.88 A gas mixture has a total pressure of 779 mm Hg; the partial pressure of N_2 is 266 mm Hg, and the partial pressure of He is 176 mm Hg; the remaining gas in the mixture is CO. Determine the partial pressure of CO in the mixture.

12.89 If 65% of the moles of a gas mixture is Ar gas and the remaining gas in the mixture is He, what is the partial pressure of each component, given that the total pressure is 1.851 atm?

12.90 A 83.4-mL sample of N_2 is collected over water at 298 K and 762 mm Hg. What is the volume of N_2, at the same conditions, without the water vapor?

12.91 What is the dry volume of Ar gas if a 0.732-L sample of Ar collected over water is at 21°C and 94.5 kPa?

12.92 When magnesium metal, Mg, is placed into hydrochloric acid, hydrogen gas, H_2, and aqueous magnesium chloride, $MgCl_2$, are produced. If 419 mL H_2 gas is collected, over water, at 23.6°C and 1.05 atm:
(a) What mass of H_2 gas is collected?
(b) What masses of Mg and HCl are combined to produce this mass of H_2?

12.93 Oxygen gas is collected over water by heating $KClO_3$ with a catalyst.

$$2KClO_3 \xrightarrow[\Delta]{MnO_2} 2KCl + 3O_2$$

It is found that 68.3 mL O_2 is produced at 28°C and 777 mm Hg. What mass of $KClO_3$ was initially heated to produce this quantity of oxygen?

Atmosphere

12.94 What scientist is credited with the discovery of each of the following gases: (a) oxygen, (b) nitrogen, (c) argon, (d) helium?

12.95 What is the principal means for fixing nitrogen (a) in the natural world, and (b) in industry?

12.96 Compare the physical properties of N_2 and O_2. How are they similar, and how are they different?

12.97 In Table 12.3, a range of values rather than a single value is given for the percentage volume of carbon dioxide. What could account for carbon dioxide's variability?

12.98 Write a short description of the nitrogen cycle, tracing nitrogen from the atmosphere to production of proteins in living systems.

12.99 How are N_2 and O_2 isolated from the atmosphere (look at their physical properties)?

12.100 List three uses for oxygen gas.

12.101 Specifically, why is O_2 needed by all animals, and why does a brief loss of O_2 result in death?

12.102 What could account for the rather recent discovery of the noble gases?

12.103 What trends are observed in the physical properties of the noble gas when proceeding from the lowest to highest atomic mass?

12.104 List uses for the noble gases.

12.105 Ar is most commonly used in light bulbs. If Kr was used in place of Ar, the bulb's life would be extended. What one reason could account for not using Kr in place of Ar?

General Problems

12.106 Dry ice is solid carbon dioxide, $CO_2(s)$. On heating, dry ice is totally converted to $CO_2(g)$. If 25 g of dry ice is placed in a 2.0-L container and then sealed, what pressure would result inside the container at room temperature, 25°C?

12.107 (a) Liquid mercury changes to a vapor at 357°C. Calculate the density of mercury vapor at 357°C and 1.0 atm.
(b) By what factor did the density decrease in the gas phase compared with the liquid phase if liquid mercury's density is 13.6 g/mL?

12.108 During the winter it is often necessary to add air to your tires to maintain the proper inflation.
(a) Why is the added air needed?
(b) Give an explanation employing the kinetic molecular theory.

· C H A P T E R ·

13

Liquids, Solids, and State Changes

Study Guidelines

After completing Chapter 13, you should be able to:
1. Describe the basic properties of liquids and solids
2. Explain the differences between gases, liquids, and solids, using the kinetic molecular theory
3. Describe the process of evaporation of liquids and factors that influence evaporation
4. Define a dynamic equilibrium in terms of evaporation and condensation
5. Discuss a liquid's equilibrium vapor pressure and how it is measured
6. List factors that influence a liquid's vapor pressure
7. Define the boiling point of a liquid, and list factors that influence boiling points
8. Explain surface tension of liquids and factors that alter surface tension
9. Describe what is meant by the viscosity of a liquid
10. Discuss the three categories of intermolecular forces in liquids (dipolar interactions, hydrogen bonding, and dispersion forces) with respect to how they are formed, strength, and examples of liquids exhibiting the force
11. List and explain the basic particles, forces, and examples of the four classes of solids: ionic, covalent, molecular, and metallic
12. Explain each segment found on a heating curve of a substance, including the points at which the substance is changing state, transition temperatures, and energy considerations
13. Predict the strength of intermolecular forces, given the heat of fusion and the heat of vaporization
14. Discuss the structure and properties of (a) individual water molecules, (b) liquid water, and (c) ice
15. Explain the unique physical properties of water
16. Outline primary parts of the water and carbon cycles, and explain how these cycles are important in nature
17. Discuss differences in the properties of the allotropic forms of carbon: graphite, diamond, and amorphous carbon

Liquids have a fixed volume and a variable shape, taking the shape of the bottom of their containers. Liquids are less fluid than gases, but they are significantly more fluid than solids. On an average, liquids are more dense than gases but less dense than solids.

Kinetic Molecular Theory

The kinetic molecular theory also applies to liquids. Like gases, the particles within liquids are in constant motion. However, the movement of liquid molecules and atoms is greatly restricted because the molecules are more closely packed together. Only a tiny proportion of the volume of liquids is empty space.

Why are liquid molecules more closely packed together than gas molecules? Attractive forces among liquid molecules are greater than those among gas molecules. Collectively, the forces that hold particles in the liquid state are called **intermolecular forces.** Compared with chemical bonds, intermolecular forces are weak. We do not encounter these forces in gases because gas molecules are widely separated and move quickly.

Intermolecular forces are the attractive forces that hold molecules together in the liquid state.

Because of the increased intermolecular forces, molecules in liquids have a more orderly arrangement than do the molecules in gases. Gas molecules are totally random and devoid of any definite structure. Liquid molecules, in contrast, stick together. This gives them a short-range order; i.e., there are small areas with an organized regular molecular pattern. While liquid molecules are more ordered than gases, they are less ordered than the particles in solids (see Figure 13.1).

Solid

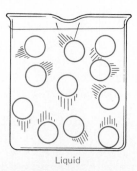
Liquid

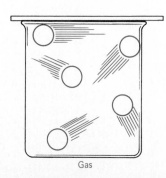
Gas

FIGURE 13.1
Solids have the most ordered arrangement of particles. Liquids possess structures that are more random than solids; only small segments are ordered. Gas particles are distributed randomly throughout their volume.

We can use the kinetic molecular theory to account for the principal liquid properties.

1. A liquid's incompressibility is explained in terms of the lack of empty spaces among molecules. If a force is applied to the surface of a liquid, there are few empty spaces for the molecules to fill.
2. Since the intermolecular forces are not strong enough to keep the molecules rigidly affixed to each other, they slide by each other. This accounts for a liquid's fluidity.
3. The higher density of liquids compared with gases results from the intermolecular forces that hold the molecules together in a relatively small volume with virtually no empty spaces among molecules.

With these general properties of liquids in mind, what are some of the specific properties associated with liquids?

Evaporation (Vaporization)

Evaporation occurs when liquids change to a vapor below the boiling point.

If a liquid is placed into an open container and left undisturbed, the liquid level steadily lowers until no liquid remains. What is happening, and where did the liquid go? If the liquid is gasoline, it is not difficult to understand that the gasoline is undergoing a phase change, liquid to vapor. Whenever gasoline is placed in an open container, the fumes are rapidly detected in the air.

Evaporation is the process in which liquid molecules break free from the surface and enter the gas phase. Evaporation is explained in terms of the energy possessed by the molecules on the liquid's surface. Surface molecules whose kinetic energies are large enough to overcome the intermolecular forces that bind them to the liquid break free and enter the gas phase.

Many higher animals regulate their body temperature through evaporation of water. Heat from their bodies evaporates water through pores in the skin.

In open containers (open systems), evaporation continues until all of the liquid is gone. If a liquid is placed into a closed container (closed system), the amount of liquid decreases for a period of time, and then does not change. In closed containers, the vapor cannot escape; as the vapor concentration increases, some of the vapor molecules lose energy and return to the liquid state. When a vapor returns to the liquid state, it is called **condensation.**

Evaporation and condensation are opposing processes. Evaporation involves molecules leaving the liquid, and condensation occurs when a vapor changes back to a liquid. Initially, after a liquid is placed in a closed container, it begins to evaporate at a constant rate; very little condensation takes place at this time. But as the concentration of the vapor increases above the liquid, the rate of condensation increases. At some point in time, the rate of condensation equals the rate of evaporation. When the rates are equal, the number of molecules entering the gas phase equals the number returning to the liquid phase in a given time interval. Consequently, the level of the liquid does not change; each liquid molecule lost is replaced by a molecule from the condensing vapor (see Figure 13.2).

A dynamic equilibrium results when two physical or chemical processes have equal but opposite rates.

When the rates of two opposing processes are equal, the system is said to be in a **dynamic equilibrium.** An equilibrium is represented in an equation by two arrows pointing in opposite directions.

FIGURE 13.2

(a) In an open system, the level of a liquid in a container continually drops as a result of evaporation until there is no more liquid remaining. (b) In a closed system there is a point when the rate of evaporation equals the rate of condensation. At this time the liquid level remains constant.

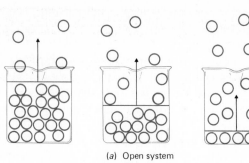

(a) Open system

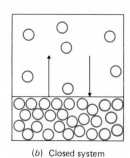

(b) Closed system

$$\text{Liquid} \underset{\text{condensation}}{\overset{\text{evaporation}}{\rightleftharpoons}} \text{vapor}$$

Vapor Pressure

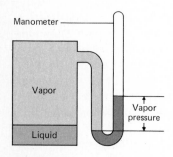

Manometer

Vapor

Vapor pressure

Liquid

FIGURE 13.3

The vapor pressure of a liquid at a specific temperature is determined by measuring the pressure exerted by a vapor in equilibrium with its liquid phase. A closed-end manometer is the pressure-measuring device in this figure.

A measure of the degree to which a liquid enters the vapor state is called **equilibrium vapor pressure.** If a manometer, a mercury-filled U tube used for measuring gas pressure, is attached to a closed vessel containing a liquid in equilibrium with its vapor, the manometer measures the pressure of the vapor above the liquid (see Figure 13.3). The pressure exerted by a vapor in equilibrium with its liquid state is called the liquid's vapor pressure. Vapor pressure is measured in the same units used for gas pressure: atmospheres, millimeters of mercury, torr, and pascals.

Vapor pressure of a liquid is independent of the amount of liquid or size of the container. Factors that affect a liquid's vapor pressure are its temperature and its intermolecular forces.

Consider the graph of the vapor pressure of water versus temperature shown in Figure 13.4. As the temperature increases, the vapor pressure increases—a direct relationship. Water's vapor pressure remains fairly low until about 50°C, when it starts to rise rapidly. At 100°C, the vapor pressure of water increases to 760 mm Hg, or 1 atm. All liquids exhibit similar vapor pressure curves. Figure 13.5 shows the vapor pressure curves for diethyl ether, ethanol, and water.

FIGURE 13.4

The vapor pressure of water is quite low at temperatures below 50°C. Above 50°C, water's vapor pressure increases rapidly. At 100°C, the vapor pressure of water is 760 mm Hg.

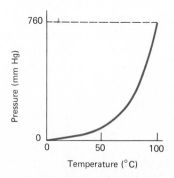

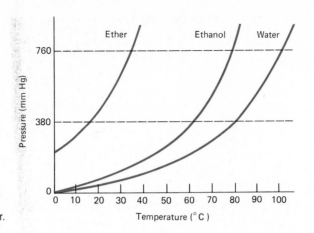

FIGURE 13.5
At all temperatures, the vapor pressure of ether is higher than the vapor pressure of either ethanol or water; thus, ether is the most volatile of the three liquids, and it has the lowest boiling point (34.6°C). Since the vapor pressure of ethanol is higher than the vapor pressure of water at all temperatures, ethanol is a more volatile liquid and boils at a lower temperature (78.3°C) than water.

Diethyl ether, $(CH_3CH_2)_2O$, is a very combustible volatile liquid. Ether, as it is frequently called, was once one of the primary general anesthetics—a substance used to put a person to sleep and block pain during medical operations.

The vapor pressure of diethyl ether (185 mm Hg) at 0°C is much greater than that of ethanol (12 mm Hg) or water (4.6 mm Hg) (see Figure 13.5). Ether's high vapor pressure is attributed to its weak intermolecular forces; thus, the surface ether molecules require less energy to break free and enter the gas phase. In a similar manner, ethanol's intermolecular forces are not as strong as those of water, and thus the vapor pressure of ethanol is greater than that of water at all temperatures.

Boiling Point

When a liquid is heated, there is a temperature at which the vapor pressure of the liquid equals the pressure of the gases above the liquid. This temperature is known as a liquid's **boiling point.** At the boiling point, bubbles of vapor form throughout the liquid and break free into the gas phase.

Most frequently, chemists are interested in a liquid's **normal boiling point,** i.e., the temperature at which a liquid's vapor pressure equals exactly 1 atm. In Figure 13.5, we see that each of the three vapor pressure curves cross the line corresponding to 1 atm at a different temperature. Diethyl ether's curve crosses at 35°C—ether's normal boiling point; the vapor pressure curves for ethanol and water cross at 78°C and 100°C, respectively.

If the pressure of gases above a liquid changes, there is a corresponding change in boiling point. When the pressure is increased, the liquid must be heated to a higher temperature before the vapor pressure equals the increased pressure. Likewise, if the pressure above the liquid is decreased, the boiling point is lowered.

The boiling point of water at various pressures is shown in Table 13.1. Notice that if the pressure is lowered to 24 mm Hg, water boils at 25°C, room temperature. Here is boiling water that you would not mind thrusting your hand into! On the other end of the scale, at 2026 mm Hg water boils at 130°C, or 30°C above its normal boiling point.

Water boiling at greatly reduced pressures is not hot enough to burn skin.

TABLE 13.1   BOILING POINT OF WATER AT
SELECTED PRESSURES

Pressure, mm Hg	Boiling point, °C
5	1
24	25
93	50
600	94
760	100
2026	130

Surface Tension

Surface tension supports small insects that walk on water!

Surface tension is the property of the surface of a liquid to act as if there is a membrane stretched across it. To understand surface tension, consider the unique properties of the surface molecules. All molecules below the surface of a liquid are surrounded in all directions by other liquid molecules. Thus, the forces exerted on subsurface molecules are balanced in all directions. Surface molecules are surrounded by other liquid molecules on all sides but one. This results in an unbalanced force pulling the surface molecules inward. Figure 13.6 shows the forces on surface and subsurface molecules.

As a result of the net inward attractive forces, surface molecules tend to minimize the amount of surface area exposed. A drop of liquid, outside of a gravitational field, takes the shape of a sphere, a geometric figure with minimum exposed surface area per unit volume.

To demonstrate the membranelike property that surface tension imparts, carefully place a needle (a metal with a density greater than water) on the surface of water. Unless the needle is pushed through the surface, it will remain suspended by the water's surface tension.

While surface tension is mainly thought of in terms of the interface between a liquid and a gas, similar generalizations can be made when a liquid is in contact with a solid. If a small quantity of liquid spreads out uniformly and

FIGURE 13.6
Subsurface molecules of a liquid have intermolecular forces exerted on them from all directions. Hence, the forces on subsurface molecules are balanced. Surface molecules have intermolecular forces in all directions except for directly above them. Hence, the forces on surface molecules are unbalanced.

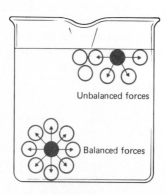
Unbalanced forces

Balanced forces

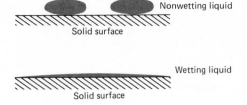

FIGURE 13.7
A liquid that wets solid surfaces spreads out and covers the surface. A nonwetting liquid, one that does not wet solid surfaces, forms beads of liquid on a surface.

Substances called wetting agents decrease the cohesive forces in liquids, reducing surface tension and allowing water to flow more readily over solid surfaces. Detergents are examples of wetting agents.

covers the surface of a solid, we say the liquid **wets** the surface (see Figure 13.7). Some liquids, like mercury, when placed on certain solid surfaces, bead up and do not spread out; a liquid of this nature does not wet the surface.

Within liquids that wet solid surfaces, the forces of attraction between the liquid and solid (adhesive forces) are larger than the internal intermolecular forces (cohesive forces). In contrast, the cohesive forces of a liquid that does not wet a surface are greater than the adhesive forces; therefore, nonwetting liquids attempt to expose a minimum surface to the solid.

Viscosity

Viscosity is the resistance of a liquid to flow, the opposite of fluidity. Liquids that flow readily are said to be fluid and have low viscosities. "Thick" liquids, those that flow slowly, have high viscosities.

A liquid's viscosity is dependent on the shape and size of molecules as well as on their intermolecular forces. Generally, high-molecular-mass substances with strong intermolecular forces have high viscosities. Examples of such liquids are molasses, lubricating oils, and asphalts.

REVIEW PROBLEMS

13.1 How does the kinetic molecular theory explain the following liquid properties: (a) higher density than gases, (b) incompressibility, (c) fluid properties?

13.2 Explain what is happening to the molecules within a liquid when it is evaporating.

13.3 What is a dynamic equilibrium and how is it different from a static equilibrium? Give an example of a dynamic equilibrium.

13.4 What variable factors influence a liquid's vapor pressure?

13.5 State specifically what is meant by the normal boiling point of a liquid. Give an example.

13.6 What is the surface tension of a liquid?

13.2 INTERMOLECULAR FORCES IN LIQUIDS

Before we complete our discussion of liquids, let us look at the intermolecular forces (cohesive forces) that hold molecules together in the liquid state. There are three primary intermolecular forces:

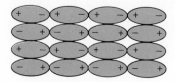

1. Dipolar (dipole-dipole) interactions
2. Hydrogen bonding
3. Dispersion forces

Each class of intermolecular force results from the attraction of unlike charged particles, called electrostatic attractions.

Dipolar Interactions

A bond is polar covalent when the two atoms bonded together have different electronegativities.

Dipolar, or dipole-dipole, interactions are the forces of attraction among molecules that naturally exist as dipoles. A dipole exists when there is a charge separation within a molecule (polar covalent molecules).

Figure 13.8 shows how dipolar interactions bind liquid molecules together. Positive ends of molecules attract the negative ends of others. Dipolar interactions are short-range forces, which means that the molecules have to be close together to produce a significant force of attraction. Only about 1 kcal/mol is needed to break dipolar interactions; this is low compared with the approximately 100 kcal/mol needed to cleave average covalent bonds.

Examples of liquids that contain molecules held together by dipolar interactions include $PCl_3(l)$, $HI(l)$, $H_2S(l)$, and $CH_2Cl_2(l)$. As a result of the relatively weak nature of dipolar interactions, most of these are volatile liquids and must be cooled to low temperatures to remain in the liquid state.

Hydrogen Bonding

Hydrogen bonds are also found in solids. They are especially important in biological compounds, e.g., proteins and nucleic acids.

Hydrogen bonding is a special case of dipolar interactions, so special, in fact, that it is classified separately. **Hydrogen bonding** results when the molecules that make up the liquid have a H atom covalently bonded to F, O, or N (elements in the second period with the highest electronegativities). As with any dipolar interactions, there is a charge separation within molecules that exhibit hydrogen bonding because of the attraction of the more electronegative atom for hydrogen's electron. A hydrogen atom consists only of a proton and an electron; when its electron is somewhat withdrawn, the compact positively charged proton is all that remains. Consequently, the force of attraction among such molecules is greater in magnitude than that among other combinations of atoms. Experimentally, hydrogen bonds are the strongest of the intermolecular forces, requiring about 5 to 6 kcal/mol to cleave.

Water is a hydrogen-bonded liquid. Water molecules have two hydrogen atoms bonded to the more electronegative oxygen atom. In liquid water, a water molecule can be hydrogen-bonded to two other water molecules, as pictured in Figure 13.9.

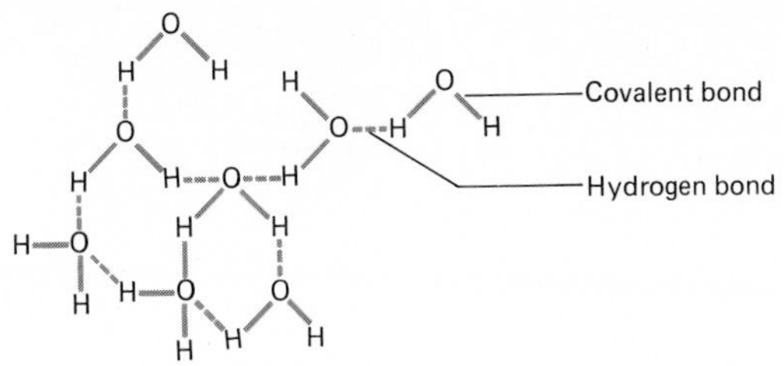

FIGURE 13.9
Hydrogen bonding in water results when the partially negative oxygen atom from one water molecule attracts the partially positive hydrogen atoms from one or two adjacent water molecules. The structure of liquid water is a network of associated water molecules. (Hydrogen bonds are indicated by dashed lines, and covalent bonds are indicated by solid lines.)

As a result of the large degree of association of water molecules produced by the hydrogen bonding, water has many unusual properties. For example, it has a high boiling point. Most low-molecular-mass molecules without hydrogen bonds are usually gases at room conditions, but water has a high boiling point because of its hydrogen bonds and is therefore a liquid at room temperature. Table 13.2 lists the boiling points of other chalcogen hydrides.

Dispersion Forces

Of the three classes of intermolecular forces, dispersion forces are the weakest. Unlike the other two, dispersion forces exist in all molecules, but in small polar molecules they are so weak they are totally overshadowed by dipolar and hydrogen bonding forces.

Dispersion forces are the weakest intermolecular forces because no permanent charge separation is normally found within the molecules.

Dispersion forces are the primary forces that attract nonpolar molecules to one another. Nonpolar molecules are those without a charge separation within the molecule. Also, dispersion forces exist in liquids of monatomic atoms, i.e., in the liquefied noble gases.

How then can molecules stick together, if there is no charge separation? Figure 13.10 shows two nonpolar molecules that are brought together. In (*a*) there is no interaction, since they are too far away. As the molecules approach each other, in (*b*), the electron clouds interact, producing a dipole for the brief instant they are very close to each other. By the time they have passed, in (*c*), they are no longer interacting. A dispersion force exists only in (*b*) when the two molecules are near each other.

Dispersion forces are also known as London forces and as van der Waals forces. Fritz London and Johannes van der Waals were two scientists who investigated and helped elucidate the nature of nonpolar interactions.

When one atom or molecule produces a momentary or instantaneous dipole in another atom, this is referred to as an *instantaneous induced dipole*. You can think of the dispersion force as a weak force of attraction of the nucleus of one atom (or molecule) for the electrons of another as the two atoms pass close

TABLE 13.2 **BOILING POINTS OF CHALCOGEN HYDRIDES**

Compound	Boiling point, °C
H_2O	100
H_2S	−61
H_2Se	−42
H_2Te	−2

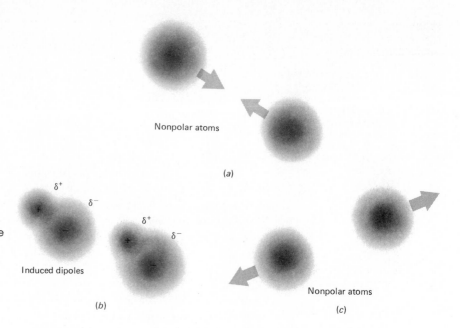

FIGURE 13.10
(a) When two nonpolar atoms are widely separated, they do not interact and dispersion forces do not exist. (b) When the nonpolar atoms are very close together, the electrons from one atom repel the electrons in the other atom, producing a momentary dipole and force of attraction. This attractive force is a dispersion force. (c) As the two atoms separate, they no longer interact. Dispersion forces are very short-range interactions.

Dispersion forces are strong enough to hold molecules in the solid state if the molecules are sufficiently large.

together. The strength of these forces depends mainly on how many electrons are distributed around the atom, and how tightly they are held. Stronger dispersion forces are found in atoms with a large number of loosely held electrons. Table 13.3 illustrates the increasing boiling points of the noble gases, indicating stronger dispersion forces.

REVIEW PROBLEMS

13.7 Write the three primary intermolecular forces found in liquids in order of increasing strength.

13.8 How is an intermolecular force different from a true chemical bond?

13.9 What accounts for the strength of hydrogen bonds relative to dipolar interactions?

13.10 Consider the nonpolar halogens, Cl_2, Br_2, and I_2. At room temperature, 25°C, Cl_2 is a gas, Br_2 is a liquid, and I_2 is a solid. In terms of their intermolecular forces, explain the trend in their physical states. Predict the physical states of F_2 and At_2.

TABLE 13.3 **BOILING POINTS OF NOBLE GASES**

Noble gas	Number of electrons	Boiling point, °C
He	2	−269
Ne	10	−246
Ar	18	−186
Kr	36	−153
Xe	54	−108
Rn	86	−61

13.3 SOLIDS

Recall from Section 4.2 that solids have a constant volume and shape, and have immense viscosities; thus, they exhibit no fluid properties. Solids have the highest average densities, melting points, and boiling points of the three physical states. Particles within solids are bound by strong intermolecular forces that inhibit the particles from moving from place to place. Solid particles are arranged in orderly geometric patterns.

Kinetic Molecular Theory

Molecular motion of molecules in solids is restricted to vibrations about a fixed position in space.

Application of the kinetic molecular theory is quite different for solids because of their strong intermolecular forces. While there is virtually no movement from place to place, the molecules and atoms in solids are in constant motion in a fixed position. Their motions are mainly vibrational in nature. Solid particles move back and forth rapidly, oscillating about a fixed position in space.

A solid's fixed shape results from the strong cohesive forces between particles; each particle is in a set position and cannot move to another position without breaking its cohesive forces. High average density is explained in terms of the closeness of the particles to each other. Once again, they are closely packed as a result of the strong intermolecular forces. Since applied pressure cannot push the particles any closer than they already are, the volume of a solid is constant.

Classes of Solids

Glass is actually a supercooled liquid. When silica, sodium carbonate, and calcium carbonate (components of glass) are heated and then cooled, instead of reforming with a highly organized regular crystalline pattern of particles, the structure resembles the short-range order of liquids.

Solids are classified as being either *crystalline* or *amorphous*. Crystalline solids are the true solids; the particles are in a regular, recurring three-dimensional pattern called a *crystal lattice*. Amorphous solids lack the regular microscopic structure of crystalline solids. Actually, their structure more closely resembles that of liquids than solids (many are actually liquids with high viscosities). Examples of amorphous solids include glass, tars, and high-molecular-mass polymers, such as Plexiglas.

Crystalline solids are further classified according to the type of forces that hold the particles in the rigid crystal lattice. Most frequently, these solids are grouped into four different classes: (1) ionic, (2) covalent, (3) molecular, and (4) metallic.

Ionic Solids

Structural analysis of crystalline solids is accomplished by placing crystals in the path of an x-ray beam. The beam is bent, depending upon the spacing and arrangement of particles within the crystal. This method of analysis is called x-ray crystallography.

You may recall the discussion of ionic solids from Chapter 8. Within **ionic solids,** oppositely charged ions are alternately arranged in the crystal's structure. Since the ions are bonded by strong ionic bonds, ionic solids have high melting and boiling points and are fairly hard.

Because the electrons in ionic solids are bound tightly to the ions, they cannot flow when a voltage is applied; hence, ionic solids are nonconductors (insulators) of electricity. However, when ionic solids are melted, most of the ionic bonds are broken; the ions then act as charge carriers and do conduct an electric current.

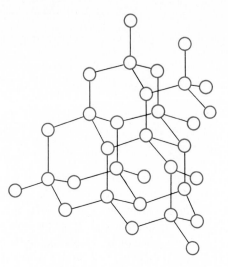

FIGURE 13.11
Diamond is composed of a network of carbon atoms each bonded to four other carbon atoms. Quartz, SiO_2, has the same structure as diamond, but there is a network of silicon and oxygen atoms.

Examples of ionic solids include sodium chloride, NaCl; magnesium oxide, MgO; potassium nitrate, KNO_3; and ammonium bromide, NH_4Br.

Covalent Solids

Covalent solids are sometimes called macromolecular (literally, "large molecule") solids, because the entire crystalline solid is one gigantic molecule held together by covalent bonds.

Atoms are the units that are bonded together in covalent solids. For example, diamond (pure carbon) is a covalent solid in which one carbon atom is bonded to four other carbon atoms, producing a huge array of carbon atoms (see Figure 13.11). We shall discuss the properties of diamond later in this chapter.

Covalent bonds, on an average, are the strongest bonds; thus, covalent solids are held together tightly. Melting points of covalent solids are among the highest known, many are in excess of 1000°C. Because the electrons are held tightly in the covalent bonds, they do not move when a voltage is applied; thus covalent solids are nonconductors. Also, it is not surprising that covalent solids are the hardest substances known.

Another common example of a covalent solid is quartz, SiO_2. In a quartz crystal, Si atoms are each bonded to four O atoms in a network of alternating Si and O atoms (see Figure 13.11). Quartz has a melting point of around 1700°C.

Molecular Solids

Molecular solids have molecules as the particles in the crystal lattice positions. These molecules are bonded by the same intermolecular forces that hold liquids together (Section 13.2): (1) dipolar interactions, (2) hydrogen bonding, and (3) dispersion forces.

Solids, under the proper conditions, can directly enter the gas phase without becoming liquid; this process is called sublimation.

Since the intermolecular forces are weak relative to covalent and ionic bonds, molecular solids tend to be relatively soft, volatile, and low-melting. There are no loosely held electrons in a molecular solid; accordingly, molecular solids are nonconductors of electricity.

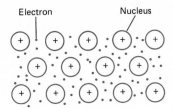

FIGURE 13.12

In metallic solids, positive metal nuclei are surrounded by loosely held delocalized electrons. The force of attraction between the metal nuclei and electrons is the metallic bond.

Three examples of molecular solids are ice, $H_2O(s)$; dry ice, $CO_2(s)$; and iodine, $I_2(s)$.

Metallic Solids

Metallic solids (metals) range from soft to hard, have a metallic luster, are malleable, and are good conductors of heat and electricity. These properties result from the fact that metallic solids possess **delocalized electrons,** i.e., electrons under the influence of more than one nucleus.

In the other solids we saw ions, atoms, and molecules as the basic units of structure. In metallic solids, positively charged metal nuclei exist in the crystal lattice positions. The structure of a metal is often described as "metal nuclei in a sea of electrons." A regular array of metal nuclei are surrounded by electrons that are attracted, not by one, but by many nuclei. This somewhat unique manner in which metal atoms are bound is called a *metallic bond.* Figure 13.12 illustrates the structure of a metal and metallic bonding.

The hardness and malleability of metals are explained by the strong electrostatic attractions between positive nuclei and negative electrons. Hammering and bending metals displaces atoms but does not break the attractive forces in metals. Electric conductivity results from the delocalization of outer electrons; the outer electrons move from one atom to another, if pushed by a voltage.

Common examples of metals are copper, Cu; silver, Ag; iron, Fe; and tungsten, W. Table 13.4 summarizes the four classes of solids.

TABLE 13.4   SUMMARY TABLE OF CLASSES OF SOLIDS

	Ionic	Covalent	Molecular	Metallic
Particles	Cations and anions	Atoms	Molecules	Metal nuclei and electrons
Strongest forces	Ionic bonds	Covalent	H bonds, dipolar, dispersion	Metallic bonds
Properties	Hard, insulators, high mp and bp	Hard, insulators, highest mp and bp	Soft, insulators, low mp and bp	Soft to hard, conductors, full range of mp and bp
Examples	NaCl, CaF$_2$, CaO	Diamond, quartz, carborundum	Ice, dry ice, SO$_2(s)$	Iron, copper, chromium, gold

REVIEW PROBLEMS

13.11 How does the structure of solids differ from that of liquids?

13.12 Describe the restricted movement of particles in the solid state.

13.13 Use the kinetic molecular theory to explain the following properties of solids: (a) high density, (b) constant volume, and (c) constant shape.

13.14 List the four general classes of solids. How do they differ?

13.15 Which class of solids has the following properties: (a) highest average melting point, (b) lowest average melting point, and (c) best conductors of electricity?

13.4 STATE CHANGES

If a solid is heated, it becomes warmer. As it heats, the solid generally undergoes two changes of state—solid to liquid, and liquid to gas. The solid-to-liquid transition is called *melting*, while the liquid-to-gas change is called *boiling*.

Heating Curve

A heating curve is a plot of temperature versus uniform addition of heat energy. Figure 13.13 is a heating curve for a hypothetical substance. Temperature of the substance is plotted on the vertical axis, and the passage of time during which heat is added to the substance is plotted on the horizontal axis.

Initially, the substance exists in the solid state (point *I*), and the addition of heat increases the temperature (segment *IJ*). When a substance's temperature increases, the particles within move faster; in other words, the average kinetic energy of the particles increases. The temperature increase of the solid depends on the solid's heat capacity, i.e., the amount of heat required to raise the temperature of a fixed amount of the solid by one degree.

Solid-to-Liquid Transition

Point *J* on the heating curve represents the time at which the first drop of liquid appears. At this time the substance is beginning to melt. From *J* to *K*, heat is

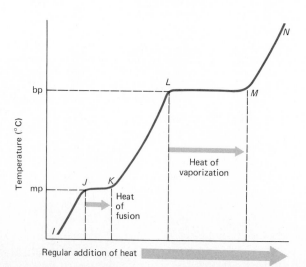

FIGURE 13.13
A heating curve for a substance is a plot of the temperature of the substance as heat is added uniformly to it.

added but there is no increase in temperature. Added heat no longer increases the kinetic energy of the molecules; instead it increases their potential energy. During the melting process, the heat energy breaks the forces that hold the solid's crystal lattice together. As long as bonds remain to be broken within the solid, the temperature remains constant at the melting point.

How much heat is needed to break the bonds in a solid to convert it to a liquid? There are two principal factors to consider: (1) the amount of substance, and (2) the strength of the bonds. A greater quantity of substance requires more heat than a lesser amount; a substance with stronger bonds also requires more heat than one with weaker bonds. Chemists measure the heat of fusion for melting solids. **Heat of fusion** is defined as the amount of heat required to change a specified amount of solid to a liquid at its melting point. Table 13.5 lists the molar heats of fusion (kJ/mol), the amount of heat energy required to change one mole of a solid to one mole of liquid at their melting points.

After all of the solid has melted (point K), added heat increases the kinetic energy of the liquid molecules, causing the temperature to rise—segment KL in Figure 13.13. The temperature increase is determined by the **heat capacity of the liquid,** i.e., the amount of heat required to raise the temperature of a fixed quantity of the liquid by one degree.

Certain substances have intermediate "states" that are not exactly solid or liquid. When melting, the solid changes first to the intermediate state and then to the liquid state. This intermediate state is less ordered than a solid but much more ordered than a liquid and is called a liquid crystal state. Liquid crystals are used in numeric displays (LCD) in watches and calculators.

Liquid-to-Gas Transition

At point L, another change of state occurs, called boiling. As previously discussed, the boiling point of a liquid is the temperature at which the vapor pressure of the liquid equals the pressure of the gases above the liquid. At the boiling point the temperature remains constant; the heat is increasing the potential energy, breaking the intermolecular forces binding the liquid together.

Notice that the segment LM on the graph is longer than segment JK, where melting occurs. Why is this? When the liquid is changed to a gas, all of the intermolecular forces must be broken for the liquid to enter the gas phase. When the solid is melting, fewer intermolecular forces are broken to enter the liquid phase; thus, less heat energy is normally required to melt a solid than is needed to boil the same substance.

In Table 13.5 the molar heat of vaporization is listed in addition to the molar heat of fusion. A liquid's molar **heat of vaporization** is the amount of heat

TABLE 13.5   MOLAR HEATS OF FUSION AND VAPORIZATION OF SELECTED SUBSTANCES

Substance	Melting point, °C	Molar heat of fusion, kJ/mol	Boiling point, °C	Molar heat of vaporization, kJ/mol
H_2O	0.0	6.01	100	40.7
Cl_2	−101	6.40	−34	20.4
NaCl	801	30	1413	—
Al	660	10.9	2467	284
CH_4	−182	0.94	−161	9.201
O_2	−219	0.44	−183	6.82
NH_3	−78	5.65	−33	23.4

required to change one mole of liquid to vapor at a specified temperature (generally, the boiling point, as in this case). Normally, the molar heat of vaporization is many times larger than that of the molar heat of fusion.

Magnitudes of the molar heats of fusion and vaporization give an indication of the strength of forces holding the molecules together in the solid and liquid states. For instance, the low heats of fusion and vaporization of oxygen indicate weak intermolecular forces. Solid oxygen is held together by weak dispersion forces (a molecular solid). On the other end of the scale, sodium chloride's rather strong ionic bonds are reflected in its high heat of fusion, 30 kJ/mol.

When point M is reached on the heating curve, all of the liquid has been converted to vapor. Additional heat increases the kinetic energy of the gas particles, raising the vapor's temperature. Hence, the temperature increase depends on the vapor's heat capacity.

Water's specific heats of fusion and vaporization in calories are 80 cal/g and 540 cal/g, respectively.

REVIEW PROBLEMS

13.16 Consider the heating curve in Figure 13.13 to answer the following. (a) In what physical state is the substance from point K to point L on the graph? (b) What phase change occurs at L? (c) What quantity of heat is required to move from point J to point K? (d) What determines the rate of increase in temperature from point M to point N?

13.17 Why is the magnitude of the heat of vaporization of a substance greater than its heat of fusion?

13.18 Of the substances listed in Table 13.5, which has the strongest intermolecular forces in the: (a) liquid state and (b) solid state?

13.5 WATER

Water is by far the most important liquid. Water covers the greatest percentage of the earth's surface, about 75%, and composes approximately 65% of the human body. All living things require a constant supply of relatively pure water.

Structure of Water

In 1781, Cavendish became the first person to show that water is produced from H_2 and O_2. A few years later Lavoisier experimentally determined the percent composition of H_2O. More recently, water was found to consist of H_2O molecules in which an O atom is covalently bonded to two H atoms that are separated by an angle of approximately 104 degrees:

$$H \overset{O}{\underset{\sim 104°}{\diagup \diagdown}} H$$

The higher electronegativity of an oxygen atom compared with that of a hydrogen atom makes water a polar covalent molecule with the "O side" more negative than the "H side."

In the liquid state, water molecules are hydrogen-bonded to each other. A certain percent of the O atoms in liquid water molecules have four H atoms

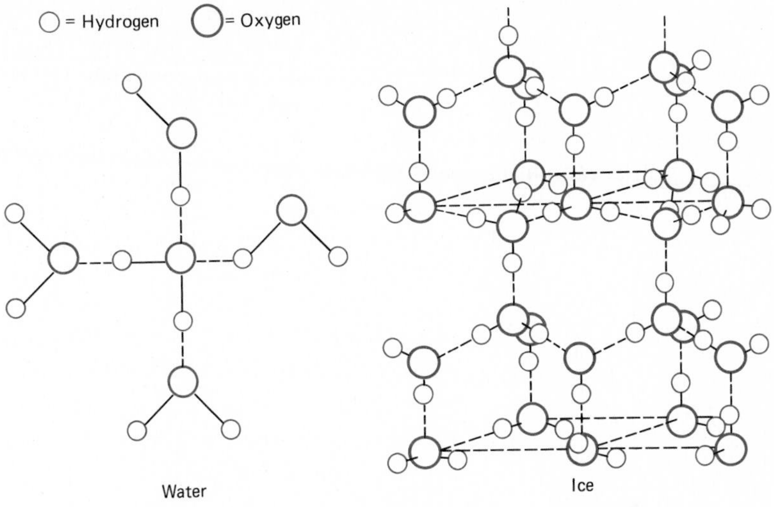

○ = Hydrogen ◯ = Oxygen

Water

Ice

FIGURE 13.14
As a result of hydrogen bonding, there are some organized regions of water molecules in liquid water. In this figure, one water molecule is hydrogen-bonded to four water molecules. The structure of ice is more organized and is like honeycomb.

associated with them. Two of the H atoms are covalently bonded, and two other H atoms are hydrogen-bonded to the O from other water molecules (see Figure 13.14). Water's structure is believed to contain some highly organized regions (similar to the regular structure in ice) and some more disordered regions of individual water molecules.

Depending on pressure, six different ice structures are known. You are familiar with ice I; there are five other forms (II–V and VII). Ice VII exists at pressures in excess of 24,000 atm, has a density of 1.7 g/mL, and is stable above the normal boiling point of water!

The crystal structure of ice is more regular than the structure of liquid water. As shown in Figure 13.14, all the O atoms in ice are surrounded by four H atoms (two covalently bonded and two hydrogen-bonded). Ice is often described as having a "honeycomb" structure. Groups of water molecules are attached in a ring structure with a large number of open spaces.

When liquid water freezes, and forms the honeycomb structure of ice, the empty spaces in the honeycomb cause ice to have a larger volume than liquid water. The increase in volume when liquid water freezes is highly deviant from the behavior of just about all other substances. Most substances occupy less volume (are more dense) in the solid state.

Liquid water's maximum density is at 4°C.

Because ice has a lower density than liquid water, ice floats on the water's surface. Antifreeze is placed into an auto's radiator to prevent the formation of ice, which can crack an engine block.

Physical Properties of Water

Hydrogen bonding gives water a set of unique properties. Water has one of the highest heat capacities of all liquids. Consequently, water is used as a coolant. Its high heat capacity means that a large amount of energy is needed to raise the temperature of water relative to other liquids. Additionally, water has both a large heat of vaporization and a high degree of heat conductivity, good properties for a coolant.

In the biological world it is fortunate that water's surface tension is large. The surface tension of water allows it to rise in small tubes, as it does in plants. Water rising in narrow tubes through surface tension is called **capillary action**.

Water, as we shall see in Chapter 15, is an excellent solvent (dissolving medium) for polar and ionic substances. Blood and the fluids in living systems are aqueous solutions, containing many dissolved substances.

Chemical Properties of Water

Water is a thermally stable molecule; it does not easily decompose when heat is applied. However, water is quite reactive and combines with many substances.

Under the proper conditions, water molecules are split by many substances. These reactions are called **hydrolysis** reactions. Consider the following hydrolysis reactions:

$$SiCl_4(l) + 4H_2O \longrightarrow Si(OH)_4(s) + 4HCl(aq)$$
$$Cl_2(l) + 2H_2O(l) \longrightarrow H_3O^+(aq) + Cl^-(aq) + HOCl(aq)$$

Water can combine with substances without being split apart. A reaction of this type is called a **hydration** reaction. Various salts combine with water to form hydrated salts:

$$Salt + xH_2O \longrightarrow Salt \cdot xH_2O$$

A salt without the attached water is called an **anhydrous** salt. After combining with water, it is referred to as a **hydrated salt** or just a **hydrate**. Two examples of the formation of a hydrated salt are:

$$Na_2B_4O_7 + 10H_2O \longrightarrow Na_2B_4O_7 \cdot 10H_2O \qquad \text{(borax)}$$
$$KAl(SO_4)_2 + 12H_2O \longrightarrow KAl(SO_4)_2 \cdot 12H_2O \qquad \text{(alum)}$$

A large quantity of borax is obtained commercially from dry lakes in California. Borax is used as a water softener and a flux in solder to dissolve oxide coatings on metals.

Anhydrous salts and hydrates exhibit three interesting properties. Various anhydrous salts absorb water directly from the air and become hydrated. Such salts are called **hygroscopic**. Some salts take water from the air so readily that they absorb enough water to dissolve. Salts that exhibit such behavior are called **deliquescent**. Certain hydrates have a higher water vapor pressure than the atmosphere, and therefore release water to the air. When this happens, a salt is said to undergo **efflorescence**.

Water Cycle

Seawater's mineral content is 55% Cl^-, 30% Na^+, 8% SO_4^{2-}, 4% Mg^{2+}, 1% Ca^{2+}, 1% K^+, 0.4% HCO_3^-, and many others.

Good estimates indicate that there is approximately 10^9 km^3 (1 km^3 = 0.24 mi^3) of water on earth. About 97% of the water on earth is salt water; only 3% is fresh water. Of the fresh water on earth, most (75%) is in the form of ice in the polar regions and glaciers. Groundwater, found below land surfaces, represents 20% of the fresh water, and the remaining 5% is in lakes, in rivers, in the soil as moisture, and in the air.

Water, like many substances on earth, is constantly on the move (Figure 13.15). Energy to fuel the water cycle comes in the form of radiant energy from the sun, which evaporates large quantities of water from the seas and oceans. After the atmosphere becomes saturated with water vapor, the water falls to earth as some form of precipitation: rain, snow, sleet, or hail. Water that falls on land

FIGURE 13.15
Water is constantly on the move. Water, rain or snow, that falls to earth travels to the oceans where it evaporates and begins the cycle again. (*Peter Miller/Photo Researchers*)

Relative humidity is a measure of the amount of water vapor in the air. It is calculated by taking the actual vapor pressure of water in the air and dividing it by the equilibrium vapor pressure, the maximum amount of water that could be in the air.

becomes a part of rivers and lakes or filters down through various ground layers, becoming groundwater. Ultimately, all of this water ends up in the oceans, evaporates back into the atmosphere, or freezes in the polar regions.

As part of the water cycle, a small percentage enters living systems. Green plants take carbon dioxide, CO_2, and combine it with water in the presence of sunlight to produce carbohydrates (sugars and starches) and oxygen, O_2.

$$n CO_2 + n H_2O \longrightarrow (CH_2O)_n + n O_2$$
Carbohydrates

Animals produce water as an end product of cellular respiration, their mechanism for producing energy.

$$2H^+ + \tfrac{1}{2}O_2 \longrightarrow H_2O$$
Final reaction in cellular respiration

REVIEW PROBLEMS

13.19 Describe the structure of: (a) an individual water molecule, (b) liquid water molecules, and (c) water molecules in ice.

13.20 What accounts for the fact that ice is less dense than liquid water?

13.21 List two ways in which water is essential to living things.

13.22 Briefly describe the movement of water through the water cycle.

13.6 CARBON

As an example of a solid, we shall consider the element carbon. Carbon exists in two different crystalline forms, graphite and diamond. In addition, there are various semiamorphous forms of carbon—charcoal, carbon black, and coke. Different forms of an element in the same physical state are called **allotropes**; thus, graphite and diamond are allotropes of carbon.

Graphite

Graphite, also known as "black lead," is a soft, black solid. Its density is 2.3 g/mL. It is slippery, melts at 3527°C, and is an electric conductor. Graphite's somewhat special properties are related to its structure, as properties always are. Graphite is composed of layers of carbon atoms arranged in six-membered rings that are fused together. Figure 13.16 shows a segment of graphite's structure.

Each carbon atom is bonded to three other carbon atoms in such a manner that one of carbon's electrons is delocalized and not tightly held by the carbon, making graphite an electric conductor. Graphite's layers are not held very tightly together, so they are free to slide by each other, which accounts for the slippery nature of graphite. As a result, graphite is used as a lubricant.

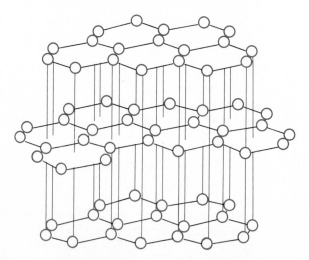

FIGURE 13.16
Graphite is composed of layers of bonded carbon atoms. Each carbon atom is covalently bonded to three other carbon atoms. The layers are not strongly bonded together, so they can slide by each other.

FIGURE 13.17
When they are found in their natural state, diamonds appear dull, and many are covered with a grayish film. They are polished, cut, and shaped to produce gems. The large diamond pictured here is the famous Hope diamond. (*Smithsonian Institution*)

Diamond

Artificial diamonds were first produced in 1955. Temperatures around 2000°C and pressures about 70,000 atm, along with a catalyst, are required to produce artificial diamonds from graphite. Those that are produced are not gem quality, but are chemically the same as natural diamonds.

Diamonds (Figure 13.17) are produced when coal and other amorphous forms of carbon are subjected to high pressures, generally from volcanic activity. Most of the world's supply of diamonds come from South Africa, Brazil, Australia, or the United States.

Diamond's properties are quite different from those of graphite. Diamond's density is 3.5 g/mL, greater than that of graphite. Higher density indicates that the carbon atoms are packed together in a smaller volume. Diamond, a network solid, is composed of carbon atoms that are bonded symmetrically to four other carbon atoms (see Figure 13.18). Since each of carbon's four electrons is a part of the four covalent bonds, diamond's electrons are not mobile, and thus diamond is a nonconductor of electricity.

Strong covalent bonds are responsible for diamond's hardness. Diamond is the hardest naturally occurring substance; on a hardness scale of 1 to 10, diamond is a 10! Since diamonds are so hard, they are placed on the ends of drill bits to bore through rock strata in search of crude oil. Diamond's melting point is 3570°C, among the highest of all substances. Extremely high melting points are a characteristic property of network solids.

Diamond is unreactive, but it does change to carbon dioxide in pure oxygen above 1000°C.

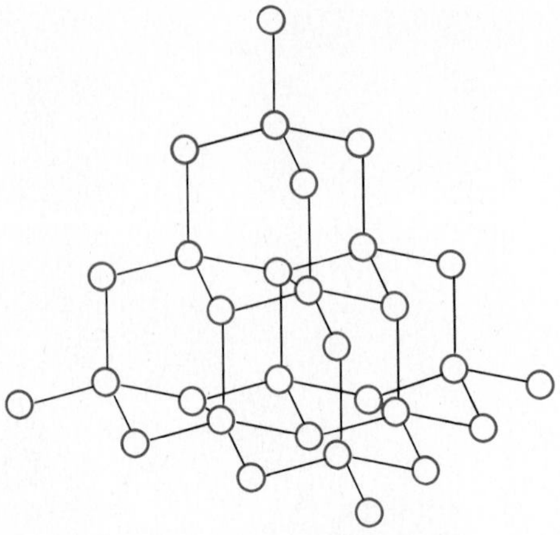

FIGURE 13.18
Diamond is a network solid of carbon atoms each bonded to four other carbon atoms. The geometric shape around each carbon atom is that of a tetrahedron. Diamond is the hardest naturally occurring substance.

$$C(\text{diamond}) + O_2(g) \xrightarrow{1000°C} CO_2(g)$$

Amorphous Forms of Carbon

Carbon black (sometimes called lampblack), a pure form of carbon, is one of the amorphous forms of carbon. Carbon black is normally prepared by heating carbon-hydrogen compounds, called hydrocarbons, in a flame. It is used primarily as a black pigment and to reinforce rubber; a large percentage of the mass of rubber tires is carbon black.

Coal is an impure form of carbon that results when dead animal and plant matter is compacted in the earth for long periods of time. Coal's structure somewhat resembles that of graphite, but is less regular. Also, many impurities exist in coal. One such impurity, sulfur, is of primary concern. When coal is combusted, sulfur is oxidized to unwanted gaseous sulfur oxides, which are major air pollutants. Coal mining, especially strip mining, also has an impact on the environment; see Figure 13.19.

Steel contains about 0.5% carbon. If the concentration of C is too large, the steel loses its strength and hardness.

When coal is heated in the absence of oxygen, volatile substances are driven off, leaving a substance called coke. Coke is extremely important in industry, since it is converted to graphite. Coke is an industrial fuel, and it is combined with metals to increase their strength.

If wood is heated in the absence of air, wood charcoal results. Most people know that charcoal is placed in grills to cook foods, but this is not the main use of charcoal. Finely powdered charcoal is commonly used to purify other substances. It adsorbs other substances onto its surface. **Adsorption** is the process in which an adsorbing substance attracts other substances to its surface. When water is filtered through charcoal, the charcoal removes many impurities that cause the water to smell bad or to have a bad taste.

Carbon Cycle

Even though carbon compounds are the central substances in living things, carbon and its compounds make up less than 0.1% of all substances on earth. Like other substances we have discussed (N_2, O_2, and H_2O), carbon and carbon compounds are linked in a complex cycle on earth.

FIGURE 13.19
Coal and other substances in the earth are obtained by strip-mining procedures. (*U.S. Department of Energy*)

Figure 13.20 is a diagram of the carbon cycle. During photosynthesis, plants and microorganisms convert atmospheric CO_2 to organic compounds. As previously described, photosynthesis is the process whereby CO_2, H_2O, and a green plant pigment called chlorophyll are combined in the presence of light, in a complex series of reactions, to produce carbohydrates and O_2.

Plants metabolize some of the sugars produced by photosynthesis for their

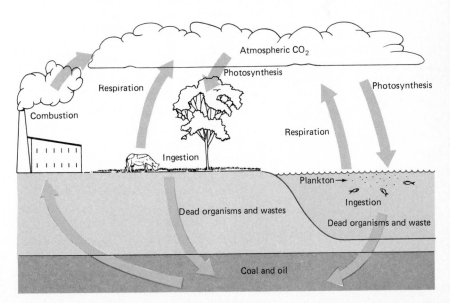

FIGURE 13.20
Carbon in many different forms travels through the atmosphere, lithosphere, and biosphere. This is known as the carbon cycle.

own energy needs; in doing so, the sugars are decomposed, releasing CO_2 to the atmosphere. Some of the sugars are stored in plants, and others are converted to other classes of biologically significant molecules. Animals eat plants, and thus carbon compounds are transferred to animals. Animals derive energy from the sugars and incorporate the carbon compounds into their bodies. After plants and animals die, their remains are decomposed by microorganisms, which transfers the carbon compounds to the soil. Ultimately, carbon compounds are oxidized in the soil to CO_2, starting the cycle again.

A large percentage of carbon compounds are found in oceans, lakes, and rivers, and as part of rocks. CO_2 from the atmosphere dissolves in the oceans, where it is converted to dissolved carbonates and hydrogen carbonates:

Limestone, $CaCO_3$, is the second most common rock in the earth's crust. Coral reefs are composed mainly of calcium carbonate from marine organisms. Pearls are layers of calcium carbonate that form inside the shell of an oyster.

$$CO_2(g) + H_2O(l) \longrightarrow [H_2CO_3] \longrightarrow H^+(aq) + HCO_3^-(aq)$$

Carbonates then precipitate out of solution in the form of carbonate-containing rocks such as limestone, calcite, chalk, and marble.

REVIEW PROBLEMS

13.23 What are allotropes? Give an example.

13.24 Describe the structure of: (a) graphite, (b) diamond, and (c) coal.

13.25 What accounts for the hardness and high melting point of diamond?

13.26 What is the significance of the photosynthetic process in the carbon cycle?

13.27 Describe the pathway of carbon from the atmosphere to a human being.

SUMMARY

Liquids are a more condensed state of matter than gases. Liquid particles are held together primarily by three intermolecular forces: (1) dipolar interactions, (2) hydrogen bonding, and (3) dispersion forces. As a result of the intermolecular forces, the particles are held closely together and their freedom of movement is restricted. Very few empty spaces exist within the structure of liquids. This renders liquids incompressible but allows space for molecules to slide by each other, which gives liquids their fluid properties.

Fast-moving liquid particles on the surface break free into the vapor state. This process is called evaporation. The opposite of evaporation is condensation, the process by which vapor molecules above a liquid cohere and fall back into the liquid state. In a closed system, at a constant temperature, an equilibrium between the rates of evaporation and condensation produces a fixed amount of vapor above the liquid. To measure the amount of vapor in equilibrium with a liquid, the vapor pressure—the amount of pressure exerted by the vapor over a liquid—is determined.

The temperature at which a liquid's vapor pressure equals the pressure of gases above the liquid is the liquid's boiling point. A liquid's boiling point is directly related to the gas pressure above the liquid; higher pressures produce higher boiling points, and lower pressures produce lower boiling points.

Other properties of liquids are surface tension, wetting, and viscosity. Surface tension is the property of the surface of a liquid to act as if it is covered by a membrane. Liquid surfaces expose the smallest amount of surface possible. Surface tension effects are observed whenever there is an interface between a liquid and another physical state, either a gas or solid. On solid surfaces, liquids either wet the surface by spreading out, or bead up and form droplets. Viscosity is a liquid's resistance to flow, the opposite of fluidity.

Two of the three intermolecular forces in liquids result from molecules that normally contain a charge separation. Dipolar interactions occur among molecules that exist as dipoles; the positive end of one molecule attracts the negative end of another, and vice versa. Hydrogen bonds (the strongest of the intermolecular forces) are a special type of dipolar interaction. All substances that exhibit hydrogen bonding have molecules with a hydrogen atom bonded to F, O, or N. All other forces between molecules are called dispersion forces or nonpolar interactions. Dispersion forces result when one nonpolar molecule induces a dipole in another nonpolar molecule; dispersion forces are generally the weakest forces in liquids.

Solids are the most condensed form of matter and possess the greatest forces between particles. Most solids have a regular geometric pattern of particles within their structure; they are called crystalline solids. A small group of solids lack the regular crystal structure and have a more random pattern; they are called amorphous solids.

Crystalline solids are normally grouped into four categories: (1) ionic, (2) covalent, (3) molecular, and (4) metallic solids. Ionic solids have ions in the crystal lattice positions and are held together by ionic bonds. Covalent solids contain atoms in the lattice positions, and are held together by covalent bonds. Molecular solids contain complete molecules at the lattice positions. Intermolecular forces that hold liquids together are the same ones that bind molecules in molecular crystals. Finally, in metallic solids, metal nuclei are the basic units. The metal nuclei are surrounded by a "sea of electrons," the electrons of the metal.

When heat is added to a solid, the temperature of the solid increases until it reaches its melting point, at which point the temperature remains constant until all of the solid has melted to a liquid. In a similar manner, addition of heat to the liquid raises the temperature of the liquid until the boiling point is reached. At the boiling point, the temperature remains constant until the liquid is totally changed to vapor. Further addition of heat raises the temperature of the vapor.

Water is a unique liquid with a set of special properties. Water has an unusually high boiling point, heat capacity, thermal conductivity, and surface tension for such a low-molecular-mass liquid. Most of the special properties are a result of water's strong hydrogen bonding. Another anomalous property of water is that liquid water has a higher density than ice.

Allotropes are different forms of an element in the same physical state. Carbon exists in several allotropic forms: diamond, graphite, and different amorphous forms. Graphite is a soft, slippery, black form of carbon, while diamond is a hard, high-melting solid.

QUESTIONS AND PROBLEMS

13.28 Define the following terms: intermolecular forces, evaporation, vaporization, condensation, dynamic equilibrium, vapor pressure, boiling point, normal boiling point, surface tension, wetting, viscosity, dipolar interactions, hydrogen bonding, dispersion forces, induced dipole, cohesive force, adhesive force, crystalline solid, amorphous solid, crystal system, delocalized electrons, heat capacity, heat of fusion, heat of vaporization, ionization of water, capillary action, hydrolysis, hydration, hygroscopic, deliquescent, efflorescence, photosynthesis, allotropes, and adsorption.

Liquids

13.29 Use the kinetic molecular theory to explain the following properties of liquids: (a) liquids flow, (b) they take the shape of the bottom of their container, (c) they have a fixed volume, (d) liquid molecules are not free to move from place to place, and (e) they have a greater density than gases.

13.30 Compare liquids with gases in terms of the following properties: (a) density, (b) fluidity, (c) intermolecular forces, (d) compressibility, and (e) shape.

13.31 Explain why only surface particles in liquids are involved in evaporation.

13.32 What is the principal difference between two liquids, at the same temperature, one of which quickly evaporates and the other does not noticeably evaporate?

13.33 What is the purpose of rubbing an alcohol on the skin of a person who has a fever?

13.34 Why is an equilibrium usually not established between evaporation and condensation in an open system?

13.35 At a fixed temperature, determine which of the described liquids would have a greater vapor pressure:
 (a) Liquid A with strong intermolecular forces or liquid B with weaker intermolecular forces
 (b) Liquid C with a higher boiling point than liquid D
 (c) Liquid E, a hydrogen-bonded liquid, or liquid F with dispersion forces
 (d) Liquid G, a thick viscous oil, or liquid H, a volatile liquid.

13.36 Consider the vapor pressure curves shown in Figure 13.21.
 (a) Which liquid has the greatest vapor pressure at 25°C?

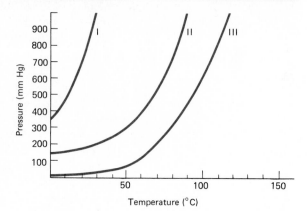

FIGURE 13.21

 (b) What is the normal boiling point of liquid I?
 (c) Rank the liquids in order of strength of intermolecular forces.
 (d) At what temperature does each liquid have a vapor pressure of 600 mm Hg?
 (e) At what temperature would each liquid boil, if the pressure above the liquids was reduced to 500 mm Hg?

13.37 What is contained in the bubbles that are observed in boiling liquids?

13.38 If the pressure above a container of water is lowered so that the water boils at room temperature, explain why this boiling water causes no burns.

13.39 (a) Would it take longer to hard-boil eggs in Denver, Colorado (the Mile High City, at an altitude of 5280 ft) or in Tampa, Florida (almost at sea level), assuming that the eggs are placed in the same size container with the same volume of boiling water? Write a complete explanation for your answer. (b) Where would it take longer to fry eggs, Denver or Tampa? Explain.

13.40 Explain how a pressure cooker boils foods faster than if the foods were cooked in an open pan of boiling water.

13.41 What is different about surface molecules of a liquid to result in the property called surface tension?

13.42 What could explain the fact that water droplets outside a gravitational field are spherical and not cubic?

13.43 Why does water spread out and cover many of the surfaces that it comes in contact with, but mercury forms small beads on similar surfaces?

13.44 (a) When water is placed into a glass tube, a concave meniscus (a liquid surface that is lower in the center than near the glass tube) is observed. Explain in terms of the cohesive forces in water, and adhesive forces between the water and glass. (b) Mercury forms a convex meniscus (rises in the center) in glass tubes. Explain in terms of cohesive and adhesive forces.

13.45 What effect should temperature have on the viscosity of liquids? Give an example.

13.46 Select from the following pairs the liquid that would be expected to have the greater viscosity:
 (a) High-molecular-mass liquid K or low-molecular-mass liquid L
 (b) Liquid M, a hydrogen-bonded liquid, or liquid N, a liquid that has dispersion forces.

Intermolecular Forces

13.47 Show that the three types of intermolecular forces in liquids result from electrostatic attractions.

13.48 What accounts for the fact that hydrogen bonds are stronger than dipolar interactions, even though both involve the attraction of a positive region of one molecule for the negative region of another, and vice versa?

13.49 Draw the dot formula for bromine monochloride, $BrCl(l)$, and show the dipolar attractions among a few BrCl molecules.

13.50 What possible atoms must H be bonded to in a molecule if the molecule exhibits hydrogen bonding?

13.51 The following three liquids, with similar molecular masses, are known to be held together by three different intermolecular forces. Given their boiling points, explain which intermolecular forces you would expect for each:

Liquid	Boiling point, °C
Q	115
R	−95
S	−2

13.52 (a) What is a dispersion force? (b) Why is it classified as an extremely short-range force?

13.53 Rank the following sets of nonpolar liquids in order of increasing boiling points:
 (a) $Ne(l)$, $Kr(l)$, and $Ar(l)$
 (b) $C_5H_{12}(l)$, $C_{10}H_{22}(l)$, and $C_8H_{18}(l)$.

13.54 Alcohols, which are carbon and hydrogen compounds containing a —OH group in the molecule, are hydrogen-bonded liquids. What could account for the increasing trend in their boiling points when a C atom and two H atoms are successively added to the previous alcohol?

Alcohol	Boiling point, °C
CH_3OH	64.7
C_2H_5OH	78.3
C_3H_7OH	97.2

Solids

13.55 How is the structure of an amorphous solid different from that of a crystalline solid?

13.56 Give two examples each of crystalline and amorphous solids.

13.57 What would happen to the structure of solids if they could be compressed?

13.58 Write two examples of each of the following classes of solids: (a) metallic, (b) ionic, (c) molecular, (d) covalent.

13.59 Ionic solids are nonconductors of electricity, but if they are melted they are capable of conducting an electric current. Write an explanation to account for this behavior of ionic substances.

13.60 Account for the extremely high melting points of covalent solids.

13.61 Dry ice is composed of carbon dioxide molecules with the structure

$$O=C=O$$

Two strong carbon-oxygen double bonds hold the molecule together; however, dry ice is a volatile solid and readily becomes a vapor, $CO_2(g)$. Write an explanation to account for this apparent inconsistency.

13.62 What accounts for the fact that metals are the only class of solids that are good conductors of electricity?

13.63 Metals exhibit the property of malleability, i.e., the ability to be hammered into different forms and shapes. Write an explanation, in terms of the structure of solids, of what happens to a metal when it is hammered into a thin foil.

State Changes

13.64 Consider the cooling curve of a hypothetical substance shown in Figure 13.22. A substance that is initially at 200°C and in the gas phase is cooled to −150°C.
 (a) What physical state does the substance exist in at each of the following temperatures: 65°C, 160°C, and −75°C?
 (b) What are the freezing and boiling point temperatures?

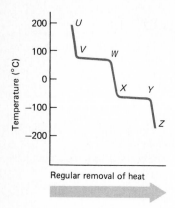

FIGURE 13.22

(c) Explain, in terms of the kinetic and potential energy of the molecules, what is happening to the substance along the following segments of the curve: UV, VW, WX, XY, and YZ.

13.65 Draw a graph of the heating curve of water, starting at −25°C and ending at 125°C. Label the axis and each segment of the curve, and indicate the heats of fusion and vaporization on the graph.

13.66 Why does the temperature remain constant when heat is continually added to a block of ice?

Use Table 13.5 to solve Problems 13.67 through 13.69.

13.67 Determine the amount of heat needed to change 0.53 mol $H_2O(l)$ at 100°C to steam at the same temperature.

13.68 How much heat must be removed from 17.3 g of water at 0.0°C to produce ice at the same temperature?

13.69 How much heat energy is required to change 0.50 mol of ice from −15°C to steam at 100°C?

Water

13.70 How would water's properties change if the water molecule was linear (H—O—H) instead of angular?

13.71 If water froze from bottom to top, what effect would that have on the animals that live in small ponds during the winter months?

13.72 What properties does water have that makes it good for: (a) a coolant in car engines, and (b) living systems?

13.73 Distinguish between efflorescence and deliquescence.

13.74 Write an equation showing: (a) the decomposition of a hydrate to an anhydrous salt and water, (b) the forma-

tion of another hydrate from an anhydrous salt and water, (c) the hydrolysis of $Cl_2(g)$, and (d) the hydrolysis of SiI_4.

13.75 Water vapor that enters the air remains in the atmosphere for a relatively short period of time, an average of 10 days. What factors tend to limit the amount of time water remains in the air?

13.76 List the three main sources of fresh water on earth.

13.77 With an ever-dwindling supply of potable fresh water, it has been proposed that icebergs, a source of a large quantity of fresh water, be towed to U.S. cities for use as drinking water. What problems might be encountered in utilizing the fresh water in icebergs?

Carbon

13.78 What properties of graphite make it a good substance to use in a pencil?

13.79 Contrast the physical properties of graphite and diamond.

13.80 What is carbon black, and how is it used commercially?

13.81 Why is graphite a conductor of electricity and diamond a very poor one?

13.82 Very little heat energy is needed to convert graphite to diamond. However, it is very expensive and difficult to change graphite to diamond. What could account for this expense and difficulty?

13.83 How are coal and coke produced naturally?

13.84 In their production, the colorless alcoholic beverages such as vodka are normally filtered through charcoal. What is the purpose of passing vodka through purified charcoal?

13.85 Briefly describe the role of carbon in living systems, starting with photosynthesis and ending with carbon compounds in animals.

13.86 Describe what happens to atmospheric CO_2 that dissolves in the ocean.

13.87 Since the industrial revolution, each year more and more CO_2 is given off to the atmosphere through the burning of fossil fuels (oil, natural gas, and coal). There has been only a small increase in the amount of atmospheric CO_2. Considering the carbon cycle, propose reasons why the CO_2 level has not increased significantly.

General Questions

13.88 If equal masses of steam at 100°C and boiling water at

100°C contact your skin, the burn is more severe from the steam. Write an explanation to account for this fact.

13.89 Copper's heat capacity is 25 J/(mol · °C). What mass of steam at 100°C must be converted to water at 100°C

to raise the temperature of 2.1 kg of copper from 5.6°C to 19.1°C?

13.90 Explain why hydrogen bonds in HF(l) are stronger than those in water.

· C H A P T E R ·

14

Descriptive Inorganic Chemistry

Chapter Guidelines

After completing Chapter 14, you should be able to:

1. Distinguish between representative metals and transition metals
2. Apply your knowledge of elements, compounds, reactions, and states of matter in discussing the physical properties, chemical properties, sources, and uses of (a) alkali metals, (b) alkaline earth metals, (c) aluminum, (d) halogens, (e) sulfur, and (f) phosphorus, and their compounds
3. Identify trends in properties of the elements discussed in the chapter
4. Write equations for the formation and reactions of the elements discussed in this chapter
5. Account for differences in properties of the alkali metals and alkaline earth metals
6. List examples of commercially important alloys and their uses
7. List examples of elements that exist in allotropic forms, and discuss differences in structure and properties
8. Distinguish between oxidizing and reducing agents, and discuss how they behave
9. Explain how chlorinated pesticides move through the food chain, creating problems for animals
10. Discuss reasons why some rain is becoming more acidic

Throughout our study of chemistry, each new chapter has brought new ideas and concepts regarding chemical principles. In Chapter 14, we deviate somewhat from that plan to reach the heart of chemistry—descriptive chemistry. **Descriptive chemistry** is the study of the properties, relationships, reactions, uses, and distribution of elements and compounds. Within the study of descriptive chemistry, an attempt is made to explain trends in properties and chemical behavior by considering microscopic structure, intermolecular forces, chemical bonding, and energy factors.

Chapter 14 is an applied chapter, in that many chemical principles that you have already learned are used to explain the properties and behavior of selected elements and their compounds. You shall apply your knowledge of atomic and molecular structure, periodic properties, stoichiometry, nomenclature, equations, and states of matter.

14.1 REPRESENTATIVE METALS

Representative metals are those metals that belong to groups IA through VA, excluding the transition metals—the group B metals. Therefore, the representative metals are (1) the alkali metals (IA), (2) the alkaline earth metals (IIA), (3) all elements in group IIIA (Al, Ga, In, and Tl) except boron, B, (4) Sn and Pb from group IVA, and (5) Bi from group VA. We shall investigate the metals in groups IA and IIA as well as one metal from group IIIA, aluminum.

Alkali Metals

Cesium's name comes from the Latin *caesius,* meaning "sky blue." Cesium was discovered by Bunsen and Kirchhoff in 1860.

Alkali metals are the elements in group IA of the periodic table: Li, Na, K, Rb, Cs, and Fr. Each of the alkali metals has an inner-core electronic configuration of a noble gas and one valence electron in the s sublevel, s^1. In chemical changes, the alkali metals tend to lose their outer s^1 electron, producing a monopositive ion.

$$M \longrightarrow M^+ + e^- \quad (M = \text{alkali metal})$$

Rubidium is derived from the Latin *rubidus,* "dark red." Rb was also discovered by Bunsen and Kirchhoff a year after they discovered cesium.

Alkali metals are widely distributed throughout the earth. However, because of their reactivity, they are always chemically combined with other elements in compounds. Sodium and potassium are two of the most abundant elements in the earth's crust and oceans. Sodium exists in rock salt, $NaCl$; saltpeter, $NaNO_3$; borax, $Na_2B_4O_7 \cdot 10H_2O$; and as dissolved ions in the oceans. Potassium is located in minerals such as sylvite, KCl; kainite, $KCl \cdot MgSO_4$; and carnallite, $KCl \cdot MgCl_2$. Li, Rb, and Cs are much less abundant than Na and K.

Table 14.1 summarizes the physical properties of the alkali metals. Alkali metals are soft and have the lowest densities of metals. Their low density is attributed to their large atomic radii (they have the largest atomic radii within a period) and their rather widely spaced metallic crystal packing pattern. Densities of alkali metals increase with increasing atomic mass.

TABLE 14.1   PROPERTIES OF THE ALKALI METALS

Properties	Li	Na	K	Rb	Cs
Atomic number	3	11	19	37	55
Atomic mass	6.941	22.99	39.10	85.47	132.9
Valence electronic configuration	$2s^1$	$3s^1$	$4s^1$	$5s^1$	$6s^1$
Density, g/mL	0.53	0.97	0.86	1.53	1.88
Melting point, °C	181	98	63	39	28
Boiling point, °C	1347	883	774	688	678
Heat of fusion, kJ/mol	3.01	2.6	2.3	2.2	2.1
Heat of vaporization, kJ/mol	135	97.9	79.0	75.8	68.3
Ionization energy, kJ/mol	526	502	425	409	382
Atomic radius, nm	0.152	0.186	0.231	0.244	0.262
Electronegativity	1.0	0.9	0.8	0.8	0.7
Formula of halide	LiX	NaX	KX	RbX	CsX
Formula of oxide	Li_2O	Na_2O	K_2O	Rb_2O	Cs_2O

Melting points of the alkali metals are the lowest among metals. Cesium's melting point, 28°C, is only a little above room temperature. Low melting points indicate rather weak metallic bonding. Similarly, the boiling points of the alkali metals are low, ranging from 1347°C for Li to 678°C for Cs. They have a large temperature range between their melting and boiling points (called the liquid range), indicating that metallic bonding must predominate in these liquids.

The electric and heat conductivities of alkali metals are among the highest of all elements. Only silver, gold, copper, and aluminum are better conductors. Good electric and thermal conductivity are explained in terms of the highly mobile, loosely held electrons. Alkali metals have the lowest ionization energies of all elements.

Chemically, the alkali metals are very reactive and combine directly with nonmetals:

$$2M + X_2 \longrightarrow 2MX \quad \text{(halide)} \quad (X = \text{halogen})$$
$$2M + H_2 \longrightarrow 2MH \quad \text{(hydride)}$$
$$4M + O_2 \longrightarrow 2M_2O \quad \text{(oxide)}$$
$$6M + N_2 \xrightarrow{\text{spark}} 2M_3N \quad \text{(nitride)}$$

This is expected, considering that, after losing an electron to a higher electronegative nonmetal, alkali metals become monopositive ions with the stable noble gas configuration.

Oxygen combines with the alkali metals to produce oxides.

$$4M + O_2 \longrightarrow 2M_2O \quad \text{(alkali metal oxide)}$$

In addition, some of the alkali metals combine in such a manner to produce peroxide ions, O_2^{2-}. Each oxygen atom in a peroxide ion has an oxidation number of -1; oxygen atoms normally exist in the -2 oxidation state.

$$2M + O_2 \longrightarrow M_2O_2 \quad \text{(peroxide)}$$

When alkali metals are placed into water, they react violently, producing an alkaline (basic) solution, hydrogen gas, and energy. Potassium and the alkali metals of higher atomic mass give off so much heat energy that the hydrogen gas liberated is ignited!

$$M + H_2O \longrightarrow 2MOH(aq) + H_2(g) + \text{energy}$$

Alkali metals combine even more vigorously if they contact acidic solutions.

Alkali metals have many uses. Lithium is alloyed with magnesium, and is the major component in Li batteries. Lithium salts help the mentally ill, especially those with manic-depressive illness. Sodium is the starting material in the manufacture of numerous substances, for example, dyes, soaps, and lead "antiknock" compounds for gasoline. Potassium compounds are major components of fertilizers and of some explosives. Photoelectric cells contain rubidium and cesium. An electric current is produced when light hits the surface of these metals.

Previously, sodium chloride, NaCl, was discussed as an alkali metal–bearing compound. Four other common alkali metal compounds that we will now discuss are (1) sodium carbonate, Na_2CO_3; (2) sodium hydrogen carbonate, $NaHCO_3$; (3) sodium hydroxide, NaOH; and (4) potassium nitrate, KNO_3.

Sodium carbonate, Na_2CO_3, is frequently called by its common name, soda ash. Soda ash is used in manufacturing soap, paper, water softeners, glass, and petroleum products. Sodium carbonate exists naturally in the mineral trona, $Na_2CO_3 \cdot NaHCO_3 \cdot 2H_2O$. Sodium carbonate is produced by the **Solvay process,** a procedure in which a solution of sodium chloride, NaCl(aq), is reacted with aqueous ammonia, NH_3(aq), and carbon dioxide, CO_2, to produce initially sodium hydrogen carbonate, $NaHCO_3$, and ammonium chloride, NH_4Cl.

$$NaCl + NH_3 + CO_2 + H_2O \longrightarrow NaHCO_3 + NH_4Cl$$

$NaHCO_3$ is then removed and heated, driving off CO_2 and H_2O and yielding sodium carbonate.

$$2NaHCO_3 \xrightarrow{\text{heat}} Na_2CO_3 + CO_2 + H_2O$$

Sodium hydrogen carbonate, $NaHCO_3$, or baking soda, is used in cooking. Baking powders contain $NaHCO_3$ along with an acidic substance (generally, cream of tartar, potassium hydrogen tartrate). As long as the mixture is dry, no

Sodium peroxide, Na_2O_2, is a yellowish white powder that is used to bleach and oxidize other substances.

Lithium carbonate, Li_2CO_3, marketed under names such as Lithane, Eskalith, and Lithonate, is an alternative drug to tranquilizers for the treatment of various psychoses.

Ernest Solvay (1838–1922), a Belgian chemist, was motivated toward industrial chemistry by his father, a salt refiner. After developing his process to manufacture $NaHCO_3$ in 1863, he founded a company that ultimately became the world's largest supplier of $NaHCO_3$.

reaction occurs. But as soon as the mixture is combined with water, $NaHCO_3$ and the acid react, releasing CO_2 (g), which causes breads and cakes to rise.

Sodium hydroxide (caustic soda and lye are two common names), $NaOH$, is a valuable commercial chemical. Sodium hydroxide exists as white crystals or pellets that readily absorb H_2O and CO_2 from the air. $NaOH$ is soluble in water, producing a strongly basic solution that feels slippery when touched. Industrially, $NaOH$ is used to produce soaps, detergents, and rayon and to extract Al from ores.

Potassium nitrate, KNO_3, is a superior fertilizer since it contains two plant nutrients when dissolved, K^+(aq) and NO_3^-(aq). KNO_3 is a component of gunpowder (black powder). KNO_3 serves as a source of oxygen to oxidize the other components. Gunpowder is a mixture of 75% KNO_3, 15% charcoal, and 10% sulfur.

> Sodium hydroxide is in the white pellets used for unclogging drains. NaOH releases a large quantity of heat when it contacts water and dissolves fats and other normally insoluble substances.

REVIEW PROBLEMS

14.1 What topics are considered in the study of descriptive chemistry?

14.2 What elements are considered the representative (a) metals and (b) nonmetals?

14.3 In what minerals does potassium exist in the earth's crust?

14.4 For the following properties, explain how the alkali metals compare with other metals: (a) density, (b) melting point, (c) electric conductivity, and (d) reactivity.

14.5 Write equations for the reaction of rubidium with: (a) $H_2(g)$, (b) $N_2(g)$, (c) $O_2(g)$.

14.6 How are the following alkali metals used in industry: (a) K, (b) Li, and (c) Na?

Alkaline Earth Metals

Alkaline earth metals are the elements that belong to group IIA of the periodic table, including beryllium, Be; magnesium, Mg; calcium, Ca; strontium, Sr; barium, Ba; and radium, Ra.

Alkaline earth metals are about as widely distributed as the alkali metals. Of all the elements in the earth's crust, calcium ranks fifth and magnesium ranks eighth in abundance. A large percentage of Ca is bonded to carbonate ions in calcium carbonate, $CaCO_3$. Calcium carbonate rock is also called limestone. Other forms of $CaCO_3$ include calcite, chalk, marble, and pearl (see Figure 14.1). Calcium is also found naturally bonded to sulfate ions in calcium sulfate, $CaSO_4$. When $CaSO_4$ is hydrated, $CaSO_4 \cdot 2H_2O$, it is called *gypsum*.

A large variety of minerals contain magnesium, including asbestos, $H_4Mg_3Si_2O_9$; talc or soapstone, $Mg_3Si_4O_{10}(OH)_2$; dolomite, $MgCO_3 \cdot CaCO_3$; and meerschaum, $Mg_2Si_3O_8 \cdot 2H_2O$. In addition, large quantities of Mg^{2+}(aq) are dissolved in the oceans. Beryllium is isolated from a mineral called beryl, $Be_3Al_2Si_6O_{18}$. Beryl is found in different colors; when beryl is a pale green, it is called aquamarine, and when it has a deep green color, it is an emerald. Strontium is quite rare; it exists primarily in celestite, $SrSO_4$, and strontianite, $SrCO_3$. Barium is obtained from barite, $BaSO_4$.

> Meerschaum is a soft, whitish mineral that is used primarily to make tobacco pipes. It has a low density and floats on water.

> Strontium is named after Strontian, a town in Scotland. Sr is a rather hard, white, metallic solid that closely resembles Ca.

The properties of the alkaline earth metals, summarized in Table 14.2, parallel those of the alkali metals.

Electronic configurations of the alkaline earth metals are such that all group members have two s electrons in their valence level, with an inner noble gas core. In chemical reactions, they tend to lose their two outermost electrons, forming 2+ ions that are isoelectronic with noble gases.

$$M \longrightarrow M^{2+} + 2e^- \qquad (M = \text{alkaline earth metal})$$

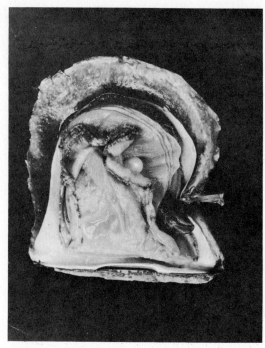

FIGURE 14.1
Limestone is composed of calcium carbonate, $CaCO_3$. Limestone makes up the stalactites hanging from the ceiling and the stalagmites growing from the floor of underground caverns (*National Park Service, photo by George Grant*). A pearl inside an oyster shell is also composed of $CaCO_3$. (*Field Museum of Natural History, Chicago*)

TABLE 14.2   PROPERTIES OF THE ALKALINE EARTH METALS

Property	Be	Mg	Ca	Sr	Ba
Atomic number	4	12	20	38	56
Atomic mass	9.012	24.31	40.08	87.62	137.3
Valence electronic configuration	$2s^2$	$3s^2$	$4s^2$	$5s^2$	$6s^2$
Density, g/mL	1.9	1.7	1.5	2.6	3.5
Melting point, °C	1280	650	842	770	725
Boiling point, °C	2970	1090	1490	1384	1640
Heat of fusion, kJ/mol	12	9.0	8.8	9.2	7.5
Heat of vaporization, kJ/mol	294	132	150	139	151
Ionization energy, kJ/mol	905	744	596	556	509
Atomic radius, nm	0.112	0.160	0.197	0.215	0.217
Electronegativity	1.5	1.2	1.0	0.9	0.9
Formula of halide	BeX_2	MgX_2	CaX_2	SrX_2	BaX_2
Formula of oxide	BeO	MgO	CaO	SrO	BaO

Densities of alkaline earth metals are higher than those of the alkali metals, as a direct result of the larger masses and smaller atomic radii of alkaline earth metals. However, their densities are low compared with those of other metals (most metals have densities in excess of 5 g/mL). They have higher melting and boiling points and are harder than corresponding alkali metals because they have two outer electrons instead of one. More electrons among metal nuclei produce stronger intermolecular forces.

Properties of the alkaline earth metals that are different from those of the alkali metals generally result from the smaller atomic radii and greater forces of attraction among atoms in alkaline earth metals. Their heats of fusion and vaporization are significantly higher than those of corresponding alkali metals. Electric and thermal conductivities of alkaline earth metals are high, but they are less than those of the alkali metals.

Chemically, the alkaline earth metals are reactive, but less so than the alkali metals. Of the alkaline earth metals, beryllium has the highest electronegativity and is the smallest in size, properties which give it metalloid characteristics. Beryllium compounds normally share more properties with covalent compounds than with ionic compounds. Beryllium actually resembles Al in chemical properties more than it resembles the alkaline earth metals.

Beryllium and its compounds are toxic and produce a degenerative, often fatal, lung disease called *berylliosis*.

Except for Be, most alkaline earth metals combine directly with nonmetals.

$$M + X_2 \longrightarrow MX_2 \quad \text{(halide)} \quad (X = \text{halogen})$$
$$M + H_2 \longrightarrow MH_2 \quad \text{(except when M = Be or Mg)}$$
$$3M + N_2 \longrightarrow M_3N_2$$

Alkaline earth metals produce oxides and peroxides when combined with oxygen. Oxide formation is only observed with Be through Ca. Strontium and barium produce both peroxides (SrO_2 and BaO_2) and oxides (SrO and BaO).

$$2M + O_2 \longrightarrow 2MO \quad \text{(oxide)}$$
$$M' + O_2 \longrightarrow M'O_2 \quad \text{(peroxide)} \quad (M' = \text{Sr or Ba})$$

Most alkaline earth metals (except Be and Mg) combine with water to produce a basic solution and liberate $H_2(g)$.

$$M(s) + 2H_2O(l) \longrightarrow M(OH)_2(aq) + H_2(g)$$
$$\text{(except when M = Be or Mg)}$$

In general, alkaline earth metals combine much less vigorously with water than do the alkali metals. Hot water is required for Ca to react at an appreciable rate.

In industry, Be and Mg metals are the most important metals in group IIA. Since Be does not absorb x-rays, it is placed into x-ray tubes as transparent "windows." Beryllium does absorb neutrons, particles given off in nuclear reactions; consequently, Be is used in shielding nuclear power plants.

Magnesium ribbon or wire gives off a blinding white light when ignited.

Economically, Mg is a most valuable metal. Magnesium's low density makes it a good metal to use in the manufacture of lightweight metal alloys.

Magnesium alloys are employed in the construction of wheels, cameras, airplane parts, tools, and luggage.

Calcium ions, Ca^{2+}, are biologically significant because they are a primary component of hydroxyapatite, $Ca_{10}(PO_4)_6(OH)_2$. Hydroxyapatite is the crystalline compound in both bone and teeth. Dental caries, or tooth decay, are thought to result when microorganisms decompose foods, especially sugars, to acids. These acids accelerate the decomposition of hydroxyapatite to Ca^{2+}(aq), $PO_4{}^{3-}$(aq), and OH^-(aq). Dental research has shown that the addition of small quantities of F^-(aq) to the drinking water helps to prevent dental caries in young people. Fluoride ions substitute for some of the OH^- in the crystal structure of hydroxyapatite, producing fluorohydroxyapatite—a compound that is more resistant to acid attack.

The ion Ca^{2+} is linked to K^+ in other biological systems; a proper Ca^{2+}-K^+ balance is required for normal heart function. Calcium ions are also involved in the blood-clotting mechanism. If Ca^{2+} is removed from blood, the blood does not clot.

Calcium carbonate, limestone, is the main source of elemental Ca. On heating limestone, calcium oxide (lime), CaO, results.

Lime, CaO, is the chemical that is ranked third in terms of amount produced in the United States; in excess of 35 billion pounds of CaO was produced in 1980.

$$CaCO_3(s) \longrightarrow CaO(s) + CO_2(g)$$

Since antiquity CaO has been known to be a good mortar when mixed with sand and water. After reacting with water, CaO yields calcium hydroxide (slaked lime, quicklime), $Ca(OH)_2$.

$$CaO + H_2O \longrightarrow Ca(OH)_2$$

As the mortar dries, $Ca(OH)_2$ combines with CO_2 in the air, re-forming the insoluble $CaCO_3$, a good binding substance.

$$Ca(OH)_2 + CO_2 \longrightarrow CaCO_3 + H_2O$$

Calcium oxide has the interesting property of giving off a brilliant white light when heated strongly. We refer to this bright light when we use the expression "in the limelight." Hundreds of industrial chemicals are produced from CaO or $Ca(OH)_2$; few other substances are used more by industry.

Let's turn our attention to magnesium compounds. Magnesium oxide (magnesia), MgO, is placed in electric heating elements, such as those in hot plates and stoves. Unlike most other substances, MgO is an excellent heat conductor but a rather poor conductor of electricity. Thus, MgO can be used to insulate an inner high-voltage heating element wire from the outside metal and, at the same time, transfer the heat generated.

Milk of magnesia, $Mg(OH)_2$, is a commonly used antacid. $Mg(OH)_2$ is an alkaline substance that neutralizes excess stomach acid secreted by the stomach lining as a result of disease or overeating. Magnesium sulfate, $MgSO_4 \cdot 7H_2O$ (Epsom salts) has varied medical uses.

Asbestos, a magnesium silicate compound, has been widely used whenever a heat-resistant substance is required. Asbestos is noncombustible and is a strong fiber; hence, automobile brake linings, construction materials, and building insulation contain asbestos. Recently, it has been discovered that microscopic asbestos particles can cause lung cancers when inhaled. Asbestos is gradually being replaced with ceramic compounds to minimize the cancer risk.

REVIEW PROBLEMS

14.7 Compare each of the following properties of alkaline earth metals with those of alkali metals: (a) electronic configuration, (b) density, (c) melting point, (d) heat of fusion.

14.8 From what minerals are the following isolated: (a) Ca, (b) Mg, (c) Be?

14.9 Write the equations for the reactions of Ba with: (a) $N_2(g)$, (b) H_2O, (c) $O_2(g)$.

14.10 List two functions that Ca^{2+} has in living systems.

Aluminum

All of the members of group IIIA are metals, except for B, which is a metalloid. We shall consider only one group member, Al. **Aluminum** is the most abundant metal in the earth's crust, and is the third most abundant element, exceeded only by O and Si.

Aluminum is in many clays as aluminum silicates, for example, feldspar, $KAlSi_3O_8$. Bauxite, $Al_2O_3 \cdot xH_2O$, a hydrated form of aluminum oxide, is the best source of Al.

Only a small percentage of Al is found in bauxite; most Al is a component of silicate clays.

Corundum is an anhydrous type of aluminum oxide that is widely distributed. Impurities in corundum create colored crystals, many of which are classified as gemstones (see Figure 14.2). If the impurities are chromium oxides in the

FIGURE 14.2
Topaz is a form of corundum that contains iron oxide impurities. Topaz is hard and exhibits a wide range of colors. (*Field Museum of Natural History, Chicago*)

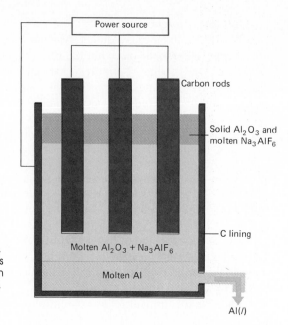

FIGURE 14.3
A Hall cell is used to extract Al from bauxite, Al_2O_3. Bauxite is first dissolved in molten cryolite, Na_3AlF_6. As an electric current is passed through the cell, molten Al is liberated from the bauxite. The molten Al then flows from the bottom of the cell.

Aluminum was once considered a precious metal (rare and valuable), like gold, platinum, and silver.

Charles Hall (1863–1914) began his search for Al after hearing a lecture by his professor, who stated that anyone who could discover an inexpensive means for extracting Al from its ore would become wealthy and famous. In 1886, in his home laboratory, Hall discovered the procedure for obtaining Al from bauxite.

corundum structure, then the crystal has a red color and is called a ruby. If the impurities are cobalt and titanium, then the crystal is blue, and is called a sapphire. If iron oxides are the impurities, the crystal is called oriental topaz. Amethyst results when manganese oxide is the impurity in corundum.

Even though it is the most abundant metal in the earth's crust, aluminum was once a rare and expensive metal that people valued highly. Its rarity stemmed from the lack of an inexpensive method for extracting Al from its ore. Charles M. Hall, while an undergraduate chemistry student at Oberlin College, solved this perplexing problem of getting to metallic Al relatively inexpensively. In his procedure, aluminum is obtained by dissolving bauxite in a molten aluminum compound, cryolite (Na_3AlF_6), at high temperatures. An electric current is then passed through the solution, producing molten Al, which flows out of the container. Figure 14.3 is a diagram of a Hall cell. This procedure is now known as the **Hall process,** and much of the world's Al is produced in this manner.

Selected properties of Al are listed in Table 14.3. Al, a member of group IIIA, has three valence electrons, $3s^2 3p^1$. These valence electrons are lost in chemical changes, yielding Al^{3+}. In compounds, Al exists only in the $+3$ oxidation state.

$$Al \longrightarrow Al^{3+} + 3e^-$$

The density of aluminum is 2.7 g/mL, a low density for a metal. Al is a good conductor of heat and electricity and is an excellent reflector of heat and light. Aluminum's melting and boiling points are moderate for metals.

Charles M. Hall (*Aluminum Company of America*)

A thin layer of aluminum is used to reflect light in large, visible-light telescopes.

As a result of its unique properties, Al has a wide variety of uses. Most people are familiar with Al as a foil used to wrap foods and other household items. Al is alloyed with other metals, especially Cu, Mg, and Mn, to produce metals that are strong and have a low density. Duraluminum, a solution of Al, Mg, Mn, and Cu, is a construction material that is used in buildings, boats, and airplanes. Another Al alloy is alnico, a strongly magnetic metal composed of Al, Ni, Co, and Fe (alnico is a mnemonic for Al, Ni, and Co). Because of the ever-diminishing world supply of Cu, aluminum now replaces Cu as the electric conductor in wires and cables.

Aluminum's chemical properties are of interest because they resemble those of metalloids. Aluminum, depending on the conditions, exhibits both acidic and

TABLE 14.3   PROPERTIES OF ALUMINUM

Property	
Atomic number	13
Atomic mass	26.98
Outer electronic configuration	$3s^2 3p^1$
Density, g/mL	2.7
Melting point, °C	660
Boiling point, °C	2467
Heat of vaporization, kJ/mol	284
Ionization energy, kJ/mol	577
Atomic radius, nm	0.143
Electronegativity	1.5

A substance that exhibits both acidic and alkaline properties is said to be amphoteric.

basic behavior. Aluminum metal samples, when exposed to the air, form an outer oxide coating that is relatively inert. Thus, Al appears to be an inert metal when, in fact, it is quite reactive.

Pure aluminum when heated in air at high temperature is totally converted to aluminum oxide, Al_2O_3.

$$4Al(s) + 3O_2(g) \longrightarrow 2Al_2O_3(s)$$

Al combines with iron(III) oxide, Fe_2O_3, releasing a tremendous amount of energy, enough that the resulting iron becomes molten.

$$2Al(s) + Fe_2O_3(s) \longrightarrow 2Fe(l) + Al_2O_3(l)$$

Thermite bombs were dropped during World War II because thermite fires are difficult to extinguish with water.

This reaction is known as the **thermite** reaction. Since temperatures in excess of 3000°C are obtained, metals are welded together using the thermite reaction.

Important Al compounds include aluminum hydroxide, $Al(OH)_3$; potassium aluminum sulfate, $KAl(SO_4)_2 \cdot 12H_2O$; aluminum chloride, $AlCl_3$; and aluminum sulfate, $Al_2(SO_4)_3$.

Aluminum hydroxide, $Al(OH)_3$, is an ingredient of antacids, which are used to lower the acid concentration in the stomach. Potassium aluminum sulfate, $KAl(SO_4)_2 \cdot 12H_2O$, commonly called alum, is added to soils that are too basic. Alum is acidic and neutralizes the basic components of soils. Aluminum chloride, $AlCl_3$, is frequently a reactant in laboratory syntheses, and is now produced as an intermediate in a new procedure for isolating Al from bauxite. Aluminum sulfate, $Al_2(SO_4)_3$, is used in the paper and pulp industry and in the purification of water.

REVIEW PROBLEMS

14.11 Consider the description of the Hall process for obtaining Al metal. What could account for the high price of isolating Al from bauxite?

14.12 Compare the following properties of Mg and Al: (a) melting points, (b) densities, and (c) ionization energies.

14.13 List two alloys of Al, and tell what they are used for.

14.2 REPRESENTATIVE NONMETALS

The general properties of oxygen, nitrogen, and the noble gases were discussed in Chapter 12, and carbon was considered in Chapter 13. In this section we will concentrate on some other nonmetals: the halogens, sulfur, and phosphorus.

Halogens

The halogens consist of the elements fluorine, F; chlorine, Cl; bromine, Br; iodine, I; and astatine, At. Halogens are one of the most reactive groups of elements on the periodic table. Each halogen (X) has one less electron (s^2p^5) than do the noble gases; thus, if a halogen obtains one electron from another atom, it becomes isoelectronic with a noble gas.

Astatine comes from the Greek *astatos,* meaning "unstable." Astatine was discovered in 1940 at the University of California by Corson, MacKenzie, and Segre.

$$X + e^- \longrightarrow X^-$$

Because of their reactivity, the halogens are always chemically combined with other elements in nature. Fluorine is relatively abundant, the primary sources being the minerals fluorite, CaF_2; fluoroapatite, $Ca_{10}F_2(PO_4)_6$; and cryolite, Na_3AlF_6. Chlorine is the most abundant of the halogens; it is located primarily in the oceans or in salt deposits in the ground. Bromine, to a much lesser extent, is found along with chlorine in the oceans and in salt deposits. Little dissolved iodine is in the oceans; most of the world's iodine is found in Chile saltpeter deposits where it occurs as sodium iodate, $NaIO_3$.

Some iodine is extracted from seaweed.

Liberating the halogens from compounds yields nonpolar diatomic molecules:

$$:\ddot{\text{X}}:\ddot{\text{X}}: \quad \text{(X = F, Cl, Br, I, or At)}$$

Fluorine, F_2, is a pale yellow gas, and chlorine, Cl_2, is a pale green gas; bromine, Br_2, is a volatile reddish-brown liquid; and iodine, I_2, is a grayish black solid that sublimes. Dispersion forces are the intermolecular forces that bind halogen molecules together and account for the trend in physical states of the halogens.

Table 14.4 presents other physical properties of the halogens. Trends in density, melting point, and boiling point are explained by the increasing dispersion forces present in atoms of succeedingly higher atomic mass. Stronger dispersion forces in the halogens are directly related to the number of electrons and how tightly they are held by the nucleus.

If we consider atomic properties, the halogens have the second highest set of ionization energies, exceeded only by those of the noble gases. The atomic radii

TABLE 14.4   PROPERTIES OF THE HALOGENS

Property	Fluorine	Chlorine	Bromine	Iodine
Atomic number	9	17	35	53
Atomic mass	19.00	35.45	79.90	126.9
Valence electronic configuration	$2s^2 2p^5$	$3s^2 3p^5$	$4s^2 4p^5$	$5s^2 5p^5$
Density	1.81 g/L	3.21 g/L	3.12 g/mL	4.94 g/mL
Melting point, °C	−220	−101	−7.3	114
Boiling point, °C	−188	−34	58.8	184
Heat of vaporization, kJ/mol	—	—	15.0	20.8
Ionization energy, kJ/mol	1680	1250	1140	1000
Atomic radius, nm	0.064	0.099	0.114	0.133
Electronegativity	4.0	3.0	2.8	2.5

of the halogens are among the smallest in each period. Both ionization energy and atomic radii trends indicate that the nuclei of halogen atoms strongly attract their electrons. Therefore, it follows that the halogens are highly electronegative elements; they strongly attract electrons in chemical bonds.

Halogen chemistry is best understood in terms of the halogen's ability to attract electrons from other elements. A substance that readily pulls electrons from other elements and compounds is called a strong oxidizing agent. **Oxidizing agents** are chemical species that attract electrons. A **reducing agent** is a substance that gives up electrons; thus it is the opposite of an oxidizing agent.

When discussing ionic substances (Chapter 8), we have already seen the oxidizing capacity of halogens. Halogens remove electrons from metals (reducing agents), yielding ionic salts, for example, $NaCl$, MgF_2, and LiI.

Fluorine gas is the strongest oxidizing agent of all elements. In fact, F_2 is such a strong oxidizing agent that it combines with just about all other elements, in many cases, violently. Substances like hydrogen and sulfur ignite immediately when contacted by F_2 gas. F_2 also replaces all of the other halogens in binary compounds.

$$F_2 + MX \longrightarrow X_2 + MF \qquad (X = Cl, Br, or I)$$

Fluorine is one of the few elements that combine with the high-atomic-mass noble gases (Xe and Kr), producing a multitude of noble gas compounds.

Chlorine gas is not as reactive as F_2; nevertheless, Cl_2 combines with most other elements under the proper conditions. For instance, when Cl_2 is combined with H_2 a reaction occurs only if the temperature is elevated or light is present; Cl_2 and H_2 do not combine at low temperatures in the dark.

$$H_2 + Cl_2 \xrightarrow{\text{heat}} 2HCl$$

Chlorine gas combines with other nonmetals to produce covalent chlorides.

$$S_8 + 4Cl_2 \longrightarrow 4S_2Cl_2$$
$$P_4 + 6Cl_2 \longrightarrow 4PCl_3$$

Chlorine gas replaces hydrogens in hydrocarbons, i.e., carbon-hydrogen compounds.

$$CH_4 + Cl_2 \xrightarrow{\text{light}} CH_3Cl + HCl$$

If excess Cl_2 is available, all of the hydrogens in methane can be replaced, yielding a mixture of CH_3Cl, CH_2Cl_2, $CHCl_3$, and CCl_4 (chlorinated hydrocarbons).

CHCl₃ is chloroform. Chloroform was an early anesthetic but is no longer used because of its high toxicity.

Most of the chemical properties of bromine and iodine are similar to chlorine. However, they are much weaker oxidizing agents than chlorine. Consequently, they react more slowly and require a greater quantity of energy.

Controversy surrounds some halogen compounds. For example, chlorinated hydrocarbons are a group of halogen compounds that have a detrimental effect on the environment. Many pesticides are complex, highly chlorinated hydrocarbons, which perform well in doing what they are designed to do, kill "bugs." But overuse of pesticides releases large quantities of these chemically stable substances into the environment, where they cause many problems.

Chlorinated hydrocarbons have a very low solubility in water and a significantly higher solubility in fat tissues. Thus, excess chlorinated hydrocarbons concentrate in biological organisms and are passed along in the food chain. Consumers at the high levels of the chain, predatory animals and human beings, concentrate a large quantity of these pesticides. Chlorinated hydrocarbons have been linked to the extinction of several bird species and to various medical problems in humans. Yet without these pesticides, insects decrease crop yields, which also has a deleterious effect on society.

A less controversial halogen compound is hydrogen fluoride, HF. HF combines with calcium silicate, a major component of glass, and is therefore used to etch glass,

$$CaSiO_3(s) + 6HF(l) \longrightarrow CaF_2(s) + SiF_4(g) + 3H_2O(l)$$

frost light bulbs, and etch calibration marks on thermometers and glassware.

Bromine and iodine are used much less frequently. Silver salts of bromine and iodine are the image-forming components of photographic films. Iodide ions are an essential mineral required in a person's diet. Iodide ions are removed from the blood by the thyroid gland and incorporated into a hormone called thyroxin. Iodine deficiency in humans results in an affliction called *goiter*, an enlargement of the thyroid gland.

Examples of chlorinated pesticides are DDT, chlordane, endrin, and lindane. These pesticides reside in the environment from 2 to 15 years before they are completely decomposed.

Black and white photographic film has particles of AgBr and AgI embedded in a coating on a plastic backing. When light strikes Ag^+X^- crystals, it activates them and makes the crystals more susceptible to reduction (gaining of electrons). During development these crystals are reduced to metallic silver.

REVIEW PROBLEMS

14.14 What are the main natural sources of the halogens?

14.15 In what physical state would you expect to find astatine? Explain in terms of a trend in intermolecular forces.

14.16 List the halogens in increasing order of their capacity to oxidize other substances.

14.17 Write and balance the following equations:
(a) Combination of H_2 and Br_2
(b) Formation of strontium iodide from the elements
(c) Chlorine gas combining with methane gas

Sulfur

An old name for sulfur was brimstone, from the Germanic word meaning "burning stone."

Sulfur belongs to the chalcogen group (VIA), and is directly under oxygen on the periodic table; accordingly, S has an s^2p^4 valence electronic configuration. Sulfur is a yellow solid that is composed of S_8 molecules in the form of rings containing eight sulfur atoms (Figure 14.4). Actually, there are two primary allotropes of sulfur, **orthorhombic** and **monoclinic** sulfur.

Orthorhombic sulfur is the normal form of sulfur. Orthorhombic refers to the type of crystal system to which the sulfur atoms belong. When orthorhombic sulfur is heated to its melting point, 112°C, it changes to a pale yellow liquid, which upon cooling changes to monoclinic sulfur. This second allotrope con-

FIGURE 14.4
Elemental sulfur is composed of cyclic molecules containing eight sulfur atoms, S_8.

tains sulfur atoms in a different crystal lattice pattern. Monoclinic sulfur slowly changes back to orthorhombic sulfur if left alone at room temperature. Table 14.5 lists the physical properties of sulfur.

Sulfur-bearing minerals include galena, PbS; pyrite, FeS_2; zinc blende, ZnS; and bornite, Cu_3FeS_3. Many sulfates, SO_4^{2-}, are dissolved in the oceans. These include sodium sulfate, Na_2SO_4, magnesium sulfate, $MgSO_4$, and the less soluble calcium sulfate, $CaSO_4$. Elemental sulfur deposits are located throughout the world; large amounts are located in the United States in Texas and Louisiana.

Underground sulfur deposits are extracted by what is known as the **Frasch process,** named for its developer Herman Frasch. A special metal pipe, which is actually three pipes in one, is inserted into the sulfur deposit (see Figure 14.5). By separately pumping both hot water and compressed air down two of the pipes, the underground sulfur is dissolved and forced up the third pipe as a bubbly, frothy mixture of sulfur, air, and water. When the water evaporates, relatively pure sulfur is obtained.

Like oxygen, sulfur combines with both metals and nonmetals. If sulfur is heated with Na metal, sodium sulfide, Na_2S, results.

$$16Na + S_8 \xrightarrow{\Delta} 8Na_2S$$

Herman Frasch's (1851–1914) interest in S developed out of his work in the field of petroleum chemistry. Frasch was concerned with S contamination of oil. This prompted him to apply principles of petroleum chemistry to tap the vast supplies of underground S in Texas.

TABLE 14.5   PROPERTIES OF SULFUR AND PHOSPHORUS

Property	Sulfur	Phosphorus
Atomic number	16	15
Atomic mass	32.05	30.97
Valence electronic configuration	$3s^2 3p^4$	$3s^2 3p^3$
Density, g/mL	2.07, rhombic	1.82, white
	1.96, monoclinic	2.34, red
Melting point, °C	112, rhombic	44, white
	119, monoclinic	600, red
Boiling point, °C	444, rhombic	280, white
Heat of fusion, kJ/mol	1.23	2.51
Heat of vaporization, kJ/mol	10.5	12.4
Ionization energy, kJ/mol	1000	1058
Electronegativity	2.5	2.1

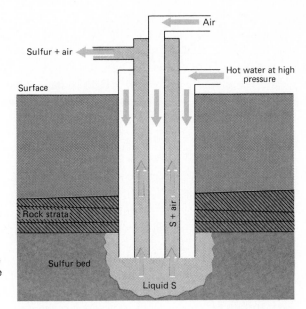

FIGURE 14.5
The Frasch process is used to bring underground sulfur to the surface. Air and hot water are forced under pressure into underground sulfur deposits. The sulfur melts and is pushed to the surface, where it dries out and solidifies.

Iron and calcium combine with sulfur as follows:

$$8Fe + S_8 \xrightarrow{\Delta} 8FeS$$

$$8Ca + S_8 \xrightarrow{\Delta} 8CaS$$

Sulfur combines with carbon, phosphorus, and hydrogen (nonmetals).

$$4C + S_8 \xrightarrow{\Delta} 4CS_2$$

$$8P_4 + 10S_8 \xrightarrow{\Delta} 8P_4S_{10}$$

$$8H_2 + S_8 \xrightarrow{\Delta} 8H_2S$$

Most of the halogens and oxygen combine with sulfur.

$$S_8 + 4Cl_2 \xrightarrow{\Delta} 4S_2Cl_2$$

$$S_8 + 8O_2 \xrightarrow{\Delta} 8SO_2$$

Uses of elemental sulfur are varied. Sulfur is a component of gunpowder and is used in rubber manufacturing. Rubber is strengthened by the addition of sulfur; this process is called **vulcanization.** However, the main use of sulfur is the formation of sulfur compounds, especially sulfuric acid, H_2SO_4. Sulfuric acid is commercially the most important of all chemicals. It is synthesized by combining sulfur trioxide, SO_3 (an acid anhydride) with water.

Vulcanization, heating rubber with sulfur at high temperatures, was developed by Charles Goodyear (1800–1860) in 1839. The term "vulcanization" comes from the Roman god of fire, Vulcan. Goodyear's fame, however, stems from the tire company that uses his name.

$$SO_3 + H_2O \longrightarrow H_2SO_4$$

The top 10 chemicals produced in the United States are:

Chemical	Amount produced, billions of pounds
H_2SO_4	81
NH_3	38
CaO	35
O_2	34
N_2	34
$CH_2\!\!=\!\!CH_2$	28
NaOH	23
Cl_2	22
H_3PO_4	22
NH_4NO_3	17

Sulfur trioxide is produced by oxidizing elemental sulfur. Commercial sulfuric acid is generally sold as a 98% H_2SO_4 solution. It is a viscous, dense (1.85 g/mL) liquid that has strong dehydrating properties. Sulfuric acid dehydrates many substances instantaneously upon contact.

There are many uses of H_2SO_4, most of which are in the phosphate fertilizer industry. Ammonia is combined with sulfuric acid to yield ammonium sulfate, $(NH_4)_2SO_4$, a good fertilizer.

$$2NH_3 + H_2SO_4 \longrightarrow (NH_4)_2SO_4$$

Synthetic fibers, pigments, petroleum products, and storage batteries (lead-acid) all require sulfuric acid in their production.

Other interesting sulfur compounds are hydrogen sulfide, H_2S, and sulfur dioxide, SO_2. In contrast to water, hydrogen sulfide is a gas at room temperature. The lower electronegativity of sulfur (2.5) compared with that of oxygen (3.5) does not allow for the formation of hydrogen bonds; accordingly, H_2S molecules are bound by the weaker dipolar interactions. H_2S is a foul-smelling gas with the odor of rotten eggs. Hydrogen sulfide gas is highly toxic, about as poisonous as hydrogen cyanide, HCN; two parts per thousand parts of air are normally fatal. Decaying animals and the refining of petroleum are sources of H_2S in the atmosphere.

Sulfur dioxide, SO_2, is a major component of air pollution. Most fossil fuels contain a small quantity of sulfur (crude oil contains about 0.2 to 0.8% sulfur) that is oxidized to SO_2 when the fuel is combusted. The three main sources of SO_2 in the air are power plants, industry, and transportation. Power companies burn immense quantities of fossil fuels in generating electricity. Industry, especially the metal industry, generates SO_2 from burning fuels and manufacturing goods. Fuels combusted in automobiles, planes, trains, and boats give off a small percentage of SO_2.

Sulfur oxides are the main type of atmospheric pollutants in what is called London smog. On occasion, London has dense fogs that last for a week or more. During these episodes, high levels of sulfur oxides build up from burning fossil fuels.

What happens to the SO_2 once it enters the atmosphere? There are two principal possibilities: It can combine immediately with water, forming the weak acid, sulfurous acid.

$$SO_2 + H_2O \longrightarrow H_2SO_3$$

Sulfurous acid falls to earth in rain and other types of precipitation. SO_2 can also combine with oxygen in the atmosphere to yield sulfur trioxide, SO_3. Sulfur trioxide combines with water, and forms sulfuric acid, which also falls to earth in rain. Rain that contains acidic substances is called **acid rain.**

Acid rain is not only harmful to plants and animals, it also dissolves exposed metals, and even dissolves buildings that are constructed out of marble, $CaCO_3$, by the following reaction.

$$H_2SO_4(aq) + CaCO_3(s) \longrightarrow CaSO_4(aq) + CO_2(g) + H_2O(l)$$

Figure 14.6 shows the effect of air pollution on a marble statue.

FIGURE 14.6
Throughout the world, components of polluted air have destroyed or damaged irreplaceable works of art and historic buildings. In these "before" and "after" pictures, the effect of air pollution on a marble statue is illustrated. (*Field Museum of Natural History, Chicago*)

Many famous buildings, like the ancient ruins in Greece and various memorials in Washington, D.C., are dissolving at a measurable rate.

REVIEW PROBLEMS

14.18 How do the two allotropic forms of sulfur differ?

14.19 (a) What method is used to bring sulfur deposits to the surface? (b) How is this accomplished?

14.20 Write balanced equations for the combination of sulfur, S_8, with: (a) Ca, (b) Cl_2, (c) O_2.

14.21 (a) How is acid rain produced? (b) Write all appropriate equations.

14.22 List three uses of sulfuric acid, H_2SO_4.

Phosphorus

Phosphorus is another element that exists in different allotropic forms. The most common allotropes of phosphorus are generally distinguished by their colors, white and red. The properties of white phosphorus are markedly different from those of red phosphorus.

White phosphorus is a soft waxy solid with a low melting point, 44°C. After exposure to the air, white phosphorus ignites spontaneously.

Black phosphorus also exists. It results when white phosphorus is subjected to high pressures.

$$P_4(s) + 5O_2(g) \longrightarrow P_4O_{10}(s)$$

FIGURE 14.7
White phosphorus is very reactive and spontaneously ignites in the air. It is composed of P_4 molecules in the shape of a tetrahedron. Red phosphorus is much less reactive than white phosphorus. It is composed of long chains of P_4 rings bonded to each other.

White phosphorus

Red phosphorus

Consequently, white phosphorus is stored so that it never contacts the air; most white phosphorus is stored under water. White phosphorus exhibits the property of **phosphorescence**; i.e., it releases light after being exposed to various types of electromagnetic radiation.

White phosphorus is made up of P_4 molecules, not individual P atoms. Each P atom in P_4 is bonded to three other P atoms in a tetrahedral pattern (Figure 14.7). When white phosphorus is heated without air to about 250°C, red phosphorus is produced. At 250°C, it is thought that the P_4 rings combine, yielding long chains of P_4 molecules attached together (Figure 14.7). Such compounds are called **polymeric substances.** Red phosphorus melts around 600°C, is stable at room temperature, and has a low toxicity compared with white P. Table 14.5 lists the properties of phosphorus.

> Polymers are high-molecular-mass substances that are composed of long-chain molecules.

Phosphorus is the twelfth most abundant element in the earth's crust. Phosphorus is in phosphate-bearing rocks such as calcium phosphate, $Ca_3(PO_4)_2$, also called phosphate rock. Elemental phosphorus is most frequently prepared from phosphate rock by mixing the rock with coke, C, and silica, SiO_2, and then placing the mixture into a high-temperature electric furnace. At high temperatures, P_4 is given off as a vapor.

$$2Ca_3(PO_4)_2 + 6SiO_2 + 10C \longrightarrow 6CaSiO_3 + 10CO + P_4(g)$$

Elemental phosphorus itself is limited in its uses. At one time phosphorus was placed in matches, but problems arose from its toxicity, and it now has been replaced by a phosphorus compound, tetraphosphorus trisulfide, P_4S_3. Fireworks and incendiary bombs contain white phosphorus. Phosphorus is principally used to synthesize phosphorus compounds, especially tetraphosphorus decoxide, P_4O_{10}, which is converted to phosphoric acid, H_3PO_4.

$$P_4O_{10}(s) + 6H_2O(l) \longrightarrow 4H_3PO_4(aq)$$

Pure phosphoric acid (actually hydrogen phosphate) is a white solid that melts at 42°C. Hydrogen phosphate is infrequently used. Instead, phosphoric acid is normally sold as an 85% aqueous solution. Compared with nitric, sulfuric, and hydrochloric acids, phosphoric acid is a weak acid.

Phosphoric acid combines with basic compounds yielding a large variety of dihydrogen phosphates, $H_2PO_4^-$, monohydrogen phosphates, HPO_4^{2-}, and phosphates, PO_4^{3-}. Calcium monohydrogen phosphate, $CaHPO_4$, is one of the polishing substances in toothpaste. Calcium dihydrogen phosphate, $Ca(H_2PO_4)_2$, is an ingredient of baking powders and is used as a fertilizer.

Major nutrients needed by plants are C, H, O, N, P, K, Ca, Mg, and S.

Phosphates are good fertilizers for plants since plants (and animals) need a constant source of phosphorus to synthesize biologically significant compounds. A substance called adenosine triphosphate, ATP, is a high-energy molecule that breaks down, giving off energy that fuels cells.

REVIEW PROBLEMS

14.23 Describe differences in the properties of white and red phosphorus.

14.24 How is elemental phosphorus obtained from phosphate rock?

14.25 What are the three types of phosphate that result when phosphoric acid combines with bases?

14.26 How is phosphorus important to animal and plant life?

SUMMARY

Descriptive chemistry is the study of the properties, reactions, uses, and sources of elements and compounds.

Alkali metals are reactive elements that are located throughout nature in ionic compounds. As elements, they have low densities, low melting points, and low boiling points. Alkali metals are excellent conductors of heat and electricity. Their reactivity is so great that most of them combine directly with nonmetals. Higher-atomic-mass alkali metals react violently and explode when combined with H_2O. Sodium metal and its compounds are commercially most important.

Alkaline earth metals are not as reactive as the alkali metals. Their densities, melting points, and boiling points are higher than those of the alkali metals. Ca and Mg are two of the most common elements in the earth's crust; the other alkaline earths are less widely distributed. Chemically, the alkaline earth metals are similar to the alkali metals except that they lose two electrons to obtain the noble gas configuration. Beryllium is an exception, acting more like a metalloid than a metal.

Aluminum is the most abundant metal in the earth's crust. Most of this Al is bonded to oxygen and silicon in oxides and silicates. Al has three valence electrons and exists in the +3 oxidation state in compounds. It is a low-density metal that is a good conductor of heat and electricity. Al is used to manufacture strong, light metal alloys and is frequently used as an electric conductor in cables and wires.

Halogens are reactive nonmetals that exist in ionic form in oceans and salt deposits around the world. As elements, the halogens exist diatomically. F_2 and Cl_2 are gases at room temperature, Br_2 is a volatile liquid, and I_2 is a volatile solid. Halogens have large ionization energies and the highest set of electronegativities. They are good oxidizing agents; halogens readily remove electrons from other substances.

Sulfur is a chalcogen with six valence electrons. Sulfur exists in different allotropic forms: orthorhombic sulfur and monoclinic sulfur. Sulfur, like O_2, combines with both metals and nonmetals, yielding metallic and nonmetallic sulfides. Gunpowder and rubber are two examples of commercial products containing S.

Phosphorus normally exists in two allotropic forms, white and red phosphorus. White phosphorus, P_4, is a reactive solid that reacts violently with O_2 in the air; it is toxic and must be handled carefully. Red phosphorus is unreactive and bears little resemblance to white phosphorus. Phosphorus is obtained from phosphate rock, $Ca_3(PO_4)_2$, which is chemically converted to many phosphate compounds and phosphoric acid.

QUESTIONS AND PROBLEMS

Alkali Metals

14.27 In the alkali metals, what trends are observed in the following properties (increasing or decreasing with atomic mass): (a) density, (b) melting point, (c) heat of fusion, (d) ionization energy?

14.28 Write the complete electronic configuration for: (a) Na, (b) Li, and (c) K.

14.29 Draw the dot formulas for the following: (a) Rb, (b) NaF, (c) K_2O, and (d) Cs_3N.

14.30 List two minerals in which each of the following alkali metals exist: (a) Na, (b) K.

14.31 (a) Determine the liquid range, in Celsius degrees, of lithium. (b) How does this range compare with the other alkali metals?

14.32 Write an explanation for the increasing atomic radii with increasing mass in the alkali metals.

14.33 What mass of Rb occupies 9.40 mL? Use data in Table 14.1.

14.34 How many joules are required to melt 48.2 g Na at its melting point?

14.35 Complete and balance the following equations:
(a) $K + Cl_2 \longrightarrow$
(b) $Rb + H_2 \longrightarrow$
(c) $Li + O_2 \longrightarrow$
(d) $Cs + H_2O \longrightarrow$
(e) $Na + HCl(aq) \longrightarrow$

14.36 Write the formulas for the following: (a) lithium hydroxide, (b) rubidium sulfate, (c) potassium sulfide, (d) sodium peroxide, (e) potassium phosphate.

14.37 (a) What substance is produced as a result of the Solvay process? (b) Write balanced equations illustrating the Solvay process.

14.38 (a) Write a balanced equation for the reaction of lithium with bromine liquid, Br_2. (b) To what general class of reaction does this reaction belong? (c) If 0.373 g Li is combined with excess Br_2, what is the theoretical yield of product?

14.39 Potassium combines with water producing potassium hydroxide and hydrogen gas. What mass of potassium metal should be placed into water to produce 28.9 L H_2 at STP?

14.40 Determine the percent by mass of Na in sodium carbonate.

14.41 Write a balanced equation to show what happens when an acid, such as HCl(aq), is combined with sodium hydrogen carbonate.

14.42 About 2.0 g K is required in a person's diet each day. Calculate the number of K atoms in 2.0 g K.

Alkaline Earth Metals

14.43 What alkaline earth metals are extracted from the following minerals: (a) beryl, (b) dolomite, (c) limestone, (d) barite?

14.44 What is the difference between gypsum and pure calcium sulfate?

14.45 What is the mass percent of Mg in asbestos, $H_4Mg_3Si_2O_9$?

14.46 Determine the amount of heat, in kilocalories and kilojoules, required to vaporize 37.3 g Ca at its boiling point. Use data located in Table 14.2.

14.47 What properties of Be atoms deviate from those of the other alkaline earth metals?

14.48 Write electron dot formulas for: (a) Mg, (b) SrS, (c) $CaCO_3$, (d) $MgSO_4$.

14.49 Complete and balance the following equations:
(a) magnesium + nitrogen gas $\longrightarrow$
(b) barium + fluorine gas $\longrightarrow$
(c) strontium + hydrogen gas $\longrightarrow$
(d) barium + water $\longrightarrow$
(e) beryllium + chlorine gas $\longrightarrow$

14.50 How is each of the following used? (a) CaO, (b) Mg alloys, (c) asbestos, (d) magnesium hydroxide, (e) beryl.

14.51 Write balanced equations showing how mortar, containing lime, is produced from limestone, and how it helps hold bricks together.

14.52 Barium peroxide, BaO_2, is combined with hydrochloric acid, $HCl(aq)$, to produce hydrogen peroxide, H_2O_2. (a) Complete and balance the equation. (b) Determine the mass of H_2O_2 that could be produced by combining 883 g of barium peroxide with excess hydrochloric acid.

14.53 Plaster of paris, $CaSO_4 \cdot \frac{1}{2}H_2O$, is formed by heating gypsum, $CaSO_4 \cdot 2H_2O$.

$$CaSO_4 \cdot 2H_2O \longrightarrow CaSO_4 \cdot \tfrac{1}{2}H_2O + \tfrac{3}{2}H_2O$$

Find the mass of plaster of paris that results when 27.0 kg of gypsum is heated.

14.54 Lime, CaO, when added to water produces slaked lime, $Ca(OH)_2$. (a) Write the balanced equation for this chemical change. (b) If 66 kJ of energy is released per mole of slaked lime produced, calculate the amount of energy released, in both joules and calories, when 4.0 kg lime is added to water.

14.55 Write a chemical explanation for tooth decay; include appropriate compounds and what happens to them.

14.56 Magnesium hydroxide, an ingredient in antacid tablets, combines with stomach acid, $HCl(aq)$, to produce aqueous magnesium chloride and water. (a) What mass of hydrochloric acid would a tablet containing 324 mg of magnesium hydroxide neutralize? (b) Compare your answer to the amount of acid neutralized by the same mass of sodium hydrogen carbonate. Sodium hydrogen carbonate and $HCl(aq)$ produce aqueous sodium chloride, water, and carbon dioxide.

Aluminum

14.57 (a) What is the mineral corundum? (b) List different forms of corundum.

14.58 Draw electron dot formulas for and name: (a) AlN, (b) AlF_3, (c) Al_2O_3, (d) $Al(OH)_3$.

14.59 Some antiperspirants contain the aluminum salts (aluminum chlorohydrates) $Al(OH)_2Cl$ and $Al(OH)Cl_2$. Write two balanced equations showing how both of these salts are produced from the combination of aluminum hydroxide and hydrochloric acid.

14.60 What could account for the difficulty and high cost of obtaining such an abundant element as Al from its ores?

14.61 (a) Write the thermite reaction. (b) If 426 kJ is released per mole of Al, what quantity of heat is released when 95.7 g Al is totally reacted? (c) What mass of water could be heated from 25°C to 100°C with the energy released from 95.7 g Al in the thermite reaction [heat capacity of water is 75 J/(mol·K)]?

14.62 Aluminum fluoride, AlF_3, has a melting point of 1291°C and exhibits some characteristics of ionic substances. However, aluminum chloride, $AlCl_3$, has a much lower melting point, 194°C, and shows little ionic character. What could explain the difference in the properties of these two aluminum halides?

14.63 Aluminum oxide combines with hydrogen fluoride to produce aluminum fluoride and water. Write a balanced equation for this reaction.

14.64 List three aluminum compounds and their uses.

Halogens

14.65 Draw the electron dot formulas for: (a) ClI_4, (b) HF, (c) ClF, (d) CaI_2, (e) SCl_2.

14.66 What volume of Cl_2 gas has a mass of 20.0 g at STP?

14.67 Among the halogens, which is the best oxidizing agent and which is the best reducing agent?

14.68 Considering the atomic properties of halogens in Table 14.4, explain fluorine's capacity to oxidize most other substances.

14.69 Complete and balance the following halogen reactions:
(a) $F_2 + S_8 \longrightarrow$
(b) $Ca + Br_2 \longrightarrow$
(c) $H_2 + I_2 \longrightarrow$
(d) $F_2 + MgCl_2 \longrightarrow$
(e) $CH_4 + Cl_2 \longrightarrow$

14.70 Chlorine gas combines with sulfur dioxide gas, yielding SO_2Cl_2. Determine the mass of SO_2Cl_2 that forms when 0.123 L of chlorine gas combines with excess sulfur dioxide; assume STP conditions.

14.71 Hydrogen fluoride combines with calcium silicate, $CaSiO_3$, the major component of glass.

$$CaSiO_3 + 6HF \longrightarrow CaF_2 + SiF_4 + 3H_2O$$

Assuming that a 125-g drinking glass is 100% calcium silicate, determine the mass of HF needed to totally combine with the glass.

Sulfur

14.72 Which of the following sulfur-containing ores has the greatest percent composition of sulfur: zinc blende, ZnS; pyrite, FeS_2; or galena, PbS?

14.73 Write balanced equations for the combination of sulfur, S_8, with: (a) calcium, (b) hydrogen gas, (c) carbon.

14.74 When 2.00 mol of zinc(II) sulfide is oxidized to zinc(II) oxide and sulfur dioxide, 905 kJ of energy is released. Determine the amount of energy given off when 3.00 g zinc(II) sulfide is oxidized.

14.75 If a gasoline contains 0.2%, by mass, of sulfur (S_8), and 1 gal of gasoline has a mass of 2.6 kg, approximate the mass of H_2SO_4 produced from burning one tank (15 gal) of gasoline, using the equations for the oxidation of S_8 to SO_2, SO_2 to SO_3, and finally SO_3 to H_2SO_4. Assume that all of the sulfur in the gasoline is initially oxidized to sulfur dioxide.

14.76 What volume of CO_2 at STP is produced when 4.99 g $CaCO_3$ combines with 6.11 g of sulfuric acid?

$$CaCO_3 + H_2SO_4 \longrightarrow CO_2 + CaSO_4 + H_2O$$

Phosphorus

14.77 Name the following phosphorus compounds: (a) PF_5, (b) H_3PO_3, (c) P_4O_6, (d) NaH_2PO_4, (e) AlP.

14.78 Find the percent of phosphorus in $Ca(H_2PO_4)_2$.

14.79 If 76.5 kg of phosphate rock, containing 75.0% calcium phosphate, is treated with silicon dioxide and carbon monoxide, determine the mass of P_4 obtained.

14.80 Determine the mass of phosphoric acid that results when 16.21 g P_4O_{10} combines with excess water.

14.81 Phosphine, PH_3, a colorless, foul-smelling poisonous gas is prepared by combining calcium phosphide with water, producing phosphine and calcium hydroxide. (a) Write and balance the equation for the reaction. (b) What volume of PH_3 at STP is produced when 35 g of calcium phosphide is placed into excess water?

14.82 Three phosphorus sulfides have the following formulas: (a) P_4S_{10}, (b) P_4S_3, and (c) P_4S_7. Write balanced equations that show the formation of each of the three sulfides, starting with P_4 and S_8.

· C H A P T E R ·

Solutions

Study Guidelines

After completing Chapter 15, you should be able to:
1. Describe how a solution is prepared
2. Distinguish between a solute and a solvent
3. List and give examples of the primary classes of solutions
4. Distinguish between miscible and immiscible liquids
5. List factors that influence a solute's solubility in a solvent
6. Explain the meaning of "like dissolves like" in terms of solute and solvent structure
7. State the relationship between solubility and temperature, given a solubility curve
8. Discuss energy requirements for a solid to dissolve in a liquid
9. Describe how each of the following affects the dissolving rate of a solute: (a) particle size, (b) temperature, (c) concentration, and (d) stirring
10. List and define the primary units of concentration
11. Determine the percent by mass (% m/m), molarity, or molality of a solution, given all necessary data
12. Explain how a specific solution concentration (% m/m, molarity, or molality) is prepared
13. Determine the exact quantity of solvent needed to dilute a more concentrated solution to a stated lower concentration
14. Give examples of strong electrolytes, weak electrolytes, and nonelectrolytes
15. Write overall and net ionic equations, given the equations in undissociated form
16. Describe how the addition of a nonvolatile solute lowers the vapor pressure of a solution
17. Explain why a solvent's boiling point is increased and its freezing point is decreased by the addition of a solute
18. Determine the increase in boiling point and decrease in freezing point, given the amounts of solute and solvent in the solution

Solutions are homogeneous mixtures of pure substances.

Solutions are homogeneous mixtures of substances that have a uniform composition throughout their volume. When a solution is prepared, one substance is added to another substance in such a manner that, after mixing, only one physical state is observed.

A solvent is the component of a solution that is present in the greatest amount. Solutes are dissolved in the solvent.

In a solution, the substance in larger amount, the one whose physical state is observed, is called the **solvent.** The substance in lesser amount, the one that becomes incorporated into the solvent, is called the **solute.** When a solute mixes with a solvent, producing a solution, the process is called **dissolving.** Solutes dissolve in solvents, yielding solutions.

Glucose is an important sugar in the blood.

Solute particles (atoms, molecules, or ions) interact with solvent particles and become a part of the structure of the solvent during the dissolving process. For instance, when a sugar such as glucose, $C_6H_{12}O_6$ (solute), dissolves in water (solvent), the solid crystalline structure of glucose is broken down by the water molecules. When totally dissolved, glucose molecules are evenly distributed throughout the water (see Figure 15.1).

$$\text{Glucose}(s) \xrightarrow{\text{water}} \text{glucose}(aq)$$

If the conditions remain constant, glucose molecules remain in solution and do not settle out on standing. Any equal-volume portion of the solution contains the same number of glucose molecules. At any time, the dissolved glucose could be recovered by evaporation of the water.

Types of Solutions

Solutions are categorized by the physical states of the solute and solvent. For example, at room conditions glucose is a solid and water is a liquid, and a glucose-water solution is therefore a solid-liquid solution. Solid-liquid solutions

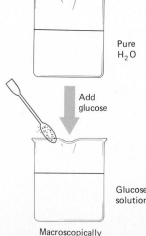

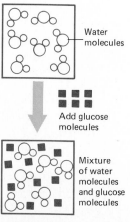

FIGURE 15.1
When glucose (solute) dissolves in water (solvent), glucose molecules become a part of the structure of water. Glucose molecules are surrounded by and bond to water molecules.

Pure H_2O

Add glucose

Glucose solution

Macroscopically

Water molecules

Add glucose molecules

Mixture of water molecules and glucose molecules

Microscopically

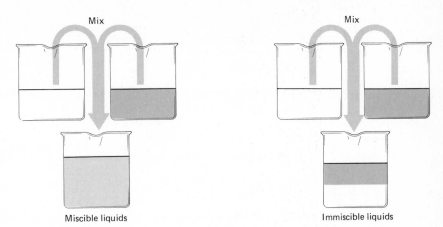

FIGURE 15.2
If two miscible liquids are poured together they dissolve in each other; they are mutually soluble. If two immiscible liquids are poured together, they do not mix; they form two separate layers.

Miscible liquids

Immiscible liquids

Average solubility values for dissolved ions in ocean water:

Ion	Solubility (g solute/kg H_2O)
Cl^-	19
Na^+	11
SO_4^{2-}	2.7
Mg^{2+}	1.3
Ca^{2+}	0.41
K^+	0.39
Br^-	0.067

are common since many solids dissolve in water. Oceans contain many dissolved substances, including sodium chloride, NaCl(aq); sodium bromide, NaBr(aq); magnesium chloride, $MgCl_2$(aq); and many others.

Another commonly encountered type of solution is a liquid-liquid solution in which two liquids are dissolved in each other. For example, an alcohol, like ethanol, $C_2H_6O(l)$, dissolves in water to produce a solution of alcohol and water. It is difficult to differentiate the solute from the solvent in a liquid-liquid solution because the solution is in the same physical state as the two components. Which component is the solvent, and which is the solute? If a small quantity of ethanol is added to a large quantity of water, then the ethanol is the solute and water is the solvent. If a small quantity of water is mixed with a larger quantity of ethanol, then water is the solute and ethanol is the solvent.

When dealing with liquid-liquid solutions, two additional terms are used. They are "miscible" and "immiscible." Two liquids are categorized as **miscible** if they are mutually soluble in each other. Ethanol and water are miscible liquids. **Immiscible** liquids are not soluble and do not mix. When combined, immiscible liquids form two layers. For example, oil and water are immiscible liquids (see Figure 15.2). There are all degrees of miscibility in between these examples; such solutions are classified as **partially miscible.**

A third class of solutions is gas-liquid solutions. In these solutions, the gas is the solute and the liquid is the solvent. Carbonated beverages are aqueous solutions of gaseous carbon dioxide, CO_2(aq). Nitrogen and oxygen gas, primary components of the atmosphere, are dissolved to a small degree in most water samples.

In addition to liquids, both gases and solids can be solvents. Gas-gas solutions are those that contain two or more nonreacting gases. Air is a gaseous solution composed of $N_2(g)$, $O_2(g)$, $Ar(g)$, $H_2O(g)$, $CO_2(g)$, and many other gases. All nonreacting gases are miscible with all other gases. Terms such as "solvent" and "solute" have little or no meaning when applied to gas-gas solutions.

TABLE 15.1   TYPES OF SOLUTIONS

Solute	Solvent	Examples
Solid	Liquid	NaCl(aq); sugar water, $C_{12}H_{22}O_{11}$(aq)
Liquid	Liquid	Vinegar, $HC_2H_3O_2$(aq); antifreeze in water, $C_2H_6O_2$(aq)
Gas	Liquid	CO_2(aq), O_2(aq), Ar(aq), NH_3(aq)
Gas	Gas	Atmosphere, any mixture of nonreacting gases
Solid	Solid	Solder (Sn and Pb), brass (Zn and Cu)
Gas	Solid	H_2 in Pd

Amalgams are used to fill dental cavities. Giovanni of Arcoli (1412–1484) was one of the first to suggest that dental cavities be filled with a metal; he proposed gold as the best material for fillings. Modern dental amalgams contain silver, copper, tin, zinc, and mercury.

Less frequently, solids are solvents. Alloys are solid-solid solutions of two or more different metals. Sterling silver is a solution of 7.5% Cu and 92.5% Ag. Eighteen carat gold is an alloy of 75% Au and variable amounts of Ag (10–20%) and Cu (5–14%). Solids can dissolve liquids and produce a solution. A good example of such solutions are amalgams. Amalgams contain varying amounts of liquid mercury, Hg(l), dissolved in metals.

Most other combinations of physical states are possible, but they are rather rare compared with those already discussed. Gases can be dissolved by solids— $H_2(g)$ forms a solution with the metal palladium, Pd. Table 15.1 lists examples of selected classes of solutions.

Factors Affecting Solubility

Solubility is the amount of solute that dissolves in a given amount of solvent. Generally, chemists measure solubility by determining the amount of solute that dissolves in 100 g of solvent. Three factors are most significant in predicting the solubility of one substance in another: (1) the nature of the solute and solvent, (2) temperature, and (3) pressure. Let's look at these factors individually.

Solubility depends on intermolecular forces (1) within the solvent, (2) within the solute, and (3) between solute and solvent particles.

The chemistry saying that **"like dissolves like"** is a guiding rule when considering the nature of solute and solvent, and solubility. Substances that have similar structures and intermolecular forces tend to be soluble; whereas, substances that have dissimilar structures and intermolecular forces are less soluble or are insoluble (don't dissolve).

Methanol, $CH_3OH(l)$, is totally miscible with water. What explains water-methanol miscibility? Looking at the structure of a water molecule, we see two H atoms bonded to an O atom; thus, we would expect hydrogen bonding to be the intermolecular force among water molecules.

Methanol is commonly called wood alcohol because when wood is heated without oxygen, methanol is one of the products formed. Methanol is a highly toxic substance if ingested.

$$\begin{array}{c} O \\ \diagup \ \diagdown \\ H \qquad H \end{array}$$

Methanol molecules are also polar covalent and exhibit hydrogen bonding among molecules in the liquid state.

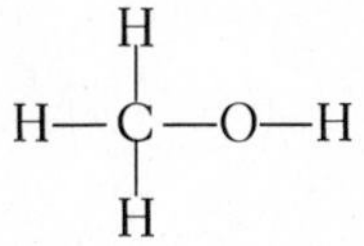

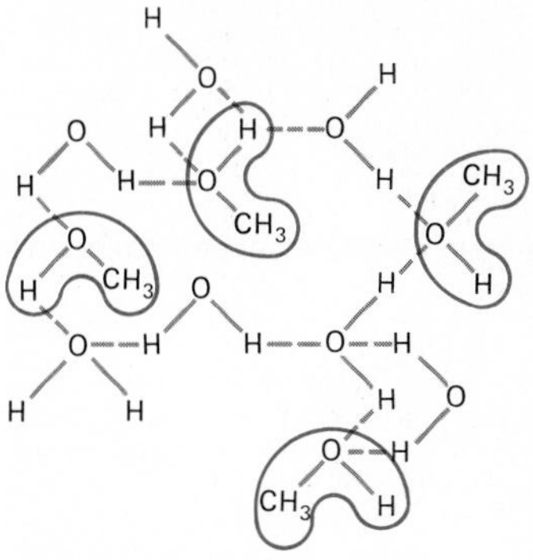

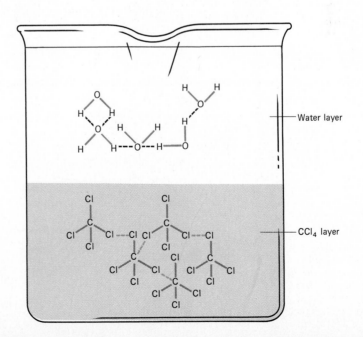

$$H-\overset{\overset{\displaystyle H}{|}}{\underset{\underset{\displaystyle H}{|}}{C}}-O-H$$

FIGURE 15.3
When methanol molecules dissolve in water, they form hydrogen bonds with water molecules. Normally, liquid molecules with similar sizes and intermolecular forces are miscible.

Carbon tetrachloride was once the main component in household cleaning fluids. It was found to cause severe liver damage and to be carcinogenic. Thus, it is no longer used as a cleaner.

Since water and methanol have similar structures and are each hydrogen-bonded liquids, they can form hydrogen bonds with each other when they are mixed (see Figure 15.3). Thus methanol and water are miscible.

As a second illustration, consider the miscibility of carbon tetrachloride, CCl_4, and benzene, C_6H_6. Carbon tetrachloride molecules are nonpolar, with dispersion forces existing among the molecules in the liquid state. Likewise, benzene is composed of nonpolar covalent molecules, also with dispersion forces in the liquid state. CCl_4 and C_6H_6 molecules attract each other with dispersion forces, which hold the two liquids together. These are miscible liquids.

What about the solubility of CCl_4 in water? Water has strong hydrogen bonds among molecules, and CCl_4 molecules have much weaker dispersion forces. In a mixture of water and CCl_4, the water molecules hold tightly together and exclude the CCl_4 molecules. Because there is no net charge separation in CCl_4 molecules, water molecules have only a small force of attraction for CCl_4 molecules (see Figure 15.4). CCl_4 is therefore only partially miscible in water— less than 0.1 g CCl_4 dissolves in 100 mL of water at 20°C.

The above cases are illustrations of liquid-liquid solutions. What predictions could be made for solids dissolving in liquids? For example, sodium nitrate, $NaNO_3(s)$, is an ionic solid. In which solvent, H_2O or CCl_4, would $NaNO_3$

FIGURE 15.4
Polar covalent water molecules are held together by strong hydrogen bonds. Nonpolar covalent carbon tetrachloride molecules are bonded by weaker dispersion forces. Water molecules, therefore, exclude the CCl_4 molecules. In general, nonpolar solvents have minimal solubility in polar solvents.

FIGURE 15.5
(a) When ionic substances dissolve in water, the polar water molecules attract the anions and cations, weakening and then breaking the ionic bonds that hold the ions to the crystal lattice. (b) After breaking free, the ions are surrounded by water molecules, and then they diffuse away from the undissolved crystal.

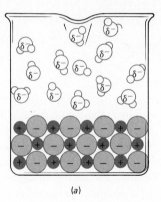

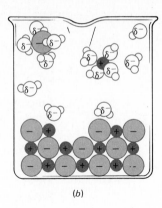

(a)   (b)

Polar covalent and ionic substances tend to dissolve in polar solvents. Nonpolar substances tend to dissolve in nonpolar solvents.

have a higher solubility? Ionic solids have a crystal lattice structure composed of oppositely charged ions that are attracted by a polar covalent substance. Accordingly, water breaks apart the crystal lattice of $Na^+NO_3^-$ and then surrounds the resulting Na^+ and NO_3^- ions (see Figure 15.5). Nonpolar CCl_4 molecules are unable to attract the ions in $NaNO_3$ and cannot break apart the crystal lattice; hence, the solubility of $NaNO_3$ in CCl_4 is minimal.

Temperature is the second factor that influences solubility. Figure 15.6 is a graph of the solubility of selected solutes in water as a function of temperature. Generally, an increase in temperature is accompanied by an increase in solubility; at higher temperatures greater masses of solutes dissolve in a fixed mass of water than at lower temperatures (see Figure 15.7). However, this is not always the case. For a small number of substances, notably gas-liquid solutions, there is

FIGURE 15.6
The solubility of a substance in water is usually measured in terms of the number of grams of solute that dissolve per 100 g of water. A solubility curve normally shows the solubility of a substance from 0°C to 100°C.

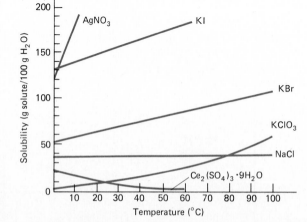

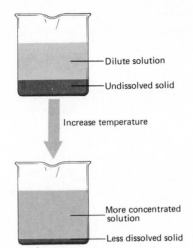

FIGURE 15.7
For many solutes, as the temperature of the solvent increases, the solubility of the solute increases. This is a direct relationship. For solutes that give the solution a color, the increased solubility is recognized by a deeper color of the solute in the solution.

Dilute solution
Undissolved solid

Increase temperature

More concentrated solution
Less dissolved solid

FIGURE 15.8
Lattice energy is the amount of energy required to break the bonds in the crystal lattice structure of a solute. Whenever bonds are broken, energy must be added; that is, it is an endothermic process. Hydration energy is the amount of energy released when water molecules surround and bond to the solute particles. Whenever bonds are formed, energy is released; that is, it is an exothermic process.

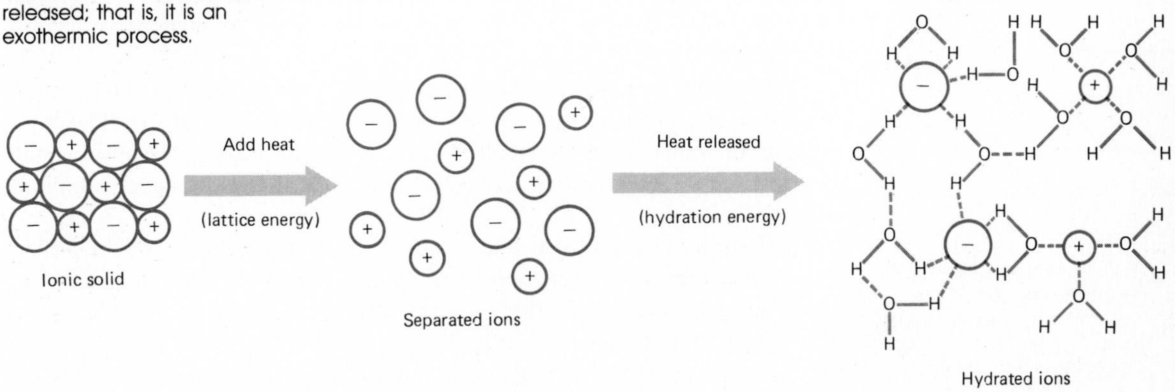

a decrease in solubility with increasing temperature. Other substances, including NaCl, have a relatively constant water solubility with increasing temperature.

An explanation for the effect of temperature on solubility requires information that we have not discussed yet. For now, we will consider only the energy effects of dissolving solids. There are two steps involved in dissolving solids: (1) the breaking of the crystalline structure of the solid, and (2) the surrounding of the solid particles by water molecules. The first step requires an input of energy (an endothermic process) to break the bonds that hold the solid together. The added energy is called the **lattice energy**. In the second step, energy is released (an exothermic process) when bonds are formed between solid particles and water molecules. The energy released is called the **hydration energy** (see Figure 15.8).

Ionic solid

Add heat

(lattice energy)

Separated ions

Heat released

(hydration energy)

Hydrated ions

If the hydration energy is greater than the lattice energy, more energy is released than consumed, resulting in a net loss of heat to the surroundings. In such cases, the temperature of the solution rises. If the opposite is true, i.e., the hydration energy is less than the lattice energy, energy is removed from the solution and the solution becomes colder. If the hydration energy equals the lattice energy, no temperature change is observed on dissolving. The amount of heat transferred when substances dissolve is called the **enthalpy of solution.**

> The amount of heat absorbed or released when a mole of substance dissolves is its enthalpy of solution, ΔH_{soln}.

Pressure also influences solubility. Only one type of solution is significantly affected by pressure changes: gas-liquid solutions. As the pressure over the solution of a dissolved gas is increased, the solubility of the gas increases. William Henry (1774–1836) was the first to describe the direct relationship between gas pressure and solubility, now known as **Henry's law.**

Divers are concerned with Henry's law. High-pressure conditions beneath the surface of water increase the solubility of N_2 and O_2 in the blood. If the diver comes to the surface too rapidly, bubbles of gas form in the blood and block small blood vessels, causing severe pain, fainting, and other unpleasant symptoms.

REVIEW PROBLEMS

15.1 Describe the difference between homogeneous and heterogeneous mixtures.

15.2 Give an example for each of the following types of solutions: (a) solute is a gas and solvent is a gas, (b) solute is a solid and solvent is a liquid, and (c) solute is a liquid and solvent is a liquid.

15.3 Write two examples of pairs of miscible liquids.

15.4 Knowing that oil and water are immiscible liquids, what could be inferred about the intermolecular forces in water and oil?

15.5 On an average, what effect does each of the following changes have on the solubility of a solute in a solvent: (a) lowering the temperature, and (b) decreasing the pressure of a gas over a solution containing that gas?

15.6 What is observed when the lattice energy of a solute dissolving in water is greater than its hydration energy?

Dissolving Rates

Dissolving rate refers to the speed with which a solute dissolves in a solvent. Dissolving rates can be measured in terms of the mass of solute that dissolves in a time interval.

A very soluble substance, like sugar, can be placed into water and then not dissolve at an appreciable rate. What is done to decrease the time that it takes for sugar or any other solute to dissolve? Four factors influence the rates at which substances dissolve: (1) particle size, (2) temperature, (3) solution concentration, and (4) stirring.

If a sugar cube is placed into water, the cube dissolves more slowly than does an equal mass of finely granulated sugar. A sugar cube exposes less surface area to the water than do the tiny sugar crystals of granulated sugar; inner regions of the sugar cube dissolve only after the outer layers are in solution. There are fewer unexposed particles when sugar is granulated, so the sugar crystals enter solution faster.

To increase the dissolving rate of large chunks of solids in liquids, it is only

> A cube that is 1 cm on each edge has a surface area of 6 cm². If the same cube is divided up into 1000 cubes of 1 mm on each edge, the total surface area is 60 cm², a tenfold increase.

necessary to grind them up into small particles. Mortar and pestles are frequently used in laboratories for just this purpose.

Temperature is another factor that changes the rate at which solutes dissolve in solvents. Increasing the temperature of the solvent generally decreases the time needed for solids to dissolve. At higher temperatures, increased molecular motion increases the number of interactions of solute and solvent particles, which increases the dissolving rate.

Solution concentration affects the rate at which substances dissolve. Solution concentration is a measurement of the amount of solute that is dissolved in a solvent. A **dilute** solution is one in which a small amount of solute is dissolved in a solvent. As more solute is added, the concentration of the solute increases. As the concentration of solute in solution increases, the amount of time needed for more solute to dissolve increases. Initially, when the first solute is added to a pure solvent, the dissolving rate is at a maximum. With each addition of solute, the dissolving rate decreases until no more solute is observed to dissolve; at this time the solution is said to be **saturated.** Any addition of solute to a saturated solution remains undissolved and falls to the bottom of the container. Prior to saturation, the solution is **unsaturated,** and more solute dissolves in the solution.

Some substances, when dissolved at a higher temperature, do not come out of solution as the temperature is lowered. Thus, these solutions contain more solute than the maximum solubility. This phenomenon is called supersaturation. Few solutions supersaturate.

When a solution is saturated, the number of solute particles entering solution equals the number of particles that leave the solution and attach to the undissolved solute. In other words, an equilibrium is established between the rate at which particles break free from the undissolved solute and the rate at which they attach to the undissolved solute. Consequently, no more solute is experimentally found to dissolve (see Figure 15.9). Prior to saturation, the number of particles breaking free from the solute is greater than the number bonding with the solute; therefore, the solute continues to dissolve.

Stirring a solution increases the rate at which a solid dissolves by decreasing the concentration of solute in the immediate region surrounding the solid, and also by increasing the amount of exposed solute surface to the solvent.

REVIEW PROBLEMS

15.7 List three ways that the dissolving rate of a sugar cube is increased.

15.8 (a) What happens when additional salt is added to a saturated salt solution? (b) What effect would stirring have on the amount of dissolved salt in this solution? (c) How could added salt be dissolved in a saturated salt solution?

FIGURE 15.9
When a solute is placed into a solvent, the rate of solute particles entering the solution is at first greater than the rate of solute particles leaving the solution and bonding to the undissolved solute. Therefore, the solute continues to dissolve. At equilibrium, the rate of solute particles entering the solution is equal to the rate of those leaving the solution; thus, there is no change in the mass of the undissolved solute.

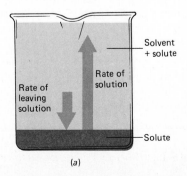
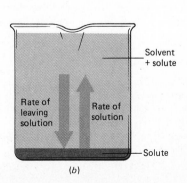

(a)   (b)

15.2 CONCENTRATION OF SOLUTIONS

Terms such as "concentrated" or "dilute" refer to the amount of solute contained in a solvent. Various solutions, when saturated, are quite dilute. A saturated solution of magnesium hydroxide, $Mg(OH)_2$, contains only 9×10^{-4} g $Mg(OH)_2/100$ mL H_2O at 25°C. On the other hand, a dilute solution of magnesium chloride hexahydrate, $MgCl_2 \cdot 6H_2O$, might contain 1 to 10 g of magnesium chloride (solubility is 167 g/100 mL H_2O at 25°C). A quantitative means is required to express exactly the amount of solute in a solution.

Many units of concentration are used by chemists. Table 15.2 lists the major units of concentration that are employed in chemistry. In this section, we will discuss percent by mass and molarity. Later in the chapter, molality is discussed.

Percent by Mass

Percent by mass, % m/m, is the unit of concentration that expresses the mass of solute per 100 g of solution (mass of solute + mass of solvent).

$$\% \text{ m/m} = \frac{\text{mass of solute}}{\text{mass of total solution}} \times 100$$

In percent concentration problems the identity of the solute is not used in the calculation.

By definition, a 5% m/m sugar solution is one that contains 5 g of sugar in every 100 g of solution. A 5% m/m sugar solution is prepared by dissolving 5 g of sugar

TABLE 15.2 COMMON CONCENTRATION UNITS

Unit	Symbol	Definition
Percent by mass	% m/m	$\% \text{ m/m} = \dfrac{\text{mass of solute}}{\text{mass of solution}} \times 100$
Percent by mass to volume	% m/v	$\% \text{ m/v} = \dfrac{\text{mass of solute}}{\text{volume of solution}} \times 100$
Percent by volume	% v/v	$\% \text{ v/v} = \dfrac{\text{volume of solute}}{\text{volume of solution}} \times 100$
Parts per thousand	ppt	$\text{ppt} = \dfrac{\text{mass of solute}}{\text{mass of solution}} \times 1000$
Parts per million	ppm	$\text{ppm} = \dfrac{\text{mass of solute}}{\text{mass of solution}} \times 1{,}000{,}000$
Molarity	M	$M = \dfrac{\text{moles of solute}}{\text{volume of solution (L)}}$
Normality	N	$N = \dfrac{\text{equivalents of solute}}{\text{volume of solution (L)}}$
Molality	m	$m = \dfrac{\text{moles of solute}}{\text{kilogram of solvent}}$

in 95 g of water. Example Problem 15.1 shows the calculations required to determine the mass of solute and solvent when using percent by mass.

Example Problem 15.1

How would 185.00 g of a 1.39% NaCl(aq) solution be prepared?

Solution

1. *What is unknown?* g NaCl and g H_2O
2. *What is known?*   185.00 g of solution, and 1.39% NaCl(aq)
   From the information given, we know that the total mass of the solution is 185.00 g, which is the mass of both the solute, NaCl, and the solvent, water. Also, a 1.39% NaCl solution contains 1.39 g NaCl per 100 g of solution.

$$1.39\% \text{ NaCl} = \frac{1.39 \text{ g NaCl}}{100 \text{ g solution}}$$

   The percent by mass is a conversion factor that relates mass of solute to the mass of total solution; accordingly, we apply the factor-label method to solve the problem (see Figure 15.10).
3. Apply the factor-label method to find the mass of NaCl.

$$185.00 \text{ g solution} \times \frac{1.39 \text{ g NaCl}}{100 \text{ g solution}} = 2.57 \text{ g NaCl}$$

4. Determine the mass of water. Knowing the mass of NaCl, 2.57 g, we subtract its mass from the total mass of the solution to find the mass of water.

$$\text{Mass of water} = \text{mass of solution (solute + solvent)} - \text{mass of solute}$$
$$\text{g } H_2O = 185.00 \text{ g solution} - 2.57 \text{ g NaCl}$$
$$= 182.43 \text{ g}$$

FIGURE 15.10
The percent by mass of a solution is a conversion factor that relates the mass of solute per 100 g of solution. To calculate the mass of solute contained in a solution, multiply the percent by mass times the mass of the solution. To calculate the mass of solution that contains a fixed mass of solute, invert the percent-by-mass conversion factor and multiply it by the mass of solute.

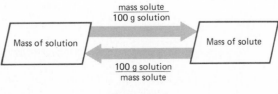

Mass of solution

$\frac{\text{mass solute}}{100 \text{ g solution}}$

$\frac{100 \text{ g solution}}{\text{mass solute}}$

Mass of solute

Percent by mass

To prepare a 1.39% NaCl solution, 2.57 g NaCl should be dissolved in 182.43 g H_2O.

Many times in the laboratory a solution is available with a known concentration, and a specific mass of solute is needed. Example Problem 15.2 illustrates how the mass of solute is determined from the concentration expressed in percent by mass.

Example Problem 15.2

Alchemists of the eighth century called nitric acid *aqua fortis*, which means "strong water." Today, nitric acid is used to produce fertilizers, explosives, plastics, dyes, and drugs.

Concentrated nitric acid, HNO_3, is normally sold as a 70% by mass aqueous solution. Determine the mass of concentrated nitric acid that contains 23 g HNO_3.

Solution

1. *What is unknown?*   Mass, g, of concentrated HNO_3 solution
2. *What is known?*   23 g HNO_3, and 70% nitric acid solution
3. Apply the factor-label method (see Figure 15.10).

$$23 \text{ g } \cancel{HNO_3} \times \frac{100 \text{ g } HNO_3 \text{ (solution)}}{70 \text{ g } \cancel{HNO_3}} = 33 \text{ g } HNO_3 \text{ solution}$$

Our answer of 32.9 g is rounded to 33 g to display the correct number of significant figures. Within a 33-g 70% nitric acid solution there is 23 g of pure HNO_3; the remaining 10 g is water.

More often than not, when dealing with percent by mass, volumes of solutions must also be considered. Study Example Problem 15.3.

Example Problem 15.3

Hydroiodic acid, HI(aq), is commercially available as a 57% by mass aqueous solution that has a density of 1.70 g/mL. Determine the mass of HI in 1.0 L of 57% HI solution.

FIGURE 15.11
Three conversion factors are needed to determine the mass of solute in a given number of liters of solution. These factors are the number of milliliters per liter, the density of the solution, and the percent by mass of the solution. First convert the number of liters of solution to milliliters. Then calculate the mass of the solution using the density, and finally multiply by the percent by mass to obtain the mass of solute.

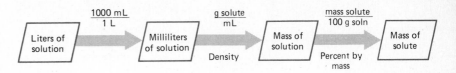

Solution

1. *What is unknown?*   Mass of HI, g HI
2. *What is known?*   1.0 L of 57% HI, with a density of 1.70 g/mL
   In this problem, the volume of the solution is given; thus, by using the density, we can find the total mass of the solution (see Figure 15.11).
3. Apply the factor-label method.

$$1.0 \text{ L solution} \times \frac{1000 \text{ mL}}{1 \text{ L}} \times \frac{1.70 \text{ g solution}}{1 \text{ mL}} \times \frac{57 \text{ g HI}}{100 \text{ g solution}}$$

$$= 969 \text{ g HI}$$
$$= 9.7 \times 10^2 \text{ g HI}$$

A solution of 1.0 L of 57% HI contains 9.7×10^2 g HI.

Other percent concentration units are percent by volume and percent by mass to volume.

$$\% \text{ v/v} = \frac{\text{volume of solute}}{\text{total volume}} \times 100$$

% m/m = % mass/mass
% m/v = % mass/volume
% v/v = % volume/volume

$$\% \text{ m/v} = \frac{\text{mass of solute}}{\text{total volume}} \times 100$$

Problems involving these concentration units are similar to those for percent by mass.

REVIEW PROBLEMS

15.9 Name the unit of concentration represented by the following: (a) moles of solute per liter of solution, (b) mass of solute per 100 mL of solution, (c) moles of solute per kilogram of solvent, (d) mass of solute per 100 g of solution.

15.10 Explain how 675 g of a 22.0% by mass solution of aqueous acetone is prepared.

15.11 What mass of $CaCl_2$ is contained in 46.4 g of a 8.1% $CaCl_2$(aq) solution?

Molarity   Molarity is the concentration unit that is most frequently encountered in beginning chemistry. **Molarity** (M) is the number of moles of solute per liter of solution. When using or determining molarity, moles of solute particles are compared with the total volume of the solution, i.e., both the solute *and* solvent.

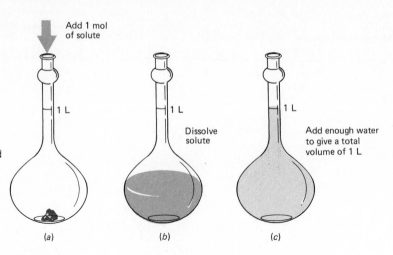

FIGURE 15.12
One way to prepare a 1 *M* aqueous solution is to (*a*) weigh 1 mol of solute on a balance and transfer the solute to a 1-L volumetric flask, (*b*) add water and dissolve the solute, and (*c*) continue adding water until the total volume of the solution is 1 L and shake to produce a homogeneous mixture.

Molarity is the number of moles of solute per liter of solution.

$$\text{Molarity} = \frac{\text{moles solute}}{\text{liter solution}}$$

One way to prepare a one molar (1 *M*) aqueous solution is to add 1 mol of solute to a 1-L volumetric flask and then add enough water so that the total volume is 1 L. See Figure 15.12.

A one molar solution can be prepared in many other ways. If a 250-mL volumetric flask is all that is available, 0.25 mol solute is placed into the 250-mL flask, and then enough water is added to yield a total volume of 250 mL.

A unit closely related to molarity is formality (*F*). Formality is the number of moles of formula units of an ionic solute per liter of solution. Formality is used for ionic substances that do not contain discrete molecules.

$$1.0 \text{ M} = \frac{0.25 \text{ mol}}{0.25 \text{ L}}$$

Example Problem 15.4 shows the calculation required before a specific volume of a given molar solution can be prepared.

Example Problem 15.4

Explain how 175 mL of a 0.320 M KI(aq) solution is prepared.

Solution

1. *What is unknown?* Grams of KI in 175 mL of 0.320 M KI(aq)

2. *What is known?* 175 mL, 0.320 M KI $\left(\dfrac{0.320 \text{ mol KI}}{1 \text{ L}} \right)$, and $\dfrac{166 \text{ g KI}}{1 \text{ mol}}$

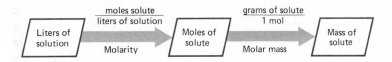

FIGURE 15.13
To find the mass of solute in a given volume of solution, multiply the volume of the solution in liters times the molarity of the solution $\left(\dfrac{\text{moles of solute}}{\text{liters of solution}}\right)$, and then multiply by the molar mass $\left(\dfrac{\text{grams of solute}}{1\text{ mol}}\right)$.

To calculate the mass of KI required to prepare the solution, we use conversion factors to change the volume to liters and then use the molarity, ratio of moles of solute to volume, and the molar mass of KI (see Figure 15.13).

3. Apply the factor-label method.

$$175 \text{ mL} \times \frac{1 \text{ L}}{1000 \text{ mL}} \times \frac{0.320 \text{ mol KI}}{1 \text{ L}} \times \frac{166 \text{ g KI}}{1 \text{ mol KI}} = 9.30 \text{ g KI}$$

After we find the volume in liters, we calculate the number of moles, using the molarity. Finally, we obtain the mass by multiplying by the molar mass of KI.

4. To prepare 175 mL of 0.320 M KI, add 9.30 g KI to a container with a graduation mark at 175 mL. Place some water into the container to dissolve the KI, and continue adding water until the total volume is 175 mL.

Example Problem 15.5

What is the molarity of a solution prepared by adding 18.3 g of methanol, CH_4O, to a container, and then adding enough water to yield a total volume of 58.6 mL?

Solution

1. *What is unknown?* $\quad M\ CH_4O\text{(aq)} = \dfrac{\text{mol } CH_4O}{1 \text{ L solution}}$

2. *What is known?* $\quad$ 18.3 g CH_4O, total volume = 58.6 mL, $\dfrac{32.0 \text{ g } CH_4O}{1 \text{ mol}}$

Since molarity is the ratio of moles of solute, CH_4O, to liters of solution, we must find the number of moles of CH_4O and then divide that by the number of liters of solution (see Figure 15.14).

3. Find moles of solute and liters of solution.

$$\text{Mol } CH_4O = 18.3 \text{ g } CH_4O \times \frac{1 \text{ mol } CH_4O}{32.0 \text{ g } CH_4O} = 0.572 \text{ mol } CH_4O$$

$$\text{L solution} = 58.6 \text{ mL} \times \frac{1 \text{ L}}{1000 \text{ mL}} = 0.0586 \text{ L}$$

FIGURE 15.14
To find the molarity of a solution given the mass of solute and total volume of the solution, convert the mass of solute to moles and divide this number of moles by the total volume of solution.

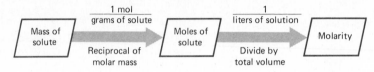

$$M_{CH_4O} = \frac{0.572 \text{ mol CH}_4\text{O}}{0.0586 \text{ L}}$$

$$= 9.76 \text{ } M \text{ CH}_4\text{O}$$

When 18.3 g CH_4O is dissolved in enough water to give a total volume of 58.6 mL, the molarity of the solution is 9.76 M.

At times, the concentration of a laboratory stock solution may not be given in the units that you want. Therefore, these units must be converted to the desired concentration units. Example Problem 15.6 shows how percent by mass is converted to molarity.

Example Problem 15.6

What is the molarity of an 85.0% by mass phosphoric acid (H_3PO_4) solution? An 85.0% H_3PO_4 solution has a density of 1.70 g/mL.

Solution

1. *What is unknown?* $\text{M H}_3\text{PO}_4 \left(\dfrac{\text{mol H}_3\text{PO}_4}{\text{L solution}} \right)$

2. *What is known?* 85.0% by mass H_3PO_4, density (solution) = 1.70 g/mL, and molar mass = 98.0 g/mol
To solve this problem, first consider what we are starting with: % by mass, or the ratio of mass of solute per 100 g of solution.

$$85.0\% \text{ H}_3\text{PO}_4 = \frac{85.0 \text{ g H}_3\text{PO}_4}{100 \text{ g solution}}$$

Thus, the mass of the solute should be converted to moles, and then the volume of the solution is calculated from its mass (see Figure 15.15).
3. Determine the number of moles of H_3PO_4 and the volume of solution.

$$\text{mol H}_3\text{PO}_4 = 85.0 \text{ g H}_3\text{PO}_4 \times \frac{1 \text{ mol H}_3\text{PO}_4}{98.0 \text{ g H}_3\text{PO}_4}$$

$$= 0.867 \text{ mol H}_3\text{PO}_4$$

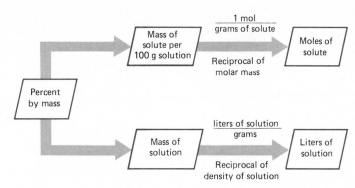

FIGURE 15.15
To change percent by mass of a solution to molarity, calculate the number of moles of solute per 100 g of solution. Then determine the volume in liters of the 100 g of solution, using the density of the solution. Obtain the molarity by dividing the number of moles of solute by the volume of the solution.

$$L \text{ solution} = 100 \text{ g solution} \times \frac{1 \text{ mL}}{1.70 \text{ g solution}} \times \frac{1 \text{ L}}{1000 \text{ mL}}$$

$$= 0.0588 \text{ L}$$

4. Determine the molarity.

$$M \text{ H}_3\text{PO}_4 = \frac{0.867 \text{ mol H}_3\text{PO}_4}{0.0588 \text{ L}}$$

$$= 14.7 \text{ } M \text{ H}_3\text{PO}_4$$

An 85.0% by mass H_3PO_4 solution is 14.7 M, or 14.7 mol H_3PO_4 is dissolved per liter.

Often, more concentrated solutions are diluted to obtain less concentrated solutions. Such dilutions are accomplished after calculating the amount of water that is mixed with the more concentrated solution to give the desired lower concentration. For example, if 100.0 mL of water is mixed with 100.0 mL of 6.00 M NaOH, the concentration of the solution is diluted to 3.00 M, assuming that the volumes are additive (100.0 mL + 100.0 mL = 200.0 mL). Initially,

When two liquids or solutions are poured together, their total volume does not always equal the sum of the individual volumes.

0.600 mol NaOH $\left(0.1000 \text{ L} \times \dfrac{6.00 \text{ mol NaOH}}{\text{L}}\right)$ is dissolved in 100.0 mL.

By increasing the total volume to 200.0 mL (doubling the volume), the concentration becomes half the original concentration,

$$\frac{0.600 \text{ mol NaOH}}{0.2000 \text{ L}} = 3.00 \text{ } M \text{ NaOH}$$

See Figure 15.16.
Example Problem 15.7 illustrates a dilution problem.

FIGURE 15.16
If 100.0 mL of 6.00 *M* NaOH is mixed with 100.0 mL of water, the concentration of the diluted solution becomes 3.00 *M* NaOH because the total volume of the solution doubles. The same number of moles of solute, NaOH, is contained in a volume twice as large as the initial volume.

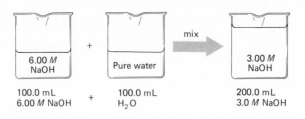

Example Problem 15.7

When diluting concentrated acids, never add water to the acid. The acid is likely to splatter as water hits it. Instead, **always** add concentrated acids to water.

What volume of water is required to dilute 235.0 mL of 12.00 M HCl to 1.000 M HCl (assume that the volumes are additive)?

Solution

1. *What is unknown?*   mL of water
2. *What is known?*   235.0 mL of 12.00 M HCl to be diluted to 1.000 M HCl
   First, calculate the number of moles of HCl present in 235.0 mL of 12.00 M HCl. Knowing the number of moles of HCl, we can determine the total volume of the diluted solution (1.000 M HCl). Knowing the final volume and the initial volume, we can obtain the volume of water for the dilution.
3. Find the number of moles of HCl.

$$\text{mol HCl} = 235.0 \; \text{mL} \times \frac{1 \, \text{L}}{1000 \; \text{mL}} \times \frac{12.00 \; \text{mol HCl}}{1 \, \text{L}}$$

$$= 2.820 \; \text{mol HCl}$$

4. Find the total volume of 1.000 M HCl containing 2.820 mol HCl.

$$\text{Volume of 1.000 M HCl} = 2.820 \; \text{mol HCl} \times \frac{1 \, \text{L}}{1.000 \; \text{mol HCl}}$$

$$\times \frac{1000 \; \text{mL}}{1 \, \text{L}}$$

$$= 2.820 \times 10^3 \; \text{mL}$$

5. We have computed that the initial 235.0 mL of 12.00 M HCl should be diluted to 2.820×10^3 mL in order to lower the concentration to 1.000 M HCl. Accordingly, the difference between 2.820×10^3 mL and 235.0 mL is the volume of water needed for the dilution.

$$\text{Volume of H}_2\text{O} = 2820 \; \text{mL} - 235.0 \; \text{mL} = 2585 \; \text{mL H}_2\text{O}$$

Properly mixing 2585 mL water with 235.0 mL of 12.00 M HCl dilutes the solution to 1.000 M HCl.

REVIEW PROBLEMS

15.12 Explain how 375 mL of 1.5 M NaOH(aq) is prepared.

15.13 What mass of $Mg(NO_3)_2$ is required to prepare 3.6 L of 0.52 M $Mg(NO_3)_2$(aq)?

15.14 Determine the molarity of a solution that is prepared by dissolving 4.22 g C_2H_6O in enough water to have a total volume of 176 mL.

15.15 What is the molarity of a 90.0% by mass formic acid (CH_2O_2) solution? A 90.0% formic acid solution has a density of 1.20 g/mL.

15.16 Glacial acetic acid, $HC_2H_3O_2$, is 99.8% acetic acid. If glacial acetic acid has a molar concentration of 17.4 M, calculate the amount of water that must be mixed with 275 mL of glacial acetic acid to produce a 6.00 M acetic acid solution. Assume the volumes are additive.

15.3 ELECTROLYTES AND NONELECTROLYTES

Solutions can also be classified according to the solute's behavior after it enters the solvent. When placed into water, the crystal structure of ionic compounds is broken apart and the ions enter solution.

$$M^+X^-(s) \xrightarrow{\text{water}} M^+(aq) + X^-(aq)$$

Jacobus van't Hoff (1852–1911) was the first to show that solutions can conduct an electric current. He advanced chemistry by showing that the solute particles in a solution follow laws similar to those governing gas molecules.

Experimentally, the existence of ions in solution is verified using a conductivity apparatus which is nothing more than a light bulb wired to a source of electricity and two electrodes. The light bulb will glow only if there is an electric conductor between the two electrodes. When the electrodes of the conductivity apparatus are immersed in a solution containing dissolved ionic substances, as in Figure 15.17, the light bulb glows. This shows that dissolved ions act as electric current carriers and complete the electric circuit. Solutes that produce solutions that conduct electric currents are called **electrolytes.** Depending on the dissolved solutes, some solutions (those with higher concentrations of dissolved ions) con-

FIGURE 15.17
A conductivity apparatus can be constructed by attaching electrodes to a light bulb and power source in such a manner that the light bulb will light only if there is a solution that conducts an electric current between the electrodes.

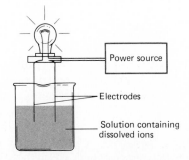

TABLE 15.3 ELECTROLYTES

Strong electrolytes	Weak electrolytes	Nonelectrolytes
HCl	HF	Water
NaCl	H_2S	Ethanol
KNO_3	HNO_2	Sugars
HNO_3	HCN	O_2
NaOH	$Ca(OH)_2$	Acetone

duct large electric currents, and others (those with lower dissolved ion concentrations) conduct smaller electric currents.

We further classify dissolved ionic substances as strong and weak electrolytes. **Strong electrolytes** are those that allow a large electric current to flow through water, indicating that a large percent of the dissolved solute particles are ions. **Weak electrolytes** allow only a small current to flow through water, indicating that a small percent of the solute is broken up into ions. Table 15.3 lists examples of strong and weak electrolytes.

Ionic solids are generally strong electrolytes. Covalent substances that are water soluble are normally weaker electrolytes or nonelectrolytes (except strong acids). A **nonelectrolyte** is a substance that does not ionize or ionizes to a very small extent when dissolved. Water itself is a nonelectrolyte; a light bulb in a conductivity apparatus does not glow when the electrodes are immersed in pure water. Table 15.3 also lists nonelectrolytes.

Ionic equations represent what happens when ionic substances dissolve. Up to this point, when a substance dissolved in water, we just wrote (aq) next to the formula. An **ionic equation** includes all of the dissolved ionic species in a chemical reaction. An ionic equation representing what happens when solid sodium chloride, NaCl(s), dissolves is as follows:

Nearly 100% of a strong electrolyte exists as ions in solution. Weak electrolytes are only partially ionized in solution.

$$Na^+Cl^-(s) \xrightarrow{\text{water}} Na^+(aq) + Cl^-(aq)$$

Sodium chloride, a strong electrolyte, dissolves in water, producing hydrated sodium cations and chloride anions.

Hydrogen chloride gas, HCl(g), when dissolved in water also ionizes totally; therefore, HCl is a strong electrolyte. In this case the polar covalent HCl molecule ionizes, producing $H^+(aq)$ and $Cl^-(aq)$.

$$HCl(g) \xrightarrow{\text{water}} H^+(aq) + Cl^-(aq)$$

Only a small percent of the dissolved particles in solutions of weak electrolytes are broken up into ions. The ionization of nitrous acid, HNO_2, a weak electrolyte, is represented as follows.

$$HNO_2(aq) \rightleftharpoons H^+(aq) + NO_2^-(aq)$$

A dynamic equilibrium results when the rate at which HNO_2 dissociates to its ions equals the rate at which the ions combine to re-form HNO_2.

Two single-barbed arrows indicate an equilibrium system; the small arrow pointing to the right indicates that only a small number of the HNO_2 molecules are ionized.

If a chemical reaction occurs in solution, ionic equations indicate what happens. Let's consider the combination of NaCl(aq) and $AgNO_3$(aq) solutions. When these two solutions are mixed, a white insoluble solid (a precipitate), AgCl(s), forms. The overall ionic equation for the reaction is

$$Na^+(aq) + Cl^-(aq) + Ag^+(aq) + NO_3^-(aq) \longrightarrow$$
$$AgCl(s) + Na^+(aq) + NO_3^-(aq)$$

On the left side of the equation, all chemical species are written in ionic form, since both NaCl and $AgNO_3$ are strong electrolytes. After the solutions combine, only Na^+ and NO_3^- remain in solution, and silver chloride, AgCl, precipitates out of solution as a result of its low solubility (see Table 10.4).

When NaCl and $AgNO_3$ solutions combine, the Na^+(aq) and NO_3^-(aq) are unchanged in the reaction; they appear as both reactants and products. We call the ions that do not change **spectator ions** since they "watch" but do not take part in the chemical change. If the spectator ions are eliminated from the overall ionic equation, the **net ionic equation** results.

Spectator ions are unchanged in aqueous ionic reactions. Thus, they are canceled out to obtain the net ionic equation.

$$\cancel{Na^+}(aq) + Cl^-(aq) + Ag^+(aq) + \cancel{NO_3^-}(aq) \longrightarrow$$
$$AgCl(s) + \cancel{Na^+}(aq) + \cancel{NO_3^-}(aq)$$
Overall ionic equation

$$Ag^+(aq) + Cl^-(aq) \longrightarrow AgCl(s)$$
Net ionic equation

Another example of an ionic reaction is the combination of HCl and NaOH solutions. Overall and net equations for this neutralization reaction are as follows:

$$H^+(aq) + Cl^-(aq) + Na^+(aq) + OH^-(aq) \longrightarrow H_2O(l) + Na^+(aq) + Cl^-(aq)$$
Overall ionic equation

$$H^+(aq) + OH^-(aq) \longrightarrow H_2O(l)$$
Net ionic equation

Both HCl and NaOH are strong electrolytes, so they exist as ions in solution. When they combine, the H^+ from HCl and the OH^- from NaOH combine to yield liquid water. The remaining ions, Na^+ and Cl^-, are spectator ions and are unchanged after the reaction.

Example Problem 15.8

Write the overall and net ionic equations for the following reaction:

$$(NH_4)_2S(aq) + FeBr_2(aq) \longrightarrow FeS(s) + 2\ NH_4Br(aq)$$

Solution

1. Overall ionic equation.

$$2NH_4^+(aq) + S^{2-}(aq) + Fe^{2+}(aq) + 2Br^-(aq) \longrightarrow$$
$$FeS(s) + 2NH_4^+(aq) + 2Br^-(aq)$$

All chemical species are dissolved and ionized except for the precipitate, FeS.

2. Net ionic equation. Identify the spectator ions, NH_4^+ and Br^-, and eliminate them from the overall equation. This gives the net ionic equation.

$$\cancel{2NH_4^+}(aq) + S^{2-}(aq) + Fe^{2+}(aq) + \cancel{2Br^-}(aq) \longrightarrow$$
$$FeS(s) + \cancel{2NH_4^+}(aq) + \cancel{2Br^-}(aq)$$

$$S^{2-}(aq) + Fe^{2+}(aq) \longrightarrow FeS(s)$$
Net ionic equation

REVIEW PROBLEMS

15.17 How is a solute identified as being an electrolyte or nonelectrolyte?

15.18 Give an example of: (a) a nonelectrolyte, (b) a weak electrolyte, (c) a strong electrolyte.

15.19 Write the overall ionic equation for the reactions where KOH(aq) solutions are reacted with: (a) HNO_3(aq), (b) H_2SO_4(aq), (c) H_3PO_4(aq).

15.4 PROPERTIES OF SOLUTIONS

Osmotic pressure is another colligative property. It concerns the ability of solvents to pass through semipermeable membranes.

Once a solute is dissolved in a solvent, the properties of the resulting solution differ from those of the solute and solvent. Solutions exhibit special properties, depending on the concentration of dissolved solute particles. These special properties are called colligative properties.

A **colligative property** is one that is directly related to the number of dissolved solute particles in the solvent. To a large extent, colligative properties are independent of the nature of the solute; they depend only on the concentration

FIGURE 15.18
In a pure solvent, all the surface molecules are solvent molecules. In a solution containing a nonvolatile solute, both solute and solvent molecules occur on the surface. The solute molecules decrease the number of solvent surface molecules, which decreases the number of solvent molecules that can evaporate.

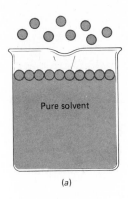

(a)

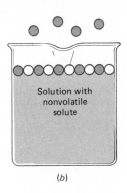

(b)

of dissolved solute particles. Three colligative properties are considered in this section: (1) vapor pressure, (2) boiling point, and (3) freezing point.

Vapor Pressure

The lowering of the vapor pressure of solutions was first proposed by Francois Marie Raoult (1830–1901). His work with solutions led directly to the development of the theory of dissociation of solutes in solution.

In Chapter 13, we discussed the equilibrium that exists between the evaporation of liquid molecules and the condensation of vapor molecules above a liquid in a closed container. A measure of the amount of vapor above the liquid is called the vapor pressure of the liquid. You may want to reread Section 13.1 for review.

At a specific temperature, each liquid has a fixed vapor pressure. In a pure liquid, all of the surface molecules are that of the liquid. However, a solution would contain both solute and solvent molecules on the surface (see Figure 15.18). Decreasing the number of surface solvent molecules decreases the rate of evaporation of the solvent, which decreases the vapor pressure. The vapor pressure of a solution containing a nonvolatile solute is always less than the vapor pressure of the pure solvent.

Boiling Point

Addition of a nonvolatile solute to a liquid increases the liquid's boiling point.

Molality (*m*) is the number of moles of solute per kilogram of solvent.

A liquid's **boiling point** is the temperature at which the vapor pressure of the liquid equals the pressure of the gases above the liquid. **Normal boiling point** is defined as the temperature at which the vapor pressure of a liquid equals one atmosphere. Water's normal boiling point is 100°C.

The addition of nonvolatile solute molecules to a solvent decreases the number of surface solvent molecules, and thus lowers the vapor pressure. After the addition of a nonvolatile solute, water's vapor pressure is lower than 1 atm at 100°C. The solution's temperature would have to be increased above 100°C in order to increase the vapor pressure to 1 atm. Accordingly, the boiling point of the solution is raised with the addition of nonvolatile solute particles.

Higher concentrations of solute particles result in higher boiling points. An increase in a solution's boiling point is directly related to the molality of the solution. **Molality** is the concentration unit that is defined as the number of moles of solute per kilogram of solvent.

A solution's molality is determined by first calculating the number of moles of solute and then dividing by the mass of the solvent expressed in kilograms.

$$\text{Molality } (m) = \frac{\text{moles of solute}}{\text{kilograms of solvent}}$$

Example Problem 15.9 shows how a solution's molality is calculated.

Example Problem 15.9

A solution is prepared by dissolving 105 g of ethylene glycol, $C_2H_6O_2$, in 649 g of water. What is the molality of the solution?

Solution

1. *What is unknown?*   Molality $\left(\dfrac{\text{mol } C_2H_5O_2}{\text{kg water}} \right)$

2. *What is known?*   105 g $C_2H_6O_2$, 649 g water, $\dfrac{62.0 \text{ g } C_2H_6O_2}{1 \text{ mol } C_2H_6O_2}$

3. Apply the factor-label method to find moles of $C_2H_6O_2$ and kilograms of H_2O (see Figure 15.19)

$$\text{mol } C_2H_6O_2 = 105 \text{ g } C_2H_6O_2 \times \frac{1 \text{ mol } C_2H_6O_2}{62.0 \text{ g } C_2H_6O_2}$$

$$= 1.69 \text{ mol } C_2H_6O_2$$

$$\text{kg } H_2O = 649 \text{ g } H_2O \times \frac{1 \text{ kg}}{1000 \text{ g}} = 0.649 \text{ kg } H_2O$$

4. Find molality.

$$m = \frac{1.69 \text{ mol } C_2H_6O_2}{0.649 \text{ kg } H_2O}$$

$$= 2.61 \ m$$

m is the symbol for molality; *M* is the symbol for molarity.

When 105 g of ethylene glycol is mixed with 649 g of water, the solution's molality is 2.61 *m*.

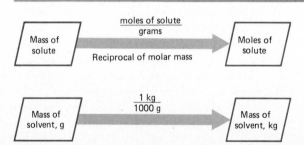

FIGURE 15.19
To find the molality of a solution, determine the number of moles of solute and divide by the mass of the solvent expressed in kilograms.

TABLE 15.4   MOLAL FREEZING AND BOILING POINT CONSTANTS
FOR SELECTED SOLVENTS

Solvent	Boiling point, °C	K_b, °C/m	Freezing point, °C	K_f, °C/m
Water, H_2O	100	0.512	0	1.86
Acetic acid, $HC_2H_3O_2$	118	3.1	16.6	3.9
Benzene, C_6H_6	80	2.5	5.5	5.1
Nitrobenzene, $C_6H_5NO_2$	211	5.2	5.7	8.1
Phenol, C_6H_6O	182	3.6	41	7.4

At what temperature would the ethylene glycol solution from Example Problem 15.10 boil? **Boiling point elevation** is calculated using the following expression:

$$\Delta T_b = K_b \times m$$

where ΔT_b is the elevation of the boiling point above the normal boiling point, K_b is the molal boiling point elevation constant, and m is the solution's molality. We have already calculated the solution's molality, and now must obtain the boiling point elevation constant. Table 15.4 contains the boiling point elevation constants for selected solvents. A **boiling point elevation constant** gives the ratio of increase in boiling point per 1 molal solution. Water's K_b value is $0.512°C/m$, or a 1.00 m solution (if the solute is nonvolatile and a nonelectrolyte) boils $0.512°C$ above the normal boiling point of water ($100°C$). A 2.00 m solution would boil 2.00 $m \times 0.512°C/m$, $1.02°C$ above the normal boiling point. Thus, the increase in boiling point of the solution from Example Problem 15.9 is calculated by multiplying 2.61 $m \times 0.512°C/m$.

$$\Delta T_b = K_b \times m$$
$$= \frac{0.512°C}{m} \times 2.61 \ m$$
$$= 1.34°C$$

A 2.61 m aqueous solution boils $1.34°C$ above the normal boiling point, or at $101.34°C$ ($100.00°C + 1.34°C$).

Freezing Point   At the freezing point, the vapor pressures of the solid and liquid phases of a substance are the same. The addition of a solute to a solvent lowers the vapor pressure of the liquid phase more than it does that of the solid phase. Consequently, in a solution the liquid freezes at a lower temperature, since the vapor pressure of the solid phase equals the vapor pressure of the liquid phase at a lower temperature.

Freezing point depression is determined much like boiling point elevation.

378 • SOLUTIONS

We multiply the molality of the solution by a constant, K_f, the **molal freezing point depression constant.**

$$\Delta T_f = K_f \times m$$

Table 15.4 presents K_f values for various solvents. Water's K_f value is 1.86°C/m. A 1.00 m solution that contains a nonvolatile, nonelectrolyte solute would freeze at -1.86°C.

$$\Delta T_f = \frac{1.86°C}{m} \times 1.00 \ m = 1.86°C$$

Camphor has a K_f value of 40°C/m. A 1 m camphor solution freezes 40°C below camphor's freezing point, 176°C.

and

$$0.00°C - 1.86°C = -1.86°C$$

Example Problem 15.10 illustrates a freezing point depression problem.

Example Problem 15.10

Ethylene glycol's structural formula is

Ethylene glycol is a common ingredient in antifreeze solutions. It is extremely toxic.

What is the freezing point of a solution prepared by adding 95.0 g of ethylene glycol, $C_2H_6O_2$, to 365 g of water?

Solution

1. *What is unknown?* °C, freezing point of solution

2. *What is known?* 95.0 g $C_2H_6O_2$, 365 g of water, $\dfrac{62.0 \text{ g } C_2H_6O_2}{1 \text{ mol } C_2H_6O_2}$,

$K_f = 1.86°C/m$, freezing point (water) = 0.00°C
Depression of the freezing point is a function of the molality of the solution. Thus, the molality should be determined first, and then the molality is multiplied by the K_f to obtain the freezing point depression.

3. Determine the solution's molality.

$$m = \frac{\text{mol } C_2H_6O_2}{\text{kg water}}$$

$$\text{mol } C_2H_6O_2 = 95.0 \text{ g } \cancel{C_2H_6O_2} \times \frac{1 \text{ mol } C_2H_6O_2}{62.0 \text{ g } \cancel{C_2H_6O_2}}$$

$$= 1.53 \text{ mol } C_2H_6O_2$$

$$\text{kg water} = 365 \text{ g water} \times \frac{1 \text{ kg water}}{1000 \text{ g water}} = 0.365 \text{ kg water}$$

$$m = \frac{1.53 \text{ mol } C_2H_6O_2}{0.365 \text{ kg water}} = 4.20 \ m$$

4. Determine the solution's freezing point.

$$\Delta T_f = K_f \times m$$
$$= \frac{1.86°C}{\not{m}} \times 4.20 \ \not{m}$$
$$= 7.81°C$$

Our calculation has determined that the freezing point is lowered by 7.81°C. Water's freezing point is 0.00°C; hence, 7.81°C is subtracted from 0.00°C, giving −7.81°C as the solution's freezing point.

$$\text{Freezing point} = 0.00°C − 7.81°C$$
$$= −7.81°C$$

A solution prepared by adding 95.0 g of ethylene glycol to 365 g of water freezes at −7.81°C.

REVIEW PROBLEMS

15.20 What determines if a property of a solution is a colligative property or not?

15.21 Explain why the addition of a nonvolatile, nonelectrolyte solute to water lowers the water's vapor pressure.

15.22 What are the expected boiling and freezing points of a 4.0 m aqueous solution?

15.23 Determine the molality of a solution prepared by dissolving 63 g CH_4O in 321 g of water.

15.24 What is the boiling point of a solution that contains 175 g $C_2H_6O_2$ in 2.80 kg of water?

SUMMARY

Solutions are homogeneous mixtures. A solution is composed of a solute, the component in lesser amount, dissolved in a solvent, the component in larger amount. Solute molecules become incorporated into the structure of the solvent.

Solutions are classified according to the physical states of the solute and solvent. Solid-liquid, liquid-liquid, gas-liquid, gas-gas solutions are the classes of solutions most frequently encountered.

Solubility is the extent to which a solute dissolves in a solvent. Solutions that have larger quantities of solute dissolved in the solvent are called concentrated solutions. Those with a small quantity of solute dissolved in a solvent are called dilute solutions. If no more solute can dissolve in a solvent, without changing the conditions, the solution is said to be saturated. If any quantity of solute less than the amount needed to saturate the solution is present, the solution is said to be unsaturated.

There are three primary factors that influence a solute's solubility: (1) the nature of solute and solvent, (2) temperature, and (3) pressure. Generally, substances that have similar structures and intermolecular forces are more soluble in each other than those that are different: like dissolves like. On an average, higher temperatures result in higher solubilities. Pressure affects mainly gaseous solutions. Higher pressures of gases above solvents result in a greater gas solubility.

Four factors affect the rate at which a solute dissolves in a solvent: (1) particle size, (2) temperature, (3) concentration, and (4) stirring. A decrease in particle size increases surface area, and thus increases the rate of dissolving. Higher temperature increases the dissolving rate. The higher the concentration of the solute, the slower a solute dissolves. Stirring solutes increases the rate of dissolving by lowering the concentration of surrounding dissolved solute particles.

Solution concentrations are measured using many different units. Concentration units used frequently in chemistry are (1) molarity, moles of solute per liter of solution, (2) percent by mass, mass of solute per 100 g of solution, and (3) molality, moles of solute per kilogram of solvent.

Solutions containing dissolved ions are called electrolyte solutions; they conduct an electric current. If a large quantity of the solute ionizes when dissolved, the solute is classified as a strong electrolyte—it conducts a large electric current. Solutes that form few ions in solution are called weak electrolytes, and nonelectrolytes are those that essentially produce no ions in solution.

Solutions exhibit special properties that depend on the concentration of dissolved solute particles. These special properties are called colligative properties. Three of the colligative properties are (1) vapor pressure, (2) boiling point, and (3) melting point. As the solute concentration increases, a solution's vapor pressure decreases, its boiling point increases, and its freezing point decreases.

QUESTIONS AND PROBLEMS

15.25 Define the following terms: solution, solute, solvent, dissolving, miscible, immiscible, solubility, hydration energy, lattice energy, dilute, concentrated, saturated solution, unsaturated solution, concentration unit, dilution, strong electrolyte, weak electrolyte, nonelectrolyte, overall ionic equation, net ionic equation, colligative property.

Solutions

15.26 Identify the solute and solvent:

(a) 1 L of water and 1 g NaCl
(b) 1 L of alcohol and 50 mL of water
(c) 1 L of alcohol and 1 L of water
(d) 1 L of water and 1 mL $O_2(g)$

15.27 (a) How can the NaCl in an aqueous NaCl solution be recovered? (b) Can the same method be used to isolate alcohol from an alcohol and water solution?

15.28 What is the difference between a substance's solubility in a solvent and its dissolving rate?

15.29 How are immiscible liquids distinguished from those that are miscible?

15.30 How could you test a solution to decide if the solution is saturated or unsaturated?

15.31 Classify the following solutions according to the physical state of the solute and solvent: (a) sugar water, (b) air, (c) brass, (d) coffee, and (e) vinegar.

15.32 In which liquid of each of the following pairs would you expect potassium chloride, KCl, an ionic solid, to be more soluble? Explain your predictions.
(a) H_2O or CCl_4?
(b) CH_3OH or $CH_3CH_2CH_2CH_2CH_3$?
(c) $CH_3CH_2CH_2CH_2CH_3$ or CH_3OCH_3?

Percent by Mass

15.33 How are each of the following solutions % m/m prepared?
(a) 451 g of 3.44% $NaNO_3$(aq)
(b) 70.0 g of 21.4% KOH(aq)
(c) 8.1 kg of 0.55% NH_4Cl(aq)

15.34 Determine the mass of concentrated HCl(aq) (37% m/m) that contains the following masses of HCl: (a) 3.7 g, (b) 0.92 g, (c) 2.1 kg, (d) 67 mg.

15.35 Concentrated ammonia, NH_3, is sold as a 29% m/m NH_3(aq) solution. Its density is 0.90 g/mL. What mass of NH_3 is contained in the following volumes of concentrated ammonia solutions: (a) 1.0 L, (b) 59 mL, (c) 0.025 mL?

15.36 What is the maximum total mass of a 5.4% m/m KI solution that could be prepared from 103 g KI(s)?

15.37 Determine the mass of water in 92.5 g of 8.3% m/m Na_2SO_4 solution.

15.38 A solution is prepared by dissolving 26.5 g of solute in 306 g of water. What is the concentration of the solution in % m/m?

15.39 (a) Initially, 275.0 g of a 9.10% m/m ammonium nitrate solution is contained in a beaker. What is the concentration of the solution after 55.0 g of water are added? (b) What mass of ammonium nitrate solid must be added to the diluted solution to change the concentration back to the original concentration of 9.10%?

Molarity

15.40 Explain how the following solutions are prepared:
(a) 100.0 mL of 0.455 M $Ca(NO_3)_2$
(b) 500.0 mL of 1.09 M NH_3
(c) 3.25 L of 0.915 M $C_6H_{12}O_6$
(d) 83.2 mL of 2.89 M H_3PO_4

15.41 What are the molarities of solutions prepared by dissolving the following amounts of solute in enough water to yield 625 mL total volume?
(a) 2.9 mol NH_4NO_3
(b) 0.771 mol KNO_3
(c) 9.31 g H_2SO_4
(d) 1.003 mg HI

15.42 Find the molarity of the following solutions: (a) 70% m/m HNO_3, density 1.42 g/mL, (b) 36% acetic acid, $C_2H_4O_2$, density 1.045 g/mL.

15.43 What volume of water should be mixed with 1.00 L of 15.9 M HNO_3 to dilute this solution to: (a) 14.9 M HNO_3, (b) 1.00 M HNO_3, (c) 0.019 M HNO_3?

15.44 Determine the number of moles of solute particles in the following solutions:
(a) 417 mL of 0.100 M CsOH
(b) 53.8 mL of 0.296 M HBr
(c) 8.98 L of 5.70 M $NaNO_3$

15.45 What is the molar concentration of a solution that contains 118 g $AgNO_3$ per liter of solution?

15.46 A 27% m/m H_2SO_4 solution has a density of 1.2 g/mL. Find the molar concentration of the solution.

15.47 To what total volume would 90.0 mL of 1.50 M HI solution have to be diluted to produce a 0.244 M HI solution?

15.48 (a) What volume of 4.55 M NaOH is required to produce 0.935 L of 0.500 M NaOH? (b) Explain how this solution is prepared.

Electrolytes

15.49 Describe the expected light intensity (bright, dim, etc.) observed when the following solutions are placed between the electrodes in a conductivity apparatus:
(a) 0.1 M NaCl(aq)
(b) 0.1 M HF(aq)
(c) pure water
(d) 0.1 M ethanol, C_2H_6O(aq)
(e) 0.1 M $Ca(OH)_2$(aq)

15.50 Hydrofluoric acid, HF(aq), is a weak electrolyte. Compare the relative amounts of un-ionized HF(aq) to that of dissolved F^-(aq) and H^+(aq).

15.51 Write an ionic equation indicating what happens when the following strong electrolytes are dissolved in water: (a) KCl(s), (b) $NaHCO_3$(s), (c) $Pb(NO_3)_2$(s), (d) CaI_2(s), (e) $MgCrO_4$(s).

15.52 Write balanced overall and net ionic equation for:

(a) $LiBr(aq) + AgNO_3(aq) \longrightarrow$
$$LiNO_3(aq) + AgBr(s)$$

(b) $Na_3PO_4(aq) + NiCl_2(aq) \longrightarrow$
$$Ni_3(PO_4)_2(s) + NaCl(aq)$$

(c) $(NH_4)_2S(aq) + CoSO_4(aq) \longrightarrow$
$$(NH_4)_2SO_4(aq) + CoS(s)$$

(d) $H_2SO_4(aq) + KOH(aq) \longrightarrow$
$$H_2O(l) + KHSO_4(aq)$$

(e) $Hg_2(NO_3)_2(aq) + CaBr_2(aq) \longrightarrow$
$$Hg_2Br_2(s) + Ca(NO_3)_2(aq)$$

15.53 Write the balanced net ionic equations (if any) for the following:
(a) $Ba(NO_3)_2(aq) + Na_3PO_4(aq) \longrightarrow$
(b) $NH_4C_2H_3O_2(aq) + Pb(NO_3)_2(aq) \longrightarrow$
(c) $Na_2CO_3(aq) + FeBr_2(aq) \longrightarrow$
(d) Potassium chloride(aq) + silver nitrate(aq) $\longrightarrow$
(e) Magnesium acetate(aq) + sodium sulfide(aq) $\rightarrow$

Molality

15.54 Find the molality of the following solutions (water's density is 1.00 g/mL):
(a) 9.9 g $NaNO_3$ in 949 g of water
(b) 45.2 g $CaCl_2$ in 1.63 kg of water
(c) 0.831 mol K_2SO_4 in 613 g of water
(d) 8.9 g NaI in 598 mL of water

15.55 Determine the molality of the following solutions:
(a) 3.4% m/m $C_6H_{12}O_6(aq)$
(b) 20% m/m $HCl(aq)$
(c) 51.5% m/m $HClO_4(aq)$

15.56 Explain how the following molal solutions are prepared:
(a) 0.094 m $Cu(NO_3)_2$ in 893 g H_2O
(b) 1.7 m $C_{12}H_{22}O_{11}$ in 6.91 kg H_2O

15.57 (a) Determine the molality of 1.54 g $I_2(s)$ dissolved in 97.7 g CCl_4.
(b) What is the % m/m of this solution?

Properties of Solutions

15.58 Explain specifically why the boiling point of a solution containing a nonvolatile solute is higher than that of the pure solvent.

15.59 A solution's vapor pressure is greater than that of the pure solvent. Write an explanation to account for this seemingly contradictory behavior.

15.60 Determine the boiling point of aqueous solutions with the following molalities (assume all solutes to be nonvolatile and nonelectrolytes): (a) 0.33 m, (b) 5.9 m, (c) 3.81 m.

15.61 What are the freezing points of the aqueous solutions in Problem 15.60?

15.62 Using Table 15.4, determine the elevation of the boiling points of 3.5 m solutions in: (a) water, (b) acetic acid, (c) benzene, (d) nitrobenzene, (e) phenol.

15.63 What are the boiling points of solutions in which benzene is the solvent with the following molalities (see Table 15.4): (a) 6.6 m, (b) 0.92 m, (c) 8 m?

15.64 What is the freezing point of each benzene solution in Problem 15.63?

15.65 What are the freezing and boiling points of a solution prepared by adding 13.3 g of ethylene glycol, $C_2H_6O_2$, to 100.0 g of water?

15.66 What are the freezing and boiling points of a solution prepared by adding 25.0 g of carbon tetrachloride, CCl_4, to 125.0 g of benzene, C_6H_6? (See Table 15.4.)

General

15.67 When cooking, what effect does adding salt to water have on the time required to boil foods?

15.68 When we discussed colligative properties, we assumed that the solute was not an electrolyte. If $NaCl(s)$ is added to water, calculate the freezing and boiling point of a 0.500 m $NaCl$ solution.

15.69 Calculate the mass of ethylene glycol, $C_2H_6O_2$, that should be added to 15 kg of water in an auto's radiator to prevent the water from freezing at 0.0°F.

15.70 Two chloride solutions are mixed: 50.0 mL of 0.115 M $NaCl(aq)$ and 25.0 mL of 0.115 M $CaCl_2(aq)$. What is the resultant molar concentration of all three ions in the solution, Na^+, Ca^{2+}, and Cl^-?

15.71 Freezing point depressions are used by chemists to determine the molecular mass of substances. Calculate the molecular mass of a solute if it is known that dissolving 0.25 g of solute in 18 g of water lowers the freezing point of water by 0.20°C.

15.72 Concentrated sulfuric acid, H_2SO_4, is sold as 96% m/m $H_2SO_4(aq)$. Its density is 1.84 g/mL. Calculate (a) the molarity and (b) the molality of the solution. (c) What quantity of water is required to dilute 25 g of the concentrated solution to 0.100 M? (d) What mass of 0.100 M H_2SO_4 reacts with exactly 254 mL of 0.925 M $NaOH(aq)$ to produce sodium sulfate and water? Assume that the density of 0.100 M H_2SO_4 is 1.00 g/mL.

· C H A P T E R ·

16

Reaction Rates and Chemical Equilibrium

Study Guidelines

After completing Chapter 16, you should be able to:
1. Explain what is meant by reaction rate
2. List the three main principles of the collision theory
3. Discuss the factors that determine whether or not a molecular collision is effective
4. List four factors that influence the rates of chemical reactions
5. State the relationship between reaction rate and (a) reactant concentration and (b) temperature
6. Explain the relationships between reaction rate and concentration and reaction rate and temperature in terms of the collision theory
7. Identify and explain the meaning of the rate-determining step for a given reaction mechanism
8. State the one condition, in terms of reaction rates, that is required for the establishment of a chemical equilibrium
9. State whether the reactants or products are favored in an equilibrium, given the value of the equilibrium constant
10. State Le Chatelier's principle
11. Describe effects of changing concentration, pressure, and temperature, and the addition of a catalyst on a chemical equilibrium

Chemical kinetics is the study of the rate with which chemical reactions take place. **Reaction rate** is the speed of a chemical reaction, i.e., how fast the products are formed from the reactants. Within the study of chemical kinetics, chemists attempt to measure accurately the rates of reactions under different conditions, and then try to account theoretically for the observed rates.

Some reactions that occur in the oceans and in the earth's crust require thousands, and in some cases millions, of years to complete.

Some reactions are nearly instantaneous—the reactants totally change to products on contact. Explosions are good examples of instantaneous reactions. Other reactions proceed at such a slow rate that years or centuries may pass before the reaction is completed. Most chemical reactions proceed at rates somewhere in between these two extremes.

Rates of chemical reactions are measured by finding either the decrease in reactant concentration or the increase in product concentration over a specific time interval.

$$\text{Reaction rate} = \frac{\text{change in concentration}}{\text{change in time}}$$

In a reaction with a high reaction rate, the time interval in which the reaction takes place is relatively short. A reaction proceeding at a slower rate requires more time to go to completion. Hence, rates of reactions are inversely proportional to the time required for the reaction to be completed.

Collision Theory

Rates of chemical reactions are theoretically explained by the **collision theory.** The basic premise of the collision theory is that in order for two substances to react the reactant particles must collide with each other. After they collide, two possibilities exist. Either the bonds in the reactant molecules are broken and the bonds in the products are formed, or the particles merely bump into each other and no bonds break and no new bonds form. If a collision occurs that results in the formation of the products, the collision is called an **effective collision,** and if the collision does not yield the products, the collision is called an **ineffective collision.**

For simplicity, let's consider the hypothetical reaction of two diatomic gas molecules, $A_2(g)$ and $X_2(g)$. In Figure 16.1a, one A_2 molecule collides with an X_2 molecule, producing two molecules of AX; here an effective collision has occurred.

$$A_2(g) + X_2(g) \longrightarrow 2AX(g)$$

In Figure 16.1b, one A_2 and one X_2 collide; but after the collision the reactant molecules remain, and no product is produced. This is an ineffective collision.

Two factors primarily determine whether or not a collision is effective: (1) energy and (2) orientation. When two particles collide, they require sufficient energy to break their bonds. For each reaction, there is a minimum energy of collision below which an effective collision cannot occur.

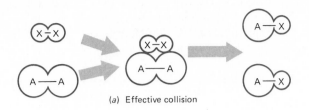

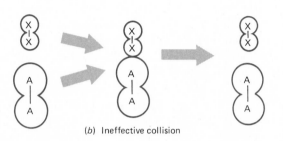

(a) Effective collision

FIGURE 16.1
(a) Molecules of A_2 and X_2 collide in such a manner that their bonds are broken and the bonds of the product, AX, are formed. This is an effective collision. (b) No bonds are broken, and molecules of A_2 and X_2 have not changed after colliding; thus, this is an ineffective collision.

(b) Ineffective collision

An effective collision occurs when the two colliding particles have the proper energy and orientation.

If the two colliding particles have sufficient energy to react, they still must collide with the proper orientation. **Orientation** refers to the alignment of the molecules as they collide. Figure 16.2 illustrates three different collision orientations. In Figure 16.2a, A_2 and X_2, on colliding, are arranged so that each A atom contacts an X atom; consequently, if they collide with the proper amount of energy, they will form a molecule of AX. In Figures 16.2b and 16.2c the orientation of the particles does not bring both A and X atoms together—an effective collision cannot occur, even with the proper amount of energy.

A higher frequency of effective collisions results in a greater reaction rate.

Rates of chemical reactions are directly linked to the frequency of effective collisions. The greater the **frequency of effective collisions,** the faster the reaction proceeds. If only a small percentage of the collisions are effective, the reaction rate is small.

REVIEW PROBLEMS

16.1 What is chemical kinetics?

16.2 Explain how the rate of a chemical reaction is measured.

16.3 List and explain the two primary factors that determine whether a molecular collision is effective or ineffective.

16.4 What is true about a reaction that has a high frequency of effective collisions?

FIGURE 16.2
(a) Proper orientation of colliding molecules to form the products. (b) and (c) Molecular orientations that do not produce the products.

(a)

(b)

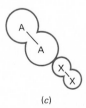

(c)

16.2 FACTORS THAT INFLUENCE REACTION RATES

There are four principal factors that influence the rates of chemical reactions: (1) the nature of the reactants, (2) concentration, (3) temperature, and (4) catalysts. We shall look at each factor individually.

Nature of the Reactants

For reactions in which two or more reactants combine, the degree to which they come in contact must be considered. Normally, the larger the surface area of reactants that are in contact, the greater the reaction rate.

If a constant set of conditions is present, the rates of chemical reactions depend on the molecular properties of the substances that are combining. If F_2 gas is placed into a container at room temperature with H_2, there is an immediate violent reaction. In contrast, if Cl_2 is combined with H_2 under the same conditions, the rate is significantly slower, and at lower temperatures, Cl_2 does not combine with H_2 at all!

The nature of reactant molecules determines whether or not there will be a reaction. As we have seen, reactants must collide before they can react. But once the reacting species collide, the interactions of the electrons in chemical bonds determine whether or not the reaction takes place and, if so, what the rate will be. For example, two oppositely charged ions combine immediately since they rapidly attract each other and it is not necessary for any bonds to be broken. Hydrogen ions immediately combine with hydroxide ions to form water:

$$H^+(aq) + OH^-(aq) \longrightarrow H_2O(l)$$

Reactants with strong covalent bonds are generally less reactive than those with weaker bonds.

Decomposition of molecules with strong bonds proceeds slowly because of the energy required to break the bonds. The decomposition of water,

$$2H_2O \longrightarrow 2H_2 + O_2$$

requires a constant source of electricity or a high temperature.

Concentration

Rates of most chemical reactions are directly proportional to the concentration of the reactants—an increase in reactant concentration produces an increase in the rate of the reaction.

When $H_2(g)$ is combined with $I_2(g)$ to produce $HI(g)$,

$$H_2(g) + I_2(g) \longrightarrow 2HI(g)$$

an increase in the concentration of either H_2 or I_2 increases the rate of reaction. For instance, if the H_2 concentration is doubled, the rate of the reaction is doubled. A similar increase is observed when the I_2 concentration is increased.

Substances that burn slowly in air (air is 20% O_2) sometimes explode when burned in pure oxygen; the increased oxygen concentration speeds up the reaction.

Collision theory explains the effect of concentration changes on reaction rates. If we assume that there are two reactants, an increase in the amount of either reactant increases the overall number of collisions that occurs in a given time interval. Earlier we mentioned that reaction rates depend on the frequency of effective collisions. Increased reactant concentration, therefore, increases the reaction rate by increasing the frequency of effective collisions.

Temperature   As the temperature of a reaction mixture is increased, the rate increases. Reaction rate is directly proportional to temperature. In chemistry laboratories, bunsen burners and hot plates are used to heat substances so they react faster. At home, foods heated to higher temperatures cook faster than those heated at lower temperatures.

At a fixed temperature, molecules have a range of velocities. Some molecules are moving rapidly, and some slowly; however, the largest percentage are moving at velocities near the average velocity. Molecular velocities are directly related to the kinetic energy of the molecules.

Temperature is directly related to the average kinetic energy of molecules. At higher temperatures, the reacting molecules are moving faster; hence, they collide more often and with greater energy. Both factors increase the frequency of effective collisions.

Let's look more closely at the energy requirements of chemical reactions. In Section 10.5, we classified chemical reactions as either exothermic or endothermic. An **exothermic** reaction is one with a net release of energy, and an **endothermic** reaction is one in which energy is absorbed. Graphs of the potential energy relationships in exothermic and endothermic reactions are shown in Figure 16.3.

In an exothermic reaction, the total potential energy stored within the reactants is greater than the potential energy in the products. The chemical potential energy stored in the reactants or products is called **enthalpy.** Consequently, in exothermic reactions the enthalpy (H) of the reactants is greater than the enthalpy of the products. In endothermic reactions, the opposite is true: the enthalpy of the products is greater than the enthalpy of the reactants.

The difference between the enthalpy of products and the enthalpy of the reactants is the heat of reaction ΔH.

FIGURE 16.3
(a) In an exothermic reaction, the enthalpy of the reactants, H_R, is greater than the enthalpy of the products, H_P. Exothermic reactions release energy to the surroundings. (b) In an endothermic reaction, the enthalpy of the reactants, H_R, is less than the enthalpy of the products, H_P. Endothermic reactions absorb energy from the surroundings.

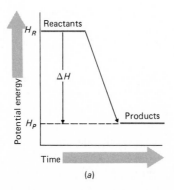

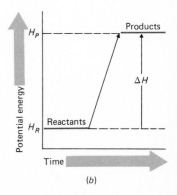

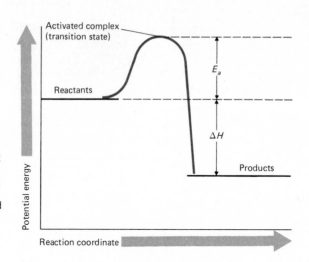

FIGURE 16.4
In an exothermic reaction, initially energy is absorbed to produce the activated complex (transition state). This energy is called the activation energy, E_a. The activated complex breaks apart, forming the products and releasing the activation energy, E_a, and the enthalpy of reaction, ΔH.

$$\Delta H = H_{\text{products}} - H_{\text{reactants}}$$

Every chemical reaction has a characteristic heat of reaction.

As a reaction proceeds from reactants to products, it follows a definite energy pathway. Figure 16.4 illustrates the energy pathway taken in an exothermic reaction. Even though the reaction is exothermic, energy is initially required to start the reaction. The minimum quantity of energy needed to get over the energy "hill" is called the **activation energy** E_a. In all reactions, an amount of energy equal to the activation energy must be present for the reaction to occur.

Activation energy depends on the nature of the reactants. Concentration and temperature have no effect on a reaction's activation energy.

Chemical examples of activation include (1) striking a match, (2) lighting a bunsen burner, and (3) sparking a H_2 and O_2 mixture. In each example, a small input of energy (activation energy) causes a self-sustaining reaction to take place.

The activated complex is a combination of the reacting molecules in which the bonds of the reactants are stretched and almost broken, and the bonds of the products are partially formed.

In terms of the collision theory, as the reactant particles approach each other, their energy increases until they reach the top of the energy hill. This energy peak is referred to as the **transition state** or **activated complex.** At this time, the reactant molecules have interacted, producing an intermediate species that has the proper orientation and energy to break apart into the products or reactants (see Figure 16.5). In other words, the activation energy of a reaction is the amount of energy needed to produce the transition state.

Catalysts

A **catalyst** is a substance that lowers the activation energy of a chemical reaction (see Figure 16.6). With a lower activation energy, a greater percentage of the particles have enough energy to produce the transition state (to have an effective collision). Accordingly, a lower activation energy increases the reaction rate. The catalyst is not consumed during the reaction, and it can be recovered unchanged after the reaction.

Catalysts are grouped into two categories: (1) homogeneous and (2) hetero-

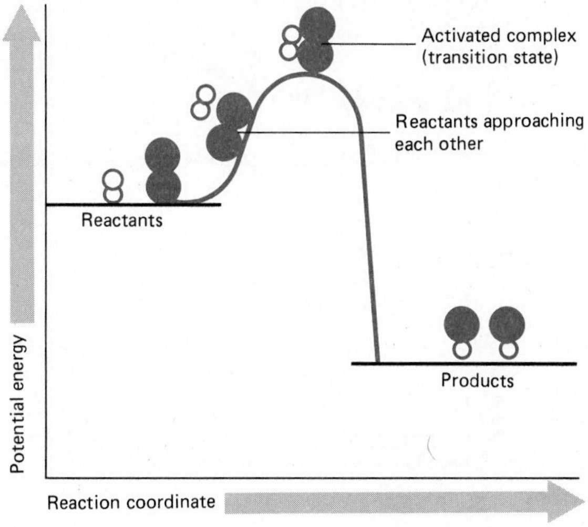

FIGURE 16.5
To form the activated complex (transition state), the reacting molecules must collide in such a way that the bonds in the reactants are partially broken and the bonds in the products are partially formed. Once the activated complex forms, it can become either the reactants or the products.

geneous. A homogeneous catalyst is in the same physical state as the reactants, and a heterogeneous catalyst exists in a different physical state than the reactants.

Reaction Mechanisms

Chemical reactions go through an exact series of steps called the reaction mechanism or reaction pathway.

Chemical reactions normally follow an ordered series of steps called a **reaction mechanism.**

For example, $H_2(g)$ and $I_2(g)$ do not simply collide to produce $HI(g)$, as implied by the equation

$$H_2(g) + I_2(g) \longrightarrow 2HI(g)$$

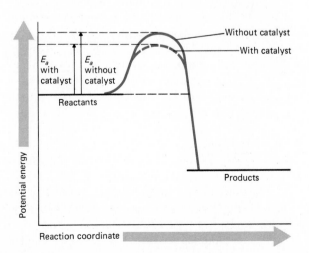

FIGURE 16.6
A catalyst is a substance that lowers the activation energy, E_a, of a chemical reaction. With a lower activation energy, the reaction proceeds at a faster rate than at a higher activation energy.

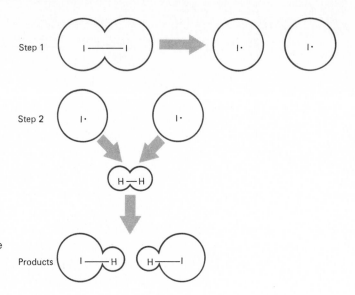

FIGURE 16.7
H_2 and I_2 follow a two-step reaction mechanism to produce HI. In the first step, the bond between the two iodine atoms is broken, yielding two free iodine atoms. In Step 2, the two iodine atoms react with a molecule of H_2, producing the products (two molecules of HI).

Instead, the reaction takes place in two steps: (1) I_2 dissociates to two I atoms, and (2) both I atoms collide with H_2 at the same time to produce two HI molecules. These steps are diagrammed in Figure 16.7.

Step 1: $\qquad\qquad I_2(g) \rightleftharpoons 2I(g) \qquad$ (fast)

Step 2: $\qquad 2I(g) + H_2(g) \longrightarrow 2HI(g) \qquad$ (slow)

Overall: $\qquad H_2(g) + I_2(g) \longrightarrow 2HI(g)$

An overall reaction is analogous to a description of the starting place and destination of a trip. The reaction mechanism is the actual route taken.

The first step, breaking the I_2 bond, occurs more rapidly than the second step. In Step 2, both I atoms collide with the H_2 simultaneously. Collisions involving three particles are slow, since there is a low probability that the three particles will collide with the proper energy and orientation.

The slowest step in a reaction mechanism is classified as the **rate-determining step,** since it determines the maximum rate of the reaction. Production of $HI(g)$ depends on the second step. No matter how fast iodine atoms are produced, HI formation depends on the slow second step. We might say that there is a "bottleneck" at this step.

An analogy to the rate-determining step is the speed at which you proceed down a single-lane highway. Your car might be capable of traveling at very high speeds; however, your speed is controlled by the slowest-moving vehicle in front of you.

In Step 1, individual iodine atoms are formed, which are then consumed in the second step. Chemical species that are in the reaction mechanism but not in the overall reaction are called reaction **intermediates.** Most intermediates are reactive and do not generally occur by themselves; such is the case of isolated iodine atoms.

REVIEW PROBLEMS

16.5 List the primary factors that influence the rates of chemical reactions.

16.6 Describe what happens to the rate of a reaction when the following changes are made: (a) concentration of the reactants is decreased, (b) a catalyst is added, (c) temperature is decreased.

16.7 (a) What is the activation energy of a reaction? (b) Will an exothermic or endothermic reaction have a greater activation energy on an average? Explain.

16.8 What is a chemical catalyst?

16.9 Consider the proposed mechanism for the decomposition of ozone, $2O_3 \rightarrow 3O_2$.

Step 1: $\qquad O_3 \rightleftharpoons O_2 + O \qquad$ (slow)

Step 2: $\quad O + O_3 \longrightarrow 2O_2 \qquad$ (fast)

(a) Which step is the rate-determining step? Explain. (b) What is the reaction intermediate in the mechanism?

16.3 CHEMICAL EQUILIBRIUM SYSTEMS

We have already encountered a physical equilibrium system, the equilibrium established by a liquid and its vapor in a closed container (evaporation-condensation). Let's review this system before discussing chemical equilibrium systems.

In the evaporation-condensation equilibrium system, an equilibrium is established when the rate of evaporation equals the rate of condensation. At equilibrium, the number of liquid molecules escaping from the surface in a particular time interval equals the number of vapor molecules that return to the liquid in the same time interval. Once the equilibrium is established, the total quantity of liquid and vapor remains constant. The level of the liquid no longer changes.

A dynamic equilibrium results when two processes have equal but opposite rates.

A chemical equilibrium system is similar to the evaporation-condensation system just described. In a chemical equilibrium there are also two opposing rates that are equal. Let us consider the gas-phase reaction of carbon dioxide, CO_2, with hydrogen gas, H_2, in a closed system.

$$CO_2(g) + H_2(g) \longrightarrow CO(g) + H_2O(g)$$

If equal moles of CO_2 and H_2 are placed into a container under the appropriate conditions, they combine, producing carbon monoxide, CO, and steam, H_2O. The products CO and H_2O can also combine, producing the reactants CO_2 and H_2.

If a reaction is in a closed system, and the products can combine to produce

A chemical equilibrium results when the forward and reverse reaction rates are equal.

the reactants, then the reaction is classified as a **reversible reaction.** Just about all chemical changes are reversible to some degree. Reversible reactions are identified by writing two arrows, one pointing in one direction and the other pointing in the opposite direction.

$$CO_2(g) + H_2(g) \rightleftharpoons CO(g) + H_2O(g)$$

As the reaction proceeds, the concentration of the products, CO and H_2O, increases and speeds up the reverse reaction (to the left). The rate of the forward reaction (to the right) decreases as the reactant concentrations decrease. If the reaction is undisturbed, eventually the rate of the forward reaction decreases to a level equal to the rate of the reverse reaction. At this time, a **chemical equilibrium** is established. A graph of the rates of the forward and reverse reactions in the above example is shown in Figure 16.8. Notice that initially, as the reactants just begin to combine, or at time t_0, the forward reaction rate is largest and the rate of the reverse reaction is zero. There are no products initially. As time passes, the rate of the forward reaction steadily decreases and the rate of the reverse reaction increases. At a certain time, t_e, the rates become equal and remain unchanged with time. At this point the system is in equilibrium. As long as the conditions remain constant, the system remains in equilibrium. A system is more stable when it is in equilibrium than when it is not in equilibrium.

When a chemical equilibrium is established, the concentrations of the reactants and products become constant. We mathematically represent the relationship between the equilibrium concentrations of the products and the equilibrium concentrations of the reactants by writing an **equilibrium expression.** For the general form of a chemical equation

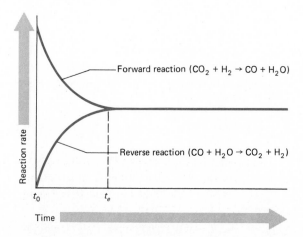

FIGURE 16.8
In the development of a chemical equilibrium, the rate of the forward reaction becomes equal to the rate of the reverse reaction. After time t_e, the rates remain equal to each other if the system is undisturbed.

TABLE 16.1   EQUILIBRIUM EXPRESSIONS FOR SELECTED EQUILIBRIA

Equilibrium	Equilibrium expression
$N_2(g) + 3H_2 \rightleftharpoons 2NH_3(g)$	$K = \dfrac{[NH_3]^2}{[N_2][H_2]^3}$
$2O_3(g) \rightleftharpoons 3O_2(g)$	$K = \dfrac{[O_2]^3}{[O_3]^2}$
$H_2(g) + C_2H_4(g) \rightleftharpoons C_2H_6(g)$	$K = \dfrac{[C_2H_6]}{[H_2][C_2H_4]}$
$I_2(g) + Cl_2(g) \rightleftharpoons 2ICl(g)$	$K = \dfrac{[ICl]^2}{[I_2][Cl_2]}$
$CO_2(g) \rightleftharpoons CO(g) + 0.5O_2(g)$	$K = \dfrac{[CO][O_2]^5}{[CO_2]}$

$$aA + bB \rightleftharpoons cC + dD$$

the equilibrium expression takes the form:

The equilibrium expression is a mathematical equation of the law of chemical equilibrium.

$$K = \frac{[C]^c[D]^d}{[A]^a[B]^b}$$

where K is the equilibrium constant, [C] and [D] are the equilibrium molar concentrations of the products, [A] and [B] are the equilibrium molar concentrations of the reactants, and a, b, c, and d are the coefficients of the reactants and products in the equation. For example, the equilibrium expression for the reaction of CO_2 and H_2 is:

$$K = \frac{[CO][H_2O]}{[CO_2][H_2]}$$

Here the coefficients for all species are 1; therefore, all concentrations are raised to the first power.

Table 16.1 provides added examples of equilibrium expressions for selected equilibria.

Equilibrium constants K are experimentally determined for an equilibrium system at a fixed temperature and pressure. The numerical value of the **equilibrium constant** indicates if the reactants or products are the primary species when equilibrium is reached. For example, the equilibrium constant for the reaction of CO_2 and H_2 is 0.14 at 550°C. Since the value for K is less than 1, this means that the product of the molar concentrations of the reactants (in the

When K is much less than 1, the reverse reaction goes nearly to completion and the forward reaction occurs to a small extent.

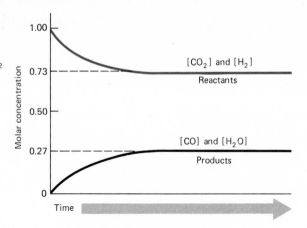

FIGURE 16.9
If initially 1.0 M CO_2 and 1.0 M H_2 are placed into a reaction vessel and the temperature is 550°C, the concentrations of both CO_2 and H_2 decrease to 0.73 M and the concentrations of CO and H_2O increase to 0.27 M. At equilibrium, the concentrations of the reactants are greater than the concentrations of the products. This equilibrium has a K value that is less than 1.0.

denominator) is greater than the product of the molar concentrations of the products (in the numerator).

$$[CO_2][H_2] > [CO][H_2O] \qquad \text{when } K < 1$$

If, for example, 1.00 M CO_2 and 1.00 M H_2 are placed into a reaction vessel and allowed to attain equilibrium at 550°C, the equilibrium molar concentrations of each of the products, CO and H_2O, is 0.27 M, and the molar concentrations of each of the reactants is 0.73 M. Figure 16.9 is a graph of the development of this equilibrium. Initially, 1.00 M CO and 1.00 M H_2 are contained in the reaction vessel; with time, their concentration drops to 0.73 M as the concentration of the products increases to 0.27 M. After the equilibrium is established, there is no further change in concentrations.

If we apply the equilibrium expression by substituting the equilibrium concentrations, we find the value of K.

$$K = \frac{[CO][H_2O]}{[CO_2][H_2]}$$

$$K = \frac{0.27 \ M \times 0.27 \ M}{0.73 \ M \times 0.73 \ M}$$

$$K = 0.14$$

Some chemical equilibria have K values that are greater than 1. For example, the combination of $H_2(g)$ and $I_2(g)$ to produce $HI(g)$ has an equilibrium constant of 51 at 440°C.

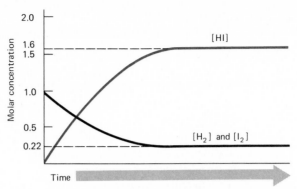

FIGURE 16.10
If initially 1.0 M H_2 and 1.0 M I_2 are allowed to come to equilibrium, the equilibrium concentration of the product, HI, is greater than the concentrations of the reactants, H_2 and I_2. This equilibrium has a K value that is greater than 1.0.

$$H_2(g) + I_2(g) \rightleftharpoons 2HI(g)$$

$$K = \frac{[HI]^2}{[H_2][I_2]} = 51$$

When K is much larger than 1, the forward reaction goes nearly to completion and the reverse reaction occurs to a small extent.

When the equilibrium constant is greater than 1, the numerator of the equilibrium expression is greater in magnitude than the denominator—the concentrations of the products are greater than those of the reactants when equilibrium is established.

If 1.00 M H_2 and 1.00 M I_2 are placed into a reaction vessel at 440°C, at equilibrium there is 0.22 M H_2, 0.22 M I_2, and 1.6 M HI. Figure 16.10 illustrates the establishment of the HI equilibrium. Exactly the same equilibrium concentrations exist for the HI equilibrium at 440°C if 2.00 M HI is initially placed in the reaction vessel. Equilibrium concentrations of all species depend only on the equilibrium constant at a specific temperature.

REVIEW PROBLEMS

16.10 What equality exists when a chemical system is in equilibrium?

16.11 Explain why the concentrations of chemical species remain constant after equilibrium is established?

16.12 For the hypothetical chemical equilibrium X(g) + Y(g) ⇌ XY(g): (a) Write the equilibrium expression. (b) What is known about this equilibrium if $K > 1$? (c) If 1 M X and 1 M Y are initially combined, draw a graph illustrating the development of the equilibrium. Assume that the value for K is greater than 1.

16.13 Write the equilibrium expressions for:
(a) $NO_2 + SO_2 \rightleftharpoons SO_3 + NO$
(b) $4HCl + O_2 \rightleftharpoons 2Cl_2 + 2H_2O$.
All substances are gases in both equilibria.

16.4 FACTORS THAT INFLUENCE CHEMICAL EQUILIBRIA

Stable systems usually will not change on their own. Less stable systems tend to undergo changes until they are more stable.

If a chemical equilibrium system is not disturbed, it will remain in equilibrium indefinitely; such behavior characterizes stable systems. A stable system is one that does not undergo spontaneous changes.

If the concentration, pressure, or temperature of a chemical equilibrium is changed, the equilibrium is disrupted and, initially, is no longer in equilibrium. Henri Le Chatelier (1850–1936) was the first to describe how a chemical equilibrium system responds to physical changes. His description of this behavior is called Le Chatelier's principle.

Henri Louis Le Chatelier (1850–1936) obtained a degree in mining chemistry. Initially he was interested in the nature of flames, and in developing ways to prevent mine explosions. He later became involved in the area of chemical thermodynamics, which led to his most valuable contribution to science, Le Chatelier's principle.

Le Chatelier's principle states that if the concentration, pressure, or temperature of a chemical equilibrium system is changed, the system shifts in such a manner to minimize the change and to bring the system back to a state of equilibrium. More simply, a chemical equilibrium attempts to remain in equilibrium by changing the concentrations of the reactants and products. If the change is such that more products are present, after reestablishing the equilibrium, the equilibrium is said to have shifted to the products (right side). If there is a net increase in reactant concentration after reestablishing equilibrium, the system has shifted towards the reactants (left side).

We shall now consider four ways in which equilibrium systems are disrupted: (1) concentration changes, (2) pressure changes, (3) temperature changes, and (4) addition of a catalyst.

Concentration Changes

When the reactant or product concentrations are changed, the equilibrium shifts to accommodate the substance added or removed. To illustrate the effect of concentration changes on equilibrium systems, we will consider the gas-phase equilibrium

A chemical equilibrium shifts to remove an added substance or replace one that has been lost.

$$CO_2 + H_2 \rightleftharpoons CO + H_2O$$

Figure 16.11 shows the concentration changes that occur when $CO_2(g)$ is added to this equilibrium system.

1. At time t_1 additional CO_2 is added to the reaction vessel, and the system is no longer in equilibrium. A higher concentration of CO_2 increases the rate of the forward reaction relative to the reverse reaction; thus, more CO_2 and H_2 combine than CO and H_2O (the reverse reaction) form.

2. As time passes, t_1 to t_2, the concentrations of CO_2 and H_2 decrease, lowering the rate of the forward reaction. At the same time, the increased concentration of products accelerates the reverse reaction. Ultimately, the two rates become equal again, and the equilibrium is reestablished at time t_2.

Once equilibrium is reattained, there are more products, CO and H_2O, and

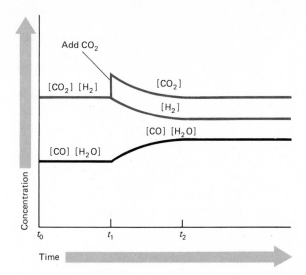

FIGURE 16.11
CO_2 is added to the water-gas equilibrium at t_1. From t_1 to t_2 the system returns to a state of equilibrium. At t_2 the concentrations of CO and H_2O have increased and the concentration of H_2 has decreased. As long as there is no change in the system, the reaction remains at equilibrium after t_2.

less reactants, CO_2 and H_2, than were present before the CO_2 was added. Consequently, the equilibrium is said to have shifted to the products (right).

In all chemical equilibrium systems, an increase in the concentration of a reactant is absorbed by shifting the equilibrium toward the formation of the product. If a product concentration is increased, the equilibrium shifts toward the reactants. A shift toward the reactants lowers the concentration of the added product.

The opposite happens when we decrease the concentration of a reactant or product. If the reactant concentration is decreased, then initially the forward reaction rate decreases while the reverse reaction rate is unchanged. Thus, the

TABLE 16.2 **CONCENTRATION EFFECTS ON A CHEMICAL EQUILIBRIUM***
$$CO_2 + H_2 \rightleftharpoons CO + H_2O$$

Concentration change	$[CO_2]$	$[H_2]$	$[CO]$	$[H_2O]$	Direction of shift
Increase $[CO_2]$	—	Dec	Inc	Inc	Products
Increase $[H_2]$	Dec	—	Inc	Inc	Products
Increase $[CO]$	Inc	Inc	—	Dec	Reactants
Increase $[H_2O]$	Inc	Inc	Dec	—	Reactants
Decrease $[CO_2]$	—	Inc	Dec	Dec	Reactants
Decrease $[H_2]$	Inc	—	Dec	Dec	Reactants
Decrease $[CO]$	Dec	Dec	—	Inc	Products
Decrease $[H_2O]$	Dec	Dec	Inc	—	Products

* Dec = decrease; Inc = increase.

equilibrium system shifts toward the reactants. Removal of product causes the system to shift toward the products, partially replacing what is removed.

Table 16.2 shows the direction the equilibrium shifts with all possible concentration changes for the CO_2-H_2 equilibrium.

Pressure Changes

Pressure changes affect mainly gas-phase equilibria.

Since the liquid and solid states are not affected by pressure changes, only equilibria that contain gases are influenced by pressure changes.

Applying Le Chatelier's principle to pressure changes, we would predict that equilibrium systems shift to decrease the pressure when the pressure is increased, and shift to increase the pressure when the pressure is decreased. Gas pressure is directly related to the number of moles of gas particles. Thus, if the pressure is increased, the equilibrium shifts in favor of the reaction producing the smaller number of moles of gas particles. If the pressure is decreased, the system shifts to favor the reaction producing the larger number of particles.

An increase in pressure on a gas-phase equilibrium favors the side of the reaction with the smallest number of particles; a decrease in pressure favors the side with the greatest number of particles.

In what direction does the equilibrium shift if the pressure is increased on nitrogen monoxide, NO, and oxygen, O_2, in equilibrium with nitrogen dioxide, NO_2?

$$\underbrace{2NO(g) + O_2(g)}_{\text{3 mol total}} \rightleftharpoons \underbrace{2NO_2(g)}_{\text{2 mol}}$$

On the reactant side of the equilibrium there are 3 mol of particles (2 mol NO + 1 mol O_2), whereas there are only 2 mol of particles on the product side (2 mol NO_2). Thus, the pressure increase is absorbed by shifting in favor of the products, the side of the equilibrium with the fewest particles (lowest concentration).

If the pressure is decreased, the NO-NO_2 equilibrium shifts toward the reactants since there are more reactant particles, thus increasing the pressure.

There are certain equilibria in which the total number of particles are the same on either side. For example,

$$\underbrace{S_2(g) + O_2(g)}_{\text{2 mol total}} \rightleftharpoons \underbrace{2SO_2(g)}_{\text{2 mol}}$$

Pressure changes do not affect equilibria that have the same number of moles of gaseous particles in the reactants and products.

Whenever the pressure is changed on such a system, neither the forward nor the reverse reaction is favored because there is the same number of moles of particles on each side (2 mol reactants and 2 mol products). There is no way for this equilibrium to decrease the pressure, and no shift is observed.

Temperature Changes

When the temperature is increased, equilibrium systems shift in favor of the reaction that absorbs added heat. When the temperature is decreased, equilibrium systems shift to replace lost heat. In a chemical equilibrium, one of the

If the forward reaction is endothermic, K increases with increasing temperature and decreases with decreasing temperature. The opposite is true for exothermic reactions; K decreases with increasing temperature and increases with decreasing temperature.

reactions (forward or reverse) is endothermic and the other is exothermic. Therefore, an increase in temperature (adding heat) favors the endothermic reaction, the one that absorbs heat energy, and a decrease in temperature (removing heat) favors the exothermic reaction.

Let us return to our model equilibrium system:

$$CO_2 + H_2 \rightleftharpoons CO + H_2O + 42.7 \text{ kJ}$$

As written, the forward reaction is exothermic, releasing 42.7 kJ of heat per mole. The reverse reaction is endothermic, requiring 42.7 kJ per mole. If the temperature of this equilibrium system is increased, the equilibrium shifts in favor of the reverse reaction; more reactants and less products are found after the temperature change. A decrease in temperature shifts the equilibrium toward the products.

When dealing with temperature changes and equilibrium systems, treat the heat energy as a reactant or product, a change in which causes the same equilibrium shifts as do concentration changes of actual reactants and products.

Addition of a Catalyst

A catalyst is a substance that lowers the activation energy of a reaction. Activation energies of forward and reverse reactions are lowered equally by a catalyst, as shown in Figure 16.12. Hence, upon addition of a catalyst both forward and reverse reactions are accelerated to the same degree, resulting in no net change in the equilibrium system. Catalysts produce no changes in the state of an equilibrium system.

Example Problem 16.1 gives examples of how to predict the direction in which an equilibrium shifts in response to a physical change.

FIGURE 16.12
Since a catalyst lowers the activation energy of the reactants and products, both the forward and reverse rates of reaction are increased equally. Thus, catalysts have no effect on the equilibrium concentrations of the reactants and products.

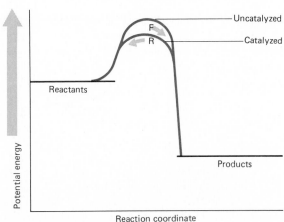
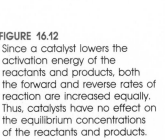

Example Problem 16.1

Consider the following gas-phase equilibrium:

$$CS_2(g) + 4H_2(g) \rightleftharpoons CH_4(g) + 2H_2S(g) + 232 \text{ kJ}$$

Predict the direction in which the equilibrium shifts for the following changes: (a) addition of $H_2(g)$, (b) removal of CS_2, (c) pressure decrease, and (d) temperature increase.

Solution

(a) Adding H_2 shifts the equilibrium to the right. An increase in the H_2 concentration (a reactant) is absorbed by shifting toward the products.
(b) Removing CS_2 shifts the equilibrium to the left. The system attempts to replace the lost CS_2 by shifting toward the reactants.
(c) Decreasing the pressure shifts the equilibrium to the left, the side of the equation with the greatest number of moles of gas.
(d) Increasing the temperature shifts the equilibrium to the left, in favor of the endothermic reaction.

REVIEW PROBLEMS

16.14 State Le Chatelier's principle.

16.15 What is meant by an "equilibrium shifting"? Explain in terms of the forward and reverse reactions.

16.16 Answer the following questions for the equilibrium system:

$$CO(g) + Cl_2(g) \rightleftharpoons COCl_2(g)$$

In what direction does the equilibrium shift (toward the reactants or products) for the following physical changes?
(a) Addition of CO.
(b) Removal of Cl_2.
(c) Removal of $COCl_2$.
(d) Doubling of the pressure.

16.17 Consider the Haber reaction:

$$3H_2(g) + N_2(g) \rightleftharpoons 2NH_3(g) + 92 \text{ kJ}$$

How does the equilibrium shift when the following changes are made?
(a) Increase in temperature.
(b) Decrease in temperature.
(c) Increase in pressure.
(d) Addition of a catalyst.

SUMMARY

Chemical kinetics is the study of the rates of chemical reactions. Rates are measured by determining the change in either reactant or product concentration over a given time interval.

Collision theory is used to explain, at the molecular level, the rates of chemical reactions. Collision theory describes atomic and molecular collisions as effective or ineffective. An effective collision occurs when two particles collide with enough energy and with the proper orientation to produce the products.

Four factors that influence the rate of chemical reactions are: (1) the nature of the reactants, (2) reactant concentration, (3) temperature, and (4) catalysts. The first factor to consider is the nature of the reactants, or more specifically, the nature of chemical bonds in the reactants. Increased concentration of reactants increases the number of effective collisions; therefore, the rate of the reaction is increased. An increase in temperature increases the average kinetic energy of the reacting particles, which also increases the number of effective collisions. Thus, higher temperatures elevate the reaction rate. Catalysts speed up chemical reactions by lowering the activation energy. A reaction's activation energy is the amount of energy needed to produce the transition state, which is a high-energy combination of colliding particles that can break apart into either the reactants or the products.

A chemical equilibrium is established when the rates of the forward and reverse reactions, in a closed system, are equal. Equilibrium systems are stable systems and do not undergo spontaneous changes. An equilibrium constant K describes the direction in which the equilibrium lies, either favoring the products or the reactants.

Various physical changes affect chemical equilibria. Le Chatelier's principle is the guiding rule used to predict the behavior of equilibria when such changes are made. Le Chatelier's principle states that if a property of a chemical equilibrium system is changed, the equilibrium shifts to minimize the change and bring the system back to equilibrium.

When there is an increase in concentration of a substance in equilibrium, the system shifts to decrease the amount of substance added. Increasing the temperature shifts the equilibrium in the direction that absorbs the heat, the endothermic reaction. An increase in pressure causes the system to shift in favor of the side with the smaller number of moles of particles. Since catalysts speed up both forward and reverse reactions equally, they have no effect on equilibrium systems.

QUESTIONS AND PROBLEMS

16.18 Define the following terms: reaction rate, effective collision, exothermic reaction, endothermic reaction, enthalpy, activation energy, catalyst, reaction mechanism, reaction intermediate, rate-determining step, equilibrium, chemical equilibrium, reversible reaction, forward reaction, reverse reaction, equilibrium shift.

Rates of Reactions

16.19 Describe the relative differences in time required to go to completion for a reaction that has a high measured rate of reaction compared with one that has a low rate.

16.20 Classify the rates of the following chemical changes as high, moderate, or low: (a) explosion of nitroglycerin, (b) hard-boiling of an egg, (c) reaction of sodium with water, (d) combination of HCl and NaOH, (e) oxidation of Fe in the air.

16.21 What is true about the rate of disappearance of reactants in a reaction that is classified as having a high rate of reaction?

Collision Theory

16.22 What is true about two colliding particles: (a) if there is an effective collision, and (b) if an ineffective collision takes place?

16.23 Discuss the orientation factors of two colliding automobiles as an analogy for the orientation effects of colliding molecules.

16.24 Determine if an effective collision will occur under the following conditions:
(a) Energy greater than the minimum energy for producing the transition state, and improper orientation
(b) Proper orientation, with energy in excess of the minimum energy to produce the transition state
(c) Energy less than the minimum energy to produce the transition state, with the proper orientation

16.25 How do the following changes alter the frequency of effective collisions for a gas-phase reaction: (a) decreased temperature, (b) increased pressure, (c) addition of a catalyst?

Factors Influencing Reaction Rates

16.26 Predict the relative rate of reactions (high, moderate, or low) by considering the nature of the reactants:
(a) $F_2 + Cl_2$
(b) $F_2 + Xe$
(c) decomposition of HF
(d) $F_2 + Na$.

16.27 Rank the following reactions according to reaction rates (highest to lowest) at 25°C:
(a) $HCl(aq) + NaOH(aq) \longrightarrow NaCl(aq) + H_2O(l)$
(b) $HCl(aq) + Na_2CO_3(aq) \longrightarrow$
$$NaCl(aq) + CO_2(g) + H_2O(l)$$
(c) $2NaCl(s) \longrightarrow 2Na(s) + Cl_2(g)$

16.28 Describe what happens to the rate of reaction for the combination of $H_2(g)$ and $Br_2(g)$,

$$H_2(g) + Br_2(g) \longrightarrow 2HBr(g)$$

when the following changes are made:
(a) Increase in $[Br_2]$
(b) Decrease in $[H_2]$
(c) Increase in temperature
(d) Addition of a catalyst
(e) Increase in the pressure

16.29 If only 1% of the reacting molecules have enough energy to form the transition state (activated complex), what would you expect the overall rate of the reaction to be? Explain.

16.30 Draw an energy-level diagram for the combination of $H_2(g)$ and $I_2(g)$ to produce $HI(g)$:

$$H_2(g) + I_2(g) + 52 \text{ kJ} \longrightarrow 2HI$$

On the graph, label energy on the vertical axis and time on the horizontal axis, and show the difference in energy, the enthalpy change, between the reactants and products.

16.31 (a) Draw an energy-level diagram for the oxidation of nickel(II) sulfide, NiS, to nickel(II) oxide, NiO, and sulfur dioxide, SO_2. Properly label both axes, and indicate the heat of reaction.

$$2NiS + 3O_2 \longrightarrow 2NiO + 2SO_2 + 936 \text{ kJ}$$

(b) On the graph draw the energy pathway that is followed by the reactants to form the products; indicate the activation energy and the position of the transition state.

16.32 After the activation energy is added, many exothermic reactions continue without any additional input of energy. Endothermic reactions cease without a constant supply of energy. Provide an explanation for these facts.

16.33 Draw a sketch of an energy-level diagram, including the reaction pathway, for an endothermic reaction that is uncatalyzed; on the same graph draw the reaction pathway if the reaction is catalyzed.

16.34 The Haber reaction is catalyzed with an iron catalyst. If a catalyst was not used, what factors would have to be changed to produce the same quantity of ammonia within a specified time interval?

16.35 What is the purpose of a catalytic converter in an automobile?

16.36 Consider the accepted mechanism for the reaction $H_2 + 2ICl \longrightarrow I_2 + 2HCl$:

(1) $H_2 + ICl \longrightarrow HI + HCl$ (slow)

(2) $HI + ICl \longrightarrow HCl + I_2$ (fast)

(a) Show that by adding the two equations in the mechanism, the overall equation is found.
(b) What is the reaction intermediate in the mechanism?
(c) What is the rate-determining step?

16.37 (a) What is the overall aqueous reaction for the following reaction mechanism?

(1) $H_2O + OCl^- \rightleftharpoons HOCl + OH^-$

(2) $HOCl + I^- \longrightarrow HOI + Cl^-$

(3) $HOI + OH^- \rightleftharpoons H_2O + OI^-$

(b) What intermediates are produced in this mechanism?

Chemical Equilibrium Systems

16.38 Correct the following incorrect statements concerning chemical equilibrium systems:
(a) At equilibrium the concentration of the reactants equals the concentration of the products.

(b) Equilibrium systems are unstable and undergo spontaneous changes.
(c) After the establishment of a chemical equilibrium the forward and reverse reactions stop.

16.39 Write an explanation to account for the requirement that a chemical equilibrium be in a closed system.

16.40 What is the purpose of writing two opposing arrows, $\rightleftharpoons$, to indicate a chemical equilibrium?

16.41 Describe what happens to the rates of the forward and reverse reactions as an equilibrium is established.

16.42 Write the equilibrium expressions for the following gas-phase reactions:
(a) $NO + SO_3 \rightleftharpoons NO_2 + SO_2$
(b) $SO_2Cl_2 \rightleftharpoons SO_2 + Cl_2$
(c) $CS_2 + 4H_2 \rightleftharpoons CH_4 + 2H_2S$
(d) $4NH_3 + 5O_2 \rightleftharpoons 4NO + 6H_2O$
(e) $4HCl + O_2 \rightleftharpoons 2Cl_2 + 2H_2O$
(f) $NOCl \rightleftharpoons NO + \frac{1}{2}Cl_2$

16.43 What is the difference between an equilibrium with a K value larger than 1 compared with an equilibrium that has a K value smaller than 1?

Le Chatelier's Principle

16.44 Predict the effect of the following concentration changes on

$$NO_2(g) + SO_2(g) \rightleftharpoons SO_3(g) + NO(g)$$

(a) lowering [NO], (b) increasing [SO$_3$], (c) increasing [SO$_2$], (d) decreasing [NO$_2$].

16.45 Describe what happens to the reaction rates of the forward and reverse reactions when a small quantity of Cl_2 is removed from the following equilibrium:

$$SO_2Cl_2(g) \rightleftharpoons SO_2(g) + Cl_2(g)$$

16.46 In which direction will the following gaseous equilibria shift (products or reactants) if the pressure is increased?
(a) $PCl_5 \rightleftharpoons PCl_3 + Cl_2$
(b) $S_2 + O_2 \rightleftharpoons 2SO_2$
(c) $NO_2 + CO \rightleftharpoons CO_2 + NO$
(d) $4HCl + O_2 \rightleftharpoons 2Cl_2 + 2H_2O$

16.47 Completely explain the behavior, in terms of Le Chatelier's principle, of the chemical equilibrium

$3O_2(g) \rightleftharpoons 2O_3(g)$ when the pressure of the system is decreased.

16.48 Describe the effect the following temperature changes have on

$$2SO_2(g) + O_2(g) \rightleftharpoons 2SO_3(g) + heat$$

(a) Temperature is decreased.
(b) Temperature is increased.

16.49 Explain why the addition of a catalyst does not cause the equilibrium to shift to minimize the addition of the catalyst.

16.50 Describe the behavior of the following gas-phase equilibria with the stated changes:

(a) Adding $I_2(g)$ to $2HI \rightleftharpoons H_2 + I_2$
(b) Decreasing the pressure on $C_2H_6 \rightleftharpoons H_2 + C_2H_4$
(c) Removing heat from $CO_2 \rightleftharpoons CO + \frac{1}{2}O_2$, $\Delta H = +284$ kJ/mol
(d) Increasing the pressure on $C_3H_8 + 5O_2 \rightleftharpoons 3CO_2 + 4H_2O$
(e) Adding a catalyst to $I_2 + Cl_2 \rightleftharpoons 2ICl$

· C H A P T E R ·

17

Acids and Bases

Study Guidelines

After completing Chapter 17, you should be able to:
1. List the four principal properties of acids and bases
2. Define and identify Arrhenius acids and bases
3. Write an equation to illustrate the ionization of water to $H_3O^+ + OH^-$
4. Define and identify Brønsted-Lowry acids and bases
5. Determine the conjugate acids of bases or conjugate bases of acids
6. Predict the relative strengths of acids and bases
7. Write equations for reactions of strong acids with strong bases, strong acids with weak bases, and strong bases with weak acids
8. Apply the K_w equilibrium expression for water to find $[H^+]$ and $[OH^-]$
9. Define and determine the pH of solutions
10. State two methods that are employed in the laboratory to determine pH of solutions
11. Explain what happens in an acid-base titration
12. Explain what is meant by the equivalence point of an acid-base titration
13. Determine the molarity of an acid, given its volume, and the volume and molarity of the base required to neutralize it
14. Determine the equivalent masses of acids and bases
15. Calculate the normality of acidic and basic solutions

Acids and bases such as hydrochloric acid, HCl(aq), nitric acid, HNO_3(aq), and potassium hydroxide, KOH(aq), were known and used by the alchemists in the eleventh century. At that time, and for many centuries thereafter, acids and bases were defined in terms of their properties. **Acids** share the following common properties. They taste sour, change the color of various indicator dyes such as litmus, react with bases to produce salts and water, and may combine with active metals liberating hydrogen gas, $H_2(g)$.

Bases have contrasting properties to acids. They taste bitter, change the color of indicator dyes opposite to the way that acids change them, feel slippery when they contact the skin, and combine with acids to produce a salt and water.

Arrhenius Definition

In 1884, Svante Arrhenius proposed the first good definitions for acids and bases. Simultaneously, he shook the world of chemistry by presenting the theory of ionic dissociation. He stated that when ionic substances are dissolved in water, they dissociate into ions and are surrounded by water molecules. A modern statement of the **Arrhenius definitions** of acids and bases is as follows:

Svante Arrhenius (*New York Library Picture Collection*)

Arrhenius taught himself to read at the age of 3 and was a brilliant student. His theory of ionic dissociation came directly from his university graduate work. Arrhenius was awarded a Ph.D. (1884) with the lowest passing grade because his mentors thought that his theory was too far-fetched. In 1903, he was awarded the Nobel prize in chemistry for this far-fetched notion!

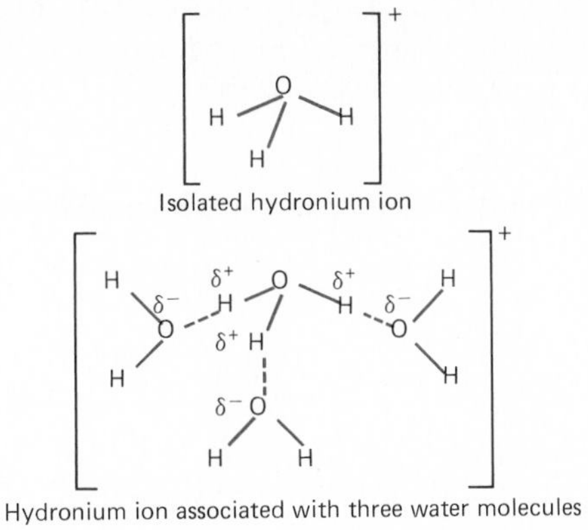

FIGURE 17.1
A hydronium ion is a hydrated proton, $H(H_2O)^+$. Most hydronium ions are hydrogen bonded to other water molecules.

Isolated hydronium ion

Hydronium ion associated with three water molecules

An **acid** is a substance that increases the hydrogen ion (H^+) concentration when dissolved in water.

A **base** is a substance that increases the hydroxide ion (OH^-) concentration when dissolved in water.

Water is an extremely weak electrolyte. Very few water molecules ionize and form H^+ and OH^- ions.

$$H_2O \rightleftharpoons H^+ + OH^-$$

But, water's ionization is not as simple as this equation indicates. Instead it should be thought of as the interaction of two water molecules, where one water molecule gives up a H^+ to another as follows:

$$H_2O + H_2O \rightleftharpoons H_3O^+ + OH^-$$

The H_3O^+ that is produced is called a **hydronium** ion, which is nothing more than a hydrated hydrogen ion (a wet proton), $H(H_2O)^+$ (see Figure 17.1). Commonly, H^+ is written instead of H_3O^+, with the understanding that the H^+ is associated with one or more water molecules and does not exist by itself.

Throughout this chapter, H^+ is written as a symbol meaning a hydrated hydrogen ion (H_3O^+).

Acids similar to HCl and HNO_3, when added to water, increase the H_3O^+ concentration as a result of their complete ionization.

$$HCl(g) + H_2O(l) \longrightarrow H_3O^+(aq) + Cl^-(aq)$$
$$HNO_3 + H_2O(l) \longrightarrow H_3O^+(aq) + NO_3^-(aq)$$

Therefore, according to the Arrhenius definition, HCl and HNO_3 are acids: they increase the number of H^+ ions in solution. Other examples of Arrhenius acids are perchloric acid, $HClO_4$; sulfuric acid, H_2SO_4; acetic acid, $HC_2H_3O_2$; and phosphoric acid, H_3PO_4.

Arrhenius bases include the metallic hydroxides because they dissociate and increase the OH^- concentration in water. Consider two strong bases KOH and NaOH:

The old name for NaOH is caustic soda.

$$KOH(s) \xrightarrow{\text{water}} K^+(aq) + OH^-(aq)$$

$$NaOH(s) \xrightarrow{\text{water}} Na^+(aq) + OH^-(aq)$$

Other examples of metallic hydroxides are calcium hydroxide, $Ca(OH)_2$; lithium hydroxide, LiOH; and barium hydroxide, $Ba(OH)_2$.

Various covalent compounds that do not contain a hydroxide ion react with water and produce hydroxide ions. Hence, they are also Arrhenius bases. For example, ammonia, NH_3, undergoes the following reaction when placed into water:

$$NH_3(g) + H_2O \rightleftharpoons NH_4^+(aq) + OH^-(aq)$$

Ammonia combines with a H^+ from water, producing an ammonium ion, NH_4^+, and a hydroxide ion, OH^-.

Brønsted-Lowry Definition

Brønsted (1879–1947) was a Danish chemist who was primarily interested in studying chemical thermodynamics. In the early 1920s, he began investigating the mechanism by which acids and bases catalyze reactions. From this work, he proposed his famous definition of acids and bases.

The Arrhenius definitions of acids and bases were extended in 1923 by two chemists, Johannes Brønsted and Thomas Lowry. Today, we call their definition the Brønsted-Lowry definition of acids and bases. A larger number of compounds are considered either acids or bases under the Brønsted-Lowry definitions than under the Arrhenius definitions.

The **Brønsted-Lowry** acid and base definitions are:

An **acid** is a substance that is a proton donor.

A **base** is a substance that is a proton acceptor.

Let's compare the Brønsted-Lowry and Arrhenius definitions. Arrhenius acids are those substances that increase the H^+ ions in water; a Brønsted-Lowry acid is a proton donor. Are hydrogen ions and protons the same? Yes, a hydrogen atom consists of a proton and an electron. Hydrogen ions form when an electron is lost, leaving a proton. Note that water is not required according to the Brønsted-Lowry definition of an acid. But without water, acids and bases could not exist according to the Arrhenius definition. Arrhenius bases increase the OH^- concentration in water, whereas Brønsted-Lowry bases accept protons. Many substances that do not contain hydroxide ions are classified as bases under the Brønsted-Lowry definition.

$$H\cdot \longrightarrow e^- + H^+$$

If we consider the reaction of hydrogen chloride gas, $HCl(g)$, and water, $HCl(g)$ gives up its H^+ ion to water producing H_3O^+, leaving a $Cl^-(aq)$.

$$\text{HCl(aq)} + \text{H}_2\text{O}(l) \longrightarrow \text{H}_3\text{O}^+(\text{aq}) + \text{Cl}^-(\text{aq})$$
$$\qquad\text{Acid}\qquad\quad\text{Base}$$

HCl is a Brønsted-Lowry acid, since it donates a proton to H_2O. Thus, H_2O is a proton acceptor, or a Brønsted-Lowry base.

After an acid has given up a proton, the resulting anion becomes a base—it could accept a proton to re-form the acid. The base that results is called the **conjugate base** of the acid. When a base accepts a proton, it forms the **conjugate acid** of that base. In the above equation, Cl^- is the conjugate base of the acid HCl, and H_3O^+ is the conjugate acid of the base H_2O.

The word "conjugate" means joined together in pairs, coupled.

Let us look at another example:

$$\text{NH}_3(\text{aq}) + \text{HBr(aq)} \longrightarrow \text{NH}_4{}^+(\text{aq}) + \text{Br}^-(\text{aq})$$
$$\quad\text{Base}\qquad\text{Acid}\qquad\qquad\text{Conjugate acid}\qquad\text{Conjugate base}$$

In this equation, HBr donates a proton to NH_3; hence, HBr is an acid and NH_3 is a base. The conjugate base of HBr is the bromide ion, Br^-, and the conjugate acid of NH_3 is the ammonium ion, $\text{NH}_4{}^+$.

Some substances are capable of both donating and accepting protons. They are called **amphiprotic,** or more generally **amphoteric,** compounds. Such substances behave as acids or bases, depending on the nature of the substances with which they are combining.

Amphoteric is the general term applied to substances that can react as acids or bases.

Water is an example of an amphiprotic substance. In the presence of stronger acids, it acts as a base. When combined with stronger bases, it is acidic.

$$\text{HI(aq)} + \text{H}_2\text{O}(l) \longrightarrow \text{H}_3\text{O}^+(\text{aq}) + \text{I}^-(\text{aq})$$
$$\quad\text{Acid}\qquad\text{Base}$$

Another acid-base definition was proposed by G. N. Lewis. He defined acids as electron-pair acceptors, and bases as electron-pair donators.

$$\text{NH}_3(\text{aq}) + \text{H}_2\text{O}(l) \rightleftharpoons \text{NH}_4{}^+(\text{aq}) + \text{OH}^-(\text{aq})$$
$$\quad\text{Base}\qquad\text{Acid}$$

In the first equation water accepts a proton from HI; thus it is classified as a base. In the second equation water donates a proton to NH_3, thus it acts as an acid.

REVIEW PROBLEMS

17.1 (a) List the four principal properties of acids. (b) List the four principal properties of bases.

17.2 Define the following: (a) Arrhenius acid, (b) Arrhenius base, (c) Brønsted-Lowry acid, and (d) Brønsted-Lowry base.

17.3 (a) What is a hydronium ion? (b) How is a hydronium ion produced?

17.4 In the following equation identify the acid and conjugate base, and base and conjugate acid.

$$\text{NH}_3(\text{aq}) + \text{HCN(aq)} \rightleftharpoons \text{NH}_4{}^+(\text{aq}) + \text{CN}^-(\text{aq})$$

17.5 What is an amphiprotic substance? Give an example by writing two equations.

17.2 RELATIVE STRENGTHS OF ACIDS AND BASES

Strong acids and bases are strong electrolytes, and weak acids and bases are weak electrolytes.

By applying the Brønsted-Lowry acid-base definition, we can predict the relative strengths of acids and bases. A **strong acid** is defined as a substance that more readily gives up a proton than does a weaker acid. A **strong base** is one that more readily accepts a proton than does a weaker base.

When a strong acid donates its proton, it produces a weak conjugate base—one that does not readily accept the proton back to re-form that acid. Thus, the conjugate bases of strong acids are weak bases. Similarly, the conjugate bases of weak acids are strong bases. To illustrate, let's look at a strong acid and a weak acid, HCl and HF, respectively.

HCl is secreted by cells in the lining of the stomach. The HCl provides the proper acidic conditions for digestion. After the contents of the stomach enter the small intestine, the acid is neutralized.

When HCl(g) is dissolved in water, it completely ionizes:

$$HCl(g) + H_2O(l) \longrightarrow H_3O^+(aq) + Cl^-(aq)$$

Hydrochloric acid is a strong acid; it has a large capacity to donate protons to water, yielding large quantities of H_3O^+ and Cl^-. The Cl^- ion has a small capacity to accept a proton from H_3O^+; consequently, Cl^- is a rather weak base.

Strong acids produce weak conjugate bases. Strong bases produce weak conjugate acids.

In contrast, hydrofluoric acid, HF(aq), is a weak acid.

$$HF + H_2O \rightleftharpoons H_3O^+ + F^-$$

Only a small percent of the HF is ionized; HF therefore has a small capacity to donate protons to water. Hydrogen fluoride's conjugate base is F^-, which is a relatively strong base. The capacity of F^- to accept protons from H_3O^+ is greater than the capacity of the weak base H_2O to accept protons from HF.

In water, the strength of HCl, HBr, and HI are equal since each is 100% ionized. All acids stronger than H_3O^+ appear to be the same strength in aqueous solutions. This phenomenon is called the *leveling effect.*

In Table 17.1, acids and their conjugate bases are listed in decreasing order of acid strength. The strongest acid listed is perchloric acid, $HClO_4$. Perchloric acid's conjugate base, perchlorate, ClO_4^-, is the weakest base listed. On the other end, methane, CH_4, is the weakest acid and its conjugate base, methanide, CH_3^-, is the strongest base listed in Table 17.1.

Strengths of binary acids can also be related to the properties of the nonmetal atoms within the acid molecules. In the periodic table, binary acid strength increases going from left to right within a period.

$$CH_4 < NH_3 < H_2O < HF$$

and

$$SiH_4 < PH_3 < H_2S < HCl$$

Increasing acid strength $\longrightarrow$

Concentrations of commercially available strong acids:

Acids	Concentration
H_2SO_4	18 M
HCl	12 M
HNO_3	16 M
$HClO_4$	12 M
HBr	9 M

What accounts for this trend? Going from left to right across a period, the electronegativity of the atoms increases. As you may recall from Section 8.4, electronegativity is the capacity of an atom to attract electrons in chemical

TABLE 17.1   RELATIVE STRENGTH OF ACIDS AND BASES

Acid		Conjugate base	
Perchloric acid	$HClO_4$	ClO_4^-	Perchlorate
Sulfuric acid	H_2SO_4	HSO_4^-	Hydrogen sulfate
Hydroiodic acid	HI	I^-	Iodide
Hydrobromic acid	HBr	Br^-	Bromide
Hydrochloric acid	HCl	Cl^-	Chloride
Nitric acid	HNO_3	NO_3^-	Nitrate
Hydronium	H_3O^+	H_2O	Water
Phosphoric acid	H_3PO_4	$H_2PO_4^-$	Dihydrogen phosphate
Hydrofluoric acid	HF	F^-	Fluoride
Nitrous acid	HNO_2	NO_2^-	Nitrite
Acetic acid	$HC_2H_3O_2$	$C_2H_3O_2^-$	Acetate
Carbonic acid	H_2CO_3	HCO_3^-	Hydrogen carbonate
Ammonium	NH_4^+	NH_3	Ammonia
Hydrocyanic acid	HCN	CN^-	Cyanide
Water	H_2O	OH^-	Hydroxide
Ammonia	NH_3	NH_2^-	Amide
Methane	CH_4	CH_3^-	Methanide

Strong acid → Weak acid (Increasing acid strength)

Weak base → Strong base (Increasing base strength)

bonds. Thus, atoms with higher electronegativities have a greater attraction for hydrogen's electron, producing molecules that have more ionic character (are more polar). Substances with greater ionic character ionize to a greater degree in aqueous solution, yielding more protons.

Within a chemical group, there is increasing acid strength with increasing atomic mass.

$$H_2O < H_2S < H_2Se < H_2Te$$

and

$$HF < HCl < HBr < HI$$

Increasing acid strength $\longrightarrow$

Increasing acidity within a family is partially explained by the increasing atomic radii. The larger the atomic radius, the weaker the bond between the nonmetal and hydrogen. Weaker bonds have a greater tendency to dissociate and give up protons.

> A concentrated acid is not necessarily a strong acid. Strengths of acids refer to their degree of ionization, and concentration is a measure of the amount of dissolved acid, i.e., moles of dissolved acid per liter.

REVIEW PROBLEMS

17.6 What is the relative strength of the: (a) conjugate acid of a weak base, (b) conjugate base of a strong acid, and (c) conjugate acid of a strong base?

17.7 Give an explanation for the fact that HF is a weaker acid than HCl in terms of their conjugate bases.

17.8 Use Table 17.1 to determine which one in each of the following pairs is a stronger acid: (a) HNO_2 or HCN, (b) H_2O or NH_4^+, (c) $HC_2H_3O_2$ or H_3O^+, and (d) H_2CO_3 or NH_3.

17.3 REACTIONS OF ACIDS AND BASES

Strong Acid–Strong Base   By far, the most important reaction of acids and bases is the combination of an acid and a base to produce a salt and water, called a **neutralization** reaction.

$$Acid + base \longrightarrow salt + water$$

To illustrate a neutralization reaction, let us consider the reaction of potassium hydroxide and hydrochloric acid.

$$KOH(aq) + HCl(aq) \longrightarrow KCl(aq) + H_2O(l)$$

In this reaction, KOH, a strong base, combines with HCl, a strong acid, to produce KCl, a salt, and water. A more accurate representation of this neutralization reaction is an overall ionic equation, showing all dissolved species. Both strong acids and strong bases are completely ionized. Therefore, they should be written as aqueous ions.

$$K^+(aq) + OH^-(aq) + H^+(aq) + Cl^-(aq) \longrightarrow$$
$$K^+(aq) + Cl^-(aq) + H_2O(l)$$

Spectator ions are unchanged in ionic reactions.   Elimination of the spectator ions from the overall equation gives the net ionic equation.

$$H^+(aq) + OH^-(aq) \longrightarrow H_2O(l)$$

Whenever a strong acid is combined with a strong base, the net ionic equation for this neutralization is

$$H^+(aq) + OH^-(aq) \longrightarrow H_2O(l)$$

In most cases, the anion in the strong acid and the cation from the strong base are spectator ions, and not a part of the net reaction.

Another example of a reaction of a strong acid and strong base is

Antacids used to relieve stomach distress are normally basic substances mixed with other ingredients. Most commercial antacids contain one or more of the following: $NaHCO_3$, $Mg(OH)_2$, $MgCO_3$, $Al(OH)_3$, or $CaCO_3$.

$$2H^+(aq) + SO_4{}^{2-}(aq) + Mg^{2+}(aq) + 2OH^-(aq) \longrightarrow$$
$$Mg^{2+}(aq) + SO_4{}^{2-}(aq) + 2H_2O(l)$$

This equation shows sulfuric acid, $H_2SO_4(aq)$, reacting with magnesium hydroxide, $Mg(OH)_2(aq)$, producing magnesium sulfate, $MgSO_4(aq)$, and water. In this equation, Mg^{2+} and $SO_4{}^{2-}$ are the spectator ions.

Strong Base–Weak Acid

A different net ionic equation is written for the reaction of a strong base and a weak acid. A strong base is completely ionized in aqueous solution; but a weak acid is a weak electrolyte, and most of the acid is un-ionized.

What happens when sodium hydroxide, NaOH, a strong base, is combined with hydrofluoric acid, HF, a weak acid?

$$Na^+(aq) + OH^-(aq) + HF(aq) \longrightarrow Na^+(aq) + F^-(aq) + H_2O(l)$$

In the overall ionic equation, NaOH is written as dissolved ions. Hydrofluoric acid, HF, is not written as ions because only a small percent is ionized. In this equation, the only spectator ion is $Na^+(aq)$, and when it is crossed out of the equation, the net equation appears as follows:

$$OH^-(aq) + HF(aq) \longrightarrow F^-(aq) + H_2O(l)$$

Hydroxide ions from strong bases accept protons from the un-ionized weak acid, yielding water and the conjugate base of the weak acid. A general net equation for the reaction of strong bases and weak acids is

$$OH^-(aq) + HA(aq) \longrightarrow H_2O(l) + A^-(aq)$$

where HA is a weak acid and A^- is its conjugate base.

As mentioned previously, the conjugate base of a weak acid is a relatively strong proton acceptor, so it can take protons from water molecules. Consequently, when equivalent amounts of strong bases and weak acids are combined, the solutions that result are basic. This is not true for the reaction of strong acids and strong bases. If equivalent amounts of strong acids and strong bases are combined, the resulting solution is neutral. Each H^+ combines with an OH^-, producing a neutral water molecule.

Other examples of weak acids combining with strong bases are:

$$K^+(aq) + OH^-(aq) + HC_2H_3O_2(aq) \longrightarrow H_2O(l) + K^+(aq) + C_2H_3O_2^-(aq)$$

$$Na^+(aq) + OH^-(aq) + HCN(aq) \longrightarrow H_2O(l) + Na^+(aq) + CN^-(aq)$$

For practice, write the net ionic equations for these two reactions.

Strong Acid–Weak Base

A strong acid combines with a weak base producing the conjugate acid of the weak base. What happens when the strong acid hydrochloric acid, HCl, combines with the weak base ammonia, NH_3?

$$H^+(aq) + Cl^-(aq) + NH_3(aq) \longrightarrow NH_4^+(aq) + Cl^-(aq)$$

A hydrogen ion from HCl is donated to the un-ionized ammonia, giving the ammonium ion. The chloride ion remains unchanged. Eliminating the chloride ion from the equation gives the net ionic equation:

$$H^+(aq) + NH_3(aq) \longrightarrow NH_4^+(aq)$$

Thus when equivalent amounts of a strong acid and a weak base react, the resulting solutions are acidic because of the formation of the acidic ammonium ion, NH_4^+. This is exactly the opposite of what was observed when a strong base reacts with a weak acid.

A general equation for a reaction of a strong acid and a weak base is

$$H^+(aq) + B(aq) \longrightarrow HB^+(aq)$$

where B is a weak base and HB^+ is the conjugate acid of the weak base.

REVIEW PROBLEM **17.9** Write the overall ionic and net ionic equations for the following neutralization reactions:
(a) $HClO_4$, a strong acid, and RbOH, a strong base
(b) NaOH, a strong base, and HNO_2, a weak acid
(c) HI, a strong acid, and NH_3, a weak base

17.4 MEASUREMENT OF H⁺ CONCENTRATION AND pH

To understand how acid-base measurements are made, we must start by considering water, the liquid in which acids and bases exist.

Water is a very weak electrolyte and ionizes to a small degree.

$$H_2O \rightleftharpoons H^+ + OH^-$$

The equilibrium of water and its ions is described by an equilibrium expression.

$$K_w = [H^+][OH^-] = 1 \times 10^{-14}$$

K_w is the ion-product equilibrium constant for water.

where K_w is the ion-product equilibrium constant for water, and $[H^+]$ and $[OH^-]$ are the molar concentrations of H^+ and OH^-. This expression shows that the product of the molar concentrations of the H^+ and OH^- ions is equal to 1×10^{-14}. Using algebra, we can determine the individual molar concentrations of both H^+ and OH^-.

If we let z equal the molar concentration of H^+, then z is also the molar concentration of OH^-, since they have equal concentrations in pure water.

$$z = [H^+] = [OH^-]$$

Therefore, $$z^2 = 1 \times 10^{-14}$$

Taking the square root of both sides of the equation, we have

$$z = 1 \times 10^{-7} \ M$$
$$z = [H^+] = [OH^-] = 1 \times 10^{-7} \ M$$

We find that, in a sample of pure water, both the H$^+$ and OH$^-$ concentrations are 1×10^{-7} M. When the concentrations of H$^+$ and OH$^-$ are equal to 1×10^{-7} M, the water is neutral, i.e., neither acidic nor basic.

When an acid is added to water, it donates H$^+$ ions, increasing the H$^+$ concentration; in contrast, bases increase the OH$^-$ concentration. Le Chatelier's principle is used to predict the effect of adding either H$^+$ or OH$^-$ to the water equilibrium system.

Le Chatelier's principle states that equilibrium systems tend to absorb changes and return to a state of equilibrium.

$$H_2O \ \rightleftharpoons \ H^+ + OH^-$$

If the [H$^+$] is increased, the equilibrium shifts to absorb the added H$^+$; the water equilibrium shifts to the left. When the equilibrium is reestablished, the concentration of OH$^-$ decreases.

If water's [OH$^-$] is increased, the equilibrium shifts to the left, decreasing the [H$^+$]. Hence, with each addition of an acid to water, the [H$^+$] increases and the [OH$^-$] decreases. Adding a base causes the [OH$^-$] to increase and the [H$^+$] to decrease.

Let's solve a problem to illustrate what happens when the [H$^+$] of water is changed by adding acid or base. What is the [OH$^-$] of a solution in which acid is added to water, producing a solution with the [H$^+$] equal to 1×10^{-3} M? In water, the product of the [H$^+$] times the [OH$^-$] is always equal to 1×10^{-14}.

$$[H^+][OH^-] = 1 \times 10^{-14}$$

Substituting the value of the concentration of H$^+$ into this expression, we get

$$(1 \times 10^{-3} \ M)[OH^-] = 1 \times 10^{-14}$$

Solve the equation for [OH$^-$] by dividing both sides by 1×10^{-3} M.

$$[OH^-] = \frac{1 \times 10^{-14}}{1 \times 10^{-3}} \ M = 1 \times 10^{-11} \ M$$

A solution with a [H$^+$] equal to 1×10^{-3} M has an [OH$^-$] of 1×10^{-11} M. Note that the product of these two numbers is 1×10^{-14}, the K_w value for water. Example Problem 17.1 illustrates another problem involving H$^+$ and OH$^-$ ions.

Example Problem 17.1

What is the $[H^+]$ of a solution prepared by adding 0.010 mol NaOH(s) to 2.0 L of water? Assume that no volume change occurs when the NaOH is added.

Solution

1. *What is unknown?* $[H^+]$ or $\dfrac{\text{mol } H^+}{1 \text{ L}}$

2. *What is known?* 0.010 mol NaOH, 2.0 L of solution

3. Determine the $[OH^-]$. NaOH is a strong base. Therefore, for each mole dissolved, a mole each of OH^- and Na^+ is added to water.

$$NaOH(s) \xrightarrow{\text{water}} Na^+(aq) + OH^-(aq)$$

If 0.010 mol NaOH is dissolved, then 0.010 mol OH^- is added to the 2.0 L of water. To calculate the molar concentration of OH^-, the number of moles of OH^- is divided by the total volume of solution.

$$M = \frac{\text{mol solute}}{\text{L solution}}$$

$$[OH^-] = \frac{0.010 \text{ mol } OH^-}{2.0 \text{ L}}$$
$$= 0.0050 \text{ M } OH^-$$
$$= 5.0 \times 10^{-3} \text{ M } OH^-$$

4. Determine the $[H^+]$ using the K_w expression.

$$K_w = [H^+][OH^-] = 1.0 \times 10^{-14}$$
$$= [H^+](5.0 \times 10^{-3} \text{ M } OH^-) = 1.0 \times 10^{-14}$$

$$[H^+] = \frac{1.0 \times 10^{-14}}{5.0 \times 10^{-3}} M$$
$$= 2.0 \times 10^{-12} M$$

A solution having an $[OH^-]$ equal to 5.0×10^{-3} M contains a $[H^+]$ equal to 2.0×10^{-12} M.

pH A commonly used means for expressing a solution's acidity is pH. Mathematically, pH is defined as follows:

$$pH = -\log [H^+]$$

In chemistry, p is an abbreviation for −log. The symbol p is also used with equilibrium constants, i.e., pK, meaning −log K.

where log is the common logarithm and $[H^+]$ is the molar concentration of the H^+.

Logarithms are nothing more than exponents. Common logarithms are exponents of 10. Accordingly, the log of a number is the exponent of 10 that gives that number. For instance, what is the log of 10 (10^1)?

$$\log 10^1 = 1$$

When taking a logarithm, ask yourself: To what exponent must I raise 10 to yield the desired number? What is the log of 100 (10^2)? What exponent of 10 equals 100? Clearly, the answer, 2, is the only exponent of 10 that gives 100:

$$10^{\log} = 10^2 = 100$$

Table 17.2 presents logarithms of numbers expressed exponentially.

Logarithms are used whenever the pH of a solution is calculated. For example, what is the pH of pure water? Pure water contains 1×10^{-7} M H^+. Thus, to determine the pH we need to determine the log of 1×10^{-7}, and then multiply by −1 (change the sign).

$$
\begin{aligned}
pH &= -\log [H^+] \\
&= -\log (1 \times 10^{-7} \, M) \\
&= -(-7.0) = 7.0
\end{aligned}
$$

The pH of pure water is 7.0. Notice that the pH is nothing more than the negative exponent of 10 for $[H^+]$.

$$1 \times 10^{-pH} = [H^+]$$

TABLE 17.2 LOGARITHMS

Number	Number expressed exponentially	Log of number
1,000,000	10^6	6
1,000	10^3	3
10	10^1	1
1	10^0	0
0.1	10^{-1}	−1
0.001	10^{-3}	−3
0.000 001	10^{-6}	−6

FIGURE 17.2
pH is the negative logarithm of the hydrogen ion concentration. Neutral solutions have a pH equal to 7. Acidic solutions have pH values less than 7, and basic solutions have pH values greater than 7. A change of one pH unit represents a tenfold change in hydrogen ion concentration.

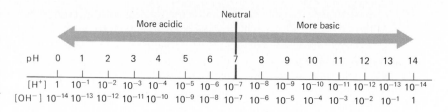

Acidic solutions have H^+ concentrations greater than 10^{-7} M. Accordingly, the values for the pH of acidic solutions are less than 7. What is the pH of an acid solution containing 0.1 M H^+?

A logarithm, such as pH, can have as many decimal places as the number of significant figures in the measurement from which it is calculated.

$$\begin{aligned} pH &= -\log [H^+] \\ &= -\log (0.1 \text{ M}) \\ &= -\log (1 \times 10^{-1} \text{ M}) \\ &= 1.0 \end{aligned}$$

A solution with 0.1 M H^+ has a pH of 1.0.

In basic solutions, the H^+ concentration is less than 10^{-7} M, so pH values of basic solutions are larger than 7.0. What is the pH of a solution with a $[H^+]$ equal to 1×10^{-12} M?

$$\begin{aligned} pH &= -\log [H^+] \\ &= -\log (1 \times 10^{-12} \text{ M}) \\ &= 12.0 \end{aligned}$$

Approximate pH of some common substances:

Ammonia cleaner	12
Detergents	10
Seawater	8.5
Egg white	8
Blood serum	7.4
Rainwater	6
Coffee	5
Wine	3.5
Vinegar	3
Soft drinks	3
Lemons	2
Gastric juice	1

Figure 17.2 illustrates the most common range of pH values. If the pH of a solution is less than 7, the solution is acidic; if it is greater than 7, the solution is basic; and if it equals 7, the solution is neutral.

$$\begin{aligned} pH < 7 \quad & [H^+] > 1 \times 10^{-7} \text{ M} = \text{acidic solution} \\ pH > 7 \quad & [H^+] < 1 \times 10^{-7} \text{ M} = \text{basic solution} \\ pH = 7 \quad & [H^+] = 1 \times 10^{-7} \text{ M} = \text{neutral solution} \end{aligned}$$

Example Problems 17.2 and 17.3 show how the pH of various solutions are calculated.

Example Problem 17.2

What is the pH of a solution that is prepared by adding 0.561 g KOH to 10.0 L of water? Assume that the total volume of the solution is 10.0 L.

Solution

1. *What is unknown?* pH of the solution
2. *What is known?* 0.561 g KOH, and 10.0 L of water
3. Find $[OH^-]$. KOH is a strong base and completely dissociates. For each mole of KOH that dissolves, there is 1 mol of dissolved OH^-.

$$mol_{OH^-} = \text{g } \cancel{KOH} \times \frac{1 \text{ mol } \cancel{KOH}}{56.1 \text{ g } \cancel{KOH}} \times \frac{1 \text{ mol } OH^-}{1 \text{ mol } \cancel{KOH}}$$

$$= 0.561 \text{ g } \cancel{KOH} \times \frac{1 \text{ mol } \cancel{KOH}}{56.1 \text{ g } \cancel{KOH}} \times \frac{1 \text{ mol } OH^-}{1 \text{ mol } \cancel{KOH}}$$

$$= 0.0100 \text{ mol } OH^-$$

$$M \text{ } OH^- = \frac{\text{mol } OH^-}{\text{L solution}}$$

$$= \frac{0.0100 \text{ mol } OH^-}{10.0 \text{ L}}$$

$$= 0.00100 \text{ M } OH^-$$

$$= 1.00 \times 10^{-3} \text{ M } OH^-$$

4. Calculate the pH of the solution. To obtain the pH, the $[H^+]$ is required. Because we now know the $[OH^-]$, we can compute the $[H^+]$ by substituting our value for $[OH^-]$ into the K_w expression.

$$K_w = 1.00 \times 10^{-14} = [H^+][OH^-]$$

$$1.00 \times 10^{-14} = [H^+](1.00 \times 10^{-3} \text{ M } OH^-)$$

$$[H^+] = \frac{1.00 \times 10^{-14}}{1.00 \times 10^{-3}} \text{ M}$$

$$= 1.00 \times 10^{-11} \text{ M } H^+$$

To obtain the pH, it is only necessary to calculate the negative logarithm of 1.00×10^{-11} M H⁺.

$$pH = -\log [H^+]$$

$$= -\log (1.00 \times 10^{-11} \text{ M})$$

$$= 11.000$$

A solution prepared by adding 0.561 g KOH to 10.0 L of water has a pH of 11.000.

> **Example Problem 17.3**

Determine the pH of a 0.055 M HCl solution.

Solution

1. *What is unknown?* pH
2. *What is known?* 0.055 M HCl
3. Find the pH of the solution. In previous problems, the $[H^+]$ was expressed with 1 as the coefficient of the exponential term. In this problem, however, we must find the log of 5.5×10^{-2} M HCl. When the coefficient is not 1, we find the log by either entering the number into a calculator or using a logarithm table. Since a large percent of students use calculators, we shall use the first method. If you do not have a calculator, use the log table in the Appendix.

   To find the log by using a calculator, simply enter the number and then press the "log" key. Don't confuse the "log" key with the "ln" key. Pressing the log key returns the common logarithm, while pressing the ln key gives the natural logarithm, that is, log base e. The common logarithm of 0.055 is -1.26.

$$pH = -\log(.055\ M)$$
$$pH = -(-1.26)$$
$$= 1.26$$

A 0.055 M HCl solution has a pH equal to 1.26.

REVIEW PROBLEMS

17.10 (a) Write the K_w expression for water. (b) State in words what this expression means.
17.11 Determine the $[OH^-]$ in solutions with the following $[H^+]$: (a) 1 M, (b) 0.081 M, (c) 1.9×10^{-4} M.
17.12 Determine the $[H^+]$ in a solution prepared by adding 100.5 g $HClO_4$ to enough water to have 25 L of solution.
17.13 What is the definition of pH?
17.14 Determine the pH of solutions with the following $[H^+]$: (a) 1×10^{-2} M, (b) 1×10^{-5} M, (c) 9.4×10^{-3} M.
17.15 Calculate the pH of a solution prepared by adding 0.0020 mol KOH to 1.00 L of water, assuming that the final volume of the solution is 1.00 L.

Acid-Base Indicators

Acid-base indicators are used to determine the approximate pH of a solution. **Acid-base indicators** are organic dyes that change color when the hydrogen ion concentration is changed. These indicators are weak acids or bases whose structures are altered by the addition and loss of protons, enough so that they change colors when protons are added or removed.

TABLE 17.3 **ACID-BASE INDICATORS**

Indicator	pH range	Color	
		Acid	Base
Methyl violet	0.1–1.5	Yellow	Blue
Thymol blue	1.2–2.8	Red	Yellow
Methyl orange	3.1–4.4	Red	Yellow
Bromthymol blue	6.0–7.6	Yellow	Blue
Phenol red	6.4–8.0	Yellow	Red
Phenolphthalein	8.2–10.0	Colorless	Red-pink
Alizarin yellow	10.2–12.0	Yellow	Red

Phenolphthalein has two different structures, depending on the [H⁺]; one compound is colorless and the other is deep pink in solution.

One of the most frequently used acid-base indicators is phenolphthalein. A few drops of phenolphthalein added to a basic solution with a pH greater than 9 changes the color of the solution to a deep pink color. If enough acid is added to the solution to lower the pH below 9, the solution becomes colorless. As we shall see, phenolphthalein is one of many indicators used in acid-base titrations (see Figure 17.3).

Table 17.3 lists other common indicators and the pH range in which they change colors.

pH Meters   For more accurate pH determinations, **pH meters** are used. While indicators can show only the general pH range, most pH meters are capable of measuring to the nearest hundredth of a pH unit. A pH meter is shown in Figure 17.4.

A pH meter functions by measuring the electric potential between two electrodes that are immersed in the solution of interest. One of the electrodes is constructed so that it is sensitive to hydrogen ions, and it changes its electric potential with changing hydrogen ion concentrations.

FIGURE 17.3
During the titration of an acidic solution, standard base is carefully added to the acid solution until reaching the end point. The end point of a titration is normally detected by the change in color of an acid-base indicator. The indicator phenolphthalein changes from colorless to pale pink at the end point.

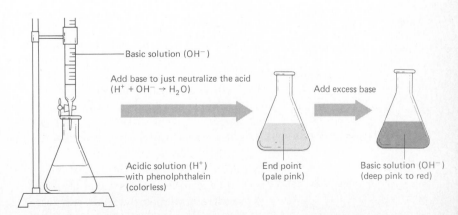

Basic solution (OH⁻)

Add base to just neutralize the acid
(H⁺ + OH⁻ → H₂O)

Add excess base

Acidic solution (H⁺)
with phenolphthalein
(colorless)

End point
(pale pink)

Basic solution (OH⁻)
(deep pink to red)

FIGURE 17.4
A pH meter is an electrical instrument that allows chemists to make quick and accurate measurements of the pH of solutions. (*Corning Glass Company*)

Acid-Base Titration

A titration is a volumetric procedure, i.e., a method that involves mainly volume measurements. Gravimetric procedures are those that involve primarily mass measurements.

An **acid-base titration** is a volumetric procedure in which a base (or acid) of known concentration is systematically added to a fixed volume of an acid (or base) of unknown concentration in order to find the point at which equivalent amounts of acid and base are present.

Acid-base titrations are conducted using burets and volumetric pipets (see Figure 17.5). Generally, a fixed volume of acid of unknown concentration is placed into a flask along with a few drops of the appropriate acid-base indicator. A standard base, one with a known concentration, is placed into the buret. The standard base is carefully added to the acid until the indicator changes color; this change of color signifies the equivalence point, i.e., the point at which equivalent amounts of acid and base are in the flask.

Let's turn our attention specifically to the titration of strong acids and strong bases. We shall not consider titrations involving weak acids or bases. A neutralization reaction occurs when acids and bases are combined. As we have already seen, the net ionic equation for the reaction of a strong acid and strong base is

$$H^+(aq) + OH^-(aq) \longrightarrow H_2O(l)$$

If a strong acid is titrated with a strong base, the solution in the flask is initially

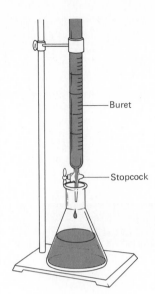

FIGURE 17.5

Titrations are performed using one or two burets, a pipet, and a flask. Normally, a known volume of acid is pipetted into the flask and the indicator is added. Then base is slowly added from the buret to the flask until the indicator just changes color.

The point at which an indicator changes color in an acid-base titration is commonly called the *end point.* Indicators are selected to change color as close as possible to the equivalence point; however, this is not always the case. The equivalence point and the end point may not be exactly the same.

acidic. With the addition of base, the amount of H^+ decreases. At a certain point, the number of moles of OH^- equals the number of moles of H^+. This is the **equivalence point.**

$$\text{mol } H^+ = \text{mol } OH^-$$

An equivalence point is detected by a change in the color of an acid-base indicator. Figure 17.6 graphically illustrates what happens to a solution's pH when a strong acid is titrated with a strong base; the graph is called a *titration curve.*

FIGURE 17.6

Most acid-base titration curves are plots of the pH of the solution versus the number of milliliters of base added. Initially upon adding a strong base to a strong acid, there is a small gradual change in the pH of the solution. As the equivalence point is approached, the pH of the solution changes very rapidly. Beyond the equivalence point, the pH again rises slowly.

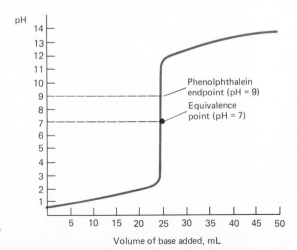

When the molarity of the H^+ of the acid is multiplied by its volume in liters, the number of moles of H^+ is obtained.

$$M_{H^+}V_{H^+} = \frac{\text{mol } H^+}{L} \times L$$

$$= \text{mol } H^+$$

Likewise, if the molarity of the OH^- of the base is multiplied by its volume in liters, the number of moles of OH^- are obtained.

$$M_{OH^-}V_{OH^-} = \frac{\text{mol } OH^-}{L} \times L$$

$$= \text{mol } OH^-$$

At the equivalence point in a titration, the number of moles of H^+ from the acid equals the number of moles of OH^- from the base. Therefore,

$$M_{H^+}V_{H^+} = M_{OH^-}V_{OH^-}$$

The product of the molar concentration of H^+ and the volume of H^+ (in liters) equals the product of the molar concentration of OH^- and the volume of OH^-. Example Problem 17.4 shows how this equation can be applied when solving titration problems.

Example Problem 17.4

A titration is performed, and it is found that 25.44 mL of 0.1212 M NaOH is required to reach the equivalence point with 30.00 mL HCl of unknown concentration. What is the molar concentration of the HCl solution?

Solution

1. *What is unknown?* molarity of HCl, [HCl]
2. *What is known?*   25.44 mL of 0.1212 M NaOH, 30.00 mL HCl
3. Determine $[OH^-]$. Before the equation can be applied, the molarity of the OH^- must be determined. Since NaOH is a strong base, and there is one mole of hydroxide per mole of NaOH, the molarity of the OH^- equals the molarity of NaOH.

$$[OH^-] = M_{NaOH} \times \frac{1 \text{ mol } OH^-}{1 \text{ mol NaOH}}$$

$$[OH^-] = \frac{0.1212 \text{ mol NaOH}}{L} \times \frac{1 \text{ mol OH}^-}{1 \text{ mol NaOH}}$$

$$= 0.1212 \text{ M OH}^-$$

4. Substitute into the equation.

$$M_{H^+}V_{H^+} = M_{OH^-}V_{OH^-}$$

Rearranging the equation, we have

$$M_{H^+} = M_{OH^-}\frac{V_{OH^-}}{V_{H^+}}$$

$$= 0.1212 \text{ M} \times \frac{0.02544 \text{ L}}{0.03000 \text{ L}}$$

$$= 0.1028 \text{ M H}^+$$

HCl is a strong acid and contains 1 mol of hydrogen per mole of HCl. There-fore, the concentration of HCl equals the H^+ concentration, 0.1028 M.

REVIEW PROBLEMS

17.16 What color is observed for the following indicators if the pH of the solution is 7.0: (a) methyl orange, (b) alizarin yellow, and (c) phenolphthalein?

17.17 What is true about the H^+ and OH^- concentrations at the equivalence point of a titration of a strong acid and strong base?

17.18 What volume of 0.100 M NaOH is required to neutralize: (a) 943 mL of 0.124 M HCl, (b) 12.2 mL of 0.438 M H_2SO_4, and (c) 1.23 L of 0.250 M HNO_3?

17.5 NORMALITY

A replaceable H^+ or OH^- is one that dissociates when added to water. Some acids contain hydrogen atoms that never dissociate. For example, in acetic acid, $HC_2H_3O_2$, there are four hydrogen atoms in the molecule; only one is replace-able.

Many chemists prefer not to use molar concentrations when solving acid-base titration problems. Instead, they use a concentration unit called normality. **Normality** is defined as the number of equivalents per liter. The primary reason for using normality is that acids such as H_2SO_4 contain 2 mol of replaceable H^+ ions per mole of acid. If H_2SO_4 is used in a titration, only half the number of moles of H_2SO_4 are required to neutralize a base compared with acids that contain only 1 mol of replaceable hydrogens per molecule. A similar situation exists for bases that contain more than one OH^- per molecule.

To avoid calculation problems associated with acids with more than one replaceable H^+ and bases with more than one replaceable OH^-, chemists deter-mine the equivalent mass for acids and bases. One **equivalent mass** for an acid is defined as the mass of acid that produces 1 mol of H^+ in solution. Sulfuric acid's

TABLE 17.4   EQUIVALENT MASSES OF SELECTED ACIDS AND BASES

Compound	Molar mass, g	Equivalent mass, g
Acid		
HCl	36.5	36.5
HClO$_4$	100.5	100.5
H$_2$SO$_4$	98.0	49.0
H$_2$C$_2$O$_4$	90.0	45.0
Base		
NaOH	40.0	40.0
KOH	56.1	56.1
Mg(OH)$_2$	58.4	29.2
Ca(OH)$_2$	74.2	37.1

Acids that donate only one proton are called *monoprotic* acids; those that donate two protons are *diprotic* acids, and those that donate three protons are *triprotic* acids.

molecular mass is 98 g. If 98 g H$_2$SO$_4$ is dissolved in water, 2.0 mol of H$^+$ are donated to the water. Accordingly, 49 g ($\frac{1}{2}$ × 98 g) H$_2$SO$_4$, when dissolved, donate 1.0 mol H$^+$.

To calculate the equivalent mass of any acid, it is only necessary to divide the molecular mass of the acid by the number of replaceable hydrogens.

$$\text{Equivalent mass of an acid} = \frac{\text{molecular mass of acid}}{\text{no. of replaceable } H^+ \text{ per mole}}$$

In a similar manner, the equivalent mass of any base is calculated by dividing the base's formula mass by the number of replaceable OH$^-$ per mole.

$$\text{Equivalent mass of a base} = \frac{\text{formula mass of base}}{\text{no. of replaceable } OH^- \text{ per mole}}$$

Table 17.4 summarizes the equivalent masses for selected acids and bases.

A 1 normal (1 N) solution can be prepared by adding one equivalent (1 equiv) of an acid or base to a flask and adding enough water to yield a total solution volume of 1 L. The term "one equivalent" is a shorthand way of stating the mass of an acid that contains 1 mol H$^+$, or the mass of a base that contains 1 mol OH$^-$. One liter of 1.0 N H$_2$SO$_4$ is prepared by adding 49 g H$_2$SO$_4$ to a flask, and then adding water to give a total volume of 1.0 L of solution.

$$1.0 \text{ N H}_2\text{SO}_4 = \frac{1.0 \text{ equiv}}{1.0 \text{ L solution}} = \frac{49 \text{ g H}_2\text{SO}_4}{1.0 \text{ L}}$$

To find the number of equivalents in a sample, the mass of the sample is divided by the mass of one equivalent.

$$\text{Number of equivalents} = \frac{\text{mass of sample}}{\text{mass of one equivalent}}$$

Example Problem 17.5 further illustrates equivalents and the normality concept.

Example Problem 17.5

Determine what mass of pure H_2SO_4 would be required to prepare 235 mL of 0.165 N H_2SO_4.

Solution

1. *What is unknown?*   g H_2SO_4

2. *What is known?*   235 mL of 0.165 N H_2SO_4, $\dfrac{49.0 \text{ g } H_2SO_4}{1 \text{ equiv}}$

3. Apply the factor-label method.

$$235 \text{ mL} \times \frac{1 \text{ L}}{1000 \text{ mL}} \times \frac{0.165 \text{ equiv}}{L} \times \frac{49.0 \text{ g } H_2SO_4}{\text{equiv}} = 1.90 \text{ g } H_2SO_4$$

First convert the number of milliliters of solution to liters so that the volume can be multiplied by the normality (equiv/L), which gives the total number of equivalents in the solution. Then multiply by the equivalent mass to obtain the mass of H_2SO_4.

The advantage of using normality is that a 1 N acid solution, no matter how many replaceable hydrogens the acid possesses, always neutralizes an equal volume of 1 N base, with any number of replaceable hydroxides. At the equivalence point in an acid-base titration, the following equation is applied

$$N_a \times V_a = N_b \times V_b$$

where N_a and N_b are the normalities of the acid and base, and V_a and V_b are the volumes of the acid and base expressed in liters. Example Problem 17.6 shows how this equation is applied for an acid-base titration.

Example Problem 17.6

Determine the volume of 0.424 N H_2SO_4 that is required to neutralize 31.4 mL of 0.593 N KOH.

Solution

1. *What is unknown?*   L of 0.424 N H_2SO_4
2. *What is known?*   0.424 N H_2SO_4, 31.4 mL of 0.593 N KOH
3. Apply the equation.

$$N_a \times V_a = N_b \times V_b$$

Rearrange the equation and substitute known values.

$$V_a = \frac{N_b \times V_b}{N_a}$$

$$= \frac{0.593 \; N \; \text{KOH} \times 31.4 \; \text{mL} \times \dfrac{1 \; \text{L}}{1000 \; \text{mL}}}{0.424 \; N \; H_2SO_4}$$

$$= 0.0439 \; \text{L} = 43.9 \; \text{mL of } 0.424 \; N \; H_2SO_4$$

Addition of 43.9 mL of 0.424 N H_2SO_4 neutralizes 31.4 mL of 0.593 N KOH.

REVIEW PROBLEMS **17.19** Determine the equivalent mass for: (a) HBr, (b) H_3PO_4, (c) H_2SO_3, (d) $H_2C_2O_4$.
17.20 How would 500 mL of 0.650 N H_2SO_4 be prepared?
17.21 Determine the normality of 35.0 mL NaOH that required 39.1 mL of 0.349 *M* H_2SO_4 to neutralize the solution.

SUMMARY

Acids are substances that have a sour taste, combine with bases to produce salts and water, change the colors of acid-base indicators, and combine with active metals to produce hydrogen gas. Bases have a bitter taste, neutralize acids, change the colors of acid-base indicators opposite to the way that acids change them, and feel slippery when touched.

Acids and bases are defined in a number of different ways. One of the earliest and most useful sets of definitions was proposed by Svante Arrhenius. He defined an acid as a substance that increases the $[H^+]$ in water, and a base as a substance that increases the $[OH^-]$ in water. Another definition, proposed by Brønsted and Lowry is that acids are proton donors and bases are proton acceptors.

According to the Brønsted-Lowry definition, once an acid gives up a proton, the resulting chemical species is the conjugate base of the acid. A conjugate base can accept a proton and re-form the acid. Bases accept protons and produce conjugate acids. Strong acids produce weak conjugate bases, and strong bases produce weak conjugate acids.

Acids react with bases to produce salts and water—a neutralization reaction. Different net ionic equations result, depending on the strength of the acids and bases combined. If equivalent amounts of a strong acid and strong base are combined, the resulting salt solution is neutral. When equivalent amounts of a strong base and weak acid are combined, the resulting salt solution is basic. A strong acid combined with a weak base gives an acidic solution.

Acid and base strength is determined by measuring the $[H^+]$ and $[OH^-]$ in water. If the $[H^+]$ is greater than 10^{-7} M, the solution is acidic. If the $[H^+]$ is less than 10^{-7} M, the solution is basic. If the $[H^+]$ equals 10^{-7} M, the solution is neutral. Commonly, acid and base strength is expressed in terms of $-\log [H^+]$, or pH.

Acid-base titrations are conducted to determine the molarity or normality of solutions with unknown concentrations of H^+ or OH^-. An acid-base titration is the systematic addition of an acid or base of known concentration to a base or acid of unknown concentration in order to determine the point at which the number of moles of H^+ equals the number of moles of OH^-.

QUESTIONS AND PROBLEMS

17.22 Define the following: Arrhenius acid, Arrhenius base, Brønsted-Lowry acid, Brønsted-Lowry base, conjugate acid, conjugate base, amphiprotic, strong acid, strong base, weak acid, weak base, neutralization, titration, pH, acid-base indicator, acid-base titration, equivalence point, equivalent mass, equivalent, normality.

Acid-Base Definitions

17.23 (a) Write two equations that illustrate the ionization of pure water. (b) Which equation better represents this ionization?

17.24 Classify each of the following substances as an Arrhenius acid or base: (a) $HClO$, (b) $Al(OH)_3$, (c) $RbOH$, (d) HI, (e) $HC_2H_3O_2$, (f) $NH_3(aq)$.

17.25 Write equations showing what happens when each of the compounds listed in Problem 17.24 is dissolved in water.

17.26 Consider the gas-phase reaction of $HCl(g)$ and $NH_3(g)$ to produce $NH_4Cl(s)$:

$$HCl(g) + NH_3(g) \longrightarrow NH_4Cl(s)$$

(a) Use the Brønsted-Lowry definitions and identify the acid and base in the equation. (b) Would this reaction be considered an acid-base reaction using the Arrhenius definition?

17.27 For each of the following equations, identify Brønsted-Lowry acids, bases, conjugate acids, and conjugate bases:
(a) $CO_2 + H_2O \rightleftharpoons H_2CO_3$
(b) $H_2SO_4 + NaF \rightleftharpoons NaHSO_4 + HF$
(c) $KOH + HCN \rightleftharpoons KCN + H_2O$
(d) $NH_3 + HC_2H_3O_2 \rightleftharpoons NH_4C_2H_3O_2$

17.28 Complete and balance the following equations:
(a) $H_2S(aq) + NaOH(aq) \rightarrow$
(b) $HBr(aq) + Ca(OH)_2 \rightarrow$
(c) $H_3PO_4(aq) + 3LiOH \rightarrow$
(d) $NH_3(aq) + H_2SO_4 \rightarrow$

17.29 Illustrate the amphiprotic nature of water by writing two equations, one in which water is an acid and another in which it is a base.

17.30 For each of the following write two equations to illustrate the substance's amphiprotic nature: (a) $H_2PO_4^-$, (b) NH_3, (c) HS^-, (d) HCO_3^-.

17.31 Write the conjugate bases of each of the following: (a) HNO_3, (b) HSO_4^-, (c) OH^-, (d) HF, (e) NH_3, (f) HPO_4^{2-}.

17.32 Write the conjugate acids of each of the following: (a) CN^-, (b) NO_2^-, (c) NH_3, (d) H_2O, (e) HNO_3, (f) HPO_4^{2-}.

Strength of Acids and Bases

17.33 Use Table 17.1 to determine which one in each of the following pairs of substances is the stronger base:
(a) $HClO_4$ or H_2SO_4
(b) F^- or NH_2^-
(c) NH_3 or NH_4^+
(d) I^- or H_2O
(e) $C_2H_3O_2^-$ or HCO_3^-

17.34 Considering the placement of the nonmetals in the periodic table, predict the strongest acid in each of the following pairs:
(a) HF or HCl
(b) H_2O or NH_3
(c) H_2Te or H_2S
(d) PH_3 or AsH_3

Reactions of Acids and Bases

17.35 Complete the following equations of strong acids combining with strong bases:
(a) $HCl + RbOH \rightarrow$
(b) $H_2SO_4 + Ca(OH)_2 \rightarrow$
(c) $HClO_4 + Ba(OH)_2 \rightarrow$
(d) $NaOH + HNO_3 \rightarrow$
(e) $HI + Sr(OH)_2 \rightarrow$

17.36 For each equation in Problem 17.35, write an overall equation and a net ionic equation.

17.37 How is the reaction of a strong base with a strong acid different than the reaction of a strong base with a weak acid?

17.38 Write the net ionic equations for the combination of the following strong bases with weak acids:
(a) $NaOH + HF \rightarrow$
(b) $KOH + HNO_2 \rightarrow$
(c) $NaOH + H_2C_2O_4 \rightarrow$

Measurements of Acids and Bases

17.39 For each of the H^+ or OH^- concentrations given, determine if the solution is acidic or basic:
(a) $[H^+] = 1 \times 10^{-3}$ M
(b) $[H^+] = 1 \times 10^{-12}$ M
(c) $[OH^-] = 1 \times 10^{-5}$ M
(d) $[OH^-] = 3.5 \times 10^{-13}$ M

17.40 Determine the $[H^+]$ in aqueous solutions containing the following concentrations of OH^-:
(a) 1×10^{-3} M
(b) 5.3×10^{-8} M
(c) 9.7×10^{-7} M

17.41 Determine the $[OH^-]$ of the following solutions:
(a) $[H^+] = 1.0 \times 10^{-1}$ M
(b) $[H^+] = 0.0001$ M
(c) $[H^+] = 1 \times 10^{-12}$ M

17.42 Find the $[H^+]$ and $[OH^-]$ of solutions prepared as follows: (a) 3.65 g HCl dissolved in 10.0 L of solution, (b) 1.63 g HNO_3 dissolved in 1.10 L of solution, (c) 6.00 g $HClO_4$ in 6.92 L of solution.

17.43 Calculate the $[H^+]$ and $[OH^-]$ of the following solutions:
(a) 0.133 g NaOH dissolved in 5.9 L of solution
(b) 0.22 mol KOH in 2.3×10^3 mL of solution
(c) 0.340 g $Mg(OH)_2$ in 121 L of solution

17.44 Determine the pH of the following solutions:
(a) $[H^+] = 1.0 \times 10^{-5}$ M
(b) $[H^+] = 1 \times 10^{-1}$ M
(c) $[H^+] = 0.1$ M

17.45 Determine the pH of the following solutions:
(a) $[OH^-] = 1.0 \times 10^{-4}$ M
(b) $[OH^-] = 1.0 \times 10^{-11}$ M
(c) $[OH^-] = 1.0 \times 10^{-7}$ M

17.46 Calculate the pH of the following solutions:
(a) 0.015 M HBr
(b) 6.1×10^{-3} M HNO_3
(c) 0.000432 M $HClO_4$
(d) 5.9×10^{-4} M NaOH
(e) 0.000033 M KOH

17.47 Find the pH of solutions prepared as follows:
(a) 0.717 g HCl dissolved in 2.11 L of solution
(b) 0.033 mol HNO_3 in 864 mL of solution
(c) 0.256 g LiOH in 9.1 L of solution
(d) 8.7×10^{-4} g $Sr(OH)_2$ in 7.3 L of solution
(e) 4.91 mg $Mg(OH)_2$ in 934 mL of solution

Acid-Base Titration

17.48 Determine the volume of 0.350 M NaOH that is required to neutralize the following amounts of HCl:
(a) 100 mL of 0.100 M HCl
(b) 1.25 L of 0.500 M HCl
(c) 14.3 mL of 0.035 M HCl
(d) 10.9 L of 1.50 M HCl

17.49 What is the molarity of a KOH solution in which 53.2 mL is the volume needed to neutralize 78.3 mL of 0.0922 M HNO_3?

17.50 What volume of the following acids can be completely neutralized with 50.0 mL of 0.345 M NaOH?
(a) 0.500 M $HClO_4$
(b) 0.0455 M HBr
(c) 0.111 M HI

17.51 How many milliliters of 0.150 M KOH are needed to reach the equivalence point for the following acid solutions?
(a) 0.0944 L of 0.370 M H_2SO_4
(b) 23.9 mL of 0.555 M H_3PO_4
(c) 0.0335 mL of 0.675 M HNO_3

17.52 A HCl solution is titrated, and 55.3 mL of 0.144 M NaOH is required to reach the equivalence point.
(a) How many moles of HCl(aq) are in the solution?

(b) What mass of HCl(aq) is contained in the solution?
(c) If the total volume of the HCl(aq) solution is 49.1 mL, what is the molarity of the H^+?
(d) What is the pH of the solution?

17.53 Calculate the $[H^+]$, $[OH^-]$, $[K^+]$, $[ClO_4^-]$, and pH of a solution prepared by mixing 75.0 mL of 0.200 M $HClO_4$ with 125.0 mL of 0.175 M KOH.

Normality

17.54 Find the gram equivalent mass for the following: (a) $Ba(OH)_2$, (b) H_2S, (c) LiOH, (d) H_2SO_3.

17.55 How many equivalents are found in each of the following samples: (a) 1.99 g $Ca(OH)_2$, (b) 45.7 g HNO_3, (c) 7.88 g H_3PO_4, (d) 0.222 g H_2SO_4, (e) 9.11 g NH_3?

17.56 What is the normality of the following solutions?
(a) 1.33 g NaOH dissolved in 100 mL of solution
(b) 0.335 g H_2SO_4 dissolved in 1.50 L of solution
(c) 1.07 g H_3PO_4 dissolved in 3400 mL
(d) 0.0019 mol $Mg(OH)_2$ in 9.95 L of solution

17.57 What mass of the following acids is required to prepare 465 mL of 0.075 N solutions: (a) HNO_3, (b) H_2SO_4, (c) H_3PO_4, (d) HBr?

17.58 Determine the volume of 0.116 N H_2SO_4 needed to neutralize the following:
(a) 9.11 L of 0.100 N NaOH
(b) 10.3 mL of 0.022 N KOH
(c) 45.6 mL of 4.5×10^{-4} N LiOH

General Problems

17.59 A titration is performed on a 1.255-g sample of an unknown acid. After the acid is dissolved, 21.59 mL of 0.2921 M NaOH is required to neutralize the acid. (a) What is the equivalent mass of the acid? (b) If it is known that the unknown acid has two replaceable hydrogens, what is the molecular mass of acid?

17.60 What volume of 0.100 M H_3PO_4 is required to combine with excess KI to produce 145 g HI in the following reaction?

$$H_3PO_4 + KI \longrightarrow KH_2PO_4 + HI$$

17.61 A beaker contains 225.0 mL of 0.1000 M HCl. A 1.000-g sample of $NaHCO_3$ is added to the beaker, and the following reaction takes place.

$$HCl(aq) + NaHCO_3(s) \longrightarrow$$
$$NaCl(aq) + H_2O(l) + CO_2(g)$$

(a) What volume of $CO_2(g)$ is liberated at 25.0°C and 1.00 atm? (b) What is the molar concentration of the dissolved $Na^+(aq)$? (c) What volume of 0.1000 M NaOH is required to neutralize the solution?

· C H A P T E R ·

Oxidation-Reduction

Study Guidelines

After completing Chapter 18, you should be able to:

1. Define oxidation and reduction in terms of electron transfer and change in oxidation numbers
2. Distinguish between an oxidizing agent and a reducing agent
3. Identify what has undergone oxidation and reduction and what the oxidizing and reducing agents are, given a chemical equation
4. Balance redox equations using the ion-electron method
5. Describe electron flow and chemical reactions that occur in electrolytic cells
6. Discuss the use of electrolytic cells in industry
7. Identify the anode and cathode in electrolytic and galvanic cells, given which substances undergo oxidation and reduction
8. Distinguish between an electrolytic and galvanic cell
9. Describe electron flow and chemical reactions that occur in galvanic cells
10. Describe commonly used galvanic cells

Redox is an abbreviation for *reduction* and *oxidation*.

Oxidation and reduction reactions (redox for short) occur when electrons move from one substance to another. Thus, one substance loses electrons and another gains electrons. Losing electrons is called *oxidation*, and gaining electrons is called *reduction*.

Oxidation occurs when a substance loses electrons.

Reduction occurs when a substance gains electrons.

Oxidation is also defined as adding oxygen to a compound or removing hydrogen from a compound. Reduction is defined as adding hydrogen to or removing oxygen from a compound.

Oxidation and reduction always occur together. When a substance loses electrons (oxidation), another substance must gain these electrons (reduction) (see Figure 18.1).

Let's consider the oxidation-reduction reaction in which metallic sodium, $Na(s)$, is combined with chlorine gas, $Cl_2(g)$, to produce sodium chloride, $NaCl(s)$.

$$2Na(s) + Cl_2(g) \longrightarrow 2NaCl(s)$$

To determine which substance has undergone oxidation or reduction, it is necessary to assign oxidation numbers for each element in the equation, and then look for those numbers that have changed. You might want to review the rules for assigning oxidation numbers given in Section 9.1.

Both reactants, Na and Cl_2, are elements, so their oxidation numbers are zero. In NaCl sodium has an oxidation number of $+1$ and chlorine has an oxidation number of -1. Frequently, the oxidation number of each element is written below the symbol within the equation, as follows:

$$\underset{0}{2Na(s)} + \underset{0}{Cl_2(g)} \longrightarrow \underset{+1\ -1}{2NaCl(s)}$$

Oxidation

Reduction

Leo the germ is a mnemonic that helps some to remember that "*lose electrons oxidation, and gain electrons reduction.*" An m is added at the end to form an English word.

Sodium's oxidation number has increased during the reaction, and chlorine's has decreased. What has happened to change the oxidation numbers? If the oxidation number of a substance increases (becomes more positive), this indicates that the substance has lost electrons. Electrons are negative particles; consequently, when an electron is lost, the substance becomes more positive.

$$2Na \longrightarrow 2Na^+ + 2e^- \quad \text{(oxidation)}$$

At exactly the same time, the electrons lost by Na are picked up by Cl_2, resulting in the decrease in the oxidation number of Cl_2.

FIGURE 18.1
Oxidation occurs when a substance loses electrons, and reduction occurs when a substance gains electrons.

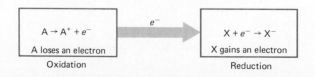

FIGURE 18.2
Reducing agents bring about the reduction of other substances by undergoing oxidation. Oxidizing agents bring about the oxidation of other substances by undergoing reduction.

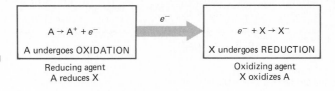

$A \rightarrow A^+ + e^-$

A undergoes OXIDATION

Reducing agent
A reduces X

e^-

$e^- + X \rightarrow X^-$

X undergoes REDUCTION

Oxidizing agent
X oxidizes A

$$2e^- + Cl_2 \longrightarrow 2Cl^- \quad \text{(reduction)}$$

Therefore, in our example reaction, Na undergoes oxidation and produces Na^+, whereas Cl_2 undergoes reduction, yielding Cl^-.

Oxidation and reduction can be redefined in terms of oxidation numbers:

Oxidation occurs when the oxidation number of a substance increases in a chemical reaction.

Reduction occurs when the oxidation number of a substance decreases in a chemical reaction.

An easy way to remember these definitions is to keep in mind that the oxidation number decreases or is *reduced* when the process of reduction occurs. Another example to illustrate oxidation and reduction is:

$$2Ca(s) + O_2(g) \longrightarrow 2CaO(s)$$

First write the oxidation numbers of all atoms.

$$\underset{0}{2Ca(s)} + \underset{0}{O_2(g)} \longrightarrow \underset{+2 \ -2}{2CaO(s)}$$

Oxidation

Reduction

Then, identify the atom whose oxidation number increases, in this case Ca, and the atom whose oxidation number decreases, O_2.

$$Ca \longrightarrow 2e^- + Ca^{2+} \quad \text{(oxidation)}$$

$$O_2 + 4e^- \longrightarrow 2O^{2-} \quad \text{(reduction)}$$

Reducing agents lose electrons, undergo oxidation, and bring about the reduction of other substances. Oxidizing agents gain electrons, undergo reduction, and bring about the oxidation of other substances.

Two terms used to describe the substances undergoing oxidation and reduction are oxidizing agent and reducing agent. An **oxidizing agent** is the reactant that gains electrons, and the **reducing agent** is the reactant that gives up electrons. An oxidizing agent takes electrons from another substance, resulting in the oxidation of that substance. A reducing agent gives up electrons to another substance, bringing about its reduction. In other words, the substance undergoing reduction is the oxidizing agent, and the substance undergoing oxidation is the reducing agent (see Figure 18.2).

Oxidizing agents undergo reduction.

Reducing agents undergo oxidation.

Oxidizing agents are used in a number of medical applications, especially as disinfectants. A 3% solution of hydrogen peroxide is used to destroy microbes in cuts and abrasions. Iodine is a weak oxidizing agent and is used to kill bacteria. It is generally applied as tincture of iodine, an alcohol solution of I_2.

Another example should help clarify these definitions. Consider the reaction of C and HNO_3:

$$C + 4HNO_3 \longrightarrow CO_2 + 4NO_2 + 2H_2O$$

In this reaction, carbon's oxidation number is initially 0 and increases to +4; consequently, C has undergone oxidation, or lost electrons. These electrons bring about the reduction of the N in HNO_3, so C is the reducing agent in this reaction. Nitric acid, HNO_3, accepts the electrons from C, which means that HNO_3 is the oxidizing agent.

When considering the elements, metals tend to be good reducing agents and nonmetals are generally better oxidizing agents. Alkali metals and alkaline earth metals are among the strongest reducing agents known. Halogens and oxygen, O_2, are the strongest of the oxidizing agents. They can take electrons from most other elements.

REVIEW PROBLEMS

18.1 What occurs in oxidation-reduction reactions?

18.2 Define and give an example for each of the following: (a) oxidation, (b) reduction, (c) oxidizing agent, (d) reducing agent.

18.3 In the following equations determine what substances are undergoing oxidation and reduction and what substances are the oxidizing and reducing agents:
(a) $H_2 + I_2 \longrightarrow 2HI$
(b) $2CO + O_2 \longrightarrow 2CO_2$
(c) $3Mg + N_2 \longrightarrow Mg_3N_2$
(d) $PCl_3 + Cl_2 \longrightarrow PCl_5$
(e) $6Fe^{2+} + 14H^+ + Cr_2O_7^{2-} \longrightarrow 2Cr^{3+} + 6Fe^{3+} + 7H_2O$

18.2 BALANCING OXIDATION-REDUCTION EQUATIONS

To better understand redox reactions and the electron transfers that occur, we will now describe how to balance redox equations. Earlier you learned the inspection method for balancing equations. Such a method is fine for simple equations, but it is inadequate for more complex redox equations.

One of the most common procedures used to balance redox equations is the ion-electron method. This method is based on the fact that the number of electrons lost by the substance undergoing oxidation must equal the number of electrons gained by the substance undergoing reduction. The **ion-electron method** is most commonly used to balance ionic oxidation-reduction equations in solutions of acids or bases. Two half-reactions are written, one for the oxida-

Half-reactions are more properly called half-equations. In themselves they are not equations but representations of either the oxidation or reduction part of a chemical reaction.

tion and one for the reduction. After the number of electrons in each half-reaction is balanced, the half-reactions are added together to give the net ionic equation.

To use the ion-electron method for balancing redox equations in acidic solutions, five steps are followed.

Step 1. For each substance that undergoes oxidation and reduction, write the substance in its initial form, then draw an arrow, and write the substance in its final form.

Step 2. Balance all atoms, except H and O, by placing the proper coefficients in front of the substances.

Step 3. Add enough water molecules to each equation to balance the oxygen atoms, and add H^+ to balance the hydrogen atoms.

Step 4. Add electrons to the appropriate side of the equation so that the charges on each side of the half-reaction are equal.

Step 5. Equalize the electrons transferred in each half-reaction, and then add the two half-reactions together, canceling all electrons. The resulting equation is the correctly balanced net ionic equation.

As an example, let's balance the following equation:

$$Cr_2O_7^{2-} + Fe^{2+} \longrightarrow Cr^{3+} + Fe^{3+} \quad \text{(acid)}$$

Potassium dichromate, $K_2Cr_2O_7$, is commonly used as an oxidizing agent. In solution, dichromate ion gives the solution a deep orange color. After oxidation in the presence of acid, the solution turns green to violet, indicating that Cr^{3+} has formed.

Step 1. Write the formula of each substance that undergoes oxidation and reduction separated by an arrow.

$$Fe^{2+} \longrightarrow Fe^{3+}$$
$$Cr_2O_7^{2-} \longrightarrow Cr^{3+}$$

Step 2. Balance all atoms except H and O. Balancing the Fe half-reaction (Steps 2 to 4) merely involves adding an electron to the right side of the half-reaction to balance the charges:

$$Fe^{2+} \longrightarrow Fe^{3+} + e^- \quad \text{(oxidation)}$$

Therefore, we shall concentrate on the dichromate half-reaction.

$$Cr_2O_7^{2-} \longrightarrow 2Cr^{3+}$$

a 2 is placed as the coefficient of Cr^{3+} to balance the chromium atoms.

Step 3. Add enough H_2O molecules to balance the oxygen atoms, and then add H^+ to balance the hydrogen atoms.

There are seven oxygen atoms on the left, therefore, seven water molecules are added to the right to balance the oxygen atoms—always balance the oxygen atoms first.

$$Cr_2O_7{}^{2-} \longrightarrow 2Cr^{3+} + 7H_2O$$

Adding 7 H_2O molecules to the right side adds 14 hydrogen atoms, which are balanced by placing 14 H^+ on the left side.

$$14H^+ + Cr_2O_7{}^{2-} \longrightarrow 2Cr^{3+} + 7H_2O$$

Step 4. Add electrons to the appropriate side of the equation so that the charges on each side of the half-reaction are equal.

The total charge now on the left side is 12+ (14 − 2), and the charge on the right side is 6+ (2 × 3+). To balance the charge, six electrons are added to the left side of the equation.

$$6e^- + 14H^+ + Cr_2O_7{}^{2-} \longrightarrow 2Cr^{3+} + 7H_2O \qquad \text{(reduction)}$$

At this point, we have two correctly balanced half-reactions.

Step 5. Equalize the electrons given off and taken in, and add the two half-reactions together.

As written, one electron is released in the oxidation half-reaction, and six electrons are gained in the reduction half-reaction. To equalize the electrons, the oxidation half-reaction is multiplied by 6.

$$6Fe^{2+} \longrightarrow 6Fe^{3+} + 6e^-$$
$$6e^- + 14H^+ + Cr_2O_7{}^{2-} \longrightarrow 2Cr^{3+} + 7H_2O$$

When the two half-reactions are added, the six electrons cancel, giving the following equation:

$$6Fe^{2+} + 14H^+ + Cr_2O_7{}^{2-} \longrightarrow 2Cr^{3+} + 6Fe^{3+} + 7H_2O$$

Always check to see that both the atoms and charges balance. There are 6 Fe, 2 Cr, 14 H, 7 O, and a total charge of 24+ on either side of the equation.

To use the ion-electron method to balance equations for redox reactions in basic (alkaline) solutions requires that the appropriate number of OH^- and H_2O are added to the half-reactions in Step 3. Some confusion results when deciding on the proper side to add the OH^- and H_2O. A simple way to avoid the problem is to follow the same steps used to balance an equation in acid solution. After adding H^+ and H_2O in Step 3, add OH^- to each side of the equation to "neutralize" ($H^+ + OH^- \longrightarrow H_2O$) all H^+, and then complete the balancing of the equation as above. The following illustrates how this little "trick" works.

Balance the following equation in basic solution:

Potassium permanganate, $KMnO_4$, is frequently used as an oxidizing agent in chemistry laboratories. It is also used as a disinfectant. Solutions of MnO_4^- have a purple color.

Step 1.

$$S^{2-} + MnO_4^- \longrightarrow S + MnO_2 \quad \text{(base)}$$

$$S^{2-} \longrightarrow S$$
$$MnO_4^- \longrightarrow MnO_2$$

Sulfur is easily balanced by adding two electrons to the right side.

$$S^{2-} \longrightarrow S + 2e^-$$

All remaining steps before adding the half-reactions together are performed for the reduction of MnO_4^-.

Step 2. For the reduction half-reaction, Step 2 is not needed because the Mn atoms are balanced.

Step 3.

$$4H^+ + MnO_4^- \longrightarrow MnO_2 + 2H_2O$$

First, balance the half-reaction with H_2O and H^+ as if it occurs in an acidic solution. Eliminate the four H^+ by adding four OH^- to both sides of the equation.

$H^+ + OH^- \longrightarrow H_2O$

$$
\begin{array}{l}
4H^+ + MnO_4^- \longrightarrow MnO_2 + 2H_2O \\
+4OH^- \qquad\qquad\qquad\qquad\qquad +4OH^- \\
\hline
4H_2O + MnO_4^- \longrightarrow MnO_2 + 2H_2O + 4OH^-
\end{array}
$$

Step 4. Subtracting two H_2O from both sides and balancing the charges on each side gives

$$3e^- + 2H_2O + MnO_4^- \longrightarrow MnO_2 + 4OH^-$$

Step 5. Two electrons are released by the oxidation half-reaction, and three electrons are taken in by the reduction half-reaction. Thus, the oxidation half-reaction is multiplied by 3, and the reduction half-reaction is multiplied by 2 to balance the electrons.

$$3 \times (S^{2-} \longrightarrow S + 2e^-) = 3S^{2-} \longrightarrow 3S + 6e^-$$

$$2 \times (3e^- + 2H_2O + MnO_4^- \longrightarrow MnO_2 + 4OH^-) =$$
$$6e^- + 4H_2O + 2MnO_4^- \longrightarrow 2MnO_2 + 8OH^-$$

Adding the two half-reactions gives

$$3S^{2-} + 2MnO_4^- + 4H_2O \longrightarrow 3S + 2MnO_2 + 8OH^- \quad \text{(balanced)}$$

Check to see that the equation is balanced. There are 3 S, 2 Mn, 12 O, 8 H, and a total charge of 8− on either side.

REVIEW PROBLEM **18.4** Balance the following skeleton equations using the ion-electron method:
(a) $PH_3 + I_2 \longrightarrow I^- + H_3PO_2$ (acid)
(b) $Zn + NO_3^- \longrightarrow Zn^{2+} + NH_4^+$ (acid)
(c) $Zn + NO_3^- \longrightarrow Zn(OH)_4^{2-} + NH_3$ (base)

18.3 ELECTROCHEMISTRY

Electrochemistry is the study of the interaction of matter and electricity. Two major areas are studied in electrochemistry: (1) the conversion of electric energy into chemical energy, and (2) the conversion of chemical energy into electric energy. Electric energy is converted to chemical energy in **electrolytic cells,** and chemical energy is converted to electric energy in **galvanic cells.** First, we shall look at electrolytic cells and then proceed to galvanic cells.

Electrolytic Cells

Two types of electric conduction are known: (1) metallic conduction, and (2) electrolytic conduction. **Metallic conduction** occurs when electrons flow through metals and is a direct result of the rather weak forces that hold electrons to metal nuclei.

Electrodes are normally strips of inert metals, such as Pt, that provide the proper conducting surface for redox reactions to occur.

Electrolytic conduction is the movement of ions in a liquid that is brought about by an external source of electricity. In Section 15.3 we discussed that solutions are classified as either electrolytes or nonelectrolytes. **Electrolyte** solutions are capable of conducting an electric current. This is accomplished by placing a solution into a container along with two electrodes that are attached to a source of direct electric current (see Figure 18.3). This system is called an electrolytic cell. When a voltage is applied, one of the electrodes develops a positive charge and the other a negative charge. Ions in electrolyte solutions are attracted to the electrodes. Cations are attracted to the negative electrode, called the **cathode,** and anions are attracted to the positive electrode, called the **anode.**

Chemical reactions occur at the electrodes in an electrolytic cell. When a cation contacts the cathode, under the proper conditions, the cation picks up one or more electrons and undergoes reduction. Similarly, anions interact with the anode and give up electrons, or undergo oxidation. In electrolytic cells, oxidation always occurs at the anode, and reduction always occurs at the cathode.

Oxidation occurs at the anode.

Reduction occurs at the cathode.

As long as oxidation and reduction occur in an electrolytic cell, the cell conducts an electric current.

During the time an electric current is flowing through the cell, electrical

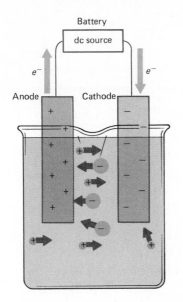

FIGURE 18.3
Cations (positive ions) diffuse toward the cathode and anions (negative ions) diffuse toward the anode in electrolytic cells.

Electrical neutrality must be maintained in electrolytic cells.

neutrality is maintained. For each electron picked up by a cation, an electron must be lost by an anion. In any region of the liquid, for each anion that migrates toward the anode, a cation must migrate toward the cathode or another anion must take its place. Positive or negative charge never builds up in an electrolytic cell.

Let us look at some specific examples of electrolytic cells. One of the easiest cells to understand is one containing molten or liquid sodium chloride, $NaCl(l)$. Figure 18.4 is a diagram of the $NaCl(l)$ electrolytic cell, illustrating electron flow and the reactions that occur. Molten NaCl contains Na^+ and Cl^-, which are free to migrate to the electrodes. At the cathode, Na^+ is reduced to Na, and at the anode, Cl^- is oxidized to Cl_2.

Cathode: $\quad Na^+ + e^- \longrightarrow Na \quad$ (reduction)
Anode: $\quad 2Cl^- \longrightarrow 2e^- + Cl_2 \quad$ (oxidation)

Equalizing the electrons and adding these two half-reactions gives the overall cell reaction:

$$2Na^+ + 2Cl^- \longrightarrow Cl_2 + 2Na$$

During the time an electric current is passed through the molten NaCl, metallic sodium coats the cathode and chlorine gas bubbles up from the anode (see Figure 18.4). Collectively, the overall chemical reaction that occurs in this or in any other electrolytic cell is called **electrolysis**, which means "electrical splitting."

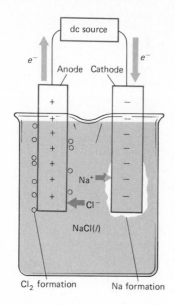

FIGURE 18.4
In the electrolysis of molten NaCl, $Cl_2(g)$ is produced at the anode and $Na(s)$ is produced at the cathode. The anode half-reaction is $2Cl^- \rightarrow Cl_2 + 2e^-$. The cathode half-reaction is $2Na^+ + 2e^- \rightarrow 2Na$.

Electrolysis of aqueous solutions is more complex because water, as well as the solute, undergoes oxidation or reduction. Water is oxidized to O_2 in the following reaction:

$$2H_2O \longrightarrow O_2 + 4H^+ + 4e^-$$

Hydrogen is the product of the reduction of water:

$$2H_2O + 2e^- \longrightarrow H_2 + 2OH^-$$

A variety of anode and cathode reactions are possible in cells containing aqueous solutions.

Whenever aqueous solutions undergo electrolysis, there are a number of possibilities as to what undergoes oxidation or reduction. Depending on the conditions within the cell, either water or the solute, or both, are involved in the cell reaction.

What happens when an aqueous solution of sodium chloride, NaCl(aq), is placed into an electrolytic cell? Two possibilities exist for both the cathode and the anode reactions. At the cathode, either Na^+ is reduced to Na or water is reduced to hydrogen gas. At the anode, either Cl^- is oxidized to Cl_2 or water is oxidized to oxygen. The overall cell reaction depends on which substances undergo oxidation or reduction most readily. In concentrated NaCl solutions, the following overall cell reaction is observed:

$$2H_2O + 2Cl^- \longrightarrow Cl_2 + H_2 + 2OH^-$$

In this electrolytic cell, water more easily undergoes reduction than Na^+, and the Cl^- more readily undergoes oxidation than water. Electrolysis of brine,

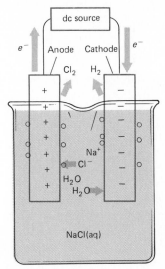

FIGURE 18.5
In the electrolysis of aqueous NaCl, chlorine gas, $Cl_2(g)$, is produced at the anode and hydrogen gas, $H_2(g)$, is produced at the cathode.

Anode: $2Cl^- \rightarrow Cl_2 + 2e^-$

Cathode: $2H_2O + 2e^- \rightarrow H_2 + 2OH^-$

an aqueous solution of NaCl, is used in the commercial preparation of $H_2(g)$ and $Cl_2(g)$. Figure 18.5 illustrates the electrolysis of NaCl(aq).

Industrial Electrolytic Cells

In Chapter 14, the Hall process for isolating aluminum was discussed. Aluminum metal is obtained in the Hall process by passing an electric current through a molten mixture of Al_2O_3 and cryolite, Na_3AlF_6. Figure 18.6 shows how the

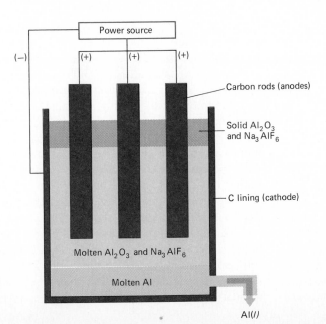

FIGURE 18.6
The Hall cell is used to commercially produce Al metal. Molten Al forms at the cathode, and then flows from the cell.

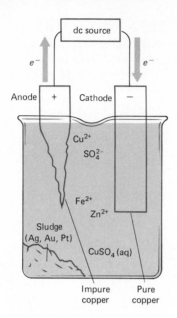

FIGURE 18.7
The final purification of Cu is an electrolytic process. As electricity passes through the cell, pure copper, Cu(s), plates onto the cathode, separating it from the metal impurities.

Corrosion of metals takes place as a result of redox reactions. Corrosion is hastened if a metal comes in contact with oxygen or oxygen dissolved in water. Metals generally corrode faster near oceans, since moisture in sea air has a relatively high salt content and is an electrolyte solution.

electrolytic cell is constructed. Both electrodes are made out of carbon; the anodes are carbon rods inserted into the molten mixture, whereas the cathode is a carbon coating on the inside of a steel container. Anode and cathode reactions in the Hall cell are as follows:

Anode: $3O^{2-} \longrightarrow \frac{3}{2}O_2 + 6e^-$
Cathode: $2Al^{3+} + 6e^- \longrightarrow 2Al$

giving an overall cell reaction of

$$2Al^{3+} + 3O^{2-} \longrightarrow \frac{3}{2}O_2 + 2Al$$

Electrolysis is also used in the final step of the purification of copper. Copper is first separated from the ore in which it naturally exists, and then it is purified until it is 99% Cu. In the remaining 1%, Fe, Zn, Au, Ag, and Pt are the major impurities. Figure 18.7 is a diagram of the electrolytic cell used to purify Cu. The anode is constructed from impure Cu, the cathode is a very pure sample of Cu, and both are immersed in a solution of $CuSO_4$.

An electric current is passed through the cell so that Cu, Fe, and Zn are the only metals that undergo oxidation at the anode. Au, Ag, and Pt do not dissolve, and ultimately fall to the bottom of the cell. The Au, Ag, and Pt are collected and sold, which greatly defrays the cost of Cu purification. At the cathode, the most easily reduced substance, Cu^{2+}, picks up two electrons and deposits on the cathode, leaving Zn^{2+} and Fe^{2+} behind. Copper purified in this manner is generally more than 99.9% pure.

If the cathode in this electrolytic cell is a metal other than Cu, when the Cu^{2+} is reduced, the resulting Cu coats or plates the metal's surface. This process is called **electroplating.** Many metals are electroplated onto the surface of other metals. Silver- and gold-plated objects are used in place of much more expensive pure silver or gold. In the automotive industry, chromium is plated on the surface of steel bumpers to protect the steel from corrosion. Tin cans actually are steel cans that have a thin layer of Sn plated on the surface. Tin is being replaced by Cr on cans. The process used to electroplate Cr is very fast, and an extremely thin Cr layer, about a millionth of a millimeter, is possible.

REVIEW PROBLEMS

18.5 (a) What is electrochemistry? (b) What are the principal topics investigated in this discipline?

18.6 What is the difference between metallic and electrolytic conduction?

18.7 Sketch a diagram of an electrolytic cell. Label the following on the diagram: (a) anode, (b) cathode, (c) power source, (d) direction of electricity flow.

18.8 For an electrolytic cell containing molten sodium chloride: (a) write the oxidation and reduction half-reactions, and (b) write the overall cell reaction.

Galvanic Cells

Galvani (1737–1798) first studied theology but turned to medicine later in life. After becoming a professor of anatomy at the University of Bologna, Galvani discovered that electricity caused the muscles in frog legs to contract. He also noticed that two different metals in contact produced the same effect in frog muscle. From these simple experiments the modern battery developed.

Galvanic cells (also called voltaic cells) are electrochemical cells that produce electricity. Galvanic cells are therefore the opposite of electrolytic cells. A galvanic cell is constructed in such a manner that the substance undergoing oxidation is separated from the substance undergoing reduction. A reaction occurs only when the cells are interconnected by a wire to carry the electrons produced by the substance undergoing oxidation to the substance undergoing reduction.

A diagram of a galvanic cell is shown in Figure 18.8. In this cell there are two electrodes, one is $Cu(s)$ immersed in a $CuSO_4(aq)$ solution, and the other is $Zn(s)$ immersed in a $ZnSO_4(aq)$ solution. Both electrodes are connected by a

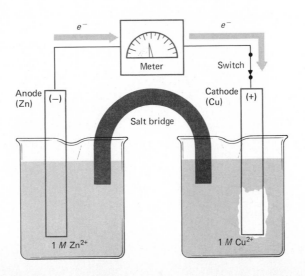

FIGURE 18.8
A galvanic cell produces an electric current through redox reactions. After the switch is closed, electrons released by the oxidation of Zn flow to the Cu electrode. These electrons reduce $Cu^{2+}(aq)$ to metallic $Cu(s)$. The electric circuit of a galvanic cell is completed with a salt bridge.

Alessandro Volta (1745–1827) was a friend of Galvani. On hearing of Galvani's study of metals and frog muscle, Volta attempted to see if the metals produced electricity without the frog muscle. He discovered that it was the metals and not the tissue that created the current. A bitter battle began between him and Galvani. Later Volta was the first to produce electro-chemical cells as we know them today. He used mainly Cu and Zn.

wire, switch, and voltmeter. The electric circuit is completed by either a salt bridge [a glass tube filled with either KNO_3(aq) or KCl(aq)] or a porous barrier.

When the switch is closed in a galvanic cell, the voltmeter registers a **voltage** (also called **electromotive force,** emf), indicating that electrons are being forced through the circuit. Since Zn is a better reducing agent than Cu, the Zn undergoes oxidation and gives up its electrons, which flow through the wire and meter to the Cu electrode. These electrons are picked up by dissolved Cu^{2+} ions, which reduces them to metallic Cu.

$$\text{Anode:} \qquad Zn \longrightarrow Zn^{2+} + 2e^- \qquad \text{(oxidation)}$$
$$\text{Cathode:} \qquad Cu^{2+} + 2e^- \longrightarrow Cu \qquad \text{(reduction)}$$

In galvanic cells, as in electrolytic cells, oxidation occurs at the anode and reduction occurs at the cathode. However, the cathode is the positive electrode, and the anode is the negative electrode.

Following the electron flow, electrons move through the wire and voltmeter to the cathode, where they are immediately picked up by Cu^{2+}. To complete the electric circuit, cations from the salt bridge diffuse out to replace cations that are reduced, and in a similar manner, anions from the salt bridge replace those that are oxidized.

Galvanic cells are more commonly called *batteries*. Today we rely heavily on electric energy from battery sources. Calculators, watches, toys, and innumerable other devices are powered by batteries. Practical batteries are normally not like the galvanic cell just described; many are "dry cells," ones that do not contain a liquid electrolyte solution.

Figure 18.9 illustrates a typical **zinc-carbon dry cell** that might power a portable radio. Inside the outer metal or cardboard jacket is a zinc container, the anode. Inside the zinc container is a moist paste of ammonium chloride,

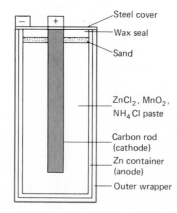

FIGURE 18.9
A carbon-zinc dry cell has Zn metal at the anode and graphite (carbon) at the cathode. Carbon-zinc dry cells are inexpensive and are used in many different electrical devices.

NH_4Cl; zinc chloride, $ZnCl_2$; and manganese dioxide, MnO_2. In the middle of the paste is a graphite rod, the cathode.

At the Zn anode, electrons are released.

Anode: $$Zn \longrightarrow Zn^{2+} + 2e^-$$

A complex reaction occurs at the cathode. It is thought that the following reaction is one of the primary reductions.

Cathode: $$2e^- + 2NH_4^+ + 2MnO_2 \longrightarrow 2NH_3 + Mn_2O_3 + H_2O$$

A buildup of NH_3 in the cell stops the flow of electricity. This buildup does not occur because most of the ammonia combines with Zn^{2+} in the paste, yielding a complex ion. Hydrogen gas is also produced at the cathode, and it is removed by combining with the MnO_2 in the following reaction.

$$H_2 + MnO_2 + 2H^+ \longrightarrow Mn^{2+} + 2H_2O$$

Another commonly encountered battery is the **nickel-cadmium** (nicad) **battery,** a battery superior to the dry cell because it is rechargeable. Cadmium metal is at the anode and NiO_2 is at the cathode. The anode and cathode reactions in a nicad battery are:

Anode: $$Cd + 2OH^- \rightleftharpoons Cd(OH)_2 + 2e^-$$

Cathode: $$2e^- + NiO_2 + 2H_2O \rightleftharpoons Ni(OH)_2 + 2OH^-$$

Lead storage batteries (Figure 18.10) are used as automobile batteries. Anodes of lead storage batteries are composed of lead, Pb, and the cathodes contain lead(IV) oxide, PbO_2. Oxidation occurs when Pb gives up two electrons and becomes Pb^{2+}, which immediately combines with SO_4^{2-} in the electrolyte, 30% H_2SO_4. The anode reaction is

Anode: $$Pb + SO_4^{2-} \longrightarrow PbSO_4(s) + 2e^-$$

Electrons from the oxidation reaction flow to the cathode where they combine with PbO_2 and H^+ from the electrolyte to produce Pb^{2+} and water. Again the Pb^{2+} immediately combines with SO_4^{2-} from the solution, producing $PbSO_4$.

Cathode: $$2e^- + PbO_2 + 4H^+ + SO_4^{2-} \longrightarrow PbSO_4 + 2H_2O$$

Adding the two equations together gives the overall reaction for lead storage batteries:

To recharge a battery, electricity is pumped back into the electrodes, reversing the reactions that liberated the electric current. After it is recharged, the battery can be used again.

The $PbSO_4$ produced when a lead-acid battery discharges is loosely held to the electrodes. If the $PbSO_4$ flakes off the electrodes, the life of the battery is shortened.

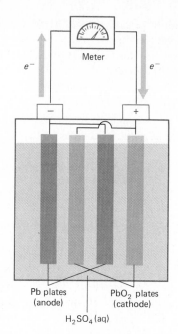

FIGURE 18.10
Lead-acid storage batteries are one of the most widely used types of batteries. The anode is composed of Pb plates, and the cathode is coated with PbO_2. Lead-acid batteries are rechargeable, and they can be used for many years.

$$Pb + PbO_2 + 4H^+ + 2SO_4^{2-} \longrightarrow 2PbSO_4 + 2H_2O$$

A more exotic type of battery is a fuel cell. A common fuel cell produces electricity from the combination of $H_2(g)$ and $O_2(g)$.
Fuel cells are efficient and are nonpolluting, but the cost is prohibitive because of high cost of the electrodes. Fuel cells are used in space missions. In addition to producing electricity, they give off water as a by-product.

Looking at the overall reaction of the lead storage battery, we see that, as electricity is being produced, insoluble $PbSO_4$ and water are formed. Since the density of water (1 g/mL) is less than that of the H_2SO_4 solution (1.3 g/mL), measurement of the density of the liquid in the battery gives an indication of how charged the battery is. A density near 1.3 g/mL indicates a charged battery; lower densities indicate a partially discharged battery.

Lead storage batteries are recharged by applying a voltage to the electrodes. During the recharging process, the battery is not acting as a galvanic cell but as an electrolytic cell. Pumping electricity into the Pb electrode dissolves the insoluble $PbSO_4$, producing Pb^{2+} and SO_4^{2-}. Withdrawing electrons at the PbO_2 electrode also dissolves $PbSO_4$, regenerating the sulfuric acid. In each case, the reactions for recharging are exactly the reverse for discharging. Consequently, the overall cell reaction is written as follows:

$$Pb + PbO_2 + 4H^+ + 2SO_4^{2-} \underset{\text{charge}}{\overset{\text{discharge}}{\rightleftharpoons}} 2PbSO_4 + 2H_2O$$

REVIEW PROBLEMS

18.9 How is a galvanic cell different from an electrolytic cell?

18.10 (a) Draw a diagram and explain each part of a galvanic cell that has a Sn electrode immersed in a tin solution and a Ni electrode in a nickel solution. (b) Explain what happens within the cell, and discuss the electron flow. *Note:* Ni is a stronger reducing agent than Sn.

18.11 What is the main difference between wet and dry galvanic cells?

18.12 What would happen if the $NH_3(g)$ produced in the "dry" cell did not combine with Zn^{2+}?

18.13 What oxidation and reduction occur when a lead-acid battery discharges?

SUMMARY

Oxidation occurs when a substance loses electrons and increases its oxidation number. Reduction is exactly the opposite; it occurs when a substance gains electrons and decreases its oxidation number. Substances are classified according to their ability to undergo oxidation or reduction. A substance that readily undergoes oxidation is called a reducing agent because it provides the necessary electrons to bring about a reduction. An oxidizing agent is a substance that undergoes reduction and takes electrons from, or oxidizes, other substances.

The ion-electron method is one way that redox equations can be systematically balanced. Two half-reactions are written, one for the substance that undergoes oxidation and one for the substance that undergoes reduction. Each half-reaction is balanced by adding either H^+ and H_2O or OH^- and H_2O, depending on the conditions of the reaction. After the half-reactions are balanced, the equations are added together to yield the net ionic equation.

Electrochemistry is the study of the interaction of chemical and electric energy. Electric energy can be used to initiate chemical reactions. This process is called electrolysis, and occurs in a container called an electrolytic cell. Generally, electrolytic cells are constructed with a source of direct current attached to two electrodes that are immersed in a solution. Reduction occurs at the negative electrode, while oxidation occurs at the positive electrode.

Galvanic cells produce rather than consume electricity. Within a galvanic cell, the substance that undergoes oxidation is physically separated from the substance that undergoes reduction. Electrons provided by the reducing agent are connected to the oxidizing agent by a wire that is attached to a meter or some electrical device. The circuit is completed with a salt bridge or porous barrier that allows the flow of charged particles to keep the solutions electrically neutral. Practical galvanic cells, or batteries, have a wide variety of uses.

QUESTIONS AND PROBLEMS

18.14 Define each of the following: oxidation, reduction, oxidizing agent, reducing agent, oxidation number, half-reaction, electrochemistry, electrolytic cell, electrolysis, anode, cathode, electric current, overall cell reaction, galvanic cell, battery.

Oxidation-Reduction—General

18.15 Describe what happens when a reducing agent and an oxidizing agent are combined.

18.16 In each of the following equations determine what has been oxidized and reduced:

(a) $2Ag^+ + Cu \longrightarrow 2Ag + Cu^{2+}$

(b) $Fe + 2H^+ \longrightarrow Fe^{2+} + H_2$

(c) $3Cu + 2NO_3^- + 8H^+ \longrightarrow$
$$3Cu^{2+} + 2NO + 4H_2O$$

(d) $Zn + 2MnO_2 + 2NH_4^+ \longrightarrow$
$$Zn^{2+} + Mn_2O_3 + 2NH_3 + H_2O$$

(e) $Cl_2 + 2I^- \longrightarrow I_2 + 2Cl^-$

(f) $2Cl^- + 2H_2O \longrightarrow Cl_2 + H_2 + 2OH^-$

(g) $6KOH + 3Br_2 \longrightarrow 5KBr + KBrO_3 + 3H_2O$

18.17 What are the oxidizing and reducing agents in the following equations:

(a) $2Al + 3F_2 \longrightarrow 2AlF_3$

(b) $HCl + HNO_3 \longrightarrow NO_2 + \frac{1}{2}Cl_2 + H_2O$

(c) $2MnO_4^- + 16H^+ + 10Cl^- \longrightarrow$
$$2Mn^{2+} + 8H_2O + 5Cl_2$$

(d) $Pb + PbO_2 + 4H^+ + 2SO_4^{2-} \longrightarrow$
$$2PbSO_4 + 2H_2O$$

Balancing Redox Equations

18.18 Balance the following, all of which are in acidic solutions:

(a) $Cr_2O_7^{2-} + Br^- \longrightarrow Br_2 + Cr^{3+}$

(b) $I^- + H_2O_2 \longrightarrow I_2 + H_2O$

(c) $Zn + NO_3^- \longrightarrow Zn^{2+} + NH_4^+$

(d) $NO_2 + HOCl \longrightarrow NO_3^- + Cl^-$

(e) $AsH_3 + Ag^+ \longrightarrow As_4O_6 + Ag$

(f) $Zn + H_2MoO_4 \longrightarrow Zn^{2+} + Mo^{3+}$

18.19 Balance the following, all of which are in basic solutions:

(a) $ClO^- + I^- \longrightarrow Cl^- + I_2$

(b) $S^{2-} + I_2 \longrightarrow SO_4^{2-} + I^-$

(c) $Al + H_2O \longrightarrow Al(OH)_4^- + H_2$

(d) $P_4 \longrightarrow PH_3 + HPO_3^{2-}$

(e) $Fe_3O_4 + MnO_4^- \longrightarrow Fe_2O_3 + MnO_2$

(f) $Si + OH^- \longrightarrow SiO_3^{2-} + H_2$.

18.20 Balance the following, using the ion-electron method:

(a) $CrI_3 + H_2O_2 \longrightarrow$
$$CrO_4^{2-} + IO_4^- + H_2O \text{ (base)}$$

(b) $XeO_3 + I^- \longrightarrow I_3^- + Xe \text{ (acid)}$

(c) $HXeO_4^- \longrightarrow XeO_6^{4-} + Xe + O_2 \text{ (base)}$

(d) $Pt + NO_3^- + Cl^- \longrightarrow PtCl_6^{2-} + NO_2 \text{ (acid)}$

(e) $CN^- + MnO_4^- \longrightarrow MnO_2 + CNO^- \text{ (acid)}$

(f) $CrO_4^{2-} + HSnO_2^- \longrightarrow$
$$CrO_2^- + HSnO_3^- \text{ (base)}$$

Electrolytic Cells

18.21 How is metallic conduction different from electrolytic conduction?

18.22 Sketch a diagram of an electrolytic cell of molten magnesium chloride. Label the anode and cathode, and write an equation to indicate what happens during this electrolysis.

18.23 Draw a diagram of an electrolytic cell containing a concentrated aqueous sodium chloride solution. Label each electrode, indicate electron flow, and write half-reactions.

18.24 In an electrolytic cell calcium metal and fluorine gas are produced at the electrodes. (a) At which electrodes are calcium and fluorine gas produced? (b) Write half-reactions to show the reactions at the electrodes.

18.25 Describe exactly how a steel can is electroplated with tin.

18.26 (a) What possible oxidations could occur in an electrolytic cell containing aqueous potassium fluoride? (b) What possible reductions occur in the same cell?

18.27 How could an impure sample of silver be purified using electrolysis?

Galvanic Cells

18.28 Draw a diagram of a galvanic cell that is constructed with Ni and Ag electrodes immersed in Ni^{2+} and Ag^+ solutions, respectively. Label the anode, cathode, and direction of electron flow.

18.29 Draw diagrams of galvanic cells with the following net reactions. Indicate anode, cathode, and direction of electron flow, and write appropriate half-reactions:

(a) $Zn + Sn^{2+} \longrightarrow Sn + Zn^{2+}$

(b) $Cl^- + MnO_4^- \longrightarrow Cl_2 + Mn^{2+}$

(c) $Zn + H^+ \longrightarrow Zn^{2+} + H_2$

18.30 Why are salt bridges or porous barriers included in galvanic cells?

18.31 (a) Write the half-reaction at the cathode of a lead storage battery. (b) Write the half-reaction at the anode of a nicad battery. (c) Write the cathode reaction for a regular dry cell.

18.32 Why is the density of the electrolyte fluid in a lead storage battery used as an indication of how charged a battery is?

18.33 What advantage does a nicad battery have over a carbon-zinc dry cell? Explain.

18.34 Describe what happens at each electrode when a lead storage battery is recharged.

· C H A P T E R ·

19

Nuclear Chemistry

Study Guidelines

After completing Chapter 19, you should be able to:
1. Describe the composition of a nucleus, given its atomic number and mass number
2. List empirical factors that are related to nuclear stability
3. Describe the properties and characteristics of the three forms of natural radiation
4. Explain the relationship between the rate of radioactive decay and a nuclide's half-life
5. Determine how much of a substance remains after a given number of half-lives
6. Write nuclear equations for alpha and beta decay, given the parent nuclide
7. Describe what happens to a nucleus when it emits a gamma ray
8. Discuss what happens in the three natural radioactive decay series
9. Write nuclear equations, complete and short-hand notation, for nuclear bombardment reactions
10. Explain and write equations illustrating how various transuranium elements were initially synthesized
11. Describe what happens to a nucleus when it undergoes fission
12. Discuss the process of nuclear fusion and how it is different from nuclear fission
13. Explain how old objects are dated using radionuclides
14. List and describe medical uses for radionuclides
15. Describe the biological effects of high, moderate, and low exposure to radiation
16. Describe the primary sections of a nuclear power plant, and explain how nuclear energy is transformed into electric energy

Nuclear chemistry is the special area of chemistry that is concerned with changes in the nuclei of atoms. The nucleus is the region within an atom where the protons and neutrons reside. Collectively, the protons and neutrons are called the **nucleons.** Compared with the total volume of the atom, the volume of the nucleus is extremely small, and the nucleus is very dense. Nuclear density is approximately 1.8×10^{14} g/mL, an incredibly high density. A marble-size sample of nuclear material would have a mass of approximately 8×10^8 tons.

Protons and neutrons are held tightly together by the strong nuclear force. If two particles are brought to within 10^{13} cm of each other, the nuclear force binds them together. The nuclear force is much stronger than electrostatic forces.

Two values are used to describe the composition of a nucleus: the atomic number and the mass number. **Atomic number** (Z) is the number of protons in the nucleus of an atom, and **mass number** (A) is the sum of the protons plus neutrons within the nucleus. Subtracting the atomic number from the mass number gives the total number of neutrons (N) in the atom.

$$A = Z + N$$

or

$$N = A - Z$$

When you write the symbol for an atom, write the mass number as a superscript to the left of the symbol of the element and write the atomic number as a subscript, also to the left of the symbol.

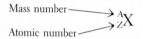

Mass number ⟶ $_Z^A X$
Atomic number ⟶

Atoms with the same atomic number but different mass numbers are called **isotopes.** In other words, these atoms have the same number of protons but different number of neutrons in the nucleus. In nuclear chemistry, the term **isotope** is normally used to refer to different forms of an individual element; the term **nuclide** generally refers to atomic forms of different elements. For example, we would speak of the primary isotopes of carbon, $_6^{12}C$ and $_6^{13}C$. However, when discussing $_6^{12}C$ compared with $_7^{14}N$, we refer to them as two different nuclides.

Nuclear Stability

Some nuclides are stable and do not undergo changes unless subjected to extreme conditions. On the other end of the scale, there are nuclides that spontaneously undergo changes. At present, modern science has not been able to unearth enough solid information to completely explain nuclear stability. It is thought that the nucleons are arranged in various energy levels. Just as there are electronic configurations that are more stable than others, some nuclear configurations are more stable than others.

It is known that nuclides with 2, 8, 20, 50, 82, or 126 protons or neutrons are usually more stable than other nuclides. For instance, there are five stable isotopes of calcium (Z = 20); but there are only two stable isotopes of potassium (Z = 19), and there is only one stable isotope of scandium (Z = 21). Four stable

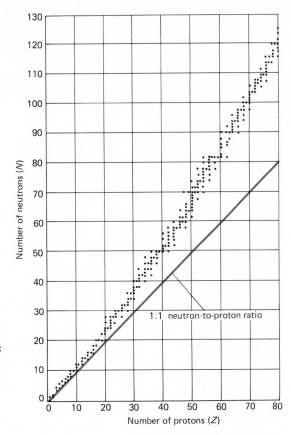

FIGURE 19.1
A plot of N versus Z reveals that the ratio of neutrons to protons increases with increasing atomic number. The most stable low-mass nuclides have the same number of neutrons and protons ($N/Z = 1$). Higher-mass nuclides have more neutrons than protons ($N/Z > 1$).

The mass of a nucleus is smaller than the sum of the masses of the protons and neutrons within the nucleus. The energy equivalent for this mass difference is calculated using Einstein's mass-energy relationship ($E = mc^2$). This energy difference is called the *nuclear binding energy*, i.e., the amount of energy needed to separate a nucleus into isolated nucleons.

nuclides with a neutron number of 20 are known; there are no stable nuclides with either 19 or 21 neutrons.

Nuclides with even numbers of nucleons are more stable than those with either an odd number of protons or an odd number of neutrons. Over 150 stable nuclides belong to the group with even numbers of protons and neutrons. There are approximately 55 stable nuclides that contain an even number of protons and an odd number of neutrons. About 50 stable nuclides have an odd number of protons and an even number of neutrons. Just five stable nuclides, all with mass numbers below 14, have odd numbers of protons and neutrons.

Finally, nuclear stability is also correlated with the ratio of the number of neutrons to the number of protons. In nuclides with small mass numbers ($A < 41$), the most stable nuclides are normally those that contain the same number of protons and neutrons ($N/Z = 1$), for example, $^{12}_{6}C$, $^{16}_{8}O$, $^{32}_{16}S$, and $^{40}_{20}Ca$. For higher-mass nuclides ($A > 41$) this is not true; the most stable nuclides have a greater number of neutrons than protons ($N/Z > 1$).

When we look at Figure 19.1, a graph of the atomic number versus the neutron number of naturally occurring stable nuclides, we see that the smaller nuclides are close to the line that represents an equal number of protons and neutrons (Z = N). Higher-mass atoms are farther from this line because they contain more neutrons than protons.

REVIEW PROBLEMS

19.1 What are nucleons, and where are they located?

19.2 What is the difference between the terms "isotopes" and "nuclides"?

19.3 Consider the nuclide $^{85}_{37}Rb$ to answer the following. (a) What is its mass number? (b) What is its atomic number? (c) How many neutrons are located in its nucleus?

19.4 (a) List the number of protons or neutrons that are normally found in more stable nuclei. (b) Are they the same numbers found for stable electronic configurations?

19.5 Considering the composition of their nuclei, predict which nuclide is the most stable in each of the following pairs: (a) $^{3}_{1}H$ or $^{4}_{2}He$, (b) $^{12}_{6}C$ or $^{14}_{6}C$, (c) $^{64}_{30}Zn$ or $^{65}_{30}Zn$.

19.2 RADIOACTIVITY

Nuclei that are unstable break down spontaneously and release various matter-energy forms. These spontaneous nuclear changes are called **radioactive disintegrations** or **radioactive decays**. In 1895, Henri Becquerel was the first scientist to observe this phenomenon. Marie and Pierre Curie later took an active role investigating the nature of radioactive substances and their emissions. Becquerel and the Curies shared the Nobel prize (1903) in physics for this pioneering work.

Three types of matter-energy forms are commonly released by naturally occurring isotopes: (1) **alpha particles, α**, (2) **beta particles, β**, and (3) **gamma rays, γ**. Table 19.1 summarizes their properties.

Alpha particles are the most massive radioactive particles; an alpha particle has the mass of a helium nucleus and a 2+ charge. Alpha particles are helium nuclei that have been shot out of the nucleus at a speed approximately one-tenth the speed of light ($c = 3 \times 10^8$ m/s). The symbol for an alpha particle is $^{4}_{2}He$ or $^{4}_{2}\alpha$. Compared with the other two emissions, alpha particles have the least capacity to penetrate matter; clothing, paper, and other thin forms of matter generally absorb a large percent of alpha radiation. As alpha particles penetrate matter, they knock electrons out of, i.e., they ionize, atoms that they encounter. An alpha particle is the radiation form with the greatest capacity to ionize matter.

Beta particles are high-energy particles that have the same mass and charge as electrons; accordingly, beta particles are symbolized as $_{-1}^{0}e$. A -1 is written as the subscript to represent the beta particle's negative charge, and a 0 is written as

TABLE 19.1   PROPERTIES OF PRINCIPAL NUCLEAR EMISSIONS

Particle	Symbol	Mass, u	Charge	Velocity*
Alpha	$^{4}_{2}He$ or $^{4}_{2}\alpha$	4.003	2+	$0.1c$
Beta	$_{-1}^{0}e$ or $_{-1}^{0}\beta$	0.0006	1-	$0.9c$
Gamma	γ	None	None	$1.0c$

*c = the speed of light.

Marie and Pierre Curie (*New York Public Library, Picture Collection*)

the mass number since a beta particle contains no nucleons. Beta particles, like alpha particles, are moving at a high velocity; some travel at a velocity that is 0.9 that of the speed of light. Beta particles penetrate a greater thickness of matter than do alpha particles before they are absorbed. A fairly thick metal is needed to stop beta particles. Their ability to ionize matter is much less than that of alpha particles, but greater than that of gamma rays.

Gamma rays have no mass or charge; therefore, the symbol for gamma rays is simply γ. Gamma rays are electromagnetic energy forms moving at the speed of light. The penetrating power of gamma rays is the greatest of the three forms discussed because of their minimal interaction with matter. Thick walls of lead or other dense substances are required to totally block out gamma radiation (see Figure 19.2) Gamma rays are also the least ionizing of the three radiation forms.

Gamma emission normally accompanies alpha and beta decay. After a nuclide releases an alpha or beta particle, the resulting nucleus is unstable and releases a gamma ray.

Half-Life Radioactive nuclides are characterized by their **half-lives,** or the amount of time required for one-half of their nuclei to decay. Let's consider a 10.0-g sample of hypothetical substance X with a half-life of 1 hr. After 1 hr elapses, only 5.00 g

FIGURE 19.2
Gamma radiation is the most penetrating form of radioactivity. Alpha radiation is the least penetrating form of radioactivity.

α, β, γ β, γ γ

Source of α, β, and γ radiation   Very thin barrier   Thick piece of metal   Thick lead wall

FIGURE 19.3
Initially, 10.0 g of radioactive substance X is present. Substance X has a half-life of 1.0 hr. Therefore, after 1.0 hr elapses, only 5.0 g of X remains. After 2.0 hr, 2.5 g of X remains. For each hour that passes, the mass of X becomes half of what it was at the beginning of the hour.

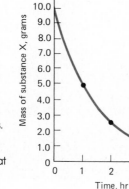

TABLE 19.2   HALF-LIVES OF SELECTED NUCLIDES

Nuclide	Half-life*	Emission
^{238}U	4.5×10^9 y	α
^{235}U	7.1×10^8 y	α
^{14}C	5700 y	β^-
^{15}O	2.1 min	β^-
^{90}Sr	28.8 y	β^-
^{139}I	2.7 s	β^-
^{17}F	66 s	β^+
^{40}K	1.3×10^9 y	α
^{106}Ag	8.6 d	β, γ
^{239}Pu	24,000 y	α
^{215}Po	1.8×10^{-3} s	α
^{250}Fm	0.5 hr	α

*y = year; d = day; hr = hour; min = minute; s = second.

The curie (Ci) is the unit of rate of radioactive decay. One Ci is the amount of substance that gives 3.7×10^{10} disintegrations per second. A curie is a very large rate of decay, and frequently the units millicuries and microcuries are used.

X would remain in the sample—half the nuclei, which represents half the mass, would change to some other substance. Analysis at 2 hr would show that only 2.50 g X remain; at 3 hr only 1.25 g X are present; and so on. Figure 19.3 shows a graph of the mass of substance X versus time.

Table 19.2 lists various nuclides and their half-lives. Half-lives can be extraordinarily long, in excess of a million years, or as short as a few milliseconds. $^{238}_{92}U$ has a half-life of 4.5 billion years, about the time span that the earth is believed to have been in existence. $^{139}_{53}I$ has a half-life of only 2.7 s; every 2.7 s, half of the nuclei in $^{139}_{53}I$ release beta particles. Most nuclides have half-lives somewhere between these two nuclides.

REVIEW PROBLEMS

19.6 Write the symbol for the radioactive emission described by the following phrases: (a) least penetrating, (b) high-energy electron, (c) most massive, (d) type of electromagnetic radiation, (e) least ionizing.

19.7 What is the half-life of a radioactive substance?

19.8 (a) From Table 19.2, find the half-life of $^{40}_{19}K$. (b) If a 25.0-g sample of $^{40}_{19}K$ is initially present, what mass remains after three half-lives?

19.3 NUCLEAR REACTIONS

Natural Nuclear Changes

Rates of nuclear reactions are not changed by temperature, pressure, or catalysts.

As we have seen, some nuclei spontaneously release alpha, beta, or gamma radiation. What happens to the nucleus in each of these changes? Let us begin with alpha decay. An alpha particle is composed of two protons and two neutrons. Thus, if a nucleus releases an alpha particle, its atomic number decreases by two and its mass number decreases by four. A nuclear equation that represents what happens in alpha decay is as follows:

$$^{A}_{Z}X \longrightarrow ^{A-4}_{Z-2}Y + ^{4}_{2}\alpha$$

A specific example is the nuclide $^{210}_{84}Po$. After undergoing alpha decay, $^{206}_{82}Pb$ is produced.

$$^{210}_{84}Po \longrightarrow ^{206}_{82}Pb + ^{4}_{2}\alpha$$

In nuclear chemistry, the nuclide undergoing decay is called the **parent nuclide,** and the nuclide that results is its **daughter nuclide.** In our example, $^{210}_{84}Po$ is the parent nuclide and $^{206}_{82}Pb$ is the daughter nuclide.

When writing nuclear equations, you must obey the law of conservation of matter. This means that the sum of the mass numbers must be the same on either side of the equation. This is also true for the atomic numbers. In other words, nucleons are conserved in nuclear reactions. Consider the alpha decay of $^{226}_{88}Ra$:

$$^{226}_{88}Ra \longrightarrow ^{222}_{86}Rn + ^{4}_{2}\alpha$$

On the left side of the equation, the total mass number is 226. On the right side, the sum of the mass numbers of $^{222}_{86}Rn$ and an alpha particle is also 226. The sum of the atomic numbers is 88 on both sides of the equation.

Beta decay is somewhat more complex than alpha decay. Beta decay occurs when a nucleus is unstable because it has too many neutrons. Neutrons undergo a spontaneous change, producing a beta particle, proton, and an antineutrino $(\bar{\nu})$.

$$^{1}_{0}n \longrightarrow {}^{1}_{1}p + {}_{-1}^{0}\beta + \bar{\nu}$$

A neutron is not composed of a proton and an electron but is transformed into a proton and an electron when it undergoes beta decay.

The resulting nucleus has the same number of total particles, or the same mass number, because one neutron changed to one proton. But, the atomic number increased by one, because there is now one more proton in the nucleus. Therefore, the general equation for beta decay is

$$^{A}_{Z}X \longrightarrow {}^{A}_{Z+1}Y + {}_{-1}^{0}\beta + \bar{\nu}$$

Besides the beta particle that is shot out of the nucleus, another matter-energy form, an antineutrino, is released. Originally thought to be a massless, chargeless particle, there is now evidence that the antineutrino possesses an infinitesimal mass. Antineutrinos belong to another class of matter called antimatter. If a "regular" neutrino encounters an antineutrino, they annihilate each other; both are transformed totally to energy.

Consider the following nuclides that undergo beta decay:

$$^{231}_{90}Th \longrightarrow {}^{231}_{91}Pa + {}_{-1}^{0}\beta + \bar{\nu}$$

$$^{14}_{6}C \longrightarrow {}^{14}_{7}N + {}_{-1}^{0}\beta + \bar{\nu}$$

$$^{36}_{17}Cl \longrightarrow {}^{36}_{18}Ar + {}_{-1}^{0}\beta + \bar{\nu}$$

In each of the above examples, a beta particle and an antineutrino are released during beta decay. The daughter nuclide that results is an isotope of the element that occupies the block to the right of the parent nuclide in the periodic table — it has one more proton in the nucleus.

Gamma emission takes place without a measurable mass change to the nucleus. Gamma rays are a more pure form of energy. Thus, nuclei that possess excess energy can become more stable by emitting a gamma ray. After gamma emission, the nucleus is in a lower energy state. A similar occurrence is the release of light energy when electrons are in higher energy states.

To represent gamma emission in an equation, an asterisk (*) is usually placed next to the symbol, indicating a nucleus with excess energy. After the gamma ray is released, the same nucleus is present, but without the excess energy, thus the asterisk is removed.

$$^{A}_{Z}X^{*} \longrightarrow {}^{A}_{Z}X + \gamma$$

Natural Radioactive Decay Series

In nature there are three primary radioactive decay series in which elements with mass numbers above 83 decay in a stepwise fashion until a stable nuclear configuration is reached. All three series end with the element Pb:

1. ^{238}U series: ^{238}U to ^{206}Pb
2. ^{235}U series: ^{235}U to ^{207}Pb
3. ^{232}Th series: ^{232}Th to ^{208}Pb

Figure 19.4 graphically illustrates the $^{238}_{92}U$ radioactive decay series that starts with the parent nuclide $^{238}_{92}U$, which initially undergoes alpha decay to produce $^{234}_{90}Th$. $^{234}_{90}Th$ gives off a beta particle, changing to $^{234}_{91}Pa$, which is also a beta emitter. After $^{234}_{91}Pa$ gives off a beta particle, $^{234}_{92}U$ is formed. After a series of five alpha emissions, the mass number is decreased by 20 and the atomic number is decreased by 10, yielding an unstable isotope of lead, $^{214}_{82}Pb$. The decay series becomes somewhat more complex at this point because more than one pathway is possible. Upon reaching $^{206}_{82}Pb$, the series ends—$^{206}_{82}Pb$ has a stable nucleus that does not undergo radioactive decay.

Artificial Nuclear Changes

In addition to the natural radioactive processes, scientists have devised procedures for initiating nuclear changes in nuclides that normally are not radioactive.

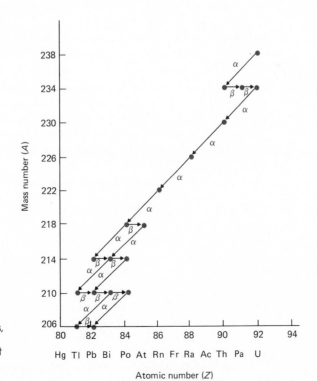

FIGURE 19.4
In the $^{238}_{92}U$ natural decay series, $^{238}_{92}U$ is the parent nuclide for a series of daughter nuclides that ultimately decay to the stable nuclide $^{206}_{82}Pb$.

The first artificial transmutation was performed by Rutherford and Soddy in 1902. They bombarded $^{14}_7N$ with alpha particles, producing $^{17}_8O$ and a proton.

$$^{14}_7N + {}^4_2\alpha \longrightarrow {}^{17}_8O + {}^1_1H$$

Irene Joliot-Curie (1897–1956) and her husband Frederic Joliot (1900–1958) were the first to demonstrate artificial radioactivity, the production of a nuclide that continuously emits radiation. They bombarded $^{10}_5B$ with alpha particles, producing $^{13}_7N$, which spontaneously decays to $^{13}_6C$:

Nuclei with too many protons can also decay by electron capture. An electron in the 1s orbital is "captured" by a proton in the nucleus and is transformed into a neutron. No radiation is emitted during this nuclear change. However, an x-ray is released when a higher-energy electron fills the vacancy left by the captured electron.

$$^{10}_5B + {}^4_2\alpha \longrightarrow {}^{13}_7N + {}^1_0n$$
$$^{13}_7N \longrightarrow {}^{13}_6C + {}^0_{+1}e^+$$

Both neutrons and positrons were emitted after this reaction. A positron is another form of antimatter. **Positrons** ($^0_{+1}e^+$) have the same properties of electrons except that they have a positive charge, and when they encounter a "regular" electron, the pair is annihilated ($e^- + e^+ \rightarrow 2\gamma$).

Unstable nuclei with an excess number of protons tend to undergo positron emission. A proton changes to a neutron and emits a positron, $^0_{+1}e^+$, and a neutrino (not an antineutrino).

$$^1_1p^+ \longrightarrow {}^1_0n + {}^0_{+1}e^+ + \nu$$

A nuclide that is a positron emitter, such as $^{13}_7N$, decays to the element with an atomic number of one less, exactly the opposite of a beta emitter.

A shorthand notation is used to simplify the writing of nuclear bombardment equations. Inside parentheses is written the bombarding particle separated by a comma from the particle that is released. The target nucleus is written to the left and the resulting nucleus is written to the right of the parentheses. For example, to simplify the nuclear equation that represents what happens when $^{10}_5B$ is bombarded with alpha particles to produce $^{13}_7N$ and a neutron, we write:

Sir James Chadwick discovered the neutron in 1932 by bombarding ^9_4Be with alpha particles, producing $^{12}_6C$ and neutrons.

$$^{10}_5B({}^4_2\alpha, {}^1_0n)^{13}_7N$$

Target nucleus ⟶   ⟵ Resulting nucleus

When $^{27}_{13}Al$ is bombarded with a neutron, $^{27}_{12}Mg$ and a proton are formed.

$$^{27}_{13}Al + {}^1_0n \longrightarrow {}^{27}_{12}Mg + {}^1_1p$$

The shorthand notation is

$$^{27}_{13}Al({}^1_0n, {}^1_1p)^{27}_{12}Mg$$

Two English scientists, Cockroft and Walton, constructed the first high-

TABLE 19.3 PREPARATION OF TRANSURANIUM ELEMENTS

Atomic number	Name	Reaction
93	Neptunium (Np)	$^{238}U(^1_0n,\beta)^{239}Np$
94	Plutonium (Pu)	$^{238}U(^2_1H,2^1_0n)^{238}Np$
		$^{238}Np \longrightarrow {}^{238}Pu + \beta$
95	Americium (Am)	$^{239}Pu(^1_0n,\beta)^{240}Am$
96	Curium (Cm)	$^{239}Pu(^4_2\alpha,^1_0n)^{242}Cm$
97	Berkelium (Bk)	$^{241}Am(^4_2\alpha,2^1_0n)^{243}Bk$
98	Californium (Cf)	$^{242}Cm(^4_2\alpha,^1_0n)^{245}Cf$
99	Einsteinium (Es)	$^{238}U(15^1_0n,7\beta)^{253}Es$
100	Fermium (Fm)	$^{238}U(17^1_0n,8\beta)^{255}Fm$
101	Mendelevium (Md)	$^{253}Es(^4_2\alpha,^1_0n)^{256}Md$
102	Nobelium (No)	$^{246}Cm(^{12}_6C,4^1_0n)^{254}No$
103	Lawrencium (Lr)	$^{252}Cf(^{10}_5B,5^1_0n)^{257}Lr$
104	Unnilquadium (Unq)	$^{249}Cf(^{12}_6C,4^1_0n)^{257}Unq$
105	Unnilpentium (Unp)	$^{249}Cf(^{15}_7N,4^1_0n)^{260}Unp$
106	Unnilhexium (Unh)	$^{249}Cf(^{18}_8O,4^1_0n)^{263}Unh$

energy linear **particle accelerator,** a machine that takes small particles, like protons, and accelerates them before they are directed into a target nuclide. Cockroft and Walton, in 1932, successfully accelerated protons and directed them into a sample of ^7_3Li, initially producing ^8_4Be, which splits and forms two ^4_2He nuclides.

$$^7_3Li + {}^1_1H \longrightarrow {}^8_4Be \longrightarrow {}^4_2He + {}^4_2He$$

Element 103 is named for the American physicist E. O. Lawrence (1901–1958), who invented the cyclotron.

Later in 1932, Lawrence and Livingston developed an even more powerful particle accelerator called a cyclotron. A **cyclotron** accelerates particles in a spiral path, which gives greater velocities, and thus allows for many more nuclear transformations.

Cyclotrons are employed to synthesize new elements that are not observed in nature. Table 19.3 lists the synthetic transuranium elements ($Z > 92$). **Transuranium elements** are those directly after uranium on the periodic table. The first transuranium element synthesized was $^{239}_{93}Np$, identified in 1940 by McMillan and Abelson in Berkeley, California. They bombarded $^{238}_{92}U$ with neutrons produced in a cyclotron, giving $^{239}_{93}Np$ and beta particles.

$$^{238}_{92}U(^1_0n,\beta)^{239}_{93}Np$$

Transuranium elements were named in a manner similar to the lanthanide elements directly above them on the periodic table. Americium was named to parallel europium. Curium was named for the Curies, who were the founders of the science of radioactivity, and gadolinium was named for the Finnish chemist Gadolin, who was a pioneer in the study of the rare earths.

Later in 1940, Seaborg, McMillan, Kennedy, and Wahl, also at Berkeley, discovered plutonium, Pu. They bombarded $^{238}_{92}U$ with deuterium, ^2_1H, initially producing an unstable daughter nuclide $^{238}_{93}Np$, which undergoes beta decay to form $^{238}_{94}Pu$.

Enrico Fermi (*Argonne National Laboratory*)

Enrico Fermi (1901–1954) was a magna cum laude Ph.D. graduate from the University of Pisa. Fermi, whose wife was Jewish, was forced to leave his native country of Italy because he refused to support the fascist government of Mussolini. He moved to the United States, where he made many valuable contributions to the understanding of the nucleus.

$$^{238}_{92}U(^2_1H, 2^1_0n)^{238}_{93}Np$$

$$^{238}_{93}Np \longrightarrow \beta + ^{238}_{94}Pu$$

Nuclear Fission

Lise Meitner (1878–1968) obtained her doctorate at the University of Vienna under the direction of Boltzmann. She traveled to Berlin in 1907 to hear Planck lecture. Here she met and joined Otto Hahn's research group. Meitner was forced to leave Germany when Hitler came to power. With the help of Neils Bohr, she obtained a position in Sweden, where she discovered nuclear fission.

Another extremely important nuclear reaction is called **nuclear fission,** in which a nucleus is split into fragments. When nuclides are fissioned, the products consist of many different smaller nuclides.

In 1934, Enrico Fermi was the first to split an atom; he bombarded $^{238}_{92}U$ with neutrons. Fermi predicted that the $^{238}_{92}U$ would absorb a neutron, undergo beta decay, and then produce the element with atomic number 93, Np. However, Fermi discovered at least four different products. Four years later, Otto Hahn and Fritz Strassmann analyzed Fermi's reaction products and isolated Ba, a substance with a mass number approximately one-half that of uranium. Hahn communicated his finding to an old friend, Lise Meitner, who immediately deduced what had happened. She realized that the U nucleus had split; Meitner then coined the term "nuclear fission." Additionally, she asked her nephew, Otto Frisch, to repeat the experiment. He discovered that the U nucleus had fragmented and produced two new atoms, barium and krypton. More importantly, Frisch discovered that the fission process released a colossal amount of energy, approximately 5×10^9 kcal/mol U.

In some cases, fission reactions even occur spontaneously without a bombardment reaction. $^{252}_{98}Cf$ spontaneously undergoes fission, producing many different products. One possible way that $^{252}_{98}Cf$ undergoes fission is:

$$^{252}_{98}Cf \longrightarrow {}^{140}_{56}Ba + {}^{108}_{42}Mo + 4{}^{1}_{0}n$$

Nuclear Fusion

When matter is heated to extremely high temperatures, like those needed to sustain fusion reactions, it is in the plasma state—a fourth state of matter.

Nuclear fusion is the reaction in which low-mass nuclides, like H, are joined, forming a more massive nucleus and liberating a large quantity of energy. A nuclear fusion reaction takes place only if a tremendous amount of energy is initially available to overcome the internuclear electrostatic repulsive forces. Temperatures in excess of 100,000,000°C are typically needed to initiate fusion reactions.

Nuclear fusion is the process that produces energy in stars. Stars generate energy by "burning" H. In the sun, the following fusion reactions take place:

$$^{1}_{1}H + {}^{1}_{1}H \longrightarrow {}^{2}_{1}H + {}^{0}_{1}e^{+}$$
$$^{2}_{1}H + {}^{1}_{1}H \longrightarrow {}^{3}_{2}He$$
$$^{3}_{2}He + {}^{3}_{2}He \longrightarrow {}^{4}_{2}He + 2{}^{1}_{1}H$$

After there is a large enough concentration of $^{4}_{2}He$ produced in a star, the $^{4}_{2}He$ is fused to produce the higher-atomic-mass elements. For example, it is thought that $^{12}_{6}C$ is produced by the fusion of three $^{4}_{2}He$:

$$3{}^{4}_{2}He \longrightarrow {}^{12}_{6}C$$

All of the heavy elements in the universe have been formed in the core of stars by nuclear fusion.

REVIEW PROBLEMS

19.9 What happens to the mass and atomic number of nuclides that undergo: (a) beta decay, (b) alpha decay, (c) gamma emission?

19.10 Write a nuclear equation for each of the following radioactive changes: (a) beta decay of $^{115}_{48}Cd$, (b) alpha decay of $^{233}_{92}U$, (c) beta decay of $^{198}_{79}Au$, (d) gamma emission of $^{89}_{38}Sr$.

19.11 List the parent nuclides of the three radioactive decay series and the stable daughter nuclides that ultimately are formed.

19.12 (a) What is a positron? List its properties. (b) Write an equation to show how a positron is produced.

19.13 Translate the following nuclear equations into the shorthand notation:
(a) $^{40}_{18}Ar + {}^{4}_{2}\alpha \longrightarrow {}^{43}_{19}K + {}^{1}_{1}p^{+}$
(b) $^{59}_{27}Co + {}^{1}_{0}n \longrightarrow {}^{56}_{25}Mn + {}^{4}_{2}\alpha$
(c) $^{12}_{6}C + {}^{2}_{1}H \longrightarrow {}^{13}_{7}N + {}^{1}_{0}n$

19.14 Using Table 19.3, translate the shorthand notation for writing nuclear equations to a complete nuclear equation for the synthesis of: (a) Np, (b) Cf, and (c) Lr.

19.15 (a) What is the difference between nuclear fission and nuclear fusion? (b) Write an equation to illustrate each.

19.4 USES OF RADIOACTIVE SUBSTANCES

Radioactive Dating   **Radioactive dating** is an indirect method of determining the age of an old rock or artifact of interest. To illustrate, let us consider *radiocarbon dating*. Ancient objects that contain carbon are dated by measuring a radioactive isotope of carbon, $^{14}_{6}C$, a beta emitter. All living things contain $^{14}_{6}C$. Plants take in $^{14}_{6}CO_2$ during photosynthesis, and animals ingest compounds containing $^{14}_{6}C$ in their foods. In nature, $^{14}_{6}C$ is produced in the atmosphere from $^{14}_{7}N$ by the following reaction:

$$^{14}_{7}N + {}^{1}_{0}n \longrightarrow {}^{14}_{6}C + {}^{1}_{1}H$$

Once formed, the $^{14}_{6}C$ is oxidized to $^{14}_{6}CO_2$:

$$^{14}_{6}C + O_2 \longrightarrow {}^{14}_{6}CO_2$$

A balance between the amount of $^{14}_{6}C$ entering and leaving a living system keeps the $^{14}_{6}C$ concentration constant. Since the concentration of $^{14}_{6}C$ in CO_2 is 1 out of 10^{12} carbon atoms, it is assumed that 1 out of 10^{12} carbon atoms in a living organism is $^{14}_{6}C$. After the organism's death, no more $^{14}_{6}C$ enters (see Figure 19.5). However, the amount of $^{14}_{6}C$ in the dead organism decreases as a result of radioactive decay. The half-life of this radioactive isotope is approximately 5700 years. Thus, with the passing of every 5700 years, the amount of $^{14}_{6}C$ decreases by one-half. Careful measurement of the amount of $^{14}_{6}C$ remaining in an old object compared with the amount of $^{14}_{6}C$ in a modern object allows scientists to estimate roughly the age of the object.

Radiocarbon dating only gives an approximation of an object's age; exact ages cannot be determined with this method. A problem with the validity of radiocarbon dating is that an assumption is made that the amount of $^{14}_{6}C$ on earth has remained constant throughout the years. Radiocarbon dating is also limited by the sensitivity of radiation-detecting devices that measure $^{14}_{6}C$. Conse-

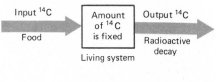

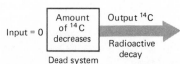

FIGURE 19.5
Within living systems, the amount of $^{14}_{6}C$ is relatively constant. After the living system dies, the amount of $^{14}_{6}C$ decreases as a result of radioactive decay.

quently, after approximately ten half-lives, $^{14}_{6}C$ activity cannot be measured as accurately.

$^{40}_{19}K$ and $^{40}_{18}Ar$ are used by geologists to date rocks. $^{40}_{19}K/^{40}_{18}Ar$ dating is different from radiocarbon dating in that $^{40}_{19}K$ decays to $^{40}_{18}Ar$, which is trapped in the rocks where it is found. Accordingly, the amount of $^{40}_{18}Ar$ accumulated is a measure of the age of the rock. Two properties of $^{40}_{19}K$ make it an excellent substance for determining the age of rocks: (1) $^{40}_{19}K$ has a long half-life, 1.3×10^9 years; and (2) $^{40}_{19}K$ has a relatively high natural abundance, in excess of 0.01%.

Medical Uses Prior to the development of medical isotopic tracers and sophisticated computer-analyzed x-ray techniques, a person suspected of having a perplexing internal problem would normally have to undergo exploratory surgery to determine the nature of the problem. In part, radioisotopes and modern x-ray techniques have diminished the need for surgical diagnostic procedures.

In excess of 100 different radionuclides have been utilized in medical diagnostic techniques. Table 19.4 lists some of the more frequently used isotopes and the problems they detect. For instance, $^{131}_{53}I$, a beta emitter, is used to diagnose diseases of the thyroid gland. The patient is given a solution of sodium iodide, $Na^{131}_{53}I(aq)$, to drink. After a certain period of time, enough for the $^{131}_{53}I^-$ to be absorbed by the blood and taken into the thyroid, the patient is placed into a scanning machine. A scanner is a machine that contains a radiation detector that transforms the intensity of radioactivity at a particular point into a visual image. Analysis of the image or "scan" by a radiologist provides valuable information about the thyroid gland.

Also listed in Table 19.4 are radionuclides that are used to treat medical problems. One of the more important therapeutic uses of radioisotopes is to destroy cancerous tumors. Radiation is either beamed through a person or a radioactive substance is implanted at the site of the tumor. $^{60}_{27}Co$, an intense

TABLE 19.4 COMMON RADIONUCLIDES USED IN MEDICINE

Nuclide	Half-life	Use*
^{137}Cs	30 y	T, implanted to destroy tumors
^{60}Co	5.3 y	T, cancers
^{198}Au	2.7 d	T, abdominal cancers
		D, liver imaging
^{32}P	14 d	T, chronic leukemia
^{131}I	8.1 d	D, thyroid and brain scans
^{24}Na	15 d	D, vascular disease
^{59}Fe	46 d	D, blood volume and iron metabolism
^{99}Tc	6.1 h	D, brain, heart, and bone scans
^{42}K	12 h	D, localizing brain tumors

*T = therapeutic use; D = diagnostic use. (Only selected uses are presented in this table; most of the above nuclides have other uses as well.)

TABLE 19.5 BIOLOGICAL EFFECTS OF RADIATION

Dose,* rem	Effect
0–25	No noticeable biological effects. A chest x-ray is 0.2 rem, and dental x-rays are approximately 0.02 rem.†
25–50	Small effect, usually a small decrease in white blood cell count.
100–200	Significant lowering of white blood cell count, nausea, and general sickness. Normally, no deaths are expected.
200–400	Vomiting and nausea first day, later diarrhea and general sickness with hair and skin loss. About 20% will die within the first month after exposure.
400–500	Same symptoms as above. About 50% will die in first month.

* Short-term exposure.
† rem = roentgen equivalent man, a unit of radiation exposure. The higher the value, the greater the damage to human tissues.

gamma and beta emitter, is used as an external source of gamma radiation that is concentrated on a tumor. $^{137}_{55}Cs$ needles are implanted near tumors until the tumors are destroyed; the $^{137}_{55}Cs$ is then removed.

Even though radioactivity is used to diagnose or treat various medical problems, exposure to radiation has a profound negative effect on human health. Once a person has been exposed to significant levels of radiation, there is nothing that the world of medicine can do to reverse the effects. Excessive exposure to high levels of radiation always results in death. Exposure to lower but significant levels initially produces nausea, anemia, vomiting, and diarrhea, as well as skin and hair loss. Later, rare blood diseases or leukemia frequently result, as well as genetic defects that can be passed to the next generation. Table 19.5 lists symptoms that are associated with different levels of exposure to radiation.

What is a safe exposure level to radiation? No one really knows the answer to this controversial question. Table 19.6 summarizes the sources and levels of

The unit rem is related to another radiation unit called the rad, radiation absorbed dose. One rad is equal to 100 ergs of energy absorbed per gram of living tissue (1 erg = 10^{-7} J). A rad is almost equivalent to a roentgen (R), a unit of the degree of ionization of dry air by radiation forms.

TABLE 19.6 RADIATION EXPOSURE IN THE UNITED STATES

Source	Average exposure, millirems/year per capita
Natural background	110
Total artificial exposure	90
Medical and dental x-rays	80
Technology and industry	4
Fallout from nuclear weapons	5
Nuclear power plants	0.3
Consumer products (watches, TV, etc.)	0.04

radiation per capita that U.S. citizens are exposed to annually. Note that natural and artificial sources of radiation are nearly equal, meaning that human activities produce approximately half of the radiation that we are exposed to. It is interesting to note that the largest percent of exposure to artificial radiation is from medical and dental x-rays!

Nuclear Energy Production

Energy from a controlled fission reaction was first harnessed by Enrico Fermi under the stadium on the campus of the University of Chicago on December 2, 1942. From this beginning, modern nuclear power plants have evolved.

A nuclear power plant (Figure 19.6) uses a fissionable material such as $^{235}_{92}U$ or $^{239}_{94}Pu$ to produce heat energy, which is then converted to electric energy. In a nuclear power plant, small fuel pellets are loaded into the core of the reactor. Located between the fuel rods are neutron-absorbing rods (usually constructed from B or Cd), which are automatically withdrawn when the neutron level decreases and reinserted when the level becomes too high. A fluid, such as water or molten sodium, is passed through the reactor core, picking up the heat liberated by the fission process. This heat is then transferred to nonradioactive water outside the core, changing the water to steam. The high-pressure steam turns a turbine that is connected to an electric generator. Other than the nuclear reactor and its associated equipment, nuclear power plants operate in a manner similar to fossil fuel plants, which burn oil, coal, and natural gas to produce heat.

Fuel rods contain an oxide of uranium, U_3O_8.

Controversy surrounds the use of nuclear power plants. Most of the controversy results from the unanswerable questions: Are nuclear power plants safe? How safe is safe? Besides the problems of operating power plants safely, there is the ever-present problem of nuclear waste storage. At present, no method has

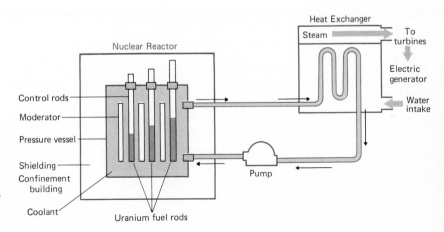

FIGURE 19.6

In a nuclear power plant, heat is generated from the nuclear fission of uranium or plutonium atoms. This heat energy converts water to steam, and the steam turns a turbine that is connected to an electric generator.

been perfected to adequately transport and store the intensely radioactive spent fuel rods. This problem could be minimized with the development of efficient, safe reprocessing facilities that could separate the "good" fissionable material from the waste.

There are no simple answers to the complex problems associated with nuclear fission power generation. Answers to these problems must take into account not only the scientific aspects of nuclear power, but social, political, and economic factors as well. We may never have to totally solve these problems because controlled fusion energy should eliminate the need for fission power generation.

Controlled fusion is a process in which low-mass elements are fused together, releasing an enormous quantity of energy, simulating the energy-producing apparatus of the sun on earth. Fuel for a fusion reactor is water, from which an unlimited supply of hydrogen could be obtained.

Many scientific and technological hurdles must be overcome to make available this unlimited supply of cheap and clean energy. At present one of the biggest problems is the temperatures required to sustain a fusion reaction, greater than 100 million degrees. No known substance can withstand such enormous temperatures. If you are an optimist, place a bet that science and technology will solve these problems in the next 50 years.

REVIEW PROBLEMS

19.16 Describe how the age of an old animal bone is determined using radiocarbon dating.

19.17 How is $^{40}_{19}K/^{40}_{18}Ar$ dating different from radiocarbon dating?

19.18 What are the two principal applications of radionuclides in medicine?

19.19 List the biological effects of exposure to sublethal doses of radiation.

19.20 Describe how the process of fission is used to generate electric energy in a nuclear power plant.

SUMMARY

The nucleus is the very small, dense region within an atom that contains the nucleons (protons and neutrons). Atomic number is the number of protons in the nucleus. Mass number is the total number of nucleons in the nucleus. Atoms with the same atomic number but different mass numbers are called isotopes.

The stability of a nuclide depends on the composition of its nucleus. Some nuclear configurations are more stable than others. Nuclides with 2, 8, 20, 50, 82, and 126 protons or neutrons are more stable than nuclides with other compositions. Atoms with both an even number of protons and neutrons are generally more stable than those with an odd number of protons or neutrons. Nuclear stability is also correlated with the ratio of neutrons to protons. Small atoms are most stable when they have an equal number of protons and neutrons, while higher-mass atoms are most stable when they contain more neutrons than protons.

There are three primary forms of natural radioactivity: (1) alpha, (2) beta, and (3) gamma radiation. Alpha radiation is observed only in more massive atoms. An alpha particle has the same mass properties as a helium nucleus. Thus, after a parent nuclide releases an alpha particle, its mass number decreases by 4 and atomic number decreases by 2. Beta decay results when a neutron spontaneously decays to a proton, releasing the beta particle and an antineutrino. Atoms that undergo beta decay produce daughter nuclei with the same mass number but with an atomic number one greater than the parent nuclei. Gamma emission results when a nucleus is in a high energy state. After the gamma ray is emitted, the nucleus returns to a lower energy state. The composition of a nucleus does not change after gamma emission.

One of the most important properties of a radionuclide is its half-life, the amount of time it takes for one-half of its nuclei to decay. A long half-life results when a substance is decaying at a slow rate. A nuclide that decays rapidly has a short half-life.

The nuclear composition of an atom can be changed artificially, through particle bombardment reactions. A target nucleus is placed into a beam of particles such as alpha particles, beta particles, neutrons, or protons. Upon bombardment, a nuclide is changed to some other nuclide, depending on the reaction. Under certain conditions, the target nucleus is split. It undergoes nuclear fission. When fission occurs, neutrons are liberated, which split other nuclei. In addition, a tremendous quantity of energy is released. The opposite of nuclear fission is nuclear fusion. Nuclear fusion results when two low-mass atoms collide and produce a larger-mass atom. As in fission, a large amount of energy is released in fusion reactions.

Radionuclides are used to date old objects. In medicine, radionuclides are used to diagnose and treat diseases. Nuclear fuels are now used to produce electric energy.

Exposure to radiation has many biological effects. A massive dose of radiation results in death. Smaller significant exposures can produce diseases like leukemia and anemia, as well as nausea, diarrhea, loss of skin and hair, and overall sickness. The biological effects of exposure to very low levels of radiation are unknown.

QUESTIONS AND PROBLEMS

19.21 Define the following terms: nucleus, atomic number, mass number, isotope, nuclide, radioactive decay, alpha particles, beta particles, gamma rays, half-life, transmutation, antimatter, neutrino, decay series, bombardment reaction, artificial radioactivity, particle accelerator, transuranium element, nuclear fission, nuclear fusion, radionuclide dating, nuclear power plant, controlled fusion.

Nucleus

19.22 Determine the volume, in liters, that the earth would occupy if earth's density was that of the nucleus. The mass of the earth is 6×10^{24} kg, and the density of the nucleus is 1.8×10^{14} g/mL.

19.23 Determine the nuclear composition for the following nuclides:

(a) $^{227}_{89}\text{Ac}$ (b) $^{77}_{33}\text{As}$ (c) $^{210}_{83}\text{Bi}$
(d) $^{51}_{24}\text{Cr}$ (e) $^{175}_{70}\text{Yb}$

19.24 Write the symbol for the nuclides with the following characteristics:
(a) A = 51, Z = 23
(b) A = 97, Z = 43
(c) N = 91, Z = 62
(d) N = 115, Z = 77

19.25 Consider the nuclear composition to determine which one of each of the following pairs of nuclides would be expected to be more stable. Explain your reasoning for each.

(a) $^{4}_{2}\text{He}$ or $^{3}_{1}\text{H}$ (b) $^{16}_{8}\text{O}$ or $^{15}_{8}\text{O}$

(c) $^{40}_{20}\text{Ca}$ or $^{40}_{19}\text{K}$ (d) $^{207}_{82}\text{Pb}$ or $^{204}_{81}\text{Tl}$

19.26 Determine the nuclear composition of each and select the most and least stable nuclide from the following:

(a) $^{141}_{57}\text{La}$, $^{184}_{74}\text{W}$, and $^{176}_{73}\text{Ta}$

(b) $^{200}_{79}\text{Au}$, $^{200}_{80}\text{Hg}$, and $^{202}_{82}\text{Pb}$

19.27 What could account for the fact that high-mass atoms (Z > 21) possess nuclei with many more neutrons than protons?

Radioactivity

19.28 Considering the mass and charge characteristics of the three forms of radiation, explain the following: (a) beta particles penetrate a greater thickness of matter than do alpha particles, and (b) alpha particles ionize matter more so than do gamma rays.

19.29 Alpha radiation has a low capacity to penetrate matter. However, nuclides that emit alpha radiation generally produce the most severe biological effects when they are inside living things. What could account for this fact?

19.30 Two radioactive nuclides of strontium are $^{89}_{38}\text{Sr}$ and $^{90}_{38}\text{Sr}$. The half-life of $^{89}_{38}\text{Sr}$ is 51 days, and the half-life of $^{90}_{38}\text{Sr}$ is 28 years. (a) Which of these two isotopes is decaying at a faster rate? Explain your answer. (b) If a sample initially contains 50.0 g $^{90}_{38}\text{Sr}$, what mass of $^{90}_{38}\text{Sr}$ remains after 112 years?

19.31 $^{131}_{53}\text{I}$ has a half-life of 8 days. How many days elapse before the $^{131}_{53}\text{I}$ atoms in a sample decrease to $\frac{1}{64}$ their initial number?

19.32 Given a 6.000-g sample of $^{250}_{100}\text{Fm}$ with a half-life of 0.5 hr. Determine the mass of $^{250}_{100}\text{Fm}$ that remains after 2.5 hr.

19.33 It is not uncommon to find pockets of He gas in uranium mines. What could account for the presence of He?

Nuclear Reactions

19.34 Write a complete nuclear equation that illustrates the alpha decay of: (a) $^{222}_{86}\text{Rn}$, (b) $^{234}_{92}\text{U}$, (c) $^{237}_{93}\text{Np}$, (d) $^{253}_{99}\text{Es}$.

19.35 Write a complete nuclear equation that illustrates the beta decay of: (a) $^{32}_{15}\text{P}$, (b) $^{144}_{58}\text{Ce}$, (c) $^{76}_{33}\text{As}$, (d) $^{191}_{76}\text{Os}$.

19.36 Explain why a nuclear equation is not generally needed to represent gamma emission.

19.37 Describe the principal characteristics of the following antimatter forms relative to their "regular" matter counterparts: (a) positron, (b) antineutrino, (c) antiproton, (d) antineutron.

19.38 In the $^{232}_{90}\text{Th}$ natural decay series, the $^{232}_{90}\text{Th}$ initially undergoes alpha decay, the resulting daughter emits a beta particle and forms the granddaughter, which also emits a beta particle. Write three nuclear equations to represent the first three steps in the $^{232}_{90}\text{Th}$ decay scheme.

19.39 What nuclide results after the following nuclear changes?
(a) $^{131}_{53}\text{I}$ releases a beta particle
(b) $^{214}_{82}\text{Pb}$ undergoes two successive beta emissions
(c) $^{218}_{85}\text{At}$ undergoes two successive alpha decays followed by a beta decay

19.40 Complete the following equations:

(a) $^{115}_{48}\text{Cd} \longrightarrow ? + ^{115}_{49}\text{In}$

(b) $? \longrightarrow ^{4}_{2}\alpha + ^{243}_{96}\text{Cm}$

(c) $^{129}_{53}\text{I} \longrightarrow ^{0}_{+1}e + ?$

19.41 Change the following shorthand notation bombardment equations to full nuclear equations:

(a) $^{32}_{16}\text{S}(^{1}_{0}n, \gamma)^{33}_{16}\text{S}$ (b) $^{130}_{52}\text{Te}(^{2}_{1}\text{H}, 2^{1}_{0}n)^{130}_{53}\text{I}$,

(c) $^{43}_{20}\text{Ca}(^{4}_{2}\alpha, ^{1}_{1}p)^{46}_{21}\text{Sc}$ (d) $^{9}_{4}\text{Be}(^{1}_{1}p, ^{4}_{2}\alpha)^{6}_{3}\text{Li}$

19.42 Change the following equations to the shorthand notation:

(a) $^{238}_{92}U + ^{16}_{8}O \longrightarrow ^{249}_{100}Fm + 5^{1}_{0}n$

(b) $^{241}_{95}Am + ^{4}_{2}\alpha \longrightarrow ^{243}_{97}Bk + 2^{1}_{0}n$

(c) $^{238}_{92}U + ^{22}_{10}Ne \longrightarrow ^{256}_{102}No + 4^{1}_{0}n$

(d) $^{242}_{94}Pu + ^{22}_{10}Ne \longrightarrow ^{260}_{104}Unq + 4^{1}_{0}n$

19.43 Complete the following equations:

(a) $^{196}_{78}Pt + ? \longrightarrow ^{197}_{78}Pt + ^{1}_{1}H$

(b) $^{94}_{42}Mo + ? \longrightarrow ^{95}_{43}Tc + ^{1}_{0}n$

(c) $^{235}_{92}U + ^{1}_{0}n \longrightarrow ^{93}_{35}Br + ^{140}_{57}La + ?$

(d) $^{24}_{12}Mg + ^{2}_{1}H \longrightarrow ? + ^{22}_{11}Na$

(e) $^{34}_{16}S + ^{1}_{0}n \longrightarrow ^{35}_{16}S + ?$

19.44 Write a complete nuclear equation that shows how the following transuranium elements were initially synthesized (see Table 19.3): (a) Am, (b) No, and (c) Unp.

19.45 What accounts for the large amount of energy produced during the fission process?

19.46 What scientists are credited with the discovery of nuclear fission?

19.47 Write the set of three equations that illustrate how hydrogen is fused to produce helium in the sun.

19.48 Why are temperatures higher than 100 million degrees required to initiate nuclear fusion reactions?

Uses of Radioactive Substances

19.49 Write equations to show how most of the $^{14}_{6}C$ on earth is formed.

19.50 If an old bone is found to contain one-half the quantity of $^{14}_{6}C$ in a modern bone, what is the approximate age of the old bone?

19.51 To date objects, using $^{14}_{6}C$, the assumption is made that the amount of $^{14}_{6}CO_2$ on earth has remained constant. What activities of human beings over the last 80 years might invalidate this assumption?

19.52 Could $^{40}_{19}K/^{40}_{18}Ar$ dating techniques be used to determine the age of a skull found in an ancient burial ground? Explain.

19.53 How have radionuclides changed the diagnostic procedures used by doctors (be specific)?

19.54 How are the following nuclides utilized diagnostically in medicine? (a) $^{131}_{53}I$, (b) $^{32}_{15}P$, (c) $^{99}_{43}Tc$, (d) $^{59}_{26}Fe$.

19.55 What might account for lack of "solid" data concerning the effects of low levels of radiation exposure on biological systems?

19.56 List two significant problems associated with nuclear power plants that are not major problems in conventional fossil fuel power plants.

19.57 How is the level of neutron radiation controlled in a nuclear power plant?

19.58 Why is a nuclear explosion, as in an atomic bomb, an impossibility in the normal operation of a nuclear power plant?

19.59 What changes might be expected in our world (political, social, economic, etc.) after the development of an economical method for harnessing fusion energy?

20

Overview of Organic and Biologically Important Compounds

Study Guidelines

After completing Chapter 20, you should be able to:
1. List the major classes of organic compounds.
2. Write IUPAC names for simple alkanes
3. Write all isomers of simple alkanes
4. Identify and give examples of unsaturated hydrocarbons
5. Identify the basic structure of aromatic hydrocarbons
6. Give examples of important hydrocarbons
7. Write the functional groups in common hydrocarbon derivatives
8. Draw the structures and name the simplest member of the major classes of hydrocarbon derivatives
9. Give examples of important compounds for each primary group of hydrocarbon derivatives
10. Identify the basic structures contained in the four principal classes of biochemicals
11. Explain the role of each class of biochemicals in living things

Modern **organic chemistry** is the study of the properties and reactions of carbon compounds. In the past, chemists did not define organic chemistry in this way. In 1807, Berzelius proposed that compounds derived from living things were "organic" and all others were "inorganic," not derived from life. Such a definition appealed to scientists of the time, especially when it was well known that organic substances could easily be converted to inorganic substances by heating or treatment with acids in the laboratory. They felt that the opposite conversion, from inorganic to organic, was not possible.

In the minds of early nineteenth century scientists, organic compounds could be produced only within living systems. At that time it was thought that living things possessed a "vital force" which changed inorganic substances to organic ones. Scientists believed that without the vital force there was no way to change an inorganic to an organic compound. In 1828, Friedrich Wöhler (1800–1882) stunned the world of science and shattered the vital force theory (also called *vitalism*) by synthesizing an organic compound from an inorganic salt.

Friedrich Wöhler studied medicine and received his degree as a physician before becoming a chemist. Besides his famous discovery, Wöhler was one of the first to investigate metabolism of compounds in living things.

Wöhler heated the inorganic salt ammonium cyanate, NH_4OCN, and analyzed the products; he discovered that urea, NH_2CONH_2, was one of the reaction products.

$$NH_4OCN \xrightarrow{\Delta} NH_2CONH_2$$

Urea is a waste product in urine. Wöhler communicated his finding to his former professor, Berzelius, who then realized that his own ideas concerning the definition of organic chemistry would have to be modified in light of Wöhler's new evidence.

Modern organic compounds are divided into two major classes, hydrocarbons (*hydro*, hydrogen) and hydrocarbon derivatives. **Hydrocarbons** are compounds that contain only the elements carbon and hydrogen. **Hydrocarbon derivatives** are those compounds that contain carbon, usually hydrogen, and some other atom. Hydrocarbon derivatives normally possess one or more of the following elements in addition to C and H: O, N, S, P, halogens, and various metals.

Some carbon compounds such as CO, CO_2, carbonates, bicarbonates, and ionic cyanides are not considered organic compounds because they have properties that more closely resemble inorganic compounds.

20.2 HYDROCARBONS

Hydrocarbons are divided into four groups: (1) alkanes, (2) alkenes, (3) alkynes, and (4) aromatics. The alkanes, alkenes, and alkynes are together known as **aliphatic hydrocarbons**. Aromatics possess a special structure called an aromatic ring that makes them chemically quite distinct. Accordingly, organic chemists place them into a separate group called **aromatic hydrocarbons**.

Alkanes

Alkanes are hydrocarbons that possess carbon atoms bonded to the maximum number of hydrogen atoms possible. There are no double or triple bonds in the alkanes. Since alkanes cannot chemically add any more hydrogen atoms, they are frequently called the **saturated hydrocarbons**—saturated with respect to the number of hydrogen atoms in the molecule. In alkanes, carbon atoms are bonded together in chains and rings. First, we shall consider alkanes that are composed of molecules with carbon chains.

Methane, CH_4, is the simplest alkane and has the following structure:

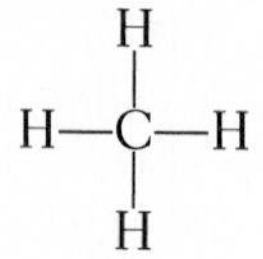

$$H-\underset{\underset{H}{|}}{\overset{\overset{H}{|}}{C}}-H$$

Methane is a colorless, odorless gas with a melting point of $-183°C$ and a boiling point of $-162°C$. Such low melting and boiling points indicate that the intermolecular forces among methane molecules are weak dispersion forces. They result because methane is a nonpolar covalent molecule. Methane, like many organic compounds, is referred to by one or more common names. For instance, methane is frequently called swamp gas. When plants decay in the absence of oxygen, a quantity of methane is produced. Such conditions exist in swamp and marsh lands where decaying plant matter is covered with water.

To alleviate the problem of having more than one name for a compound, organic chemists have adopted the IUPAC systematic procedure for naming organic compounds. IUPAC stands for the International Union of Pure and Applied Chemistry, an international organization that oversees the naming of chemical substances. Table 20.1 lists the accepted names for the first 10 alkanes.

Each name is composed of two parts: stem and suffix. The stem indicates the number of carbon atoms in a molecule. For instance, *eth* indicates there are two carbon atoms in a molecule, *prop* means three carbon atoms, and *but* indicates four carbon atoms. The suffix tells what class of organic compound the molecule belongs to. The alkanes are designated by the ending *ane*. Each different class of organic compounds has its own unique suffix.

> Methane is the principal component of natural gas. Natural gas is used for heating and cooking. Since natural gas is odorless, a small quantity of an organic sulfur compound is added so that its presence can be detected.

TABLE 20.1 NAMES OF FIRST 10 ALKANES

Molecular formula	Condensed structure*	IUPAC name
CH_4	CH_4	Methane
C_2H_6	CH_3CH_3	Ethane
C_3H_8	$CH_3CH_2CH_3$	Propane
C_4H_{10}	$CH_3(CH_2)_2CH_3$	Butane
C_5H_{12}	$CH_3(CH_2)_3CH_3$	Pentane
C_6H_{14}	$CH_3(CH_2)_4CH_3$	Hexane
C_7H_{16}	$CH_3(CH_2)_5CH_3$	Heptane
C_8H_{18}	$CH_3(CH_2)_6CH_3$	Octane
C_9H_{20}	$CH_3(CH_2)_7CH_3$	Enneane or Nonane
$C_{10}H_{22}$	$CH_3(CH_2)_8CH_3$	Decane

* A condensed formula is one in which the number of bonded hydrogen atoms is written to the right of each carbon atom. This is a shorthand way to express organic structural formulas.

Alkanes have the general formula C_nH_{2n+2}. If there are n carbon atoms within the molecule, there are $2n + 2$ hydrogen atoms attached. An alkane with 20 carbon atoms, for example, will have 42 $(2 \times 20 + 2)$ hydrogen atoms.

Alkanes are an example of a homologous series of compounds. A **homologous series** is a group of compounds in which one member differs from the compound that immediately precedes or follows it by a fixed amount, in this case by a CH_2 unit. If we select ethane, C_2H_6, it differs from both methane, CH_4, and propane, C_3H_8, by a CH_2 unit. Homologous series have observable trends in physical properties and exhibit similar chemical properties.

The first four members of the alkane series, methane to butane, are gases at room temperature. Pentane to octadecane, $C_{18}H_{38}$, are liquids, and the remaining alkanes, those with 19 or more carbon atoms, are solids. Such a trend indicates that there is a gradual increase in boiling and melting points with increased molecular mass. Chemically, the alkanes as a group are rather inert. They combine only with very reactive substances like the halogens.

In addition to the alkanes that are molecules with unbranched chains, there are alkanes with branches in the chain. For example, butane's molecular formula is C_4H_{10}. Molecules of butane contain four carbon atoms in a continuous chain.

$$CH_3CH_2CH_2CH_3$$

There is another compound with exactly the same molecular formula but with a different structural formula.

$$CH_3CHCH_3$$
$$|$$
$$CH_3$$

In this second compound there is a branch in the carbon chain. These two compounds are called **structural isomers,** compounds with the same molecular formula but with a different arrangement of carbon atoms (different structural formulas). Each has its own unique set of physical and chemical properties.

IUPAC has established specific rules for naming the alkanes with branches in the carbon chain. These rules are as follows:

Rule 1. Find the longest continuous chain of carbon atoms within the molecule.

This is accomplished by starting at one end of the molecule and counting carbon atoms that are consecutively bonded together. The name of the molecule is derived from the name of the alkane with the same number of carbon atoms in the longest continuous chain. This is called the **parent alkane.** All of the other carbon atoms in the molecule are considered branches off of the main chain and are called **substituent groups.**

In the above example the longest continuous chain is three carbon atoms; hence, the parent alkane is propane.

There are many more organic compounds than there are words in the English language. The task of properly naming such an immense number of compounds is quite difficult.

Rule 2. Number the longest continuous chain in such a way that the lowest possible number is given to the carbon atom to which the substituent group or branch is bonded.

Our example has only one branch, which is in the center of the longest chain. Thus, it does not matter in which direction the three-carbon chain is numbered.

$$\overset{1}{C}H_3\overset{2}{\underset{\underset{CH_3}{|}}{C}}H\overset{3}{C}H_3$$

Rule 3. Identify each branch (substituent group) in the chain, and the carbon to which it is attached.

A name is required for the substituent group CH_3. This group is the first in a series of what are called **alkyl groups.** An alkyl group is an alkane minus one hydrogen ($CH_4 - H = CH_3$) that is bonded to a carbon chain or other organic structure. To name alkyl groups remove the *ane* ending and replace with *yl*. Therefore, a CH_3 group is a *methyl* group (methane − *ane* + *yl*). If a hydrogen is removed from ethane, C_2H_6, an ethyl group, C_2H_5, results. Methyl and ethyl groups are the only alkyl groups that we shall consider.

In our example, there is a methyl group bonded to the second carbon of the three-carbon chain.

Rule 4. Write the name of the compound by first writing the number of the carbon to which the substituent group is bonded, followed by a hyphen that is connected to the name of the substituent group and the name of the parent hydrocarbon.

Thus, the name of the example hydrocarbon is 2-methylpropane. If we analyze the name, we see that located on the second carbon of a three-carbon chain is a methyl group.

2-methylpropane

Carbon number of bonded group ⟶   ↑   ⟵ Parent hydrocarbon

Alkyl group

To illustrate both isomers and IUPAC naming, let's identify and name all of the isomers of pentane, C_5H_{12}. Starting with pentane, the unbranched chain is:

$$CH_3CH_2CH_2CH_2CH_3 = \text{pentane}$$

Next we can write a four-carbon chain with one substituent group.

In 1892, the problem of naming compounds was first considered at the International Congress of Chemistry. Subsequently, regular meetings were held to develop a system for naming compounds. An outgrowth of these early meetings was the development of the IUPAC.

There are 75 isomers of decane ($C_{10}H_{22}$), 366,319 isomers of eicosane ($C_{20}H_{42}$), and 4,111,846,763 isomers of triacontane ($C_{30}H_{62}$).

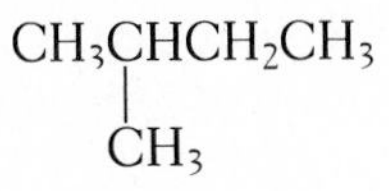

$$CH_3CHCH_2CH_3$$
$$|$$
$$CH_3$$

To name this compound, follow each of the above steps: (1) Identify the longest continuous chain: four carbon atoms are located in the longest chain. (2) Number the chain, starting from the end closest to the branch:

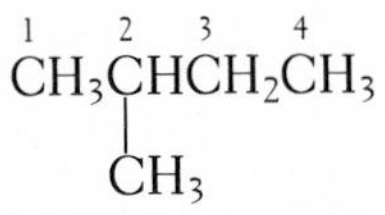

$$\overset{1}{C}H_3\overset{2}{C}H\overset{3}{C}H_2\overset{4}{C}H_3$$
$$|$$
$$CH_3$$

(3) Identify substituent groups and the carbon to which they are bonded: There is a methyl group bonded to the second carbon. (4) Write the name of the compound, starting with the number of the carbon to which the substituent group is bonded, followed by the name of the alkyl group and the parent hydrocarbon. Thus, the name of this compound is 2-methylbutane.

The compound 2-methylbutane is the only pentane isomer with butane as the parent hydrocarbon. If we write the structure with the methyl group bonded to the next carbon in the chain, it is just another way of writing 2-methylbutane. Following the rules, the chain should be numbered starting from the end closest to the substituent group.

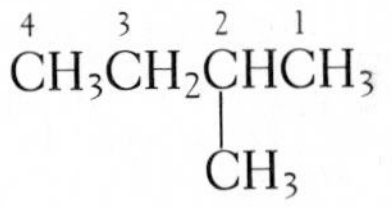

$$\overset{4}{C}H_3\overset{3}{C}H_2\overset{2}{C}H\overset{1}{C}H_3$$
$$|$$
$$CH_3$$

The remaining isomer of pentane has a parent chain of three carbon atoms with two methyl groups bonded to the second carbon.

$$CH_3$$
$$\overset{1}{C}H_3-\overset{2}{C}-\overset{3}{C}H_3$$
$$|$$
$$CH_3$$

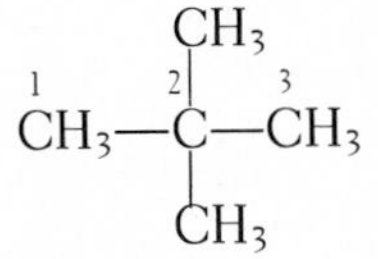

Following the rules, we find three carbon atoms as the longest continuous chain with two methyl groups attached at the second carbon. To indicate that two groups are bonded to the chain, the prefix *di* is placed in front of methyl, and to indicate that both methyl groups are bonded to the second carbon, the number 2 is written twice. The correct name for the compound is 2,2-dimethylpropane.

So far we have seen that alkanes exist as chains and branched chains. They can also exist as ring structures. An alkane ring structure, or **cycloalkane,** is one in which each carbon is bonded to two other carbon atoms and two hydrogen atoms. A three-carbon cycloalkane is the smallest ring structure:

Cyclopropane is an excellent anesthetic, but it must be handled carefully because it is explosive. It is a potent anesthetic, with minimal side effects and rather low toxicity.

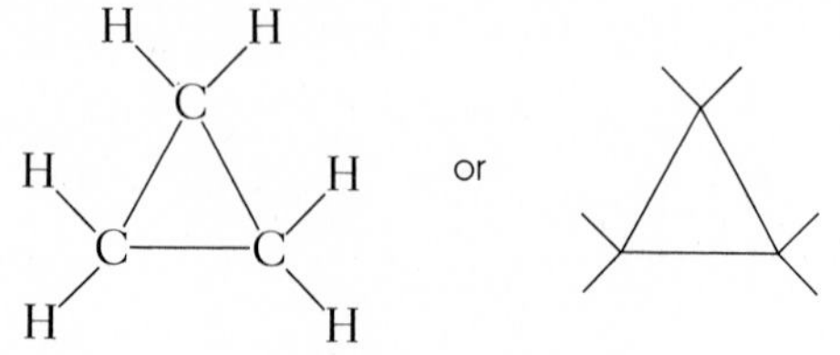

Cyclopropane

To name cycloalkanes, the prefix *cyclo* is placed in front of the name of the parent alkane. Consequently, a three-carbon cycloalkane is called cyclopropane. The next compound in the series is cyclobutane. Its structure is

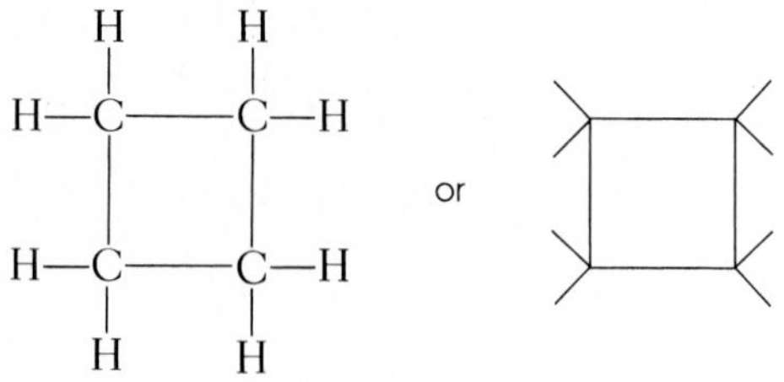

Cyclobutane

Next in the series are cyclopentane and cyclohexane, etc.

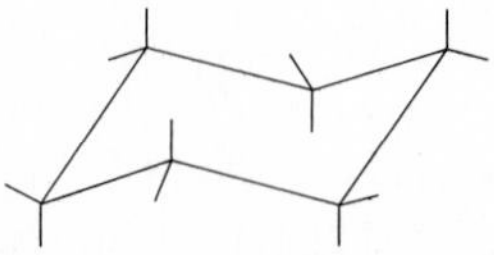

Cyclopentane   Cyclohexane

Even though cyclic structures are symbolized as being flat (planar), most rings are "puckered." For instance, cyclohexane's configuration is best represented as

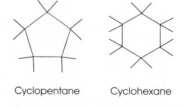

Note that the general formula of the cycloalkanes is different from that of the noncyclic alkanes. Cycloalkanes have the general formula of C_nH_{2n}. There are also cycloalkanes with alkyl groups bonded to the ring. See the following examples.

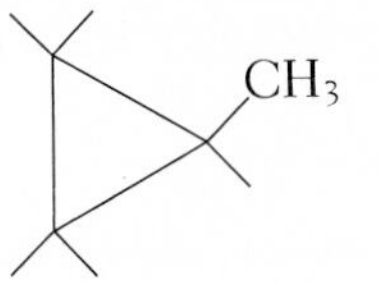

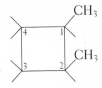 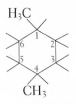

Methylcyclopropane   1,2-Dimethylcyclobutane   1,4-Dimethylcyclohexane

REVIEW PROBLEMS   **20.1** What are the four classes of hydrocarbons?

20.2 Write the names of the alkanes with the following number of carbon atoms in the molecule: (a) two, (b) four, (c) five, (d) seven, (e) nine.

20.3 What are the molecular formulas of the following: (a) methane, (b) propane, (c) pentane, (d) hexane, (e) decane?

20.4 Write the condensed formulas for the following structures:

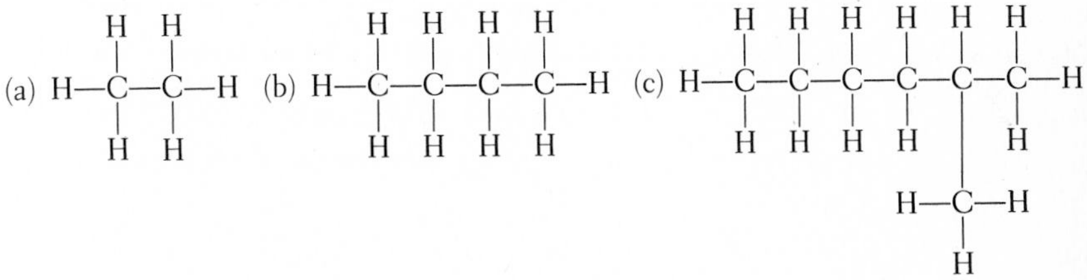

(a)
```
   H  H
   |  |
H—C—C—H
   |  |
   H  H
```
(b)
```
   H  H  H  H
   |  |  |  |
H—C—C—C—C—H
   |  |  |  |
   H  H  H  H
```
(c)
```
   H  H  H  H  H  H
   |  |  |  |  |  |
H—C—C—C—C—C—C—H
   |  |  |  |  |
   H  H  H  H  H
            |
         H—C—H
            |
            H
```

20.5 Draw the structures and write the IUPAC names for all five isomers of hexane.

20.6 (a) Draw the structure of the simplest cycloalkane. (b) How is the structure of this molecule different from its noncyclic parent molecule?

Alkenes   **Alkenes** are hydrocarbons that contain a carbon-carbon double bond. Alkenes are classified as unsaturated hydrocarbons. **Unsaturated hydrocarbons** undergo hydrogenation reactions (reaction with H_2 gas) and add at least 1 mol H_2 per mole of compound. After adding H_2, they become saturated and produce alkanes.

The general formula for alkenes is C_nH_{2n}.

The simplest alkene contains two carbon atoms joined by a double bond.

$$H_2C=CH_2$$
Ethene

All alkenes are given the ending *ene* in the IUPAC naming system. Thus, the simplest alkene is ethene, C_2H_4. (A common name, ethylene, is frequently used when discussing ethene.)

Following ethene in the alkene series is the three-carbon alkene, propene, C_3H_6:

$$H_2C=CHCH_3$$

Propene

Both ethene and propene are gases at room temperature; they have boiling points of $-102°C$ and $-48°C$, respectively.

Butene, C_4H_8, is the next alkene in the series. There are two isomers of butene. With four carbon atoms in the chain, the double bond can be either between the first and second carbons or between the second and third carbons. The structures of the two butene isomers are as follows:

$$\overset{1}{C}H_2=\overset{2}{C}H\overset{3}{C}H_2\overset{4}{C}H_3 \qquad \overset{1}{C}H_3\overset{2}{C}H=\overset{3}{C}H\overset{4}{C}H_3$$

1-Butene   2-Butene

To distinguish between these two isomers, a number that indicates the position of the double bond is placed before the name of the compound. Thus, in 1-butene the carbon-carbon double bond is located between C_1 and C_2, whereas, in 2-butene the double bond is located between C_2 and C_3.

As with the alkanes, there is no limit to the number of carbon atoms that can be in an alkene chain. Similarly, there are also branched chains and rings. An example of a cycloalkene is cyclohexene.

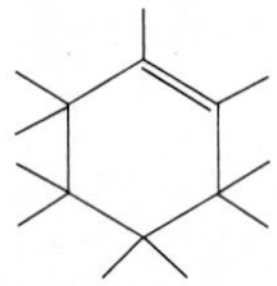

Cyclohexene

Alkenes can also possess more than one double bond within the molecule. Those containing two double bonds are called **alkadienes,** or **dienes** for short. One interesting alkadiene is 2-methyl-1,3-butadiene, commonly called isoprene.

$$\overset{1}{C}H_2=\overset{2}{C}-\overset{3}{C}H=\overset{4}{C}H_2$$
$$\overset{|}{C}H_3$$

Isoprene

Under the proper conditions, isoprene, like many alkenes, combines with itself

Polymers result when one or more monomers (small molecules) combine repeatedly to form long chains of bonded monomer units. Examples of common polymers include polyethylene; polyvinylchloride, PVC; polystyrene; nylon; and Teflon.

forming high-molecular-mass compounds, collectively called **polymers.** Substances containing long chains of connected isoprene molecules resemble natural rubber (*cis*-polyisoprene). A segment of the polyisoprene chain appears as follows:

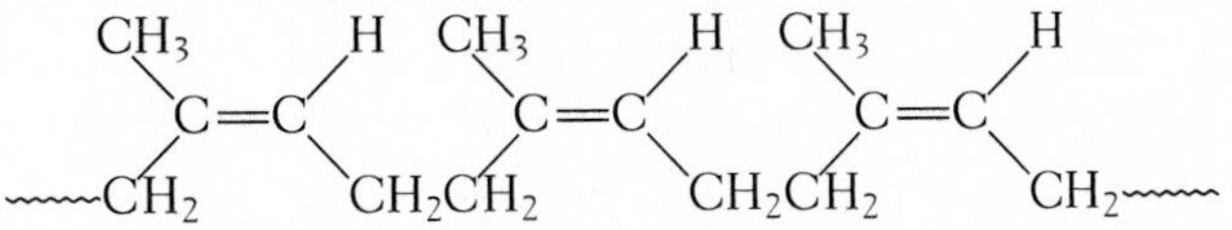

Segment of polyisoprene chain (natural rubber)

Alkynes

Alkynes contain a carbon-carbon triple bond and, like alkenes, are classified as unsaturated hydrocarbons. One mole of an alkyne combines with 2 mol $H_2(g)$ to form a saturated hydrocarbon. Ethyne is the simplest alkyne:

The general formula for alkynes is C_nH_{2n-2}.

$$HC \equiv CH$$

Ethyne

Alkynes are named in a similar manner to alkenes except that *yne* is placed after the stem to indicate that a triple bond is located in the molecule.

Ethyne (or acetylene, as it is more frequently called) is industrially the most significant alkyne. It is relatively inexpensive to synthesize. Lime, CaO, is heated with carbon to produce calcium carbide, CaC_2, which is reacted with water to produce acetylene.

$$CaO + 3C \longrightarrow CaC_2 + CO$$
$$CaC_2 + 2H_2O \longrightarrow HC \equiv CH + Ca(OH)_2$$

Acetylene is combusted in oxyacetylene torches, which are used to weld and cut through metals.

Aromatic Hydrocarbons

Aromatic hydrocarbons are those that possess the general properties and structure of benzene, C_6H_6.

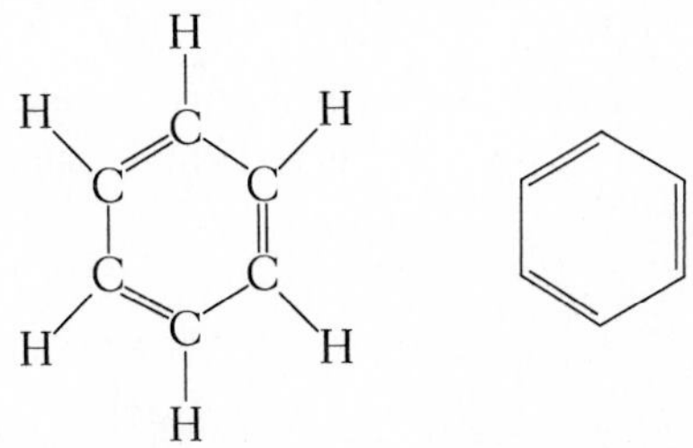

Complete structure   Shorthand notation

Structure of benzene

August Kekulé, in 1865, first proposed the cyclic structure of benzene. It was purportedly the result of a dream that Kekulé had in which he saw a snake bite its own tail.

There are many structures that are classified as aromatic hydrocarbons. However, in our cursory discussion, we shall only consider the benzenelike aromatics.

Benzene is unique with respect to the classes of hydrocarbons already discussed. In the structure that we draw for benzene, it appears as if benzene is a cyclic molecule with alternating double and single bonds. But benzene molecules exhibit **resonance.** There is more than one dot formula that can be drawn for the molecule. As with other molecules that exhibit resonance, the actual structure of benzene resembles the average of the resonance structures.

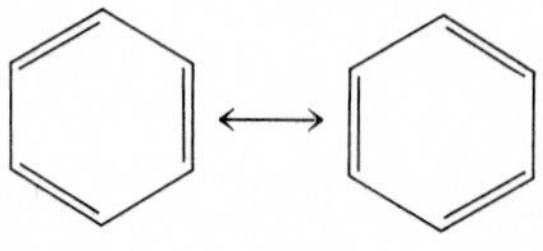

Two resonance structures of benzene

Benzene molecules actually contain no double or single bonds. Instead, each bond in benzene is intermediate between a single and a double bond, and has a bond order of 1.5. Accordingly, most chemists represent a molecule of benzene by drawing a six-member ring with a circle in the center.

Benzene

Other aromatic hydrocarbons contain the benzene structure with substituent groups attached. If a methyl group is bonded to the ring, toluene, also known

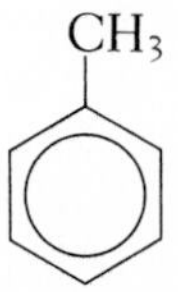

CH_3

Toluene

as methylbenzene, results. The series continues with ethylbenzene, which has an ethyl group bonded to the ring.

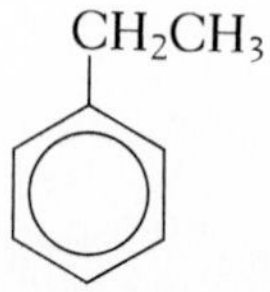

CH_2CH_3

Ethylbenzene

More than one group can be bonded to a benzene ring. Any combination of up to six substituent groups can be bonded to a benzene ring. If, for example, two methyl groups (or any substituent groups) are attached to a benzene ring, there are three different structural isomers possible.

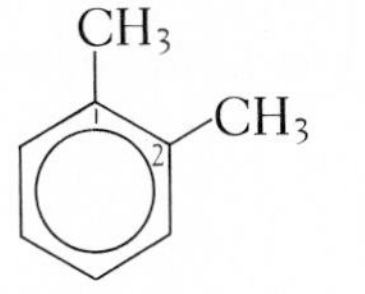

| 1,2-Dimethylbenzene | 1,3-Dimethylbenzene | 1,4-Dimethylbenzene |
| ortho-Xylene | meta-Xylene | para-Xylene |

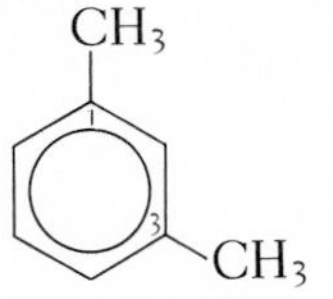

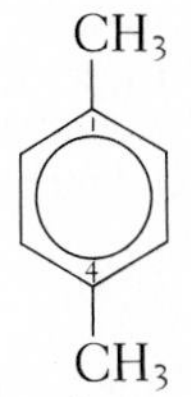

In the first structure, the two methyl groups are attached on C_1 and C_2, giving 1,2-dimethylbenzene. The second structure has the two methyl groups attached to the C_1 and C_3 positions, resulting in 1,3-dimethylbenzene. In the third structure, the methyl groups are attached to C_1 and C_4, and this compound is 1,4-dimethylbenzene. A common name for the three isomers of dimethylbenzenes is xylene. In an old naming system, three terms were used to represent the placement of two substituents on a benzene ring: (1) *ortho,* attachment at carbon atoms 1 and 2; (2) *meta,* attachment at carbon atoms 1 and 3; and (3) *para,* attachment at carbon atoms 1 and 4. Many people refer to the dimethylbenzenes as *ortho*-xylene, *meta*-xylene, and *para*-xylene.

Any number of benzene rings can be bonded together. A group of hydrocarbons called the **polycyclic aromatics** are compounds with two or more benzene rings fused together. Naphthalene is the first in the series:

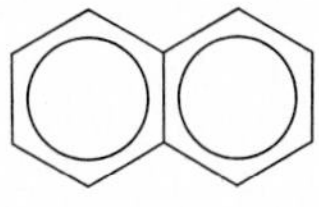

Naphthalene

Naphthalene has a distinct pungent odor and is used as an ingredient of moth balls.

Two of the three ways of fusing three benzene rings together are:

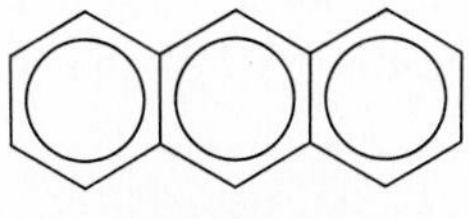

Anthracene 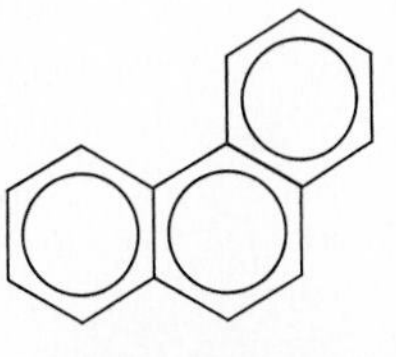   Phenanthrene

In the first compound, anthracene, the three rings are bonded in a linear fashion. In phenanthrene, the rings are bonded in a nonlinear fashion; they are angular. Both anthracene and phenanthrene are components of coal tar. Anthracene is utilized to synthesize various dyes.

Many of the polycyclic aromatics induce cancer in animals. They are classified as carcinogenic compounds. Two such compounds identified in cigarette smoke are 3,4-benzpyrene and dibenz[a,h]anthracene.

Most of the polycyclic aromatic hydrocarbons are not carcinogenic in themselves. Once they enter living systems, they are chemically altered to cancer-producing substances.

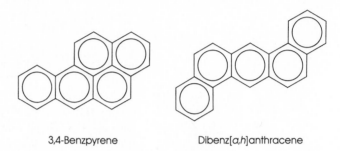

3,4-Benzpyrene   Dibenz[a,h]anthracene

REVIEW PROBLEMS

20.7 What is the difference between a saturated and unsaturated hydrocarbon? Give an example of each.

20.8 Draw the structure of a compound that represents the following groups: (a) alkyne, (b) cycloalkene, (c) alkadiene, (d) monosubstituted aromatic.

20.9 What is isoprene, and how is it important in naturally occurring substances?

20.10 Write two resonance structures that represent benzene.

20.11 Draw the structures for and name the three dimethylbenzenes.

20.12 What is a polycyclic aromatic hydrocarbon? Give an example.

20.3 HYDROCARBON DERIVATIVES

Hydrocarbon derivatives are organic compounds that also contain atoms such as oxygen, nitrogen, sulfur, phosphorus, and halogens. Each class of hydrocarbon derivatives is characterized by a functional group. **Functional groups** are a specific arrangement of atoms that give a class of organic compounds their characteristic chemical properties. Functional groups, themselves, are bonded to a carbon chain or ring. Table 20.2 lists the primary hydrocarbon derivatives with their functional groups. The symbol R is employed in organic chemistry to represent any alkyl group (hydrocarbon − H).

TABLE 20.2   HYDROCARBON DERIVATIVES

Class	Functional group	Structural formula*	Condensed formula*
Alcohol	—OH	R—OH	ROH
Ether	—O—	R—O—R	ROR
Aldehyde	$\overset{\displaystyle O}{\overset{\|}{-C-H}}$	$\overset{\displaystyle O}{\overset{\|}{R-C-H}}$	RCHO
Ketone	$\overset{\displaystyle O}{\overset{\|}{R-C-R'}}$	$\overset{\displaystyle O}{\overset{\|}{R-C-R'}}$	RCOR'
Organic acid	$\overset{\displaystyle O}{\overset{\|}{-C-OH}}$	$\overset{\displaystyle O}{\overset{\|}{R-C-OH}}$	RCOOH
Ester	$\overset{\displaystyle O}{\overset{\|}{-C-OR}}$	$\overset{\displaystyle O}{\overset{\|}{R-C-OR'}}$	RCOOR'
Amine	—NH$_2$	R—NH$_2$	RNH$_2$
Amide	$\overset{\displaystyle O}{\overset{\|}{-C-NH_2}}$	$\overset{\displaystyle O}{\overset{\|}{R-C-NH_2}}$	RCONH$_2$
Nitrile	—CN	R—C≡N	RCN
Thiol	—SH	R—SH	RSH
Thioether	—S—	R—S—R	RSR
Halide	—X (X = F, Cl, Br, I)	R—X	RX

*R′ indicates that the second alkyl group does not necessarily have to be the same as the first, although it can be.

Alcohols and Ethers

Alcohols and **ethers** can be thought of as organic derivatives of water. Water's molecular formula is H_2O. If a hydrogen atom is removed from a water molecule and replaced with an alkyl group, an alcohol results. In a similar manner, if both hydrogen atoms are removed, an ether results.

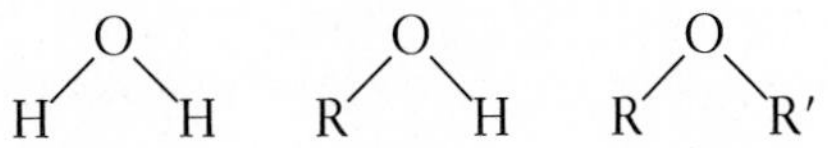

Water
HOH

Alcohol
ROH

Ether
ROR'

Methanol is the simplest alcohol, and consists of one carbon, three hydrogen atoms, and an attached —OH group.

$$CH_3OH$$
Methanol

Notice that if a hydrogen atom is removed from methane and replaced with alcohol's functional group, —OH, methanol results. IUPAC naming of alcohols requires that the parent alkane be identified; the *e* at the end of the name is removed and replaced with *ol*. As with all organic substances, there are many names for methanol, including methyl alcohol and wood alcohol.

Once ingested, methanol is converted to formaldehyde and formic acid by the liver in an attempt to rid the body of this toxic substance. Ultimately, methanol is converted to CO_2 and H_2O.

Methanol, like most of the alcohols, is a poisonous substance. Ingestion of small amounts of methanol can cause severe medical problems, including blindness. Drinking large quantities of methanol results in death. Methanol is used in industry as a solvent, and it once was a component of antifreeze.

Ethanol is the two-carbon alcohol:

$$CH_3CH_2OH$$
Ethanol

Common names for ethanol are grain alcohol, drinking alcohol, or just plain alcohol. It is called grain alcohol because it can be produced by the fermentation of grains and fruits. **Fermentation** is the process in which microorganisms are allowed to decompose the sugar components of grains and fruits, in the absence of oxygen, to ethanol and carbon dioxide.

Ethanol is metabolized in the liver to CO_2 and H_2O at a rate of about 8 g of ethanol per hour. Ethanol amounts greater than this remain in the bloodstream and cause inebriation.

Alcoholic beverages contain varying amounts of ethanol. For instance, beer has approximately 4% by volume ethanol, wines have about 12%, and whiskies contain 40 to 50% ethanol. If ethanol is ingested in low concentrations over a relatively long time period, our bodies decompose the alcohol and prevent its toxic effects. However, if a large amount of ethanol is consumed rapidly, it is deadly.

Ethanol has many other uses. It is a solvent in lacquers, varnishes, perfumes, and medicines. A mixture of ethanol and gasoline, called gasohol, is sometimes used as a fuel for internal combustion engines instead of gasoline. Ethanol is also used as the starting material in the synthesis of other organic compounds. See Table 20.3 for other examples of important alcohols.

Ethers have the general formula of R—O—R′; thus, the simplest ether contains two carbon atoms:

$$CH_3—O—CH_3$$
Dimethyl ether

Commercially, the most important ether possesses four carbon atoms with two ethyl groups:

$$CH_3CH_2—O—CH_2CH_3$$
Diethyl ether

At one time, diethyl ether, or simply ether, was one of the primary general anesthetics used in surgery. A **general anesthetic** is a substance that depresses the central nervous system, causing unconsciousness and insensitivity to pain. Many

TABLE 20.3   REPRESENTATIVE ALCOHOLS AND ETHERS

Name	Formula	Selected uses and importances
Isopropyl alcohol	CH_3CHCH_3 | OH	Rubbing alcohol, solvent
Ethylene glycol	$HO—CH_2CH_2—OH$	Permanent antifreeze for engines
Glycerol	$CH_2—CH—CH_2$ | | | OH OH OH	Lubricant, ingredient in hand lotions, sweetener, production of plastics, and nitroglycerin
Phenol (carbolic acid)	⬡—OH	Disinfectant
Divinyl ether	$(CH_2{=}CH)_2O$	Fast-acting anesthetic
Tetrahydrocannabinol		Psychoactive substance in marijuana

side effects, such as nausea, vomiting, and irritation of the lungs, occur after the administration of ether. Ether is also highly flammable. For these reasons, diethyl ether is no longer a principal anesthetic. Other ethers and a variety of halogenated hydrocarbons are used in its place.

Like other classes of organic compounds, ethers can exist in cyclic structures. Organic ring structures that contain atoms other than carbon are called **heterocyclic compounds.** Cyclic ethers contain one or more oxygen atoms in the ring. Ethylene oxide (common name) is the simplest heterocyclic ether.

Heterocyclic molecules are ring structures that contain carbon plus one or more other types of atoms.

$$\triangle$$

Ethylene oxide

Ethylene oxide's primary use is that of a starting material in the synthesis of a large variety of other organic compounds. See Table 20.3 for more examples of ethers.

Aldehydes and Ketones

Both aldehydes and ketones contain a **carbonyl group,** C=O. An **aldehyde** has the carbonyl group at the end of a carbon chain. A **ketone** has the carbonyl group located in the middle of a chain.

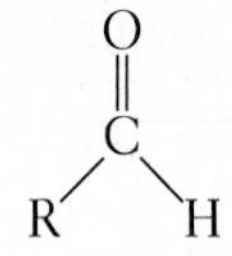

Aldehyde 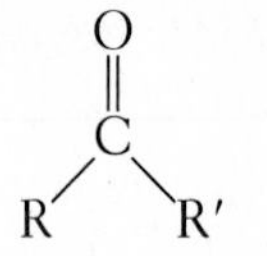   Ketone

The simplest aldehyde contains only one carbon. Its IUPAC name is methanal, but it is more commonly called formaldehyde.

$$H_2C{=}O$$

Formaldehyde

The shortest possible carbon chain in a ketone is one with three carbon atoms. Its IUPAC name is propanone, but once again the common name, acetone, prevails.

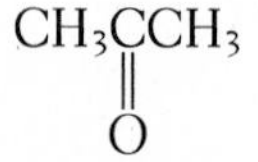

$$CH_3CCH_3$$
$$O$$

Acetone

Formaldehyde is a colorless, odorless gas that is water soluble. Commercially, formaldehyde is sold as a 37 to 40% aqueous solution called formalin. Most biology students know that formalin is used to preserve dead biological specimens.

Acetone is the most important industrial ketone. It is a low-boiling, colorless liquid that is utilized in the synthesis of varnishes, plastics, and resins. A small quantity of acetone is produced in living things as a result of the breakdown of fats. In uncontrolled diabetics, large quantities of acetone are produced. Most of it is excreted in the urine, but in severe cases of diabetes so much acetone is produced that it is detected on the person's breath.

Acetone is commonly placed into nail polish removers.

A carbonyl group can be bonded to an aromatic ring, producing aromatic aldehydes and ketones. Benzaldehyde is the simplest aromatic aldehyde. Benzaldehyde is a colorless liquid that has the odor of almonds—its common name is oil of almonds. Two other examples of aromatic aldehydes are vanillin, which gives the flavor to vanilla, and cinnamaldehyde, which has the odor of cinnamon.

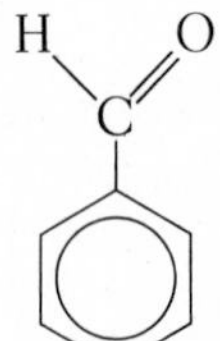

Benzaldehyde 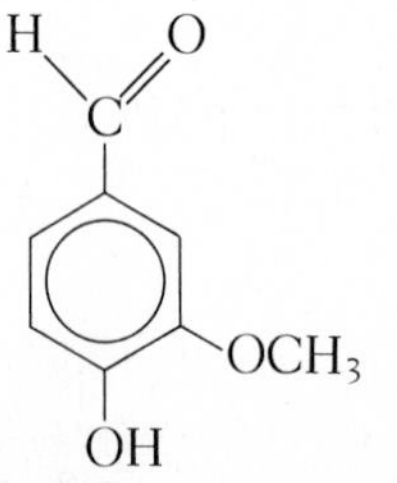   Vanillin 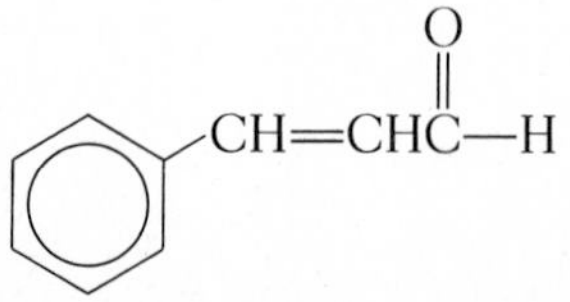   Cinnamaldehyde

TABLE 20.4 REPRESENTATIVE ALDEHYDES AND KETONES

Name	Formula	Selected uses and importance
Acetaldehyde	CH_3CHO	Biochemical intermediate
Citral	$CH_3C=CHCH_2CH_2C=CHCHO$ (with CH_3 substituents)	Lemon oil
Camphor	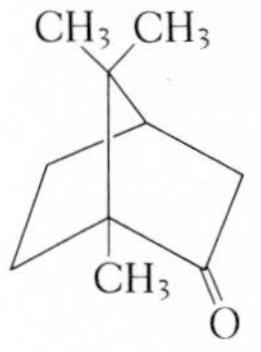	Analgesic in ointments
Jasmone	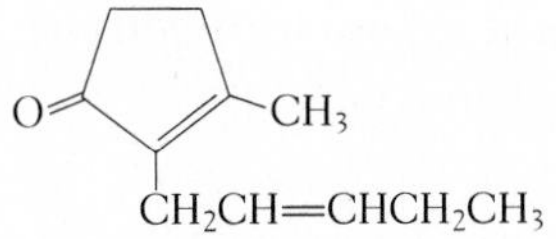	Odor of jasmine
Progesterone	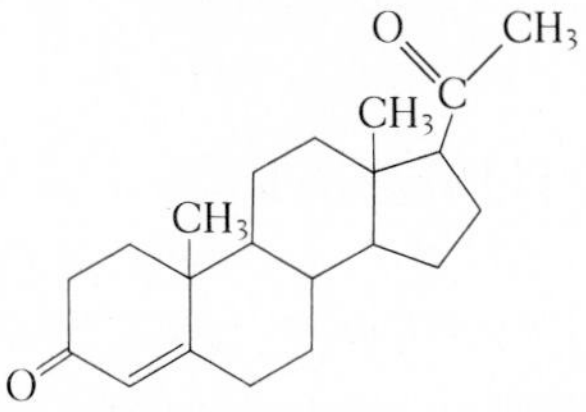	Female sex hormone

Carbonyl groups are an important functional group in the biologically significant compounds called carbohydrates. Table 20.4 lists other aldehydes and ketones.

Organic Acids (Carboxylic Acids)

A **carboxyl group** is the functional group of **organic acids.**

$$\underset{\text{Carboxyl group}}{-\overset{\overset{\textstyle O}{\|}}{C}-OH}$$

A carboxyl group is most frequently written in condensed form as either —COOH or —CO$_2$H.

H—COOH
Formic acid

The simplest carboxylic acid, as organic acids are sometimes called, is methanoic acid (IUPAC name) or formic acid (common name). Formic acid is a liquid with a pungent, irritating odor. Its name is derived from the Latin word for "ant." Ants have formic acid within their systems.

CH_3COOH
Acetic acid

Second in the carboxylic acid series is ethanoic acid (IUPAC name) or acetic acid (common name from the Latin *acetum*, meaning "vinegar"). Vinegar is a 4 to 5% solution of acetic acid. Acetic acid is used in the dye industry for its acidic properties and is a chemical intermediate in the production of various fibers and pharmaceuticals.

Butanoic acid (IUPAC name) or butyric acid (common name derived from the Latin *butyrum* for "butter"), is a four-carbon, putrid-smelling acid and is found in perspiration, rancid butter, and rancid margarine. Pentanoic acid's common name is valeric acid, derived from the Latin "to be strong." Hexanoic acid (IUPAC name) or caproic acid (from the Latin *caper*, meaning "goat") is present in excretions from the skin cells of goats.

An acid with two carboxyl groups is classified as a dicarboxylic acid. Oxalic acid is the simplest dicarboxylic acid.

$$HOOC—COOH$$

Oxalic acid

A potassium salt of oxalic acid, $HOOC—COO^-K^+$, is a component of rhubarb and spinach. Both the acid and its salts are toxic. Luckily, cooking destroys a large percent of the toxic salt.

An example of a tricarboxylic acid is citric acid. In addition to the three carboxyl groups, there is also an alcohol group bonded to the citric acid molecule. Citric acid is contained in citrus fruits such as lemons and oranges. It is very important in cellular energy production in living systems.

$$
\begin{array}{c}
H \\
| \\
H—C—COOH \\
| \\
HO—C—COOH \\
| \\
H—C—COOH \\
| \\
H
\end{array}
$$

Citric acid

Another vital acid in biological systems is lactic acid. Under low oxygen conditions, lactic acid is produced in cells from glucose. Microorganisms (lactobacilli) decompose sugars to lactic acid. Sour milk results when lactobacilli decompose milk sugars to lactic acid. Table 20.5 lists other carboxylic acids.

Bacteria on the surface of teeth convert dietary sugars to lactic acid. Increased acid on teeth causes the tooth enamel to dissolve, producing cavities.

$$
\begin{array}{c}
OH \\
| \\
CH_3CHCOOH
\end{array}
$$

Lactic acid

Esters Esters are acid derivatives. They are produced when a carboxylic acid is combined with an alcohol under the proper conditions.

$$\text{Organic acid} + \text{alcohol} \longrightarrow \text{ester} + \text{water}$$

$$RCOOH + R'OH \longrightarrow RCOOR' + H_2O$$

TABLE 20.5 REPRESENTATIVE ORGANIC ACIDS

Name	Formula	Selected uses or importance
Stearic acid	$CH_3(CH_2)_{16}COOH$	Used to make soap; coating for pills
Nicotinic acid	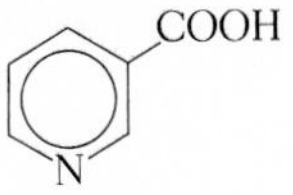 (pyridine ring with COOH)	Component of B vitamins
Pyruvic acid	CH_3CCOOH (with =O below C)	Biochemical intermediate in metabolic processes
Tartaric acid	$HOOCCHCHCOOH$ (with HO OH below)	By-product of the wine industry; used in foods, medicine, and dye industry

Esters structurally resemble organic acids except that instead of a hydrogen atom bonded to the oxygen atom, an alkyl group (R group) or aromatic group is attached.

$$RC{-}OR'$$

Ester

If the one-carbon carboxylic acid is combined with the one-carbon alcohol, the simplest ester is produced, methyl methanoate (IUPAC name) or methyl formate (common name).

$$H{-}C{-}OCH_3$$

Methyl formate

Table 20.6 lists some common esters. Many esters have pleasant, fruity odors.

TABLE 20.6 REPRESENTATIVE ESTERS

Name	Formula	Odor or flavor
Methyl butyrate	$CH_3(CH_2)_2COOCH_3$	Apples
Ethyl butyrate	$CH_3(CH_2)_2COOC_2H_5$	Pineapples
Pentyl butyrate	$CH_3(CH_2)_2COOC_5H_{11}$	Apricots
Octyl acetate	$CH_3COOC_8H_{15}$	Oranges
Ethyl formate	$HCOOC_2H_5$	Rum

In addition to its analgesic effect, aspirin lowers body temperature (is antipyretic) and has some antiseptic properties.

Three esters of the aromatic acid salicylic acid are used in medicines. If salicylic acid is combined with acetic acid, under the proper conditions, acetylsalicylic acid (aspirin) is formed. Aspirin is the most widely used painkiller (analgesic) in the world today. Another salicylic acid ester is phenyl salicylate, or salol. Salol is used to coat pills containing medicines that are irritating to the

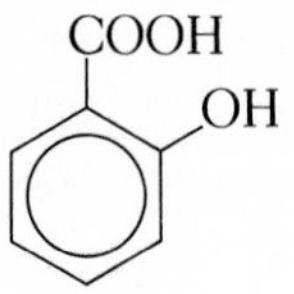

Salicylic acid

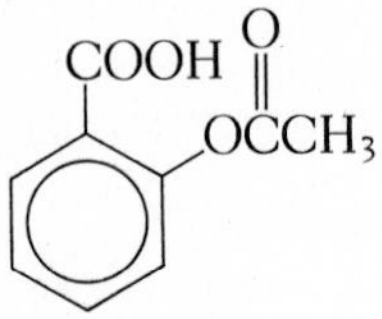

Acetylsalicylic acid
(aspirin)

stomach. Salol passes through the acidic environment of the stomach without being decomposed, but it breaks down in the alkaline conditions of the small intestine. Methyl salicylate, or oil of wintergreen, is also an ester of salicylic acid.

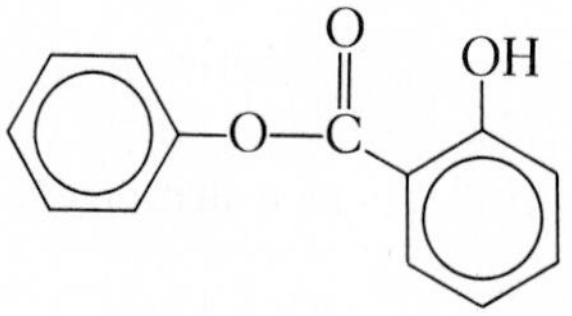

Phenyl salicylate (salol)

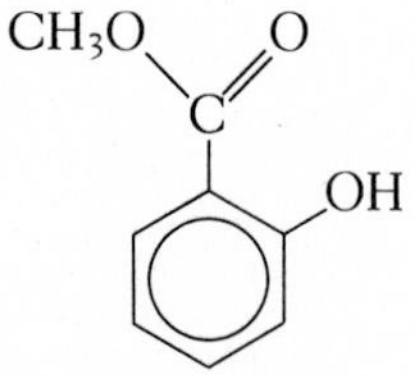

Methyl salicylate
(oil of wintergreen)

Oil of wintergreen is used in liniments and skin rubs. It is absorbed through the skin and enters sore muscles, where it is hydrolyzed to salicylic acid, an analgesic.

Amides

Amides are another class of carboxylic acid derivatives. If the —OH portion of the carboxyl group is removed and replaced with an —NH$_2$ or a nitrogen with bonded alkyl groups, an amide results. Amides have the following general formulas:

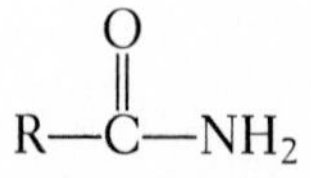

Simple amide

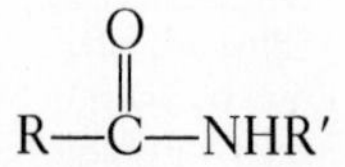

Monosubstituted amide

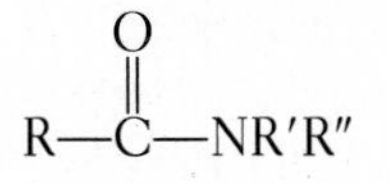

$$R\overset{\overset{\displaystyle O}{\|}}{-}C-NR'R''$$

Disubstituted amide

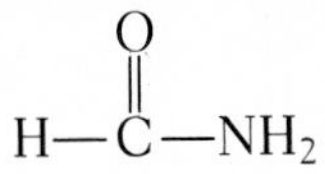

$$H\overset{\overset{\displaystyle O}{\|}}{-}C-NH_2$$

Formamide

Formamide (common name) is the simplest amide. It is a derivative of formic acid and has the structure shown in the margin.

Many amides are of biological interest. Proteins, one of the major classes of biological compounds, are amides. One of the B vitamins, niacinamide, contains a ring structure with an attached amide group.

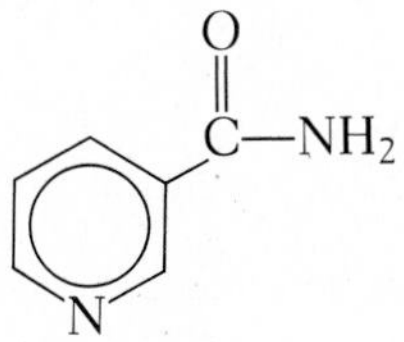

Niacinamide

Lysergic acid diethylamide, LSD, contains a disubstituted amide. LSD is a mind-altering drug (hallucinogen) that is synthesized from lysergic acid, a component of a fungus called ergot.

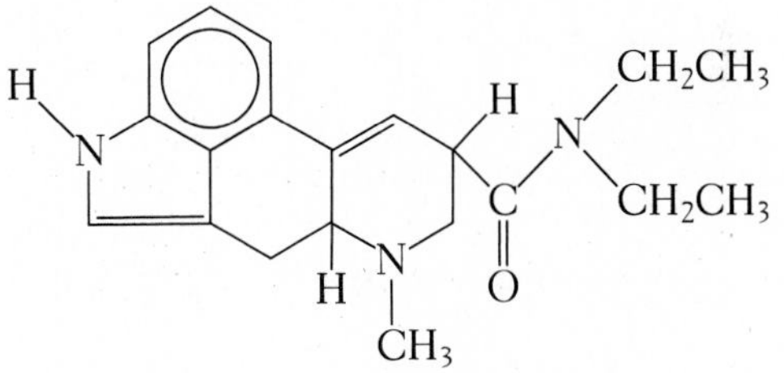

Lysergic acid diethylamide (LSD)

Acetaminophen, an aspirin substitute, is a fairly simple monosubstituted amide.

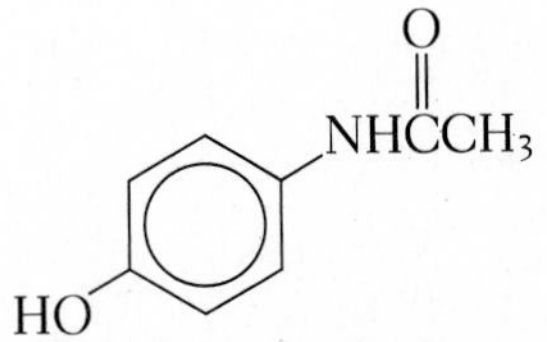

Acetaminophen

Amines

Amines are organic derivatives of ammonia, NH_3. If a hydrogen atom is removed from an ammonia molecule and is replaced by an alkyl group, a primary amine results. When two hydrogen atoms are removed and replaced with two alkyl groups, a secondary amine forms, and if three alkyl groups are bonded to a nitrogen atom, a tertiary amine results.

$$RNH_2 \qquad R_2NH \qquad R_3N$$

| Primary amine | Secondary amine | Tertiary amine |

Thus, the simplest primary, secondary, and tertiary amines are methylamine, CH_3NH_2, dimethylamine, $(CH_3)_2NH$, and trimethylamine, $(CH_3)_3N$, respectively.

Aniline is the simplest of the aromatic amines. Aniline and its derivatives are used in the synthesis of dyes. Aniline is a starting material in the production of pharmaceuticals and photographic chemicals.

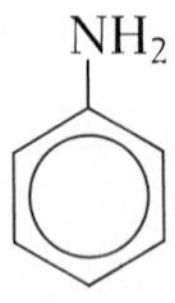

Aniline

When a nitrogen atom is a member of an organic cyclic carbon structure, the molecule is classified as a **heterocyclic amine.** Examples of heterocyclic ring structures include pyrrole, pyridine, and indole.

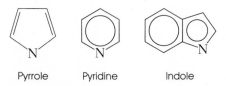

Pyrrole   Pyridine   Indole

Pyrrole is a five-member ring structure containing a nitrogen atom and four carbon atoms. Pyrrole rings are components of molecules such as hemoglobin, chlorophyll, and vitamin B_{12}. Hemoglobin is the O_2 and CO_2-carrying molecule of blood, chlorophyll is the green pigment in plants that helps convert CO_2 to O_2, and vitamin B_{12} is an essential substance in a person's diet.

Pyridine is part of the structure of two B vitamins, niacin and pyridoxine. It is also part of the structure of nicotine, a deadly compound that is contained in plants such as the tobacco plant. In small quantities, nicotine acts as a mild stimulant.

Naturally occurring amines in plants that have a physiological effect on animals are classified as alkaloids. For example, morphine is isolated from the opium poppy. Quinine is in the cinchona tree of the Andes, and tubocurarine is the active ingredient in curare.

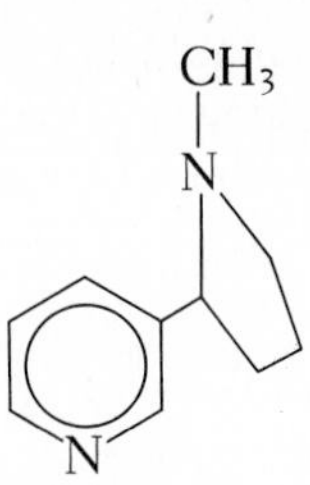

Nicotine

Indole is a common structure in nature, and is found in substances ranging from indole alkaloids such as lysergic acid, reserpine, and strychnine to important compounds in brain physiology such as serotonin. See Table 20.7 for other examples of amines.

Alkyl Halides and other Halogenated Hydrocarbons

Alkyl halides are produced when a halogen atom (F, Cl, Br, or I) replaces a hydrogen atom in a nonaromatic hydrocarbon. If methane is combined with a halogen under appropriate conditions, a halogen atom substitutes for one or more hydrogen atoms.

TABLE 20.7   **REPRESENTATIVE AMINES**

Name	Formula	Selected uses or importance
Putrescine	$H_2N(CH_2)_4NH_2$	Decomposition product of proteins
Benadryl	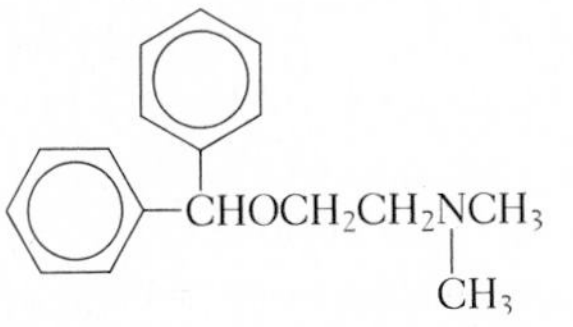	Antihistamine
Coniine	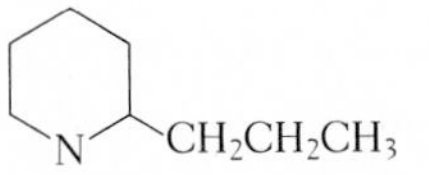	Principal alkaloid in hemlock, the chemical that killed Socrates
Caffeine	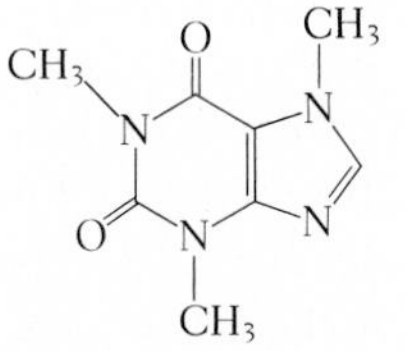	Stimulant in coffee, tea, and cola drinks
Amphetamine	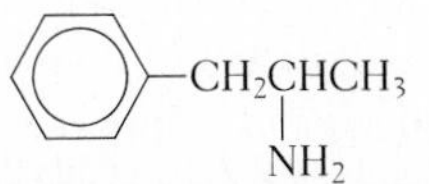	Potent synthetic stimulant

If $Cl_2(g)$ is combined with methane, chloromethane, CH_3Cl (common name, methyl chloride), results.

$$CH_4 + Cl_2 \xrightarrow{\text{light}} CH_3Cl + HCl$$

If excess Cl_2 is present, the Cl_2 combines with the newly formed chloromethane, producing a disubstituted methane, dichloromethane, CH_2Cl_2 (methylene chloride). When a third hydrogen is removed, trichloromethane, $CHCl_3$ (chloroform), results. This reaction continues until all of the hydrogen atoms are removed, finally yielding tetrachloromethane (carbon tetrachloride), CCl_4.

Each of the one-carbon chlorinated hydrocarbons are generally good solvents for organic and nonpolar substances. Chloroform, $CHCl_3$, was one of the earliest general anesthetics. It is no longer used because it is fairly toxic, and it has been identified as a cancer-causing agent. Besides its use as an anesthetic, $CHCl_3$ was once commonly placed in cough remedies until the Food and Drug Administration (FDA) banned its use in 1976.

At one time, carbon tetrachloride had a number of household uses. CCl_4 was an ingredient in spot removers and cleaning fluids until it was discovered to be too toxic for home use. CCl_4 causes serious damage to the liver and kidneys. Generally, the substances that replaced "carbon tet," as it is called, are not much better. Trichloroethylene and tetrachloroethylene are now placed in cleaning fluids, especially "dry cleaning fluids," but it is now known that they are also carcinogenic.

More than one halogen can be incorporated into a molecule. Freons are good examples of this; they contain both fluorine and chlorine atoms within their structures. Freons are fluorochlorocarbons that are used as refrigerants in refrigerators, freezers, and air-conditioning equipment. Two commercially important Freons are Freon 11 and Freon 12.

$$CCl_3F \qquad CCl_2F_2$$

Freon 11   Freon 12

When used as refrigerants, they cause minimal environmental problems because they are enclosed. But, at one time Freons were widely used as propellants in aerosol spray cans. They are no longer used as propellants because research indicated that Freons could have a significant environmental impact. Studies have shown that Freons could somehow diffuse to the upper layers of the atmosphere, in particular to the ozone layer. Under these high-energy conditions, the normally inert Freons combine with the ozone, O_3. If a small amount of the O_3 is depleted, significantly higher intensities of ultraviolet radiation would reach the surface of the earth, with very serious biological consequences.

The ozone layer is part of the stratosphere about 15 mi above the surface of the earth. A 5% depletion of the ozone layer would cause a 10% increase in the intensity of uv radiation reaching the earth.

Another class of controversial halogenated hydrocarbons are the polychlorinated pesticides. These compounds are used to control populations of insects that cause disease or consume crops. One of the first widely used polychlorinated pesticides was DDT or *dichlorodiphenyltrichloroethane*.

Trichloroethylene

Tetrachloroethylene

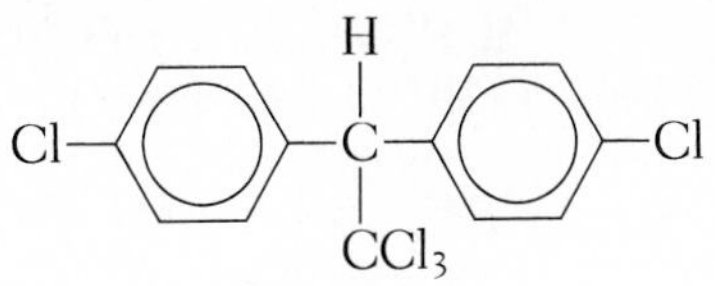

DDT

In 1972, the United States banned the use of DDT because its overuse caused many environmental problems. DDT became concentrated in the food chain and was responsible for the extinction of certain bird species.

Table 20.8 presents selected examples of organic halides and their importance and uses.

REVIEW PROBLEMS

20.13 To what class of hydrocarbon derivative does each of the following belong: (a) ROR, (b) RNH_2, (c) RCN, (d) RCHO, (e) RCOOH?

20.14 Write the functional group contained in each of the following: (a) ketones, (b) esters, (c) amides, (d) alkyl halides, (e) alcohols.

20.15 Write the formula for the simplest: (a) alcohol, (b) amine, (c) ether, (d) carboxylic acid, (e) ketone.

TABLE 20.8 REPRESENTATIVE ORGANIC HALIDES

Name	Formula	Selected uses and importance
p-Dichlorobenzene	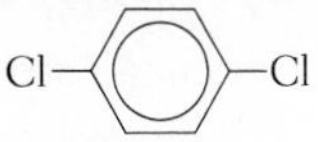	Pesticide in moth balls
Chlordane	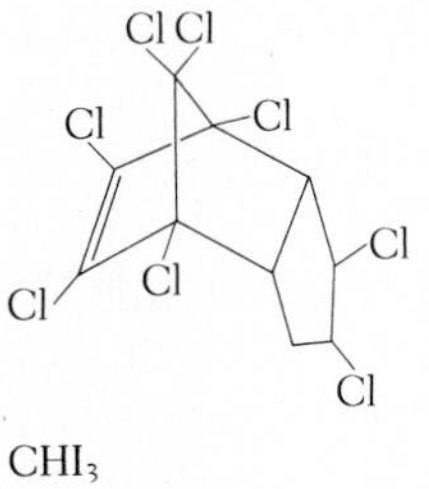	Pesticide used to control ants, termites, and lawn pests
Iodoform	CHI_3	Antiseptic
Halothane	$CF_3CHClBr$	Common general anesthetic
Polychlorinated biphenyls	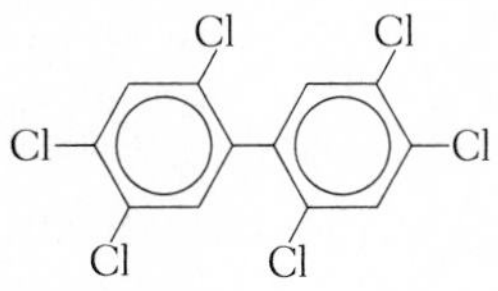	Group of compounds used as heat transfer agents and hydraulic fluids; once used in plastics until their environmental impact was found
Teflon	$-(CF_2-CF_2)_n$	High-molecular-mass polymer used as an electrical insulator and as a nonstick coating for utensils

20.16 Give an example of a specific compound that could have the following use: (a) pesticide, (b) anesthetic, (c) antiseptic agent, (d) embalming fluid, (e) flavoring agent.

20.17 Give an example of a molecule with the stated characteristics: (a) heterocyclic amine, (b) aromatic halide, (c) ester containing five carbon atoms, (d) mono-substituted amide, (e) cyclic ketone.

20.18 Describe the environmental impact of the indiscriminate "dumping" of chlorinated hydrocarbons.

20.4 CLASSES OF BIOLOGICALLY SIGNIFICANT COMPOUNDS

Living systems are composed principally of four different classes of compounds: (1) carbohydrates, (2) proteins, (3) lipids, and (4) nucleic acids. In this section we shall take a very brief look at these complex substances.

Carbohydrates

Carbohydrates are the group of biologically essential compounds that include sugars and starches. Carbohydrates are structural components of cells and are sources of energy. Chemically, carbohydrates are polyhydroxy (i.e. they contain more than one alcohol group, $-OH$) aldehydes and ketones, or they yield such substances when hydrolyzed.

Most carbohydrates are generally placed into three primary categories: (1) monosaccharides, (2) disaccharides, and (3) polysaccharides. The simplest group of carbohydrates are the **monosaccharides,** the simple sugars. Monosaccharides are combined together to produce the other two carbohydrate groups. **Disaccharides** are formed when two monosaccharide molecules are chemically combined. **Polysaccharides** are polymers of monosaccharides, i.e., long chains of monosaccharides bonded together.

Two varieties of monosaccharides exist, polyhydroxy aldehydes and polyhydroxy ketones. Normally, these molecules contain from three to six carbons. A polyhydroxy aldehyde is called an **aldose** (the ending *ose* indicates a carbohydrate), and a polyhydroxy ketone is called a **ketose.**

$$
\begin{array}{cc}
& \text{CH}_2\text{OH} \\
\text{CHO} & \text{C}=\text{O} \\
(\text{H}-\text{C}-\text{OH})_n & (\text{H}-\text{C}-\text{OH})_n \\
\text{CH}_2\text{OH} & \text{CH}_2\text{OH} \\
\text{Polyhydroxy aldehyde} & \text{Polyhydroxy ketone} \\
\text{Aldose} & \text{Ketose} \\
(n=1, 2, 3, 4, \text{ or } 5) & (n=0, 1, 2, 3, \text{ or } 4)
\end{array}
$$

Glyceraldehyde is the simplest aldose, while the simplest ketose is dihydroxyacetone.

$$
\begin{array}{cc}
\underset{|}{C}HO & \overset{H}{\underset{|}{C}}—OH \\
H—\underset{|}{C}—OH & \underset{|}{C}=O \\
H—\underset{|}{C}—OH & H—\underset{|}{C}—OH \\
H & H
\end{array}
$$

Glyceraldehyde   Dihydroxyacetone

Many of the biologically important monosaccharides contain five and six carbons. By far the most important five-carbon monosaccharides are ribose and deoxyribose. Ribose is an aldose, or aldopentose (five-carbon aldose), that exists throughout nature. It is a component of nucleic acid molecules (RNA) and molecules that generate biological energy. Deoxyribose differs from ribose by only one oxygen atom. On the second carbon of deoxyribose two hydrogen atoms are bonded; on the second carbon of ribose a H and −OH are bonded. Deoxyribose is the monosaccharide in deoxyribose nucleic acid, DNA.

$$
\begin{array}{cc}
\underset{|}{C}HO & \underset{|}{C}HO \\
H—\underset{|}{C}—OH & H—\underset{|}{C}—H \\
H—\underset{|}{C}—OH & H—\underset{|}{C}—OH \\
H—\underset{|}{C}—OH & H—\underset{|}{C}—OH \\
H—\underset{|}{C}—OH & H—\underset{|}{C}—OH \\
H & H
\end{array}
$$

Ribose   Deoxyribose

Naturally occurring monosaccharides normally exist in cyclic structures rather than in chains. The cyclic structure of glucose is

There are three major six-carbon monosaccharides, or hexoses. Each has the formula $C_6H_{12}O_6$; they differ with respect to the placement of −OH groups bonded to the carbon chain and the location of the carbonyl group in the molecule. Glucose and galactose are aldohexoses, and fructose is a ketohexose.

$$
\begin{array}{ccc}
\underset{|}{C}HO & \underset{|}{C}HO & \overset{H}{\underset{|}{C}}—OH \\
H—\underset{|}{C}—OH & H—\underset{|}{C}—OH & \underset{|}{C}=O \\
HO—\underset{|}{C}—H & HO—\underset{|}{C}—H & HO—\underset{|}{C}—H \\
H—\underset{|}{C}—OH & HO—\underset{|}{C}—H & H—\underset{|}{C}—OH \\
H—\underset{|}{C}—OH & H—\underset{|}{C}—OH & H—\underset{|}{C}—OH \\
H—\underset{|}{C}—OH & H—\underset{|}{C}—OH & H—\underset{|}{C}—OH \\
H & H & H
\end{array}
$$

Glucose   Galactose   Fructose

By a large margin, glucose is the most abundant sugar in nature. It is often called by its common name dextrose, and is sometimes referred to as blood or grape sugar. A large percent of all the sugars ingested are converted to glucose, one of the cells' primary fuels. Galactose does not exist alone in nature. Galactose is always combined with some other molecule, normally glucose. Fructose is the only biologically important ketohexose. It is one of the sugars that contribute to the sweetness of various fruits.

Disaccharides result when two monosaccharides are chemically combined. Table sugar or sucrose is an example of a disaccharide. If sucrose is decomposed, glucose and fructose remain. Sucrose's cyclic structure is

Sucrose

Also known as cane sugar, beet sugar, or just plain sugar, sucrose is the world's most frequently used sweetening agent.

Maltose (malt sugar) is another example of a disaccharide. It is composed of two glucose molecules. Maltose is rarely found in large quantities by itself in nature, but it is detected in most living cells when more complex carbohydrates are decomposed. Lactose (milk sugar) is the disaccharide made up of glucose and galactose monosaccharide units bonded together. Human milk contains about 7% lactose, a higher percentage than cow's milk, which only contains approximately 5% lactose.

Maltose   Lactose

Just about all animals lack the enzymes that decompose cellulose; accordingly, most animals, including humans, obtain no nutritional value from cellulose. Nevertheless it serves an important role in the intestines in connection with fluid retention.

Polysaccharides are the most abundant carbohydrates in nature, primarily because of the omnipresence of cellulose. Cellulose is the major component of the cell walls of all plants. After hydrolysis of cellulose only one substance is isolated—glucose. Cellulose is a polymer of glucose. Other naturally occurring polysaccharides are starch and glycogen. Both starch and glycogen are a storage reserve for glucose in plants and animals, respectively.

Proteins

Proteins are the main structural components of animal cells. A large percent of hair, skin, muscles, tendons, and connective tissues is composed of proteins. Additionally, proteins are components of nerve tissue, antibodies, and some hormones. Enzymes, the chemical controlling agents of cells, are mainly protein structures.

As the monosaccharide is the basic unit of carbohydrates, the **amino acid** is the basic unit of proteins. Amino acids have the general structure:

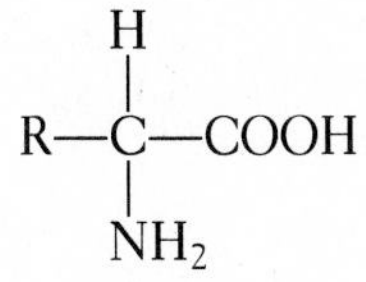

$$
\begin{array}{c}
\text{H} \\
| \\
\text{R}-\text{C}-\text{COOH} \\
| \\
\text{NH}_2
\end{array}
$$

Amino acids possess both an amino group ($-NH_2$) and a carboxyl group ($-COOH$). While an unlimited number of amino acids could exist, only about 20 to 30 amino acids are found in living things. Amino acids differ with respect to the attached R$-$ group. Examples of common amino acids are:

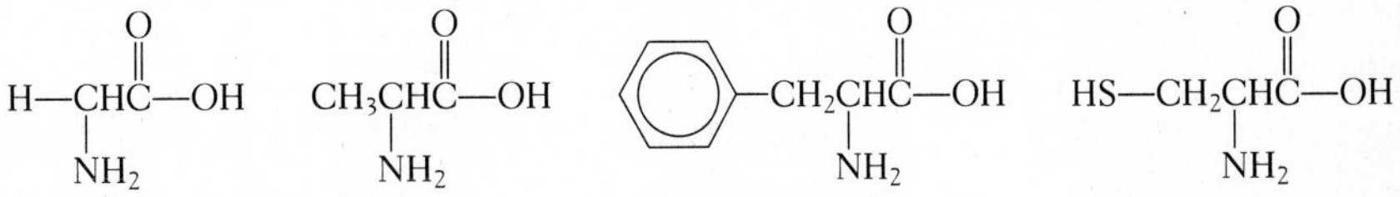

Glycine   Alanine   Phenylalanine   Cysteine

Amino acids are chemically combined within cells to produce chains of bonded amino acids. They combine in such a manner that the amino group from one amino acid combines with the carboxylic acid group from the other, producing an amide linkage that is frequently called the **peptide bond.** If two amino acids are bonded, the structure is called a **dipeptide;** three attached amino acids are **tripeptides,** etc.

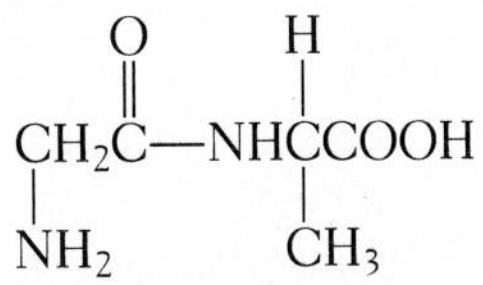

$$
\begin{array}{cc}
\text{O} & \text{H} \\
|| & | \\
\text{CH}_2\text{C}-\text{NHCCOOH} \\
| & | \\
\text{NH}_2 & \text{CH}_3
\end{array}
$$

Dipeptide (glycylalanine)

Chains of five to thirty amino acids are termed **polypeptides,** while even longer chains are called proteins. **Proteins** are high-molecular-mass polypeptides, or polymers of amino acids.

Each protein has a specific sequence of amino acids. A protein's properties are in part dependent on the sequence and number of amino acids within the

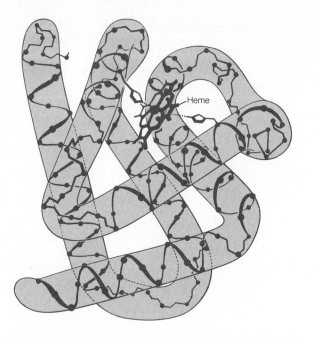

Heme

FIGURE 20.1
Myoglobin carries oxygen from the circulatory system to the muscles. Its structure is that of a helix which has a series of bends, folds, and twists. Within the protein chain is a heme group. The heme group is the component of myoglobin that bonds to the O_2 molecule. All proteins have specific three-dimensional configurations. (Modified and reprinted with permission from R. E. Dickerson, in H. Neurath, Ed., *The Proteins,* vol. 2, Academic Press, New York, 1964.)

chain. If one amino acid in a long chain is removed and replaced with another amino acid, the protein would have a different set of properties. Besides the number and sequence of amino acids in the protein, the actual three-dimensional shape of the molecule plays a significant role in determining the protein's properties. Figure 20.1 shows the complex structure of an oxygen-carrying molecule, myoglobin.

Two general classes of proteins exist: (1) simple and (2) conjugated. **Simple proteins** are composed of combined amino acids, with no other nonprotein groups attached to the molecule. **Conjugated proteins** have a nonprotein group, called a prosthetic group, attached to the protein structure.

Four principal classes of simple proteins exist: (1) scleroproteins, (2) globulins, (3) albumins, and (4) histones. **Scleroproteins** make up skin, hair, and protective and connective tissues. **Globulins** are a water-insoluble class of simple proteins found in antibodies and other blood components. **Histones** contain many basic amino acids and are generally associated with nucleic acids. Finally, the most widely distributed group of simple proteins is the **albumins.** Many blood proteins are albumins. Egg whites contain a high percentage of albumin.

Five of the most important classes of conjugated proteins are: (1) glycoproteins, (2) lipoproteins, (3) nucleoproteins, (4) phosphoproteins, and (5) chromoproteins. Each of these groups has a nonprotein group bonded to the protein molecule. **Glycoproteins** are conjugated proteins that have a carbohydrate group

TABLE 20.9   CONJUGATED PROTEINS

Class of conjugated proteins	Examples
Glycoproteins	Mucin, a component of saliva; gamma globulin, a blood protein
Lipoproteins	Blood plasma beta lipoproteins
Nucleoproteins	Component of ribosomes, cell components; tobacco mosaic virus
Phosphoproteins	Casein, a milk protein
Chromoproteins	Hemoglobin, O_2-carrying protein; cytochromes, part of cells' energy-producing apparatus

bonded to the molecule, and **lipoproteins** have a lipid structure bonded to the protein molecule. It is beyond the scope of this brief overview to discuss each of these groups. Table 20.9 presents examples of each type of conjugated protein.

Enzymes are one of the most interesting and important types of proteins. **Enzymes** are biochemical catalysts, i.e., substances that increase the rate of chemical reactions within living things. Enzymes, like catalysts, are not consumed during chemical reactions. Virtually all reactions in living cells are catalyzed by enzymes. Enzymes make life possible because most of the reactions in the cell could not take place under the mild conditions present if they were not catalyzed.

Lipids   Compared with carbohydrates and proteins, **lipids** are composed of a more diverse group of compounds. There is no basic structure from which all the classes of lipids are derived. Classes of compounds that are considered lipids include triglycerides, waxes, steroids, prostaglandins, and compound lipids.

Triglycerides are the most abundant of all the classes of lipids. Triglycerides, when decomposed, yield glycerol and fatty acids. Glycerol is a three-carbon polyalcohol,

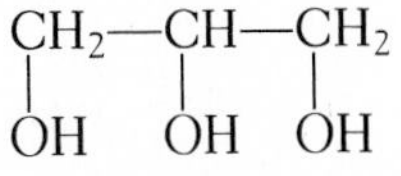

$$CH_2-CH-CH_2$$
$$|||$$
$$OHOHOH$$

Glycerol

and fatty acids are carboxylic acids with 10 to 24 carbons normally. A saturated fatty acid has the general formula

$$CH_3(CH_2)_nCOOH$$

Saturated fatty acid

where n is a number from 8 to 22. The structure of a simple triglyceride is

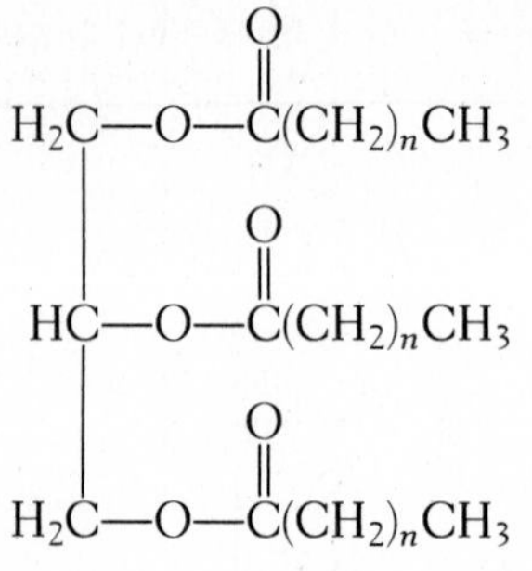

$$H_2C—O—\overset{\overset{\displaystyle O}{\|}}{C}(CH_2)_nCH_3$$

$$HC—O—\overset{\overset{\displaystyle O}{\|}}{C}(CH_2)_nCH_3$$

$$H_2C—O—\overset{\overset{\displaystyle O}{\|}}{C}(CH_2)_nCH_3$$

Triglyceride

When triglycerides are combined with NaOH, soaps are produced.

Each of the three fatty acids is connected to the glycerol molecule by an ester linkage. If all of the fatty acid molecules have saturated chains, then the triglyceride is called a **saturated fat.** Addition of unsaturated fatty acids produces an **unsaturated fat.** Triglycerides with a greater percent of saturated fatty acids tend to be solids (fats); those with more unsaturated fatty acids tend to be liquids (oils).

Waxes, another class of lipids, are also esters. They are esters of long-chain monoalcohols and fatty acids. Normally the carbon chains in both the alcohol and acid components consist of between 10 and 30 carbons. Beeswax is formed from a 30-carbon alcohol and a 16-carbon acid. Waxes are widely found throughout the plant and animal kingdoms.

Steroids are structurally different from the triglycerides and waxes. Steroid molecules have four fused rings; three of the rings contain six carbon atoms, and one ring contains five carbons. Steroid structures differ with respect to functional groups and substituent groups. For instance, the most abundant steroid in humans is cholesterol.

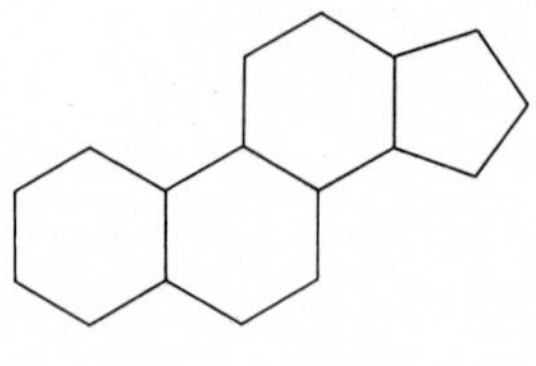

Steroid structure

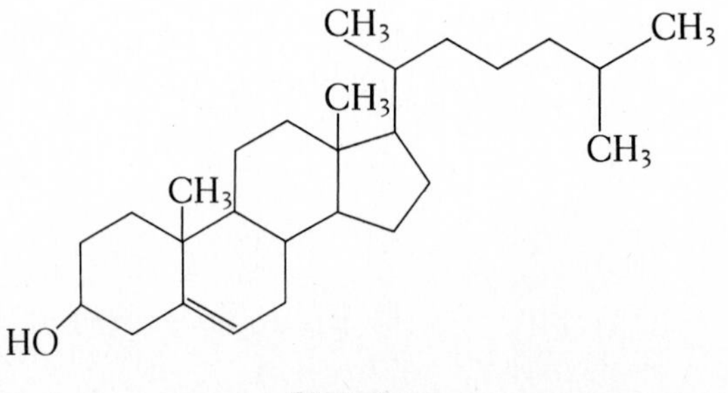

Cholesterol

Cholesterol is the most abundant steroid in humans. Most individuals contain about 225 g of cholesterol within their bodies. Cells use cholesterol to synthesize other steroids.

Steroids have widely varying functions in living things. Bile acids, such as cholic acid, aid in the digestion of dietary lipids. Many of the male and female sex hormones are steroids. Sex hormones are responsible for a person's secondary sex characteristics. Important hormones (about 30) secreted by the adrenal gland, called corticoids, are also steroids.

A class of lipids which has a biological regulatory function is the **prostaglandins.** All prostaglandins have the basic structure of prostanoic acid.

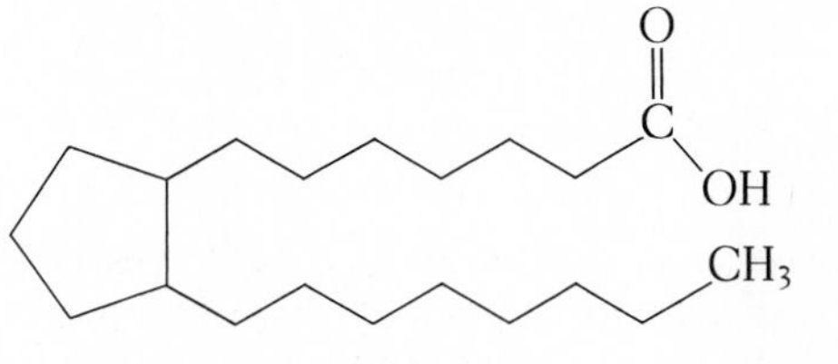

Prostanoic acid

Prostaglandins help regulate body temperature, smooth muscle contraction, and various tissue secretions. They also seem to be involved in the mechanism for controlling inflammations. Much is still to be learned about this puzzling class of lipids.

Like the proteins, lipids bond to nonlipid molecules. Two of the more common complex lipids are **phospholipids,** which contain a phosphate group, and **glycolipids,** which contain a carbohydrate molecule.

Nucleic Acids

Nucleic acids are very complex biomolecules composed of a unit called a **nucleotide.** Most **nucleic acids** are polymers of nucleotides. Two different varieties of naturally occurring nucleotides exist; thus, two different classes of nucleic acids exist. The two classes of nucleotides are ribose nucleotides and deoxyribose nucleotides.

Nucleotides have three components: (1) a sugar, either ribose or deoxyribose, (2) a heterocyclic amine, and (3) a phosphate group. An example of a deoxyribose nucleotide is as follows:

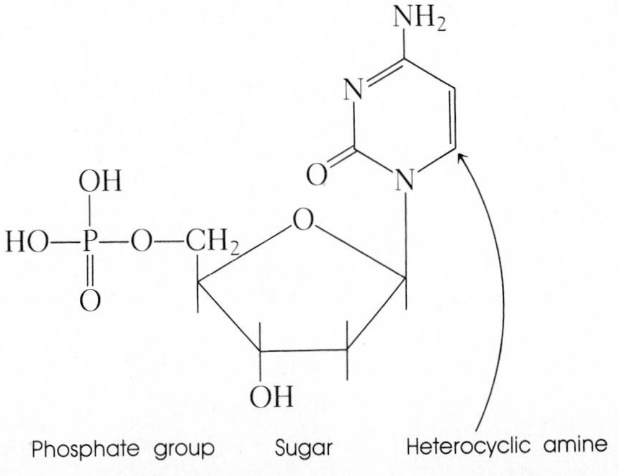

Phosphate group   Sugar   Heterocyclic amine

Five primary heterocyclic amines occur in nucleotides: (1) adenine, (2) guanine, (3) thymine, (4) cytosine, and (5) uracil.

Adenine   Guanine   Thymine   Cytosine   Uracil

Nucleic acids range from rather small transfer RNAs to the massive DNA polymers, with their double-helix structure. Refer to a biochemistry textbook for a complete discussion of the complex structure of nucleic acids. Nucleic acids have innumerable functions within cells, including the storing of genetic information and the regulation of protein synthesis.

REVIEW PROBLEMS

20.19 What are the four classes of biologically important compounds?

20.20 What is the basic molecular structure of: (a) proteins, (b) carbohydrates, (c) nucleic acids?

20.21 What functional groups are found in (a) carbohydrates, (b) proteins, (c) lipids?

20.22 Give an example of each of the following: (a) disaccharide, (b) fatty acid, (c) amino acid, (d) steroid, (e) nucleic acid, (f) polysaccharide.

20.23 What is the difference between a simple and conjugated protein?

20.24 What compounds result when the following are hydrolyzed: (a) triglycerides, (b) disaccharides, (c) nucleotides, (d) proteins, (e) waxes?

SUMMARY

All organic compounds are divided into two categories: hydrocarbons and hydrocarbon derivatives. Hydrocarbons are those compounds that contain only carbon and hydrogen. Hydrocarbon derivatives are organic substances that contain atoms such as the halogens, oxygen, nitrogen, sulfur, or phosphorus in addition to carbon and hydrogen.

There are four classes of hydrocarbons: (1) alkanes, (2) alkenes, (3) alkynes, and (4) aromatics. Alkanes are called the saturated hydrocarbons because they have the maximum number of hydrogen atoms bonded to their carbon atoms and thus contain only carbon-carbon single bonds. Alkane molecules exist in chains, branched chains, and rings. Methane, CH_4, is the simplest alkane, and ethane, C_2H_6, is the next member of the series. Each succeeding member of the alkanes differs only by a —CH_2— group. Such a series of compounds is called a homologous series.

Alkenes have a carbon-carbon double bond and, as a result, are unsaturated hydrocarbons. As with the alkanes, there is a homologous series of alkenes. Alkynes are also unsaturated; however, they contain a triple covalent bond. Aromatic hydrocarbons have a special set of properties that resemble the properties of benzene, C_6H_6. Most aromatics have a stable benzenelike ring structure as part of their molecules.

Each class of hydrocarbon derivatives is identified by its functional group. A functional group is a specific arrangement of atoms that gives a compound its characteristic chemical properties. For instance, the functional group in all alcohols is an —OH group. It does not matter how many carbon atoms are present or how they are arranged, if an —OH group is attached to a carbon group it behaves as an alcohol.

Alcohols and ethers can be thought of as organic derivatives of water, H_2O. If one of water's hydrogen atoms is replaced with an alkyl group, an alcohol results. When both hydrogen atoms are removed and replaced with alkyl groups, an ether results. Aldehydes and ketones are called carbonyl compounds; they have a $C=O$ (carbonyl group) within their molecules. If the carbonyl group is at the end of a chain and has a hydrogen atom attached, then the molecule is classified as an aldehyde. When the carbonyl is in the middle of a chain, bonded to two carbon atoms, it is a ketone.

Organic acids contain a carbon atom that has both an alcohol and carbonyl group bonded; this combination, —COOH, is called a carboxyl group. Many organic acid derivatives exist. Organic acids react with alcohols to produce esters; a —COOR is the functional group in esters. Another derivative of the organic acids is the amide—compounds that contain —$CONH_2$.

Organic derivatives of ammonia are amines, RNH_2. Amines can have up to a maximum of three alkyl groups attached to the nitrogen atom. Amines are widely distributed in the natural world, especially in molecules that have both carbon and nitrogen atoms within cyclic structures. The latter are called heterocyclic amines. The last class of organic compounds discussed was the alkyl halides, or halogenated hydrocarbons, which are organic compounds containing one or more halogens.

Biologically important organic substances are classified into four classes: (1) carbohydrates, (2) proteins, (3) lipids, and (4) nucleic acids. Carbohydrates, when broken down to their smallest units, are polyhydroxy aldehydes and ketones. Three classes of carbohydrates are monosaccharides, disaccharides, and polysaccharides. Monosaccharides are the basic unit of carbohydrates. When two monosaccharide molecules bond, a disaccharide results. If many bond together, a polysaccharide is formed.

Proteins are polymers of amino acids. An amino acid is a chemical structure that contains both an amino group (—NH_2) and a carboxylic acid group (—COOH). Approximately 20 amino acids exist in nature. These 20 amino acids are combined into chains that range from two or three amino acids to high-molecular-mass proteins with thousands of amino acids in the chain. Proteins are a major structural component of living things. A special class of proteins, called enzymes, controls the rates of most chemical reactions within living things.

Lipids are a diverse group of biological substances, ranging from the triglycerides to steroids. Triglycerides are esters of the polyalcohol glycerine. Steroids, in contrast, possess a ring system with three six-member rings and one five-member ring fused together. Other lipids include the waxes, prostaglandins, and complex lipids.

Most nucleic acids are high-molecular-mass polymeric biomolecules. Each nucleic acid is composed of a basic unit called a nucleotide. A nucleotide has three parts: (1) a five-carbon sugar (ribose or deoxyribose), (2) a heterocyclic amine, and (3) a phosphate group. The two classes of nucleic acids are the DNAs (deoxyribose nucleic acids) and RNAs (ribose nucleic acids).

QUESTIONS AND PROBLEMS

20.25 Define the following words and terms: organic chemistry, hydrocarbon, hydrocarbon derivative, saturated hydrocarbon, unsaturated hydrocarbon, homologous series, IUPAC, isomer, substituent group, cyclic structure, aromatic hydrocarbons, polycyclic compound, alkyl group, functional group, anesthetic, carbonyl group, carboxyl group, analgesic.

Hydrocarbons

20.26 (a) What scientist is credited with the discovery that organic compounds could come from nonliving things? (b) How did he prove his point?

20.27 What is the difference between a hydrocarbon and a hydrocarbon derivative?

20.28 What distinguishes each class of hydrocarbon from the others?

20.29 How many carbon atoms are in each of the following hydrocarbons: (a) heptane, (b) butane, (c) decane, (d) ethane, and (e) octane?

20.30 Draw the complete structural formulas from the following condensed formulas:
(a) $CH_3CH_2CH_2CH_2CH_3$
(b) $CH_3CH_2CH_2CH_2CH_2CH_2CH_3$
(c) 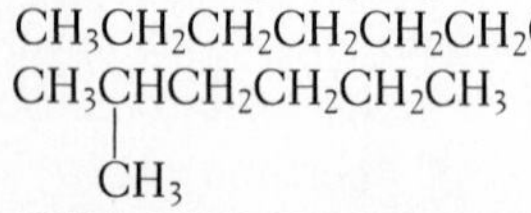 $CH_3CHCH_2CH_2CH_2CH_3$
  |
  CH_3
(d) $(CH_3)_3CCH_2CH_2CH_2CH_3$

20.31 For each of the hydrocarbons listed in Problem 20.30, write the name of the unbranched isomer.

20.32 Write the names of the following alkanes:

(a) 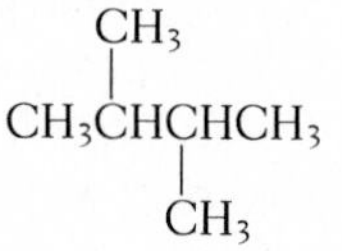
$$CH_3CHCHCH_3$$
with CH_3 groups above and below

(b) 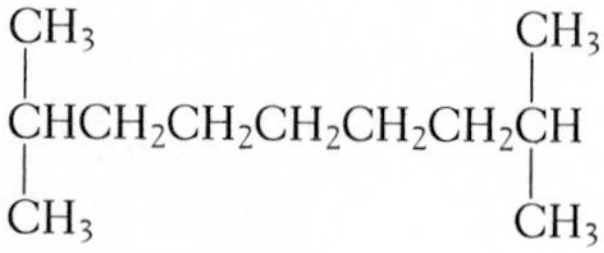
$$CHCH_2CH_2CH_2CH_2CH_2CH$$
with CH_3, CH_3 and CH_3, CH_3 substituents

(c) 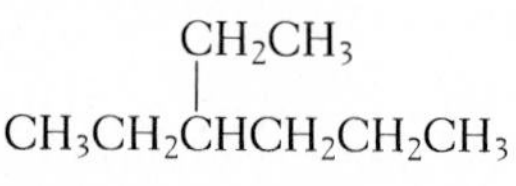
$$CH_3CH_2CHCH_2CH_2CH_3$$
with CH_2CH_3 substituent

(d) 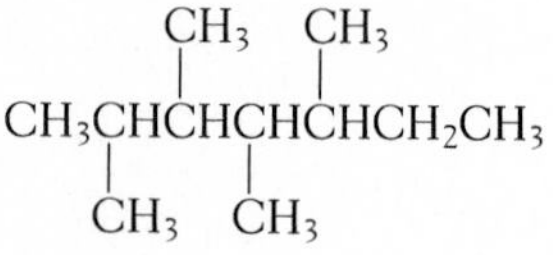
$$CH_3CHCHCHCHCH_2CH_3$$
with CH_3 CH_3 and CH_3 CH_3 substituents

(e) 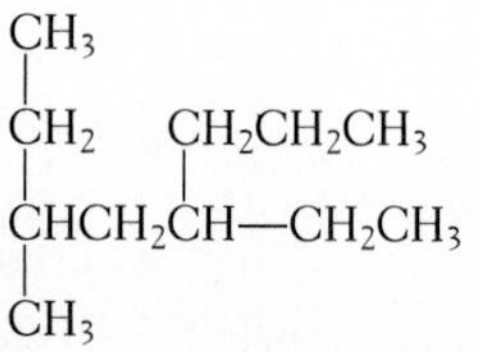
$$CHCH_2CH—CH_2CH_3$$
with $CH_3/CH_2/CH_2/CH_2CH_3$ and $CH_2CH_2CH_3$ and CH_3 substituents

(f)

$$CH_3-\overset{\overset{\displaystyle CH_3}{|}}{\underset{\underset{\displaystyle CH_3}{|}}{C}}-\overset{\overset{\displaystyle CH_3}{|}}{\underset{\underset{\displaystyle CH_3}{|}}{C}}-CH_3$$

20.33 Draw the structures of:
(a) 3-methylhexane
(b) 2,2-dimethylbutane
(c) 3-ethyl-4-methylnneane
(d) 2,2,4-trimethylpentane
(e) 2,2,3,3-tetramethylbutane
(f) 4-ethyldecane
(g) methylcyclopentane
(h) 1,2,3-triethylcyclohexane

20.34 Draw the structures and write the IUPAC names for all isomers of heptane.

20.35 Within a homologous series there is generally a trend of increasing boiling points. What structurally accounts for this trend?

20.36 Write the names of the following cycloalkanes:

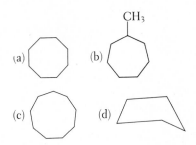

20.37 If there are *n* carbon atoms in a cycloalkane, how many hydrogen atoms are also in the molecule?

20.38 Draw the structures of the following unsaturated hydrocarbons: (a) a three-carbon alkene, (b) a five-carbon cycloalkene, (c) two unbranched five-carbon alkenes, (d) two four-carbon alkadienes, (e) all cyclic six-carbon alkadienes.

20.39 How many hydrogen atoms are contained in: (a) 2-methylpropane, (b) octene, (c) ethylcyclopentane, (d) propyne, (e) benzene, (f) 3-methylheptane?

20.40 Draw the structures of all isomers of C_5H_{10}. Don't forget about cyclic structures.

20.41 How is ethyne (acetylene) prepared commercially?

20.42 What is the general formula of any alkyne?

20.43 Draw the structures of the three diethylbenzenes, and write a name for each.

20.44 (a) Determine how many different trimethylbenzenes exist. (b) Draw the structure of each trimethylbenzene.

20.45 Draw the structure of three polycyclic aromatic compounds containing three fused benzene rings.

Hydrocarbon Derivatives

20.46 What type of hydrocarbon derivative is each of the following: (a) RCOOR, (b) RX, (c) RCOR, (d) RSH, (e) RCONH$_2$?

20.47 What functional group is in each of the following: (a) aldehydes, (b) acids, (c) amines, (d) thioethers?

20.48 Circle and identify each functional group in the following molecules:

(a)
$$CH_3-\overset{\overset{\displaystyle OH}{|}}{\underset{\underset{\displaystyle H}{|}}{C}}-\overset{\overset{\displaystyle O}{\|}}{C}-OCH_3$$

(b)
$$Cl-CH_2\overset{\overset{\displaystyle NH_2}{|}}{C}HCHCH_2-Br$$
with a $\overset{}{\underset{\displaystyle H}{C}}\diagdown O$ group below

(c) CH_3-O on a cyclopentenone ring with $C=O$ and NH_2 substituents

20.49 Draw the structures of the following: (a) a two-carbon alcohol, (b) a one-carbon aldehyde, (c) a five-carbon acid, (d) a three-carbon ketone, (e) a four-carbon amine, (f) a cyclic seven-carbon ester.

20.50 What are the IUPAC names for the following hydrocarbon derivatives: (a) grain alcohol, (b) acetone, (c) formic acid, (d) butyric acid?

20.51 What class of compounds contains: (a) a nitrogen atom within a ring, (b) a carbonyl group, (c) a carboxyl group, (d) fluorochlorocarbons?

20.52 List a use for each of the following hydrocarbon derivatives: (a) ethanol, (b) diethyl ether, (c) ethylene glycol, (d) formaldehyde, (e) vanillin, (f) acetone.

20.53 (a) What is the structural difference between an aldehyde and a ketone? (b) Give an example of each in your answer.

20.54 (a) What is oxalic acid? (b) How is oxalic acid different from acetic acid?

20.55 (a) What common property do the simple esters share? (b) Give two examples.

20.56 List three esters of salicylic acid and how they are used as medicines.

20.57 Draw the structure of: (a) a simple two-carbon amide, (b) a methyl-substituted two-carbon amide, and (c) an aromatic methyl-disubstituted amide.

20.58 List three biologically significant amides.

20.59 Draw the structures of a (a) primary, (b) secondary, and (c) tertiary aromatic amine.

20.60 (a) How is a heterocyclic amine different from other cyclic compounds? (b) Give two examples of heterocyclic amines.

20.61 Write an equation that represents what happens when 1 mol Cl_2 gas combines with 1 mol CH_4 in the presence of light.

20.62 What biological effects are observed when humans are exposed to halogenated hydrocarbons?

20.63 (a) What are Freons? (b) How are Freons used? (c) What problems result when Freons are released into the atmosphere?

20.64 (a) What problems were solved after the discovery of DDT? (b) What problems were produced by the use of DDT?

Biochemistry

20.65 What two general classes of compounds make up the carbohydrates?

20.66 Draw the structures of the simplest aldose and ketose.

20.67 How does a monosaccharide differ from a polysaccharide?

20.68 Give an example for each of the following: (a) five-carbon aldose, (b) most commonly used disaccharide, (c) plant polysaccharide, and (d) most abundant sugar in nature.

20.69 (a) What is the molecular formula for aldohexoses? (b) What is the molecular formula of a disaccharide produced from two aldohexoses? (c) Look at these molecular formulas and state what molecule is also formed when two aldohexoses combine to produce a disaccharide?

20.70 What is the main function of glycogen in animals?

20.71 In what type of tissues are proteins found?

20.72 In what class of hydrocarbon derivatives could proteins be placed?

20.73 Name and write the structures for two amino acids.

20.74 (a) What are conjugated proteins? (b) List the five main classes of conjugated proteins?

20.75 (a) What is an enzyme? (b) To what class of biochemicals do enzymes belong?

20.76 List the five principal classes of lipids, and give an example of each.

20.77 Draw the structure of a triglyceride that is produced from three 12-carbon fatty acid molecules and glycerol.

20.78 Are there any differences between fatty acids and regular carboxylic acids?

20.79 Draw the ring structure that is in all steroids.

20.80 What are three functions of steroids in humans?

20.81 How are prostaglandins important in living things?

20.82 List three classes of complex lipids.

20.83 What is the basic structure found in nucleic acids?

20.84 How are nucleic acids important within cells?

· APPENDIX ·

Review of Math Skills

I. ALGEBRAIC OPERATIONS

Algebra is an area of mathematics that deals with equations and equalities. Generally, equations are solved to determine the value of an unknown quantity. This is accomplished by applying the most basic rule of algebra: Isolate the unknown quantity in an equation by mathematically treating both sides of the equation in the same way. An equation is unchanged as long as whatever is done to one side of the equation is also done to the other side. Consider the following equation:

$$X - 10 = 12$$

If the same number is added to both sides of the equation, the equality remains unchanged. Therefore, if 10 is added to both sides of the equation, the unknown value, X, is isolated on the left side (-10 plus 10 equals zero). On the right side, 10 is added to 12, giving 22; hence, the unknown value of X is 22.

$$
\begin{array}{rr}
X - 10 = & 12 \\
+\,10 & +10 \\
\hline
X = & 22
\end{array}
$$

Always check the answer by substituting the value obtained back into the equation. In the above example: $22 - 10 = 12$.

The above equation belongs to a general class of equations having the form:

Type 1: $X + m = n$

where X is the unknown quantity and m and n are known quantities. To solve Type 1 equations, m is either added or subtracted to isolate X by itself on one side of the equation. See Example Problem 1.

Example Problem 1

Solve the following equation: $X + 23 = -100$.

Solution

To solve this equation, subtract 23 from both sides.

$$
\begin{array}{r}
X + 23 = -100 \\
\underline{-\ 23 \qquad -\ 23} \\
X = -123
\end{array}
$$

Check the answer: $-123 + 23 = -100$.

Type 2 equations are solved by multiplying and dividing appropriate quantities to isolate the unknown value. Type 2 equations have the general form:

Type 2: $mX = n$ or $\dfrac{X}{m} = n$

In the first equation, X is obtained by dividing both sides of the equation by m.

$$\frac{\cancel{m}X}{\cancel{m}} = \frac{n}{m} \quad \text{or} \quad X = \frac{n}{m}$$

In the second equation, X is determined by multiplying both sides by m.

$$\cancel{m}\left(\frac{X}{\cancel{m}}\right) = mn \quad \text{or} \quad X = mn$$

In chemistry, we use both multiplication and division of numbers to solve equations. Consider the following equation:

$$\frac{aX}{m} = n$$

To solve for X in an equation of this form, multiply each side of the equation by m/a, then cancel a/m on the left side, yielding $X = mn/a$.

$$\frac{\not{m}}{\not{a}} \times \frac{\not{a}X}{\not{m}} = \frac{mn}{a}$$

Study Example Problem 2 as an example of solving Type 2 equations.

Example Problem 2

Solve the following equation for X:

$$\frac{5X}{3} = -20$$

Solution

Multiply both sides of the equation by $\frac{3}{5}$, thereby cancelling the $\frac{5}{3}$ on the left side of the equation and isolating X.

$$\frac{\not{3}}{\not{5}} \times \frac{\not{5}X}{\not{3}} = \frac{3}{5}(-20)$$

$$X = \frac{3}{5}(-20)$$

$$X = -12$$

Check the answer:

$$\frac{5 \times (-12)}{3} = -20$$

Many algebraic equations are a combination of Type 1 and Type 2 equations. Isolating the unknown quantity in this type equation requires both addition and multiplication. To illustrate, let's solve the following equation for X:

$$a(X - m) = n$$

First divide each side by a, and then add m to both sides.

$$\frac{\cancel{a}(X - m)}{\cancel{a}} = \frac{n}{a}$$

$$X - m = \frac{n}{a}$$

$$\underline{+ m = +m}$$

$$X = \frac{n}{a} + m$$

If possible, when solving such equations, initially remove all terms from the side of the equation containing the unknown quantity by multiplying and dividing. Finally, add or subtract the remaining terms, leaving the unknown quantity by itself.

In other mixed equations the opposite procedure should be followed: Add and subtract first, and then multiply and divide (see Example Problem 3).

Example Problem 3

Solve for X:

$$\frac{18}{X} + 7 = 13$$

Solution

Subtract 7 from both sides of the equation.

$$\frac{18}{X} + 7 = 13$$

$$\underline{\quad -7 \quad -7}$$

$$\frac{18}{X} = 6$$

Multiply both sides of the equation by X, and then divide the resulting equation by 6, giving the answer 3.

$$X\left(\frac{18}{X}\right) = 6X$$

$$18 = 6X$$

$$\frac{18}{6} = \frac{\cancel{6}X}{\cancel{6}}$$

$$3 = X$$

Solving more complex equations involves applying the same general principles: Rearrange the equation and isolate the unknown quantity through additive and multiplicative operations. However, the number of operations needed to solve these equations increases as the equations become more complex. Consider the more involved equation in Example Problem 4.

Example Problem 4

Solve the following equation for X:

$$\frac{a + 1}{X + b} = \frac{m}{n}$$

Solution

1. Multiply both sides of the equation by $(X + b)$.

$$(\cancel{X + b}) \times \frac{a + 1}{\cancel{X + b}} = (X + b) \times \frac{m}{n}$$

$$a + 1 = (X + b) \times \frac{m}{n}$$

2. Multiply both sides by n/m.

$$\frac{n}{m} \times (a + 1) = (X + b) \times \frac{\cancel{m}}{\cancel{n}} \times \frac{\cancel{n}}{\cancel{m}}$$

$$\frac{n}{m} \times (a + 1) = X + b$$

3. Subtract b from both sides of the equation.

$$\frac{n}{m} \times (a + 1) - b = X + b - b$$

This yields

$$\frac{n}{m} \times (a + 1) - b = X$$

II. SCIENTIFIC NOTATION

Exponential Numbers　　Extremely small and extremely large numbers are often encountered in chemistry. Numbers such as 0.0000000000005 and 6,000,000,000,000,000,000,000 are commonplace in chemical applications. Dealing with numbers in this form is awkward and unwieldy. Consequently, large and small numbers are ordinarily expressed exponentially or in a special exponential system called **scientific notation.**

Before considering the specifics of scientific notation, let's review some basic principles concerning exponential numbers. An exponent is a number or symbol written as a superscript to the right of a base number (or symbol) indicating how many times the base number is multiplied by itself. For instance, 10^3 is $10 \times 10 \times 10$, 2^6 is $2 \times 2 \times 2 \times 2 \times 2 \times 2$, and a^4 is $a \times a \times a \times a$. The exponent of a number is called the *power*, and the number being raised to the power is termed the *base*. We will restrict our discussion to numbers with base 10. Table 1 lists examples of exponential numbers with base 10.

You must obey a couple of simple rules when you multiply and divide exponential numbers (addition and subtraction are discussed later). When multiplying exponential numbers with the same bases, add the exponents. For example, the product of 10^6 and 10^8 is

$$10^6 \times 10^8 = 10^{6+8} = 10^{14}$$

The explanation for adding exponents when multiplying is straightforward: 10^6 is $10 \times 10 \times 10 \times 10 \times 10 \times 10$, and 10^8 is $10 \times 10 \times 10 \times 10 \times 10 \times 10 \times 10 \times 10$; hence, $10^6 \times 10^8$ equals $(10 \times 10 \times 10 \times 10 \times 10 \times 10) \times (10 \times 10 \times 10 \times 10 \times 10 \times 10 \times 10 \times 10)$, or 10^{14}.

When dividing numbers with the same base, subtract the exponent in the denominator from the exponent in the numerator. Thus the quotient of 10^5 divided by 10^4 is

TABLE 1　**POWERS OF TEN**

$10^0 = 1$ (all numbers to the zero power are 1)
$10^1 = 10$
$10^2 = 10 \times 10 = 100$
$10^3 = 10 \times 10 \times 10 = 1000$
$10^4 = 10 \times 10 \times 10 \times 10 = 10,000$
$10^{-1} = \frac{1}{10} = 0.1$
$10^{-2} = \frac{1}{10} \times \frac{1}{10} = 0.01$
$10^{-3} = \frac{1}{10} \times \frac{1}{10} \times \frac{1}{10} = 0.001$
$10^{-4} = \frac{1}{10} \times \frac{1}{10} \times \frac{1}{10} \times \frac{1}{10} = 0.0001$

$$\frac{10^5}{10^4} = 10^{5-4} = 10^1$$

or, without using exponential notation,

$$\frac{100,000}{10,000} = 10$$

Scientific Notation   Numbers are expressed in **scientific notation** by separating them into two factors: (1) decimal factor and (2) exponential factor. Examples of numbers expressed in scientific notation are:

$$1.234 \times 10^9$$
$$9.87 \times 10^{-3}$$
$$3.0 \times 10^{59}$$

In each example, the first number (1.234, 9.87, and 3.0) is the decimal factor. By convention, the decimal factor is always given a value between 1 and 9.9999. The decimal factor is multiplied by the exponential factor, 10 to some power.

To convert numbers to scientific notation, adjust the decimal point so that the decimal factor has a value between 1 and 9.9999, and, depending on how many places the decimal point is moved and in what direction, give the appropriate power to the base, 10, so as not to change the value of the number.

To illustrate, let's change 23,000 to scientific notation. First, we adjust the decimal point to give a number between 1 and 9.999. To accomplish this, we move the decimal point four places to the left, giving the number 2.3000. For each place to the left we move the decimal point, we add 1 to the exponent of 10^0 (10^0 equals 1; by definition any number to the zero power is 1). So $23,000 \times 10^0$ is the same as $23,000 \times 1$ or just 23,000. Therefore, the exponent of 10^0 is increased by 4 to 10^{0+4} or 10^4.

$$23,000. \times 10^0 \qquad \text{converts to} \qquad 2.3000 \times 10^4$$

The exponent is increased because each time the decimal point is moved to the left, it is the same as dividing the number by 10, or decreasing the value by a factor of 10, and in order not to change the number, it has to be multiplied by 10. If a number is multiplied and divided by 10 at the same time, it is the same as multiplying by 1 ($\frac{10}{10} = 1$), which does not change the value of the number.

When large numbers are converted to scientific notation, the decimal point is moved to the left, thus increasing the value of the exponent. When numbers smaller than 1 are changed to scientific notation, the opposite is true—the decimal point is moved to the right, decreasing the value of the exponent.

How is 0.00000091 expressed in scientific notation? First, move the decimal point 7 places to the right, giving the value 9.1 for the decimal factor. In order not to change the numerical value of the number, 7 is subtracted from the exponent, 0 (10^0), giving -7 as the exponent. Thus

$$0.00000091 \quad \text{becomes} \quad 9.1 \times 10^{-7}$$

Each time the decimal is moved one place to the right, the magnitude of the number is increased by 10; at the same time, it must be divided by 10 (decreasing the value of the exponent) so the value remains constant. Study the following example problem to better understand changing numbers to scientific notation.

Example Problem 6

Change each number to scientific notation: (a) 390,000,000,000,000,000 (b) 0.00000000000000000000072.

Solution

(a) Move the decimal point 17 places to the left, giving 3.9.

$$390,000,000,000,000,000. \quad \text{gives} \quad 3.9$$

Add 17 to the exponent.

$$3.9 \times 10^{0+17} = 3.9 \times 10^{17}$$

(b) Move the decimal point 22 places to the right, giving 7.2.

$$0.00000000000000000000072 \quad \text{gives} \quad 7.2$$

Subtract 22 from the exponent.

$$7.2 \times 10^{0-22} = 7.2 \times 10^{-22}$$

To change a number expressed in scientific notation to a nonexponential number, the operation is reversed. If the exponent is positive, move the decimal point to the right, and if the exponent is negative, move the decimal point to the left. For example, to change 1.75×10^5 to nonexponential form:

$$1.75000 \times 10^5 \quad \text{gives} \quad 175,000 \times 10^0 = 175,000 \times 1 = 175,000$$

Each time the decimal is moved to the right, the value of the number is increased by a factor of 10; to keep the value of the number the same, 1 is subtracted from the exponent. In the case of changing from scientific notation to nonexponential form, the exponential factor is changed to 10^0, or 1, which is not written. See Example Problem 7.

Example Problem 7

Change each number to nonexponential form:
(a) 1.19×10^{-7}
(b) 6.50×10^6

Solution

(a) Since the exponent is -7, move the decimal point 7 places to the left and add 7 to the exponent in order not to change the value of the number.

$$0000001.19 \times 10^{-7} \quad \text{gives} \quad 0.000000119 \times 10^{-7+7}$$

or $\qquad\qquad\qquad\qquad$ 0.000000119

(b) Since the exponent is $+6$, move the decimal point 6 places to the right and subtract 6 from the exponent in order not to change the value of the number.

$$6.500000 \times 10^6 \text{ gives } 6,500,000. \times 10^{6-6}$$

or $\qquad\qquad\qquad$ $6,500,000$

Arithmetic Operations: Multiplication and Division

Arithmetic operations using numbers expressed in scientific notation are handled the same way as operations using any exponential numbers. The only difference is performing the proper operation on the decimal factor at the same time that the appropriate exponent operation is calculated.

To review: When multiplying, add exponents; and when dividing, subtract exponents. For example, what is the product of $(3 \times 10^4) \times (2 \times 10^6)$? It is easier to separate the decimal factors from the exponential factors, giving

$$(3 \times 2) \times (10^4 \times 10^6)$$

Multiply the decimal factors, and then add the exponents,

$$3 \times 2 = 6 \quad \text{and} \quad 10^{4+6} = 10^{10}$$

which gives the correct answer,

$$6 \times 10^{10}$$

Dividing numbers expressed in scientific notation is carried out in a similar manner except that the decimal factors are divided and the exponents are subtracted. Example Problem 8 illustrates multiplying and dividing numbers expressed in scientific notation.

Example Problem 8

Simplify the following expression:

$$\frac{8 \times 10^{12}}{4 \times 10^{15}} \times 1.5 \times 10^{-3}$$

Solution

1. First divide 8×10^{12} by (4×10^{15}).

$$\frac{8}{4} \times 10^{12-15} = 2 \times 10^{-3}$$

Separate the factors 8 and 4 from the exponential factor and divide: $\frac{8}{4} = 2$. Then subtract the exponents, 12 and 15, yielding -3.
2. Multiply the resulting number, 2×10^{-3}, by 1.5×10^{-3}.

$$2 \times 1.5 \times 10^{-3+(-3)} = 3 \times 10^{-6}$$

Multiply the decimal factors, 2 and 1.5, resulting in the product 3. Add the exponents, -3 and -3, giving -6. The same answer is obtained if 8×10^{12} is multiplied by 1.5×10^{-3} and the product is divided by 4×10^{15}.

Arithmetic Operations: Addition and Subtraction

To complete our study of numbers expressed in scientific notation, we turn our attention to addition and subtraction operations. Once again, one general rule prevails, for both operations: Only numbers with exactly the same exponent can be added or subtracted.

What is the sum of 1×10^3 (1000) and 1×10^2 (100)? To add these numbers, both exponents must be the same. Therefore, either the 3 is changed to 2 or the 2 is changed to 3. Generally, it is best to change the smaller exponent to a larger exponent, as we shall see. If the exponent is increased by 1 in (1×10^2), the decimal factor is divided by 10 (move the decimal point to the left):

$$1. \times 10^2 \quad \text{gives} \quad 0.1 \times 10^{2+1} = 0.1 \times 10^3$$

Now that both exponents are the same, the two numbers can be added together:

$$0.1 \times 10^3 + 1 \times 10^3 = 1.1 \times 10^3 = 1100$$

To add numbers in scientific notation, the decimal factors are added and the exponent of the answer remains the same (do not add them).

In the above example, if the smaller exponent was not changed to a larger exponent, the resulting answer would not have been in scientific notation initially. Commonly, answers obtained after arithmetic operations are not in scientific notation; in other words, the decimal factor is not between 1 and 9.9999. Whenever this case is encountered, it is necessary to change the answer back to scientific notation. Let's look at one final example problem illustrating scientific notation.

Example Problem 9

Evaluate the following, and express the final answer in scientific notation:

$$\frac{3 \times 10^7}{1.5 \times 10^{-2}} - (7.5 \times 10^8) + (1.25 \times 10^{10})$$

Solution

1. Divide (3×10^7) by (1.5×10^{-2}).

$$\frac{3}{1.5} = 2 \quad \text{and} \quad \frac{10^7}{10^{-2}} = 10^{7-(-2)} = 10^9$$

This yields

$$2 \times 10^9$$

2. Since the next operation is subtraction, and the numbers have different exponents, we must change one of the exponents. Change the exponent in 7.5×10^8 to 10^9.

$$7.5 \times 10^8 \quad \text{gives} \quad 0.75 \times 10^{8+1} = 0.75 \times 10^9$$

Subtract this from the first number.

$$2 \times 10^9 - 0.75 \times 10^9 = 1.25 \times 10^9$$

3. Finally, add the last number, 1.25×10^{10}, after the exponents are changed to the same value.

$$1.25 \times 10^9 \qquad \text{gives} \qquad 0.125 \times 10^{9+1} = 0.125 \times 10^{10}$$

and

$$0.125 \times 10^{10} + 1.25 \times 10^{10} = 1.375 \times 10^{10}$$

III. GRAPHING

A **graph** is a convenient means for displaying and observing trends in data. Frequently, in chemistry, collected data are graphed to clearly show patterns within the data.

Graphing involves placing or "plotting" data points on graph paper and drawing a smooth curve or straight line through the plotted points. Regular graph paper has both horizontal and vertical lines drawn symmetrically on the surface of the paper for this purpose. All data are plotted between two perpendicular axes, called the x and y axes. Normal graphing convention defines x as the horizontal axis and y as the vertical axis. In mathematics, the x axis is called the **abscissa** and the y axis is termed the **ordinate.**

The first step in graphing data is to **scale** the axes, i.e., to place appropriate values along each axis to accommodate all data points. It is a good practice to make use of as much of the graph paper as possible. If the x values range from 0 to 100, 0 is placed at the left side and 100 is placed as far to the right as possible, considering the magnitude of each division on the x axis. The size of each division depends on the collected data. For instance, if the data values are measured to the nearest 0.5 unit, the scale should allow room so that values like 23.5 or 73.5 are easily plotted.

After each axis is scaled, data points (ordered pairs) are plotted on the graph to correspond to each data pair, i.e., each x and y value. This is accomplished by moving across the x axis to the correct value, and then rising vertically until the y value is reached. At the intersection a mark is made, and in many cases is labeled with the x and y values. All data points are plotted in a similar manner before the graph is drawn.

A common error is to connect the plotted points with a jagged line, as in the children's game of "connect the dots" (Figure A.1a). Data points are not always

FIGURE A.1
(a) The *improper* way to construct a graph. Always draw a smooth curve or straight line that best fits the data points.
(b) The correct way to construct a graph.

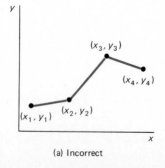

(a) Incorrect

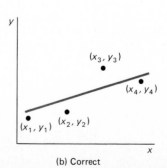

(b) Correct

connected directly. Instead, a straight edge or french curve is used to draw the best fitting line through the maximum number of data points. Usually some points do not fall exactly on the line since experimental errors are present in all measurements (see Figure A.1b).

Example Problem 10 illustrates the above procedures.

Example Problem 10

On a recent auto trip across an interstate highway the following distance and time data were collected:

Time, hr	Distance, mi
0.5	25
1.0	50
2.0	100
3.0	150

Plot a graph of the time and distance data, with time on the x axis and distance on the y axis.

Solution

Time values range from 0.5 to 3.0 hr. To include all values, scale the x axis from 0 to 4 hr in 0.5-hr intervals. Scale the y axis from 0 to 200 mi. It is a common practice to scale the axis slightly above and below the range of data values collected.

Once the axes are scaled, plot each data pair. First, plot the (0.5 hr, 25 mi) point by moving horizontally to 0.5 hr, and then rising vertically to the 25-mi line. Plot all other data pairs in a similar fashion (Figure A.2). It is apparent that

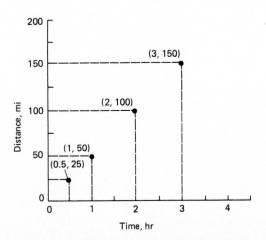

FIGURE A.2
Plotted points with correctly scaled axes.

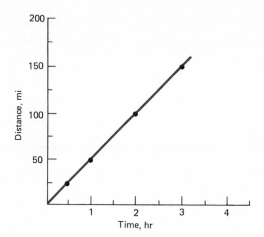

FIGURE A.3
Complete graph of distance traveled in miles versus time in hours.

the points are aligned in a linear fashion, or in a straight line. Using a ruler, draw a straight line through all plotted points, as in Figure A.3.

If you examine the graph plotted in Example Problem 10 carefully, you can readily extract additional information. The graph shows a straight line or linear relationship between the variables of time and distance. Within each unit time interval of the trip, the same amount of distance was traveled. No matter what hour interval is considered, 50 mi is traversed (check this for yourself). The change in distance per time interval is constant in the above graph—50 mi/hr. All linear relationships have this common characteristic, called the **slope,** or rate of change of the y variable for a unit change in the x variable. To compute the slope of the line, select two data points on the line, determine the change in y values, and divide this factor by the corresponding change in x values. Verify the fact that the change is 50 mi/hr in the above graph.

Slopes of linear relationships are either positive, as above, or negative. A **positive slope** indicates that for each increase in the variable plotted on the x axis there is a resulting increase in the y variable. In our example, an increase in time traveled produces an increase in distance. A **negative slope** indicates that for each increase in the x variable, there is a resulting decrease in the y variable. Figure A.4 shows linear relationships with both positive and negative slopes.

Only four data pairs were collected in Example Problem 10. Nevertheless, the distance traveled at any time interval can be determined by correctly reading the graph. How far did the auto travel after 2.5 hr? Find 2.5 hr on the x axis, and draw a perpendicular line from this point to the line plotted on the graph. Then draw a horizontal line from the intersection of the vertical line to the y axis. The point where it meets the axis is the distance travelled in 2.5 hr (125 mi). Reverse the above procedure to find time traveled given distance. For example, how long did it take to travel 75 mi?

FIGURE A.4

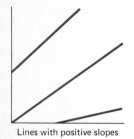

Lines with positive slopes

Lines with negative slopes

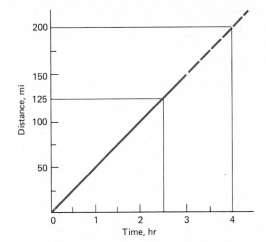

FIGURE A.5
Extrapolation of the distance versus time data indicates that in 4 hours a total distance of 200 miles is traveled.

In many instances, the limits of the data are extended, especially when there is a good fit between the line and the data points. Continuing the line on the graph above and below the range of collected data points is called **extrapolation.** Extrapolation is justified when there is a good reason to believe that the trend extends beyond the observed data. Extending the graph in Example Problem 10 to 4 hr indicates that the auto would have traveled 200 mi in 4 hr. See this in Figure A.5.

Nonlinear relationships are also frequently encountered in chemistry. As the name implies, a **nonlinear relationship** is characterized by a graph that is not a straight line. Instead, the data points are connected by a curved line. Figure A.6 presents examples of nonlinear relationships.

All graphs, no matter what type, are interpreted in the same way. Ask

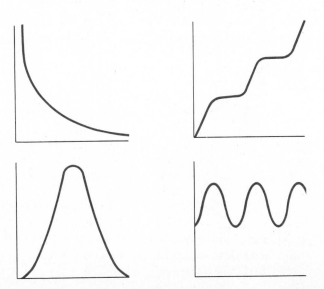

FIGURE A.6
Four types of nonlinear relationships.

yourself the following questions when you are interpreting a graph: What general trends are found? How are the variables changing (increasing or decreasing) with respect to each other? What special characteristics are there? What are the limits of the variables?

QUESTIONS AND PROBLEMS

Equations

1. Solve each of the following equations for X:
(a) $X - 55 = -22$ (b) $125 = 49 + X$
(c) $61 = -X - 78$ (d) $-9X = -34$

2. Solve each of the following equations for Z:

(a) $\dfrac{1}{Z - 9} = 0.5$

(b) $\dfrac{Z}{2Z + 5} = \dfrac{1}{3}$

(c) $(a + b)(Z + 2) = \dfrac{n}{m}$

(d) $\dfrac{a + 2}{b - Z} = n + 1$

3. If $m = 5$ and $n = 3$, solve for X:
(a) $X = 5m - 3n + 2$
(b) $X = 5(m - 3)n + 2$
(c) $X = (5m - 3n) + 2$
(d) $X = 5m - (3n + 2)$

4. Solve the following equations for X:

(a) $5X = 45$ (b) $\dfrac{X}{13} = 12$

(c) $2X + 5 = 6X$ (d) $\dfrac{10X}{30} = 24$

(e) $\dfrac{2X - 10}{8} = 15$

5. Solve the following equations for Y:

(a) $a(nY - m) = 1$ (b) $\dfrac{a + b}{Y + m} = n$

(c) $3a + b = 4Y + n$ (d) $\dfrac{2}{Y} + \dfrac{2}{Y} = \dfrac{a}{b}$

6. If $a = 10$, $b = 5$, $m = 2$, and $n = -1$, determine the value of Y in each equation in Problem 5.

Scientific Notation

7. Change the following numbers to scientific notation:
(a) 0.00049   (b) 120,000,000
(c) 0.00000000135   (d) 0.000000000000000000010

8. Change the following numbers to nonexponential form:
(a) 3.11×10^{-1} (b) 2.90×10^{-8}
(c) 7.7742×10^3 (d) 8.001×10^{18}

9. Change the following numbers in exponential form to scientific notation:
(a) 0.000436×10^3
(b) 1215×10^2
(c) 0.0001001×10^{-5}
(d) 979×10^8
(e) $6,361,000 \times 10^{-14}$

10. Add the following numbers:
(a) $(1.25 \times 10^5) + (6.5 \times 10^5)$
(b) $(5.1 \times 10^{-3}) + (8.2 \times 10^{-3})$
(c) $(9.355 \times 10^8) + (1.722 \times 10^9)$
(d) $(2.109 \times 10^{35}) + (9.64 \times 10^{34})$
(e) $(3.400 \times 10^{19}) + (4.23 \times 10^{17}) + (5.01 \times 10^{18})$

11. Subtract the following numbers:
(a) $(4.38 \times 10^5) - (7.17 \times 10^4)$
(b) $(9.921 \times 10^{-12}) - (7.22 \times 10^{-14})$
(c) $(2.55 \times 10^{45}) - (5.690 \times 10^{46})$
(d) $(6.13 \times 10^{-34}) - (1.749 \times 10^{-32})$

12. Determine the product for each:
(a) $(1.75 \times 10^3) \times (2.44 \times 10^3)$
(b) $(3.54 \times 10^{54}) \times (9.82 \times 10^{82})$
(c) $(1.11 \times 10^{-3}) \times (5.35 \times 10^{15}) \times (7.44 \times 10^{28})$

13. Determine the quotient for each:

(a) $\dfrac{4.175 \times 10^{-31}}{9.329 \times 10^{-15}}$

(b) $\dfrac{5.67 \times 10^{84}}{1.05 \times 10^{-34}}$

(c) $\dfrac{6.8 \times 10^6}{(2.79 \times 10^{-4}) \times (6.11 \times 10^4)}$

14. Perform the indicated operations:

(a) $\dfrac{1.42 \times 10^{-12}}{1.56 \times 10^{-24}} - (9.09 \times 10^{11})$

(b) $\dfrac{(6.788 \times 10^{57}) - (8.54 \times 10^{56})}{5.93 \times 10^{57}}$

Graphing

15. Graph the following data on a full sheet of graph paper.

X	Y
10	36
25	81
40	126
55	171
80	246

(a) What type of relationship is plotted? (b) What Y value corresponds to an X value of 70? (c) What X value corresponds to a Y value of 56? (d) Extrapolate the line to determine the expected Y value corresponding to an X value of 100. (e) What is the slope of the line?

16. Plot the following data, with Fahrenheit temperatures on the x axis and Celsius temperatures on the y axis.

Fahrenheit temperature, degrees	Celsius temperature, degrees
−40	−40
0	−17.8
40	4.4
80	26.7
120	48.9

(a) What is the slope of the line? (b) Using the graph, determine what Celsius temperature corresponds to the following Fahrenheit temperatures: 98.6°F, 75°F, −25°F, and 135°F.

17. (a) On a full sheet of graph paper, plot years on the x axis and the approximate population of the United States on the y axis, given the following data:

Year	Population (in millions)
1940	130
1950	150
1960	175
1970	205
1980	225

(b) Is this a linear function? Explain. (c) What was the approximate population of the United States in 1945? (d) Extrapolate the population data to the year 2000. What is the expected population of the United States in the year 2000?

18. For each of the following graphs, explain what happens to the variable written on the vertical axis when the variable listed on the horizontal axis is increased across its full range.

(a)

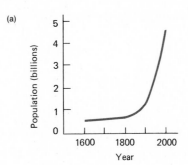

(b)

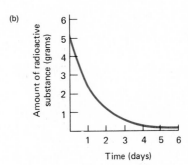

(c)

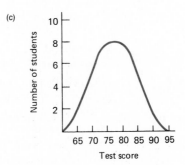

(d)

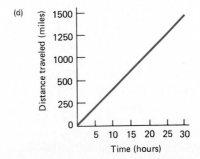

FIGURE A.7

· APPENDIX ·

2

Chemistry Calculations Using Calculators

It is a good idea to use a hand calculator for most calculations in chemistry. A calculator is a time-saver. It will allow you to spend more time actually doing chemistry and less time "crunching numbers" by hand.

Two different formats are used to enter numbers and obtain answers in modern calculators: **algebraic notation** and **reverse polish notation** (RPN). In Section I, the basics of using an algebraic notation calculator are presented, and in Section II the basics of using an RPN calculator are described. To determine what type of calculator you own, check to see whether your calculator has a key with = (equals sign) printed on it (normally near the bottom right-hand side). All algebraic notation calculators have such a key. An RPN calculator does not; it has an ENTER key instead.

Before proceeding, become familiar with your calculator's organization and the various function keys it has. Then read the appropriate section and perform the indicated exercises on your calculator.

I. USE OF ALGEBRAIC NOTATION CALCULATORS

Calculators with algebraic notation are easy to use since the procedure for entering numbers and obtaining answers is similar to the procedure used when performing algebraic and arithmetic operations on paper.

Entering Numbers   To place a number into the calculator, press the appropriate number and decimal point keys. Enter each of the following numbers, checking to see that the display corresponds to the number you think you entered. After entering each number, press the CLEAR key, which deletes the displayed digits. Enter the following: (a) 10, (b) 2395, (c) 123456789, (d) .8, (e) .00002310, and (f) 0.000000345678.

   If your calculator has an EE (enter exponent) or similar key, you can enter numbers in exponential notation. You may enter the number with any decimal and exponential factors, and most calculators will convert the number to scientific notation. To enter such numbers: (1) enter the decimal factor as above, and (2) press the EE key, which lights the two digits at the far right of the display, and then press the number that corresponds to the exponent. Enter the following exponential numbers, check display, and clear: (a) 2×10^3, (b) 5.675×10^{12}, (c) 123.4×10^{56}, and (d) 0.0034×10^{81}.

   Negative numbers and negative exponents are often encountered. Most calculators have a change-of-sign key, $+/-$. When the change-of-sign key is depressed, the displayed number is multiplied by -1. Thus, to enter a negative exponential number, enter the decimal factor, press the change-of-sign key, and then enter the exponent. To enter a positive number with a negative exponent, enter the decimal number, then the exponent, and then press the change-of-sign key. Practice using the change-of-sign key by entering the following exponential numbers, checking the display, and then clearing: (a) -4.3×10^{11}, (b) 9.11×10^{-6}, and (c) -5.51245×10^{-61}.

Arithmetic Operations   Locate the $+$, $-$, $\div$, and $\times$ keys. To perform addition, subtraction, multiplication, and division: (1) enter first number, (2) press arithmetic operator, (3) enter second number, and (4) press the $=$ key. Follow the stated steps for each of the following: (1) $23 + 92 =$, (2) $124 - 321 =$, (3) $8.533 \div 9.112 =$, and (4) $-12.003 \times -3.175 =$.

   Multiple operations of the same arithmetic procedure are accomplished in a similar manner. After you enter each number, press the arithmetic operator key. After you enter the last number, press the $=$ key to obtain the final answer. The calculator displays a new value after each arithmetic operation so that intermediate values are obtained. Perform the following multiple operations: (a) $1501 + 23 + 91.4 + 888 =$, (b) $56.99 - 1.20 - 46.9 - 0.8 =$, (c) $599 \div 23 \div 0.75 \div 34.72 =$, and (d) $-.9345 \times .05699 \times -.00344 \times 1.0101 =$.

   Care must be exercised when performing multiple arithmetic operations of more than one type. Depending on the convention followed by the calculator, the calculator may process the numbers from left to right or it may follow the algebraic operating system, which means that it will perform multiplication and division from left to right before performing addition and subtraction. An example will serve to illustrate this difference. What is the answer to the following?

$$8 + 5 - 3 \times 16 \div 8 =$$

If your calculator performs the arithmetic operations from left to right, you will obtain the answer 20. But if your calculator follows the algebraic operating system, you will obtain the answer 7. Placing parentheses into the expression will show how the answer of 7 is arrived at.

$$8 + 5 - \left(\frac{(3 \times 16)}{8}\right) = 7$$

Refer to the manual that comes with your calculator to find which method it uses. Practice multiple operations by performing and checking the following: (1) $5 \times 6 - 32 \div 8 + 12 =$, (2) $27 \div 9 - 44 \times 2 + 60 \div 20 =$, and (3) $1.15 - 0.233 \div 0.0295 - 0.010 + 15 \times 0.543 =$.

If your calculator has parenthesis keys, (and), arithmetic operations can easily be performed in any order, depending on how the parentheses group the numbers and operations. For example, consider the following:

$$\frac{9 + 2 + (4 \times 6)}{(2 + 5 - 6) \times 5} =$$

To correctly solve this problem, enclose the complete numerator in parentheses and the denominator in parentheses so that they can be divided. Also, the parentheses are used to group the operations individually in the numerator and denominator. The following key strokes are necessary to obtain the correct answer of 7.

$$\underset{\text{Numerator}}{(9 + 2 + (4 \times 6))} \div \underset{\text{Denominator}}{((2 + 5 - 6) \times 5)} = 7$$

Practice using the parentheses key with the following examples:
(a) $((23.4 - 15.4) \times 31) \div 0.22 =$
(b) $((1.934 \div 3.12 + 5.22) - 4.84) \div (1 + (3.14 \times 0.566 - 1.11)) =$

Function Keys Many calculators contain function keys that perform various mathematical functions. In each case, if a number is entered and then a function key is pressed, the resulting display presents the answer. If 2 is entered and then the reciprocal key $(1/x)$ is pressed, then the reciprocal of 2, 0.5, is displayed. If the reciprocal key is pressed again, the number 2 results.

Other function keys that are found on calculators include: square, x^2; square root, $\sqrt{x}$; SIN; COS; TAN; $\log_{10}$, LOG; $\log_e$, LN x; Y^X; 10^x; and e^x. Of these functions, the logarithm keys are probably the most important for chemistry. If the LOG key is pressed after a number is entered, the logarithm to the base 10 is calculated. However, if LN x is pressed, the natural logarithm is displayed. To illustrate this, enter the number 10 and find the common log and the natural log. For the former, the answer is 1.0, and for the latter the answer is 2.303. Antilogs are determined by pressing either 10^x for common antilogs or e^x for natural antilogs. Practice using the function keys by performing the following operations: (a) 25^2, (b) $345^{.5}$, (c) sin 45°, (d) cos 45°, (e) tan 30°, (f) log 6.45, (g) ln 1.2×10^2, (h) $.96^{11}$, (i) antilog 0.5123, and (j) antiln 4.2485.

II. USE OF CALCULATORS WITH REVERSE POLISH NOTATION

Entering Numbers   In RPN calculators, a number must first be placed into a memory location before the calculator can use it. Many RPN calculators have four principal memory locations, which are "stacked on top of each other." When a number key is pressed, the number initially occupies the lowest position in the stack. To move it to the next highest memory location, the ENTER key is pressed.

Picture the memory locations as follows:

W
Z
Y
X

where X is the memory location that is constantly displayed. Immediately above this is memory location Y. When a number is first "ENTEREd," it is moved into Y. Thus, if the number 7 key is pressed, location X receives the 7, and when the ENTER key is pressed, 7 is placed into Y, without removing it from X; 7 still remains displayed (X = 7, Y = 7, Z = 0, and W = 0). If another number is pressed, let's say 6, the 6 would occupy memory location X (X = 6, Y = 7, Z = 0, W = 0). Now when the ENTER key is pressed, the 6 is placed into Y and the 7 is pushed up to Z (X = 6, Y = 6, Z = 7, and W = 0). Pressing the 5 key places 5 in X, and then pressing the ENTER key moves 5 to Y and pushes both the 6 and 7 up in the stack (X = 5, Y = 5, Z = 6, and W = 7). One more key stroke will fill the stack with four different numbers: pressing the 4 key would result in a stack with X = 4, Y = 5, Z = 6, and W = 7

Most RPN calculators have a key that allows each stack memory location to be viewed; on many calculators this key is the **R↓** . When the **R↓** key is pressed, each member of the stack is dropped down, and the contents of X is placed into W. If the stack contains the numbers mentioned above, when the **R↓** key is pressed once, 5 is displayed, 4 goes to the top of the stack, and the 6 and 7 drop down. If the **R↓** key is pressed again, 6 is displayed, 5 goes to the top, etc.

W = 7	W = 4	W = 5	W = 6
Z = 6	Z = 7	Z = 4	Z = 5
Y = 5	Y = 6	Y = 7	Y = 4
Display → X = 4	X = 5	X = 6	X = 7
Start	Press **R↓** once	Press **R↓** again	Press **R↓** once again

With the $\mathbf{R}\downarrow$ key, the numbers can also be moved, depending on the operation being performed. Additionally, RPN calculators have an $\mathbf{X} \rightleftarrows \mathbf{Y}$ key to exchange places of numbers in memory locations X and Y.

Practice entering numbers into the stack and using the $\mathbf{R}\downarrow$ key to view the contents before continuing.

Negative numbers are obtained by using the CHS (change sign) or $+/-$ key. To place a negative number into Y, first place the number into X, then press the CHS key followed by the ENTER key.

Numbers expressed in scientific notation are usually entered into RPN calculators in the same way that they are entered into algebraic notation calculators. Refer to the appropriate paragraph in Section I.

Arithmetic Operations

All arithmetic operations are performed by entering a number into memory location Y, then placing a number into memory location X, and then pressing the arithmetic operator key. To add 2 + 5, press the 2 key, and then press the ENTER key. This moves 2 to Y. Press the 5 key, which places 5 in X, and then press the + key. At this time, 7 is displayed. Subtraction is the same as addition, except the − key is pressed. The contents of X are subtracted from the contents of Y. To multiply 2 by 5, the same procedure is followed: (1) Press 2 and then ENTER. (2) Press the 5 key. (3) Press the × key. In division, the dividend is placed into Y, the divisor is placed into X, and the ÷ key is pressed.

The following summarizes the four basic arithmetic operations on an RPN calculator:

Addition:	$y + x$
Subtraction:	$y - x$
Multiplication:	$y \cdot x$
Division:	$y \div x$

where y is the contents of memory location Y, and x is the contents of memory location X. The answer is placed into X; in other words, it is displayed. Practice simple arithmetic operations by calculating the answers to the following: (1) 24 + 91, (2) 16 − 44, (3) 3.14 × 63.69, and (4) 0.023 ÷ 451.

Multiple operations are performed exactly in the same way as are single arithmetic operations. The first number is entered into Y, the second number placed in X, and the arithmetic operator button is then depressed. To continue the process, a new number is placed into X and the operator key is pressed. For example, what key strokes are necessary to perform the following?

$$12 + 22 \div 2 - 10 \times 3$$

First, enter 12 into Y, place 22 into X, and then press the + key. Press 2, ÷, 10, −, 3, and × to give the answer of 21. Practice multiple operations by finding the answers to the following: (a) 32 − 16 + 22 ÷ 11, (b) 125 ÷ 75 × 14 − 8 ÷ 12, and (c) 9 × 5 − 5 ÷ 4 ÷ 5 + 2.

In many calculations, the order of operations must be changed and grouped. For example, how would the following calculation be performed?

$$\frac{4 \times (16 + 3)}{3 \times (12 - 5)}$$

Initially 4 and 16 are both entered, which places 4 in Z and 16 in Y. After the 3 key is depressed, 3 is located in X; the **+** key is pressed, giving 19, followed by the **×** key, giving 76. Next the denominator is determined by entering 3 and 12 (pushing 76 up to Z) followed by pressing 5, **−**, and **×**. This yields the number 21. Now the 76 has dropped back down to Y; hence, it is only necessary to press the **÷** key to obtain the final answer of 3.619.

Perform the following for practice:

(a) $(3 + 9) \div (2 + 8)$

(b) $1 + (2.455 - 1.391) \div (6 \times (0.44 \div 0.11)) + 0.65$

(c) $(((9.2 - 10.2) \div 5.66) \times -124.52) + 33 =$

Function Keys Performing functions such as reciprocals, square roots, and logarithms requires only that a number be placed into memory location X followed by pressing the appropriate function key. If 12 is placed into X and the x^2 key is pressed, the answer of 144 is displayed. Since the use of function keys in RPN calculators is the same as in algebraic notation calculators, refer to the appropriate paragraph in I for a more detailed discussion.

· APPENDIX ·

Physical Constants and Conversion Factors

I. PHYSICAL CONSTANTS

Unified atomic mass unit (u)	$1\ u = 1.66057 \times 10^{-27}\ kg$
	$= 1.66057 \times 10^{-24}\ g$
	$6.022045 \times 10^{23}\ u = 1.000\ g$
Avogadro's number	$N = 6.022045 \times 10^{23}$ entities/mol
Electron mass	$m_e = 9.10953 \times 10^{-28}\ g$
	$= 5.48580 \times 10^{-4}\ u$
Electronic charge	$e = 1.6022 \times 10^{-19}\ C$
Faraday's constant	$F = 9.6485 \times 10^{4}\ C/mol$
Ideal gas constant	$R = 0.082057\ L \cdot atm/(mol \cdot K)$
	$= 8.3144\ J/(mol \cdot K)$
	$= 8.3144\ Pa \cdot dm^3/(mol \cdot K)$
Molar volume (ideal gas)	$V = 22.41383\ L/mol$
Neutron mass	$m_{n^{\circ}} = 1.67495 \times 10^{-24}\ g$
	$= 1.00866\ u$
Proton mass	$m_{p^{+}} = 1.67265 \times 10^{-24}\ g$
	$= 1.00728\ u$
Speed of light	$c = 2.997925 \times 10^{8}\ m/s$

II. UNIT CONVERSION FACTORS

Mass
$1 \text{ g} = 10^{-3} \text{ kg}$
$1 \text{ mg} = 10^{-6} \text{ kg}$
$1 \text{ lb} = 0.453592 \text{ kg} = 453.592 \text{ g}$
$1 \text{ kg} = 2.205 \text{ lb}$
$1 \text{ oz} = = 28.349523 \text{ g}$
$1 \text{ u} = 1.66057 \times 10^{-27} \text{ kg}$
$1 \text{ metric tonne} = 1000 \text{ kg}$

Length
$1 \text{ cm} = 10^{-2} \text{ m}$
$1 \text{ mm} = 10^{-3} \text{ m}$
$1 \text{ nm} = 10^{-9} \text{ m}$
$1 \text{ Å} = 10^{-10} \text{ m} = 0.10 \text{ nm}$
$1 \text{ in.} = 2.54 \times 10^{-2} \text{ m} = 2.54 \text{ cm}$
$1 \text{ cm} = 10 \text{ mm}$
$1 \text{ mi} = 1.609 \text{ km}$
$1 \text{ yd} = 0.9144 \text{ m}$

Time
$1 \text{ day} = 8.64 \times 10^4 \text{ s}$
$1 \text{ hr} = 3.60 \times 10^3 \text{ s}$
$1 \text{ ms} = 10^{-3} \text{ s}$

Temperature
$0°C = 273.15 \text{ K}$
$0°F = 255.37 \text{ K}$
$-273.15 \,°\text{C} = 0 \text{ K}$
$-459.67°F = 0 \text{ K}$

Volume
$1 \text{ L} = 10^{-3} \text{ m}^3$
$1 \text{ L} = 1.057 \text{ qt}$
$1 \text{ qt} = 0.9463 \text{ L}$
$1 \text{ cm}^3 = 10^{-6} \text{ m}^3$
$1 \text{ mL} = 1 \text{ cm}^3$
$1000 \text{ mL} = 1 \text{ L}$

Energy
$1 \text{ cal} = 4.184 \text{ J}$
$1 \text{ cal} = 2.612 \times 10^{19} \text{ eV}$
$1 \text{ erg} = 10^{-7} \text{ J}$
$1 \text{ erg} = 2.3901 \times 10^{-8} \text{ cal}$

Pressure
$1 \text{ atm} = 101,325 \text{ Pa}$
$1 \text{ atm} = 760 \text{ mm Hg}$
$1 \text{ atm} = 14.70 \text{ lb/in.}^2$
$1 \text{ mm Hg} = 1 \text{ torr}$
$1 \text{ mm Hg} = 133.322 \text{ Pa}$

· A P P E N D I X ·

4

Table of Ions and Their Formulas

Ion	Name	Ion	Name
AlO_3^{3-}	Aluminate	IO_3^-	Iodate
AsO_4^{3-}	Arsenate	IO^-	Hypoiodite
AsO_3^{3-}	Arsenite	MnO_4^-	Permanganate
BiO_3^{3-}	Bismuthate	MnO_4^{2-}	Manganate
BO_3^{3-}	Borate	NH_4^+	Ammonium
BrO_3^-	Bromate	NO_3^-	Nitrate
BrO_2^-	Bromite	NO_2^-	Nitrite
BrO^-	Hypobromite	O_2^{2-}	Peroxide
CO_3^{2-}	Carbonate	OCN^-	Cyanate
HCO_3^-	Hydrogen carbonate	OH^-	Hydroxide
$C_2O_4^{2-}$	Oxalate	PO_4^{3-}	Phosphate
$C_2H_3O_2^-$	Acetate	PO_3^{3-}	Phosphite
CN^-	Cyanide	PO_2^{3-}	Hypophosphite
HCO_2^-	Formate	$P_2O_7^{4-}$	Pyrophosphate
ClO_4^-	Perchlorate	SCN^-	Thiocyanate
ClO_3^-	Chlorate	SeO_4^{2-}	Selenate
ClO_2^-	Chlorite	SeO_3^{2-}	Selenite
ClO^-	Hypochlorite	SiO_3^{2-}	Silicate
CrO_4^{2-}	Chromate	SnO_4^{4-}	Stannate
$Cr_2O_7^{2-}$	Dichromate	SO_4^{2-}	Sulfate
GaO_3^{3-}	Gallate	SO_3^{2-}	Sulfite
GeO_4^{4-}	Germanate	$S_2O_3^{2-}$	Thiosulfate
GeO_3^{4-}	Germanite	TeO_6^{6-}	Tellurate
H^-	Hydride	TeO_3^{2-}	Tellurite
IO_4^-	Periodate		

· APPENDIX ·

Logarithms to Base 10

Use of a Logarithm Table

I. To find the logarithm of a number between 1 and 9.99

The logarithm of a number between 1 and 9.99 is obtained directly from the table.

Step 1. Find the first two digits of the number in the first column.

Step 2. Move horizontally across the page to the column that corresponds to the third digit of the number.

Step 3. The number at the intersection is the common logarithm.

For example, what is the logarithm of the number 7.55?

1. Find the row that begins with 75.
2. Move across the table to the column labeled 5.
3. Read the number at the intersection of 75 and 5.

The common logarithm of 7.55 is .8779.

II. To find the logarithms of a number that is not between 1 and 9.99

Step 1. Express the number in scientific notation.

Step 2. Find the logarithm of the decimal factor as in part I.

Step 3. The logarithm of the exponential factor is the value of the exponent.

Step 4. Add the two logarithms to find the logarithm of the number.

For example, what is the logarithm of 3720?

1. The number 3720 expressed in scientific notation is 3.72×10^3.
2. Using the logarithm table, the logarithm of 3.72 is .5705.
3. The logarithm of 10^3 is 3.
4. Add the two numbers: $.5705 + 3 = 3.5705$

The common logarithm of 3720 is 3.5705.

	0	1	2	3	4	5	6	7	8	9
10	0000	0043	0086	0128	0170	0212	0253	0294	0334	0374
11	0414	0453	0492	0531	0569	0607	0645	0682	0719	0755
12	0792	0828	0864	0899	0934	0969	1004	1038	1072	1106
13	1139	1173	1206	1239	1271	1303	1335	1367	1399	1430
14	1461	1492	1523	1553	1584	1614	1644	1673	1703	1732
15	1761	1790	1818	1847	1875	1903	1931	1959	1987	2014
16	2041	2068	2095	2122	2148	2175	2201	2227	2253	2279
17	2304	2330	2355	2380	2405	2430	2455	2480	2504	2529
18	2553	2577	2601	2625	2648	2672	2695	2718	2742	2765
19	2788	2810	2833	2856	2878	2900	2923	2945	2967	2989
20	3010	3032	3054	3075	3096	3118	3139	3160	3181	3201
21	3222	3243	3263	3284	3304	3324	3345	3365	3385	3404
22	3424	3444	3464	3483	3502	3522	3541	3560	3579	3598
23	3617	3636	3655	3674	3692	3711	3729	3747	3766	3784
24	3802	3820	3838	3856	3874	3892	3909	3927	3945	3962
25	3979	3997	4014	4031	4048	4065	4082	4099	4116	4133
26	4150	4166	4183	4200	4216	4232	4249	4265	4281	4298
27	4314	4330	4346	4362	4378	4393	4409	4425	4440	4456
28	4472	4487	4502	4518	4533	4548	4564	4579	4594	4609
29	4624	4639	4654	4669	4683	4698	4713	4728	4742	4757
30	4771	4786	4800	4814	4829	4843	4857	4871	4886	4900
31	4914	4928	4942	4955	4969	4983	4997	5011	5024	5038
32	5051	5065	5079	5092	5105	5119	5132	5145	5159	5172
33	5185	5198	5211	5224	5237	5250	5263	5276	5289	5302
34	5315	5328	5340	5353	5366	5378	5391	5403	5416	5428
35	5441	5453	5465	5478	5490	5502	5514	5527	5539	5551
36	5563	5575	5587	5599	5611	5623	5635	5647	5658	5670
37	5682	5694	5705	5717	5729	5740	5752	5763	5775	5786
38	5798	5809	5821	5832	5843	5855	5866	5877	5888	5899
39	5911	5922	5933	5944	5955	5966	5977	5988	5999	6010
40	6021	6031	6042	6053	6064	6075	6085	6096	6107	6117
41	6128	6138	6149	6160	6170	6180	6191	6201	6212	6222
42	6232	6243	6253	6263	6274	6284	6294	6304	6314	6325
43	6335	6345	6355	6365	6375	6385	6395	6405	6415	6425
44	6435	6444	6454	6464	6474	6484	6493	6503	6513	6522
45	6532	6542	6551	6561	6571	6580	6590	6599	6609	6618
46	6628	6637	6646	6656	6665	6675	6684	6693	6702	6712
47	6721	6730	6739	6749	6758	6767	6776	6785	6794	6803
48	6812	6821	6830	6839	6848	6857	6866	6875	6884	6893
49	6902	6911	6920	6928	6937	6946	6955	6964	6972	6981
50	6990	6998	7007	7016	7024	7033	7042	7050	7059	7067
51	7076	7084	7093	7101	7110	7118	7126	7135	7143	7152
52	7160	7168	7177	7185	7193	7202	7210	7218	7226	7235
53	7243	7251	7259	7267	7275	7284	7292	7300	7308	7316
54	7324	7332	7340	7348	7356	7364	7372	7380	7388	7396

	0	1	2	3	4	5	6	7	8	9
55	7404	7412	7419	7427	7435	7443	7451	7459	7466	7474
56	7482	7490	7497	7505	7513	7520	7528	7536	7543	7551
57	7559	7566	7574	7582	7589	7597	7604	7612	7619	7627
58	7634	7642	7649	7657	7664	7672	7679	7686	7694	7701
59	7709	7716	7723	7731	7738	7745	7752	7760	7767	7774
60	7782	7789	7796	7803	7810	7818	7825	7832	7839	7846
61	7853	7860	7868	7875	7882	7889	7896	7903	7910	7917
62	7924	7931	7938	7945	7952	7959	7966	7973	7980	7987
63	7993	8000	8007	8014	8021	8028	8035	8041	8048	8055
64	8062	8069	8075	8082	8089	8096	8102	8109	8116	8122
65	8129	8136	8142	8149	8156	8162	8169	8176	8182	8189
66	8195	8202	8209	8215	8222	8228	8235	8241	8248	8254
67	8261	8267	8274	8280	8287	8293	8299	8306	8312	8319
68	8325	8331	8338	8344	8351	8357	8363	8370	8376	8382
69	8388	8395	8401	8407	8414	8420	8426	8432	8439	8445
70	8451	8457	8463	8470	8476	8482	8488	8494	8500	8506
71	8513	8519	8525	8531	8537	8543	8549	8555	8561	8567
72	8573	8579	8585	8591	8597	8603	8609	8615	8621	8627
73	8633	8639	8645	8651	8657	8663	8669	8675	8681	8686
74	8692	8698	8704	8710	8716	8722	8727	8733	8739	8745
75	8751	8756	8762	8768	8774	8779	8785	8791	8797	8802
76	8808	8814	8820	8825	8831	8837	8842	8848	8854	8859
77	8865	8871	8876	8882	8887	8893	8899	8904	8910	8915
78	8921	8927	8932	8938	8943	8949	8954	8960	8965	8971
79	8976	8982	8987	8993	8998	9004	9009	9015	9020	9025
80	9031	9036	9042	9047	9053	9058	9063	9069	9074	9079
81	9085	9090	9096	9101	9106	9112	9117	9122	9128	9133
82	9138	9143	9149	9154	9159	9165	9170	9175	9180	9186
83	9191	9196	9201	9206	9212	9217	9222	9227	9232	9238
84	9243	9248	9253	9258	9263	9269	9274	9279	9284	9289
85	9294	9299	9304	9309	9315	9320	9325	9330	9335	9340
86	9345	9350	9355	9360	9365	9370	9375	9380	9385	9390
87	9395	9400	9405	9410	9415	9420	9425	9430	9435	9440
88	9445	9450	9455	9460	9465	9469	9474	9479	9484	9489
89	9494	9499	9504	9509	9513	9518	9523	9528	9533	9538
90	9542	9547	9552	9557	9562	9566	9571	9576	9581	9586
91	9590	9595	9600	9605	9609	9614	9619	9624	9628	9633
92	9638	9643	9647	9652	9657	9661	9666	9671	9675	9680
93	9685	9689	9694	9699	9703	9708	9713	9717	9722	9727
94	9731	9736	9741	9745	9750	9754	9759	9763	9768	9773
95	9777	9782	9786	9791	9795	9800	9805	9809	9814	9818
96	9823	9827	9832	9836	9841	9845	9850	9854	9859	9863
97	9868	9872	9877	9881	9886	9890	9894	9899	9903	9908
98	9912	9917	9921	9926	9930	9934	9939	9943	9948	9952
99	9956	9961	9965	9969	9974	9978	9983	9987	9991	9996

· A P P E N D I X ·

Answers to Problems and Selected Questions

CHAPTER 2

2.4 (a) 36.7 crates   (b) 17,400 oranges

2.5 17.1 gallons

2.6 (a) 163 hours   (b) 9792 minutes   (c) 587,520 seconds
(d) 0.019 years

2.7 50 decades

2.13 470 miles

2.14 (a) 420 seconds   (b) 0.425 year   (c) 0.6 century   (d) 716.7
dozen apples   (e) 5.83 days   (f) 5.38 gross

2.15 18.2 hours

2.16 16,000 seconds

2.17 7665 days

2.18 (a) 0.0651 year   (b) 0.6849 decade   (c) 328,500 days   (d) 0.17
day   (e) 0.274 millennium

2.19 6.25 miles   33,000 feet

2.20 36 cents

2.21 (a) 28.6 miles per gallon   (b) 0.035 gallons per mile
(c) $5.41   (d) $71.30

2.22 (a) $6.67   (b) $0.0089   (c) $5.93   (d) 20 boxes   (e) 4.2
pounds

2.23 0.000155 century

2.24 2.4×10^9 seconds

2.25 40 rods

2.26 3.7×10^9 miles

CHAPTER 3

3.7 (a) 3.50×10^5 mm   (b) 1.6×10^3 cm   (c) 0.00231 lb
     (d) 7.6×10^{12} ng

3.9 (a) 398 K   (b) 49 K   (c) 11°C   (d) -122°F

3.11 (a) 1.25×10^4 mL 1.25×10^4 cm³   12.5 dm³   0.0125 m³
     (b) 5.6×10^{-4} m³ 5.6×10^{-1} L

3.12 1.7×10^2 L

3.13 (a) 0.8536 g/mL   (b) potassium

3.14 54.63 mL

3.15 0.895 g/mL

3.21 (a) 4   (b) 2   (c) 6   (d) 3   (e) 4   (f) 2   (g) 15   (h) 1

3.22 (a) 298.6 g   (b) 0.194 mL   (c) 551 m   (d) 224.077427 s
     (e) 6.60×10^{23}

3.23 (a) 0.53 g/mL   (b) 4×10^2 m²   (c) 2.40 cm³
     (d) 2.26×10^5 (e) 5.6×10^5 mg²

3.24 (a) 3 m²   (b) 2×10^{-4} g/s   (c) 9.7×10^{22}

3.30 (a) 0.012 m   (b) 5.09×10^3 mm   (c) 1.2×10^5 cm
     (d) 6.8×10^9 μm   (e) 8.5×10^6 nm   (f) 4.6×10^3 km

3.31 (a) 1 yd   (b) 3 ft   (c) 62.2 mi   (d) 20 in.
     (e) 3×10^2 fathoms

3.32 (a) 16 km   (b) 9.1×10^2 mm   (c) 1.9 m   (d) 91.4 m
     (e) 3.8×10^5 nm   (f) 8.2×10^6 mm

3.33 (a) 1.5×10^8 km   (b) 1.5×10^{11} m   (c) 1.5×10^{14} mm

3.35 (a) 59.0 kg   (b) 1.7×10^2 cm   (c) 94 cm   (d) 76 cm
     (e) 89 cm

3.38 (a) 5×10^6 mg   (b) 2.7×10^{-3} g   (c) 8.571×10^1 kg
     (d) 9×10^{12} pg   (e) 7.5×10^1 μg   (f) 5.5×10^3 Mg
     (g) 5.3 dg   (h) 3.147×10^5 cg

3.39 (a) 1115 lb   (b) 0.3214 oz   (c) 6.9×10^6 ton
     (d) 4.49×10^{-4} lb   (e) 4.171×10^{-5} grain   (f) 1 English ton

3.40 (a) 4.5×10^3 g   (b) 2.62×10^7 mg   (c) 3.8 Mg
     (d) 1.5×10^3 cg   (e) 3.0×10^4 ng   (f) 4.0×10^{12} dg

3.41 (a) 364 K   (b) 238 K   (c) 7.6 K   (d) 546 K

3.42 (a) -248°C   (b) -173°C   (c) 127.3°C   (d) 25.7°C

3.43 (a) 90.6°C   (b) -126°C   (c) 537.8°C   (d) -273°C

3.44 (a) 383°F   (b) -319°F   (c) 1832°F   (d) -31°F

3.45 (a) -1.1×10^2°F   (b) -282.3°F   (c) 1340°F   (d) 99.7°F

3.46 -40°C or -40°F

3.47 (a) 83.7°F

3.48 3.5 K   4.3 K

3.49 (a) 1.5 cm³   (b) 1.5 mL   (c) 0.0015 L

3.50 1.4×10^3 dm³ 1.4×10^3 L

3.51 (a) 1000 cm³   (b) 2.4×10^{-5} m³   (c) 8.3422×10^5 μL
     (d) 4.4×10^5 mL   (e) 840.000 m³   (f) 9.14×10^5 m³

3.52 (a) 0.0581 qt   (b) 6×10^4 in.^{3} (c) 0.106 qt   (d) 0.2472 ft^3

3.53 (a) 3.784×10^{-1} m^3 (b) 4.5×10^4 cm^3
     (c) 3.9×10^{-3} dm^3 (d) 4.2×10^6 mm^3

3.54 (a) 2.80 g/mL   (b) 0.808 g/mL   (c) 0.977 g/mL   (d) 0.9 g/mL

3.55 (a) 61 g   (b) 1.1×10^3 g   (c) 9.2×10^3 g   (d) 1.7 g
     (e) 5.8×10^3 g   (f) 5.3×10^6 g

3.56 (a) 21 mL   (b) 0.0536 mL   (c) 47.2 mL
     (d) 2.08×10^3 mL   (e) 4×10^5 mL

3.57 (a) 36.5 g/ft^3 (b) 1.29 kg/m^3 (c) 0.0805 lb/ft^3

3.58 21 g/mL

3.59 1.35 g/mL

3.60 (a) 1×10^4 g Hg   (b) 3×10^1 lb Hg

3.61 1.41 g/mL

3.62 9.68×10^4 g

3.63 1.61×10^5 g

3.71 (a) 6   (b) 5   (c) 5   (d) 6   (e) 1   (f) 3   (g) 11
     (h) 2   (i) 2   (j) 3

3.72 (a) 1   (b) 0   (c) 0   (d) 0   (e) 9   (f) 8   (g) 1
     (h) 9   (i) 5   (j) 8

3.73 (a) 2×10^2 (b) 3   (c) 2×10^3 (d) 1×10^{-5} (e) 9
     (f) 8×10^{-7} (g) 2×10^3 (h) 3×10^{-3} (i) 8×10^4
     (j) 1×10^{-4}

3.74 (a) 61.64   (b) 2.000   (c) 641.1   (d) 0.9991
     (e) 7.995×10^5 (f) 2.000×10^{11}

3.75 (a) 7×10^4 (b) 7.0×10^4 (c) 7.00×10^4

3.76 (a) 4.40 g   (b) 23 mL   (c) 16.9 g   (d) 226.71 m

3.77 (a) 16.3 mL   (b) 0.064 cm   (c) 0.23 g   (d) 10 mm

3.78 (a) 0.07 g^2 (b) 625 s^2 (c) 3×10^1 cm^2 (d) 0.527 kg^2

3.79 (a) 3.2 g/mL   (b) 630.0   (c) 2×10^{15} m/s
     (d) 3.0×10^{-2} g/cm^3

3.80 (a) 5.7×10^3 (b) 0.476   (c) 0.006

3.81 (a) 805 km/h   (b) 223 m/s   (c) 2.33×10^{-4} mm/ns

3.82 13.7 g/mL

3.83 3.9 cm

3.84 1.5×10^2 in.^{3}

3.85 (a) 160°C = 320°F   (b) −24.6°C

3.86 0.92033 g/mL

CHAPTER 4

4.1 Physical: (a), (b)   Chemical: (c), (d)

4.2 Chemical: (a), (c), (e)   Physical: (b), (d)

4.13 (a) Potassium 1, fluorine 1   (b) Beryllium 1, chlorine 2
(c) Nickel 1, sulfur 1   (d) Iron 1, oxygen 3, hydrogen 3
(e) Phosphorus 1, iodine 3   (f) Barium 1, nitrogen 2, oxygen 6
(g) Sodium 1, manganese 1, oxygen 4   (h) Nitrogen 2, hydrogen 8, carbon 2, oxygen 4

4.21 Electric to heat, light, and sound

4.25 (a) 2.2×10^4 (b) 0.14 J   (c) 34.4 kcal

4.27 (a) 7.1×10^{-2} cal/(g·°C)   (b) 0.297 J/(g·°C)

4.28 6.0×10^3 cal

4.29 14.4°C

4.30 9.0×10^7 J

4.34 Physical: (a), (b), (f), (h), (i), (k), (l), (m)   Chemical: (c), (d), (e), (g), (j), (n), (o)

4.35 Physical: (a), (d), (e), (f), (i)   Chemical: (b), (c), (g), (h)

4.36 Physical: (a), (b), (d)   Chemical: (c), (e)

4.37 Physical: color, solid phase, forms dark brown allotrope on heating to 180°C   Chemical: burns in air

4.39 (a) Vegetable oil   (b) Motor oil   (c) Pudding   (d) Molasses

4.40 Solid: (d), (f), (g), (h), (j)   Liquid: (c), (d), (e)   Gas: (a), (b), (i)

4.41 Some substances decompose before the boiling point is reached.

4.42 (a) Gas   (b) Solid

4.43 (a) Solution   (b) Element   (c) Heterogeneous mixture
(d) Compound   (e) Element

4.44 Pure substances: (c), (f), (g), (i), (j)   Mixtures: (a), (b), (d), (e), (h)

4.51 Element: (b), (c)   Compounds: (a)

4.53 Homogeneous: (a), (b), (d), (e), (f)   Heterogeneous: (c), (g), (h)

4.54 (a) Add water then filter   (b), (d) Distill   (c) Filter

4.55 (a) 2   (b) 1   (c) 2   (d) 3

4.58 Position: (b), (e)   Condition: (c), (d)   Composition: (a), (f)

4.59 (a) Bunsen burner flame   (b) Bathtub filled with 90°C water
(c) Large block of ice at 0°C   (d) Gallon of water at 50°C
(e) Neither

4.60 (a) 1.39×10^3 J   (b) 1049 cal   (c) 1.9×10^2 kcal
(d) 4.15×10^8 kJ   (e) 6.15×10^{-4} kJ

4.61 2.2 J/(g·°C)

4.62 0.40 cal/(g·°C)

4.63 2.22×10^3 cal

4.64 2.80×10^3 cal

4.65 75 cal/g   75 kcal/kg

4.66 (a) 3×10^4 g   (b) 3×10^1 L

4.67 Gold 3.1×10^2 cal   Water 1.0×10^4 cal

4.68 (a) 44 g   (b) Conservation of matter

4.69 Potential energy of water to kinetic energy of water to kinetic energy (mechanical energy) of the turbine to kinetic energy of the generator to electric energy.

Chapter 5

5.4 (a) Repel   (b) Attract

5.5 (a) Number of protons   (b) Number of protons plus neutrons   (c) Average mass of naturally occurring isotopes   (d) Number of neutrons

5.6 (a) $p^+ = 15$; $n^0 = 16$ (b) $p^+ = 34$; $n^0 = 44$ (c) $p^+ = 89$; $n^0 = 138$

5.9 63.55

5.11 (a) 4   (b) 15   (c) 25   (d) 55   (e) 54

5.14 (a) 2   (b) 8   (c) 2   (d) 6   (e) 2

5.15 (a) $1s^2$ (b) $1s^2 2s^2 2p^1$ (c) $1s^2 2s^2 2p^6$ (d) $1s^2 2s^2 2p^6 3s^2$ (e) $1s^2 2s^2 2p^6 3s^2 3p^5$ (f) $1s^2 2s^2 2p^6 3s^2 3p^6 4s^1$

5.16 (a) Na·   (b) N̈:   (c) :Ẍe: (d) Ba: (e) ·S̈e:

5.19 (a) Repel   (b) Attract   (c) No force

5.20

	Symbol	Atomic number	Mass number	No. of protons	No. of neutrons	No. of electrons
(a)	$_2^4$He	2	4	2	2	2
(b)	$_{12}^{26}$Mg	12	26	12	14	12
(c)	$_{16}^{36}$S	16	36	16	20	16
(d)	$_{22}^{48}$Ti	22	48	22	26	22
(e)	$_{34}^{80}$Se	34	80	34	46	34
(f)	$_{40}^{90}$Zr	40	90	40	50	40
(g)	$_{44}^{99}$Ru	44	99	44	55	44
(h)	$_{47}^{108}$Ag	47	108	47	61	47
(i)	$_{52}^{118}$Te	52	118	52	66	52
(j)	$_{58}^{136}$Ce	58	136	58	78	58

5.22 (a) gamma   (b) beta   (c) gamma   (d) alpha

5.25 151.96

5.26 204.4

5.27 72.7

5.28 (a) 2   (b) 6   (c) 3   (d) 8   (e) 10   (f) 27   (g) 4   (h) 2   (i) 1   (j) 5

5.30 (a) 5th energy level   (b) 3s sublevel   (c) 3s sublevel

5.31 (a) $1s^2 2s^2$ (b) $1s^2 2s^2 2p^3$ (c) $1s^2 2s^2 2p^6$ (d) $1s^2 2s^2 2p^6 3s^2 3p^4$
(e) $1s^2 2s^2 2p^6 3s^2 3p^6 3d^{10} 4s^2 4p^1$ (f) $1s^2 2s^2 2p^6 3s^2 3p^6 3d^{10} 4s^2 4p^5$
(g) $1s^2 2s^2 2p^6 3s^2 3p^6 3d^{10} 4s^2 4p^6 5s^1$
(h) $1s^2 2s^2 2p^6 3s^2 3p^6 3d^{10} 4s^2 4p^6 4d^{10} 5s^2 5p^3$
(i) $1s^2 2s^2 2p^6 3s^2 3p^6 3d^{10} 4s^2 4p^6 4d^{10} 5s^2 5p^6$
(j) $1s^2 2s^2 2p^6 3s^2 3p^6 3d^{10} 4s^2 4p^6 4d^{10} 5s^2 5p^6 6s^2$

5.32 (a) $2s^1$ (b) $3s^2$ (c) $3s^2 3p^1$ (d) $4s^2 4p^4$ (e) $5s^2 5p^5$

(f) $6s^1$ (g) $5s^25p^2$ (h) $5s^25p^1$ (i) $6s^26p^2$ (j) $7s^1$

5.33 (a) B   (b) Na   (c) S   (d) K   (e) Al   (f) Se   (g) Sr
(h) Xe

5.34 (a) $\ddot{\text{C}}\colon$ (b) Na·   (c) ·$\ddot{\text{A}}\text{s}\colon$ (d) $\colon\!\dot{\ddot{\text{I}}}\colon$ (e) $\colon\!\ddot{\text{A}}\text{r}\colon$ (f) $\colon\!\ddot{\text{S}}\colon$ (g) $\dot{\ddot{\text{G}}}\text{a}$·
(h) $\colon\!\ddot{\text{S}}\text{e}\colon$ (i) $\colon\!\ddot{\text{B}}\text{r}\colon$ (j) $\colon\!\dot{\text{I}}\text{n}$

5.35 (a) $\ddot{\text{A}}\text{l}$·   (b) $\colon\!\ddot{\text{O}}\colon$ (c) Mg$\colon$ (d) $\colon\!\text{Pb}\colon$ (e) $\colon\!\ddot{\text{X}}\text{e}\colon$ (f) K·
(g) $\colon\!\dot{\text{S}}\text{i}$·   (h) Ra$\colon$ (i) Rb·   (j) $\colon\!\ddot{\text{B}}\text{r}\colon$

5.36 (a) Na   (b) Br, I, At   (c) Y, Zr, Nb, Mo, Tc, Ru, Rh, Pd
(d) Be, Mg, Ca   (e) P, S, Cl, Ar   (f) H, He, Li, Be

5.37 (a) $1s^22s^22p^6$ (b) $1s^22s^22p^6$ (c) $1s^22s^22p^6$

CHAPTER 6

6.6 (a) F   (b) Al   (c) Mg   (d) Be   (e) Ga

6.11 Hydrogen's melting point, boiling point, density, and ionization energy are not consistent with those of the alkali metals.

6.12 Low density, melting point, and boiling point

6.14 Properties of the elements are periodic functions of their atomic number.

6.15 (a) Oxygen or chalcogens   (b) Noble gases   (c) Alkaline earth metals   (d) Alkali metals   (e) Transition metals
(f) Nitrogen   (g) Noble gases   (h) Boron-aluminum
(i) Carbon   (j) Transition metals

6.16 All A group elements are representative elements and B group elements are transition elements.

6.17 (a) s^1 (b) s^2p^5 (c) s^2p^4 (d) s^2p^6 (e) s^2 (f) s^2p^3

6.19 Metal: (a), (c), (h), (i)   Nonmetal: (d), (e), (f), (g)   Metalloid: (b), (j)

6.20 (a) Nonmetal   (b) Metal

6.21 (a) Be   (b) Sn   (c) Sb   (d) Mg

6.22 (a) (1) 3   (2) 2   (3) 1   (4) 2   (5) 3   (6) 1
(b) (1) 3−   (2) 2−   (3) 1−   (4) 2−   (5) 3−   (6) 1−

6.23 (a) (1) 2   (2) 1   (3) 3   (4) 3   (5) 2   (6) 4
(b) (1) 2+   (2) 1+   (3) 3+   (4) 3+   (5) 2+   (6) 4+

6.24 (a) 1+   (b) 2−   (c) 1−   (d) 2−   (e) 2+   (f) 2−
(g) 1+   (h) 3−   (i) 1−   (j) 1−

6.25 (a) $\text{N} + 3e^- \longrightarrow \text{N}^{3-}$ (b) $\text{Cl} + 1e^- \longrightarrow \text{Cl}^-$
(c) $\text{Mg} \longrightarrow \text{Mg}^{2+} + 2e^-$ (d) $\text{Li} \longrightarrow \text{Li}^+ + 1e^-$
(e) $\text{P} + 3e^- \longrightarrow \text{P}^{3-}$ (f) $\text{Ca} \longrightarrow \text{Ca}^{2+} + 2e^-$

6.26 Sb, 209

6.27 (a) Sn (b) As (c) Rb (d) Cl

6.28 (a) C (b) Kr (c) Ca (d) Ne

6.29 (a) He (b) Po (c) Tl (d) Cs (e) La (f) P

6.30 Element 118 would be in the noble gas group.
(a) Nonmetal (b) Gas (c) Poor conductor of electricity (d) Colorless (e) Large (f) Small

6.31 (a) He (b) Fr (c) F (d) I (e) Te (f) Po

6.32 Na releases $1e^-$; Cl gains $1e^-$.

6.33 NaOH, H_2, and energy

6.35 Metallic

6.38 Density increases to Al, then decreases.

6.39 (a) Ionization energy increases as atomic number increases. (b) Third row ionization energies are larger.

6.40 Nuclear charge of 2+, one full energy level of electrons.

6.41 Properties similar to other elements in group IIA.

6.42 Group IA; the second electron would be in a full energy level.

CHAPTER 7

7.4 6.022×10^{14} billion

7.5 (a) 2.0 mol He (b) 0.0100 mol K (c) 1.00×10^1 mol As (d) 0.0500 mol I

7.6 (a) 3.00×10^2 g Tc (b) 5.8 g Y (c) 1.7×10^4 g Te (d) 5.20×10^3 g Li

7.7 (a) 1.51×10^{23} atoms He (b) 2.98×10^{22} atoms Ne (c) 1.5×10^{22} atoms Ar (d) 7.2×10^{21} atoms Kr

7.8 (a) 6.6×10^{-1} mol Al (b) 2.91×10^{-2} mol P (c) 1.2×10^2 mol Zn (d) 1.00×10^{-1} mol Xe

7.9 (a) 103 g (b) 76.2 g (c) 98.1 g

7.10 (a) 5×10^2 g $AlCl_3$ (b) 1.8×10^1 g PH^3 (c) 0.31 g IBr

7.11 (a) 3.54×10^{22} molecules NH_3 (b) 4.12×10^{21} molecules SF_6 (c) 1.88×10^{21} molecules C_3Cl_8

7.12 (a) CO_2

7.15 (a) 19% Be, 81% F (b) 32.9% K, 67.1% Br (c) 1.00% H, 35.3% Cl, 63.7% O (d) 54.1% Ca, 43.2% O, 2.7% H

7.16 (a) CH_4 (b) SO_2

7.17 TiO_2

7.18 C_6H_{12}

7.19 B_2H_6

7.22 Atomic mass

7.23 (a) 14.01 g (b) 63.55 g (c) 35.45 g (d) 69.72 g (e) 137.3 g (f) 197.0 g

7.24 2×10^4 centuries

7.25

	Element	No. moles	No. atoms	Mass, g
(a)	He	1.00	6.02×10^{23}	4.00
(b)	Be	0.699	4.21×10^{23}	6.30
(c)	B	0.501	3.02×10^{23}	5.41
(d)	Ne	1.00	6.02×10^{23}	20.2
(e)	V	0.10	6.0×10^{22}	5.1
(f)	Rb	10.0	6.02×10^{24}	855.
(g)	Zn	1.00	6.02×10^{23}	65.4
(h)	Cd	1.000	6.022×10^{23}	112.4
(i)	Pd	7.000	4.215×10^{24}	744.8
(j)	Sr	0.0199	1.20×10^{22}	1.75

7.26 (a) 0.001 mol Li   (b) 0.14 mol Li   (c) 10.0 mol Li
(d) 0.01441 mol Li

7.27 (a) 0.10 mol Sc   (b) 0.1002 mol Si   (c) 0.100 mol Se
(d) 99 mol Sr

7.28 (a) 6.0×10^{22} atoms Sc   (b) 6.034×10^{22} atoms Si
(c) 6.02×10^{22} atoms Se   (d) 6.0×10^{25} atoms Sr

7.29 (a) 12.4 g He   (b) 20.0 g Ar   (c) 55 g K   (d) 0.497 g Cr

7.30 (a) 8.28×10^{23} atoms C   (b) 4.9×10^{20} atoms Al
(c) 7.155×10^{22} atoms Zn   (d) 3.31×10^{20} atoms U

7.31 (a) 0.2 mol Ne   (b) 1.2 mol Ag   (c) 3.235 mol Pb
(d) 0.0043 mol Hg

7.32 (a) 2.00 g Xe   (b) 7.6×10^3 g Mn   (c) 0.122 g Ba
(d) 2.805×10^4 g Sn

7.33 (a) 19 g Na   (b) 3.8×10^{-12} g Na   (c) 1.909×10^{-19} g Na
(d) 4×10^{-23} g Na

7.34 (a) 0.096 mol Si   (b) 0.689 mol V   (c) 0.083 g Kr
(d) 1.5×10^{18} atoms Ni   (e) 23.2 g Cs

7.35 (a) 3.70×10^{-4} mol Cd   (b) 7.537×10^{24} atoms Br
(c) 2×10^{-24} mol Os   (d) 223 mg Zr   (e) 1.9×10^1 kg Pt

7.36 Highest to lowest: (d) 1.1 g He   (b) 1.0 g H   (c) 0.80 g He
(a) 0.75 g H

7.37 Largest to smallest: (d) 3.9×10^{23} atoms Ne
(c) 3.0×10^{23} atoms H   (a) 3.7×10^{22} atoms Pu
(b) 2.9×10^{22} atoms Ra

7.38 (a) 28   (b) 46   (c) 254   (d) 16   (e) 102

7.39 (a) 220   (b) 395   (c) 176   (d) 245   (e) 108   (f) 90
(g) 154   (h) 342

7.40

	Compound	No. moles	No. molecules	Mass, g
(a)	H_2S	1.000	6.022×10^{23}	34.06
(b)	BrCl	0.50	3.0×10^{23}	58
(c)	NH_3	0.175	1.05×10^{23}	2.98
(d)	H_2SO_4	0.100	6.02×10^{22}	9.81

	Compound	No. moles	No. molecules	Mass, g
(e)	N_2O_4	0.186	1.12×10^{23}	17.1
(f)	AlF_3	1.00	6.02×10^{23}	84.0
(g)	PBr_5	0.100	6.02×10^{22}	43.1
(h)	OCl_2	9.3	5.6×10^{24}	8.1×10^2
(i)	SO_3	0.500	3.01×10^{23}	40.1
(j)	C_2H_6	2.14	1.29×10^{24}	64.2

7.41 (a) 12 g ClO_2 (b) 4.1×10^2 g HBr (c) 174.4 g SeO_2
(d) 8×10^2 g H_2SO_4

7.42 (a) 1.1×10^{23} molecules H_2O_2 (b) 2.8×10^{20} molecules ClF_3
(c) 2.35×10^{23} molecules N_2H_4 (d) 5.7×10^{22} molecules H_3PO_3

7.43 (a) 9 mol O (b) 2.2 mol O (c) 0.023 mol O
(d) 0.90 mol O

7.44 (a) 1.11 g H (b) 0.998 g H (c) 18.0 g H (d) 1.15 g H

7.45 (a) 1×10^1 g PH_3 (b) 6.7 g H_2Se (c) 5.60×10^3 g $HClO_3$
(d) 5.845×10^{-7} g UF_6

7.46 (a) 0.0201 mol O (b) 55 g HBr (c) 1.52 mol H
(d) 9.2×10^{15} molecules H_3PO_4

7.47 Largest to smallest: (c) 1.2×10^{23} atoms P
(d) 3.0×10^{22} atom P (b) 4.2×10^{21} atoms P
(a) 2.2×10^{20} atoms P

7.48 (a) 58.5 (b) 111.1 (c) 187.5 (d) 115.8

7.49 (a) 0.19 mol $NaNO_3$ (b) 3.223×10^{-4} mol $Ca_3(PO_4)_2$
(c) 5.28 mol $KClO_3$

7.50 (a) 8.7×10^2 g $CuBr_2$ (b) 2.4×10^2 g SnF_2 (c) 0.274 g $PbSO_4$

7.51 (a) 9.42×10^{23} formula units (b) 4.2×10^{24} formula units
(c) 4.0×10^{24} formula units (d) 9.822×10^{21} formula units

7.52 (a) 5.0% H, 95.0% F (b) 60.3% Mg, 39.7% O (c) 55.7% Hg,
44.3% Br (d) 60.1% Si, 39.9% N (e) 86.6% Pb, 13.4% O

7.53 (a) 54.2% Ba, 20.5% Cr, 25.3% O (b) 27.4% Na, 1.2% H,
14.3% C, 57.1% O (c) 41.4% Sr, 13.2% N, 45.4% O
(d) 34.4% Ni, 28.1% C, 37.5% N (e) 11.1% N, 3.2% H,
41.3% Cr, 44.4% O

7.54 (a) 45.95% Ag (b) 63.51% Ag (c) 64.65% Ag
(d) 66.79% Ag

7.55 (a) 48.2% K (b) 31.9% K (c) 19.7% K (d) 30.5% K

7.56 (a) 36.1% H_2O (b) 14.7% H_2O (c) 33.7% H_2O

7.57 Highest to lowest: (c) 72.3% Fe (b) 62.1% Fe (d) 36.7% Fe
(a) 18.9% Fe

7.58 (a) FeS_2 (b) NO (c) CH_3 (d) SiI_4 (e) NCl_3

7.59 (a) KCN (b) PbC_2O_4 (c) Cu_2SF_6 (d) CH_2Br_2

7.60 CrO_3

7.61 Ca_3P_2

7.62 CH_4N

7.63 $C_2F_3Cl_3$

7.64 $(NH_4)_3PO_4$

7.65 $C_{10}H_{20}$

7.66 Al_2Cl_6

7.67 HBrO

7.68 $C_3H_6O_3$

7.69 3×10^{26} atoms

7.70 93.7%

7.71 $(MgO)_2(SiO_2)_3 \cdot 2H_2O$

7.72 2.3×10^{21} atoms Au

CHAPTER 8

8.1 (a) Metal and nonmetal   (b) Nonmetal and nonmetal

8.2 (a) Lithium fluoride   (b) Sodium sulfide   (c) Magnesium oxide   (d) Potassium phosphide   (e) Calcium nitride

8.4 (a) Sulfur difluoride   (b) Nitrogen tribromide   (c) Nitrogen monoxide   (d) Bromine trichloride   (e) Tetraphosphorus decoxide

8.11 Covalent.

8.12 A bond involving transfer of electrons.

8.14 (a) Na^+, F^- (b) Rb^+, Br^-

8.16 (a) $1s^2 2s^2 2p^6$ (b) $1s^2$ (c) $1s^2 2s^2 2p^6 3s^2 3p^6$ (d) $1s^2 2s^2 2p^6 3s^2 3p^6$

8.17 (a) Sr^{2+} $\left[:\overset{..}{\underset{..}{S}}: \right]^{2-}$ (b) Mg^{2+} $2\left[:\overset{..}{\underset{..}{Br}}: \right]^{-}$ (c) $2Na^+$ $\left[:\overset{..}{\underset{..}{O}}: \right]^{2-}$

8.18 (a) AlP   (b) CsCl   (c) Ca_3N_2 (d) RbF

8.19 Electrons are shared in covalent bonds, transferred in ionic bonds.

8.21 (a) P > As > Sb   (b) B > Be > Li   (c) Sr > Rb > Ba > Cs

8.22 (a) 2   (b) 4   (c) 6

8.23 (a) H:H   (b) $:\overset{..}{O}::\overset{..}{O}:$ (c) $:\overset{..}{\underset{..}{F}}:\overset{..}{\underset{..}{F}}:$ (d) $:N⋮⋮N:$

(e) $:\overset{..}{\underset{..}{Cl}}:\overset{..}{\underset{..}{Cl}}:$

8.25 (a) $^{\delta+}H:\overset{..}{\underset{..}{I}}:^{\delta-}$ (b) $^{\delta+}:\overset{..}{\underset{..}{I}}:\overset{..}{\underset{..}{F}}:^{\delta-}$

8.27 (a) $:\overset{..}{\underset{..}{F}}:\overset{..}{\underset{..}{O}}:\overset{..}{\underset{..}{F}}:$ (b) $:\overset{..}{\underset{..}{I}}:\overset{..}{\underset{..}{N}}:\overset{..}{\underset{..}{I}}:$ (c) $H:\overset{..}{As}:H$ (d) $H:\overset{\displaystyle H}{\underset{\displaystyle H}{Si}}:H$

$:\overset{..}{\underset{..}{I}}:$ H

(e) $H:\overset{..}{Te}:H$

8.28 $:C⋮⋮O:$

8.29 (a) $:\overset{..}{\underset{..}{O}} - \overset{:\overset{..}{O}:}{S} = \overset{..}{\underset{..}{O}}$ $:\overset{..}{\underset{..}{O}} - \overset{\overset{\displaystyle :O:}{\|}}{S} - \overset{..}{\underset{..}{O}}:$ $\overset{..}{\underset{..}{O}} = \overset{:\overset{..}{O}:}{S} - \overset{..}{\underset{..}{O}}:$

8.30 (a) $\left[:\!\ddot{\text{O}}\!:\!\text{H}\right]^-$ (b) $\left[:\!\ddot{\text{O}}\!:\!\overset{\displaystyle :\ddot{\text{O}}:}{\underset{\displaystyle :\ddot{\text{O}}:}{\text{P}}}\!:\!\ddot{\text{O}}:\right]^{3-}$

8.31 (a) $:\!\ddot{\text{F}}\!:\!\text{Be}\!:\!\ddot{\text{F}}:$ (c) $:\!\ddot{\text{F}}\!:\!\overset{\displaystyle :\ddot{\text{F}}:}{\text{I}}\!:\!\ddot{\text{F}}:$

8.37 (a) Lithium bromide   (b) Magnesium sulfide   (c) Cesium bromide   (d) Calcium nitride   (e) Barium chloride   (f) Lead sulfide   (g) Silver fluoride   (h) Aluminum phosphide

8.39 (a) 4   (b) 6   (c) 2   (d) 8   (e) 3

8.40 (a) Dinitrogen oxide   (b) Phosphorus pentafluoride (c) Dinitrogen tetroxide   (d) Xenon difluoride   (e) Disulfur dichloride   (f) Arsenic pentafluoride   (g) Iodine trifluoride (h) Selenium tetrabromide

8.41 (a) Diiodine tetroxide   (b) Potassium sulfide   (c) Dichlorine heptoxide   (d) Tetraiodine ennoxide (nonoxide)   (e) Zinc sulfide   (f) Calcium chloride   (g) Ditellurium decafluoride (h) Selenium hexafluoride

8.42 (a) Ionic   (b) Ionic   (c) Ionic

8.43 Ions are free to move and thus conduct electricity when a salt is molten but are rigidly held in the crystal in the solid state.

8.44 N atoms need three electrons to achieve the noble gas structure.

8.47 Ne, Ar, and Xe have full energy levels.

8.48 (a) Xe   (b) Ar   (c) Ar   (d) Xe   (e) Kr   (f) He (g) Ne   (h) Xe

8.49 (a), (b) $1s^2 2s^2 2p^6$ (c), (d) $1s^2 2s^2 2p^6 3s^2 3p^6$ (e), (f) $1s^2 2s^2 2p^6 3s^2 3p^6 3d^{10} 4s^2 4p^6$

8.50 (a) Se^{2-} (b) N^{3-} (c) Na^+ (d) K^+

8.51 (a) $2\text{Na}^+ \left[:\!\ddot{\text{S}}\!:\right]^{2-}$ (b) $\text{Mg}^{2+} \, 2\left[:\!\ddot{\text{F}}\!:\right]^-$ (c) $2\text{Al}^{3+} \, 3\left[:\!\ddot{\text{O}}\!:\right]^{2-}$

8.52 (a) $\text{K}^+ \left[:\!\ddot{\text{Cl}}\!:\right]^-$ (b) $\text{Ca}^{2+} \, 2\left[:\!\ddot{\text{Cl}}\!:\right]^-$ (c) $\text{Sr}^{2+} \left[:\!\ddot{\text{S}}\!:\right]^{2-}$ (d) $2\text{Rb}^+ \left[:\!\ddot{\text{O}}\!:\right]^{2-}$ (e) $3\text{Mg}^{2+} \, 2\left[:\!\ddot{\text{N}}\!:\right]^{3-}$

8.53 (a) AlN   (b) CaS   (c) MgI_2 (d) Ga_2O_3 (e) KBr (f) Li_2O

8.57 (a) Br   (b) Sn   (c) Fr

8.60 (a) $1s$, $4p$ (b) $5p$, $2p$ (c) $4p$, $3p$

8.61 Polar: (b), (c), (e)   Nonpolar: (a), (d), (f)

8.62 $^{\delta+}:\!\ddot{\text{Cl}}\!:\!\ddot{\text{F}}\!:^{\delta-}$

8.63 (a) $:\!\ddot{\text{F}}\!-\!\overset{\displaystyle :\ddot{\text{F}}:}{\underset{\displaystyle :\ddot{\text{F}}:}{\text{C}}}\!-\!\ddot{\text{F}}:$ (b) $:\!\ddot{\text{Cl}}\!-\!\ddot{\text{O}}\!-\!\ddot{\text{Cl}}:$ (c) $\text{H}\!-\!\ddot{\text{Se}}\!-\!\text{H}$

(d) $:\ddot{B}r-\overset{\displaystyle :\ddot{B}r:}{\underset{\displaystyle }{P}}-\ddot{B}r:$ (e) $:\ddot{F}-\ddot{S}-\ddot{F}:$ (f) $:\ddot{C}l-\overset{..}{N}-\overset{..}{N}-\ddot{C}l:$ with $:\ddot{C}l:$ $:\ddot{C}l:$ below

(g) $\overset{\displaystyle H}{\underset{\displaystyle H}{C}}=\overset{\displaystyle :\ddot{B}r:}{\underset{\displaystyle :\ddot{B}r:}{C}}$ (h) $:\ddot{C}l-\overset{\displaystyle :\ddot{C}l:}{\underset{\displaystyle :\ddot{C}l:}{S}i}-\overset{\displaystyle :\ddot{C}l:}{\underset{\displaystyle :\ddot{C}l:}{S}i}-\ddot{C}l:$ (i) $:\ddot{O}-\overset{\displaystyle H}{\underset{\displaystyle \ddot{H}}{O}}$

(j) $H-\ddot{O}-\overset{\displaystyle H}{\underset{\displaystyle :\ddot{O}:}{\overset{\displaystyle :O:}{S}i}}-\ddot{O}-H$ with H above top :O: and H below bottom :O:

8.64 (a) $\left[:\ddot{O}-H\right]^{-}$ (b) $\left[:C\equiv N:\right]^{-}$ (c) $\left[H-\overset{\displaystyle H}{\underset{\displaystyle N}{N}}-H\right]^{+}$

(d) $\left[H-\overset{\displaystyle H}{\underset{\displaystyle H}{B}}-H\right]^{-}$ (e) $\left[:\ddot{O}-\overset{\displaystyle :\ddot{O}:}{\underset{\displaystyle :O:}{C}l}-\ddot{O}:\right]^{-}$ (f) $\left[:\ddot{O}-\ddot{O}:\right]^{2-}$

8.65 (a) $H-\overset{\displaystyle H}{\underset{\displaystyle H}{C}}-\overset{\displaystyle H}{\underset{\displaystyle H}{C}}-H$ (b) $H-\overset{\displaystyle H}{C}=\overset{\displaystyle H}{C}-H$ (c) $H-C\equiv C-H$

8.66 (a) $H-\ddot{O}-\overset{\displaystyle }{\underset{\displaystyle :O:}{C}}-\ddot{O}-H$ (b) $H-\ddot{O}-\overset{\displaystyle :\ddot{O}:}{\underset{\displaystyle :O:}{S}}-\ddot{O}-H$

(c) $H-\ddot{O}-\overset{\displaystyle :\ddot{O}:}{\underset{\displaystyle :O:}{P}}-\ddot{O}-H$ with H below bottom :O: (d) $H-\ddot{O}-\overset{..}{N}=\overset{..}{\underset{\displaystyle :O:}{O}}$

8.67 (a) $H-\overset{\displaystyle H}{C}=\overset{\displaystyle H}{C}-H$ (b) $\ddot{O}=C=\ddot{O}$ (c) $:N\equiv N:$

(d) H—C≡C—H   (e) :C̈l—B̈—C̈l: with :C̈l: above B

8.68 (b) PF_5 (c) IF_5

8.69 (a) $[H—\ddot{O}—C=\ddot{O}]^- \longleftrightarrow [H—\ddot{O}—C—\ddot{O}:]^-$ with $:\ddot{O}:$ below C

(b) $[\ddot{O}=C—\ddot{O}:]^{2-} \longleftrightarrow [:\ddot{O}—C—\ddot{O}:]^{2-} \longleftrightarrow [:\ddot{O}—C=\ddot{O}]^{2-}$

(c) $[:\ddot{O}—N=\ddot{O}]^- \longleftrightarrow [\ddot{O}=N—\ddot{O}:]^-$

(d) $:\ddot{O}—\ddot{O}=\ddot{O} \longleftrightarrow \ddot{O}=\ddot{O}—\ddot{O}:$

(e) $:N=N=\ddot{O} \longleftrightarrow :N≡N—\ddot{O}: \longleftrightarrow :\ddot{N}—N≡O:$

8.70 (a) Pyramidal   (b) Angular   (c) Tetrahedral   (d) Trigonal planar   (e) Linear

8.71 Triple bonds have six electrons and therefore more energy.

8.72 (a) 3   (b) 1   (c) 2   (d) 1

8.73 Si has a larger radius than C.

8.74 (a) . . . are *not* always . . .   (b) . . . and its *bonding* electrons . . .   (c) . . . are *smaller* for single . . . (d) *Covalent* . . .   (e) . . . sulfur *trioxide* . . .   (f) . . . *formula units* . . .   (g) . . . is *shared by* fluorine . . .   (h) . . . atoms, or *atoms with the same electronegativity* . . .

8.75 (a) H—C—C—Ö—H   (b) H—C—Ö—C—H   (c) H—C—C—Ö—H

8.76 H—C—C—C—H (propane structure with H's)

8.77 $\ddot{P}=\ddot{P}$ / $P=P$; $\ddot{P}—\ddot{P}$ / $P—P$; tetrahedral P_4 :P with P's

8.78 (a) $:N≡N:$ (b) $H:Ge:H$ with H above and below   (c) H—C≡C—C=Ö with H above

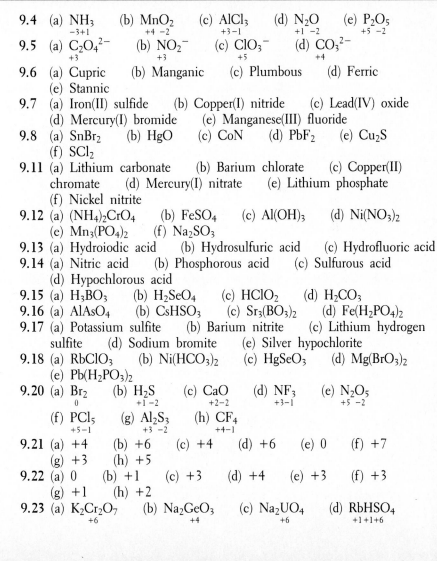

CHAPTER 9

9.4 (a) NH_3 (b) MnO_2 (c) $AlCl_3$ (d) N_2O (e) P_2O_5
       $-3 +1$ $+4\ -2$ $+3 -1$ $+1\ -2$ $+5\ -2$

9.5 (a) $C_2O_4{}^{2-}$ (b) $NO_2{}^-$ (c) $ClO_3{}^-$ (d) $CO_3{}^{2-}$
       $+3$ $+3$ $+5$ $+4$

9.6 (a) Cupric   (b) Manganic   (c) Plumbous   (d) Ferric
(e) Stannic

9.7 (a) Iron(II) sulfide   (b) Copper(I) nitride   (c) Lead(IV) oxide
(d) Mercury(I) bromide   (e) Manganese(III) fluoride

9.8 (a) $SnBr_2$ (b) HgO (c) CoN (d) PbF_2 (e) Cu_2S
(f) SCl_2

9.11 (a) Lithium carbonate   (b) Barium chlorate   (c) Copper(II)
chromate   (d) Mercury(I) nitrate   (e) Lithium phosphate
(f) Nickel nitrite

9.12 (a) $(NH_4)_2CrO_4$ (b) $FeSO_4$ (c) $Al(OH)_3$ (d) $Ni(NO_3)_2$
(e) $Mn_3(PO_4)_2$ (f) Na_2SO_3

9.13 (a) Hydroiodic acid   (b) Hydrosulfuric acid   (c) Hydrofluoric acid

9.14 (a) Nitric acid   (b) Phosphorous acid   (c) Sulfurous acid
(d) Hypochlorous acid

9.15 (a) H_3BO_3 (b) H_2SeO_4 (c) $HClO_2$ (d) H_2CO_3

9.16 (a) $AlAsO_4$ (b) $CsHSO_3$ (c) $Sr_3(BO_3)_2$ (d) $Fe(H_2PO_4)_2$

9.17 (a) Potassium sulfite   (b) Barium nitrite   (c) Lithium hydrogen
sulfite   (d) Sodium bromite   (e) Silver hypochlorite

9.18 (a) $RbClO_3$ (b) $Ni(HCO_3)_2$ (c) $HgSeO_3$ (d) $Mg(BrO_3)_2$
(e) $Pb(H_2PO_3)_2$

9.20 (a) Br_2 (b) H_2S (c) CaO (d) NF_3 (e) N_2O_5
       0 $+1\ -2$ $+2 -2$ $+3 -1$ $+5\ -2$
(f) PCl_5 (g) Al_2S_3 (h) CF_4
    $+5 -1$ $+3\ -2$ $+4 -1$

9.21 (a) $+4$ (b) $+6$ (c) $+4$ (d) $+6$ (e) 0 (f) $+7$
(g) $+3$ (h) $+5$

9.22 (a) 0 (b) $+1$ (c) $+3$ (d) $+4$ (e) $+3$ (f) $+3$
(g) $+1$ (h) $+2$

9.23 (a) $K_2Cr_2O_7$ (b) Na_2GeO_3 (c) Na_2UO_4 (d) $RbHSO_4$
     $+6$ $+4$ $+6$ $+1 +1 +6$

(e) $Ca(HS)_2$ (f) $K_4V_2O_7$ (g) $Na_2C_2O_4$ (h) $Cu(CN)_2$
    $+2+1\ -2$ $+5$ $+3$ $+2\ +2\ -3$

(i) Na_2O_2 (j) MgH_2
   $+1\ -1$ $+2\ -1$

9.24 (a) Ce   (b) Sb   (c) Cd   (d) Cr   (e) Co   (f) Cu
      $+3$ $+5$ $+2$ $+3$ $+2$ $+1$

(g) Mn   (h) Hg
    $+2$ $+1$

9.25 (a) Carbon dioxide   (b) Dinitrogen oxide   (c) Carbon tetrachloride   (d) Phosphorus tribromide   (e) Oxygen difluoride   (f) Silicon dioxide

9.26 (a) Tin(II) bromide   (b) Cobalt(III) nitride   (c) Lead(II) sulfide   (d) Copper(I) phosphide   (e) Mercury(II) oxide (f) Lithium selenide   (g) Strontium chloride   (h) Scandium oxide

9.27 (a) FeI_2 (b) Mn_2O_3 (c) $CuBr_2$ (d) CrF_3 (e) MgS
(f) Ca_2C (g) Hg_2Cl_2 (h) Sn_3N_4

9.28 (a) Tl_2O (b) UO_2 (c) Au_2O_3 (d) MoO_2 (e) Mn_2O_3
(f) RuO_4 (g) W_2O_7 (h) V_2O_5

9.29 Cations: ammonium, calcium, iron(III), mercury(II), aluminum, tin(II), cesium

Anions: acetate, phosphate, permanganate, cyanide, carbonate, chromate

Compound names have the cation followed by the anion, e.g., ammonium acetate, ammonium phosphate

9.30

$Sr(OH)_2$	$Sr_3(PO_3)_2$	$SrSO_4$	$Sr(ClO_3)_2$	$SrSeO_4$
KOH	K_3PO_3	K_2SO_4	$KClO_3$	K_2SeO_4
$Pb(OH)_2$	$Pb_3(PO_3)_2$	$PbSO_4$	$Pb(ClO_3)_2$	$PbSeO_4$
$Co(OH)_3$	$CoPO_3$	$Co_2(SO_4)_3$	$Co(ClO_3)_3$	$Co_2(SeO_4)_3$
NH_4OH	$(NH_4)_3PO_3$	$(NH_4)_2SO_4$	NH_4ClO_3	$(NH_4)_2SeO_4$
$Ga(OH)_3$	$GaPO_3$	$Ga_2(SO_4)_3$	$Ga(ClO_3)_3$	$Ga_2(SeO_4)_3$

9.31 (a) $Al_2(SO_4)_3$ (b) Calcium carbonate
(c) $NaHCO_3$ (d) Copper(II) sulfate
(e) $SrSO_4$ (f) Lead(II) chromate
(g) $Au(CN)_3$ (h) Sodium sulfate
(i) Zn_2SiO_4 (j) Barium nitrate

9.32 (a) $(NH_4)_2Se$ (b) $(NH_4)_2SO_3$ (c) NH_4NO_3 (d) NH_4IO_3
(e) NH_4ClO_4 (f) NH_4HCO_3 (g) NH_4BrO_3 (h) NH_4ClO

9.33 (a) $Ba(C_2H_3O_2)_2$ (b) $Cd(C_2H_3O_2)_2$ (c) $Fe(C_2H_3O_2)_3$
(d) $CuC_2H_3O_2$ (e) $Ga(C_2H_3O_2)_3$ (f) $In(C_2H_3O_2)_3$
(g) $Pb(C_2H_3O_2)_2$ (h) $Pt(C_2H_3O_2)_2$

9.34 (a) Hydrobromic acid   (b) Hydroiodic acid   (c) Hydroselenic acid

9.35 (a) Boric acid   (b) Hypochlorous acid   (c) Iodic acid
(d) Arsenic acid   (e) Carbonic acid   (f) Perbromic acid

9.36 Lactate ion

9.37 (a) Periodic acid   (b) Acetic acid   (c) Boric acid

(d) Hypoiodous acid   (e) Hydrocyanic acid

9.38 (a) Periodate   (b) Acetate   (c) Borate   (d) Hypoidite
(e) Cyanide

9.39 (a) Hydrogen sulfite   (b) Monohydrogen phosphite   (c) Hydrogen
sulfate   (d) Hydrogen carbonate

9.40 (a) Lower-oxidation-state oxyanion   (b) Higher-oxidation-state
oxyanion   (c) Binary compound   (d) Higher-oxidation-state cation
or acid   (e) Lower-oxidation-state cation or acid

9.41 (1) ammonium sulfide   (2) antimony triiodide   (3) arsenic
acid   (4) diarsenic trioxide   (5) barium chromate
(6) beryllium selenite   (7) bismuth(IV) chloride   (8) boron
nitride   (9) bromine dioxide   (10) cadmium bromate
(11) calcium hypochlorite   (12) disulfur decafluoride
(13) cerium(III) hydroxide   (14) cesium hydrogen carbonate
(15) dichlorine heptoxide   (16) chromium(III) sulfite
(17) cobalt(III) fluoride   (18) copper(II) selenate   (19) gallium
chloride   (20) germanium sulfide   (21) hydrogen cyanide
(22) hydrocyanic acid   (23) iodic acid   (24) iodine
pentafluoride   (25) iridium(III) sulfide   (26) iron(III) dihydrogen
phosphate   (27) lead hypoarsenite   (28) lithium chlorate
(29) magnesium nitrate   (30) manganese(III) arsenide
(31) mercury(II) bromate   (32) molybdenum(VI) sulfite
(33) trisilicon octachloride   (34) osmium(II) oxide
(35) palladium(II) nitrate   (36) phosphorus pentabromide
(37) platinum(II) hydroxide   (38) potassium dihydrogen arsenate
(39) radium carbonate   (40) rhenium(VI) fluoride
(41) rhodium(III) sulfate   (42) rubidium selenate   (43) scandium
oxide   (44) selenium tetrafluoride   (45) silver tellurite
(46) sodium dihydrogen phosphite   (47) strontium permanganate
(48) disulfur difluoride   (49) thallium(I) cyanide   (50) tin(II)
sulfate

9.42 (1) TiF_4 (2) WBr_5 (3) $ZnCrO_4$ (4) ZrO_2
(5) $(NH_4)_2SO_4$ (6) $CsOH$ (7) $CaSO_3$ (8) $Ba(BrO_3)_2$
(9) Na_2HPO_4 (10) BiI_3 (11) $Cu(ClO)_2$ (12) $Al(NO_2)_3$
(13) $FeAs$ (14) $Pb(ClO_2)_2$ (15) Li_2HPO_3 (16) $MnSO_4$
(17) Hg_2S (18) MoS_2 (19) $Ni(HCO_3)_2$ (20) N_2O_3
(21) $PdSi$ (22) $OsSO_3$ (23) $NaClO_4$ (24) H_3PO_2
(25) PN (26) KIO_2 (27) $ReCl_6$ (28) Sn_3P_2
(29) $RbClO_2$ (30) $Sc_2(SO_4)_3$ (31) Si_3N_4 (32) H_4SiO_4
(33) Ag_3PO_4 (34) $NaHSO_3$ (35) $Sr(IO_3)_2$ (36) TaN
(37) Tl_2SO_4 (38) SnI_4 (39) $Zn(CN)_2$ (40) $Co_3(PO_4)_2$
(41) NH_4BrO_3 (42) Al_4C_3 (43) $CaTe$ (44) $HClO_3$
(45) Fe_5B_3 (46) $MnCO_3$ (47) HgS (48) HCN
(49) $KMnO_4$ (50) $Ce_2(CO_3)_3$

CHAPTER 10

10.4 Solid carbon reacts with gaseous oxygen in the presence of heat to produce gaseous carbon dioxide.

10.5 (a) 1, 2 (b) 2, 3, 2 (c) 1, 6, 4 (d) 2, 3, 2, 2 (e) 5, 2, 1, 4

10.6 $Ca_3P_2 + 6H_2O \longrightarrow 3Ca(OH)_2 + 2PH_3$

10.7 $3H_2SO_4(aq) + 2Al(OH)_3(aq) \longrightarrow 6H_2O + Al_2(SO_4)_3$

10.12 (a) Decomposition (b), (d) Single replacement (c) Combination (e) Metathesis

10.13 (a), (c), (e) Combination (b), (h) Decomposition (d), (g) Metathesis (f) Single displacement

10.14 (a) $MgSO_4(s) + 7H_2O(g)$ (b) $Ag_2S(s) + NaNO_3(aq)$ (c) $P_4O_{10}(s)$ (d) NR (e) $KClO_3(aq) + H_2O(l)$

10.15 (a) $2Ca(s) + O_2(g) \longrightarrow 2CaO(s)$
(b) $2KClO_3(s) \longrightarrow 2KCl(s) + 3O_2(g)$
(c) $SO_2(g) + H_2O(l) \longrightarrow H_2SO_3(aq)$
(d) $2Cs(l) + 2H_2O(l) \longrightarrow 2CsOH(aq) + H_2(g)$
(e) $Pb(NO_3)_2(aq) + K_2SO_4(aq) \longrightarrow PbSO_4(s) \longrightarrow 2KNO_3(aq)$
(f) $H_2(g) + S(s) \longrightarrow H_2S(g)$

10.18 (a), (d) Exothermic (b), (c) Endothermic

10.22 Barium bromide + ammonium carbonate $\longrightarrow$ barium carbonate + ammonium bromide

10.23 (a) Two moles of gaseous sulfur trioxide, heated, produce two moles of gaseous sulfur dioxide and one mole of oxygen gas.
(b) One mole of liquid mercury and one mole of chlorine gas produce one mole of mercury(II) chloride.
(c) One mole of nitrogen gas and three moles of hydrogen gas in the presence of a catalyst produce two moles of ammonia gas.
(d) One mole of solid aluminum sulfide and six moles of liquid water produce two moles of solid aluminum hydroxide and three moles of aqueous hydrogen sulfide.

10.24 (a) 1, 5, 2 (b) 3, 1, 1 (c) 1, 1, 2 (d) 1, 2, 1, 2 (e) 2, 1, 2 (f) 1, 1, 1 (g) 1, 2, 2, 3 (h) 1, 3, 1, 3 (i) 1, 3, 2 (j) 2, 3, 2, 2

10.25 (a) 2, 13, 8, 10 (b) 1, 3, 1, 3 (c) 2, 2, 4, 1 (d) 3, 2, 1, 3, 3 (e) 2, 5, 2, 4 (f) 2, 3, 1, 3 (g) 4, 1, 1, 3 (h) 1, 3, 2, 3 (i) 1, 1, 1, 4 (j) 2, 31, 20, 22

10.26 (a) 2, 15, 3, 3, 6 (b) 3, 4, 3, 1 (c) 2, 2, 1, 1, 1, 4 (d) 3, 8, 9, 4 (e) 1, 1, 1, 2 (f) 2, 15, 14, 6 (g) 12, 1, 4, 6 (h) 2, 2, 2, 1, 4 (i) 1, 5, 1, 2, 3 (j) 1, 2, 1, 2

10.27 (a) $NaCl(aq) + AgNO_3(aq) \longrightarrow NaNO_3 + AgCl$
(b) $Al(OH)_3(aq) + 3HNO_3(aq) \longrightarrow Al(NO_3)_3 + 3H_2O$
(c) $Cl_2(g) + 2RbBr(aq) \longrightarrow 2RbCl + Br_2(l)$
(d) $2Fe(C_2H_3O_2)_3(aq) + 3Na_2S(aq) \longrightarrow 6NaC_2H_3O_2 + Fe_2S_3$

(e) $SiF_4(g) + 2H_2O \longrightarrow SiO_2(s) + 4HF(aq)$

(f) $MnO_2 + 4HCl \longrightarrow MnCl_2 + Cl_2(g) + 2H_2O$

(g) $N_2O_4(g) \overset{\Delta}{\longrightarrow} 2NO_2$

(h) $Ca_3P_2 + 6H_2O \longrightarrow 3Ca(OH)_2 + 2PH_3$

(i) $2AgNO_3 \overset{\Delta}{\longrightarrow} 2Ag + 2NO_2(g) + O_2(g)$

(j) $2Al + 3CuSO_4 \longrightarrow 3Cu + Al_2(SO_4)_3$

(k) $Hg_2(NO_3)_2 + 2KCl \longrightarrow Hg_2Cl_2 + 2KNO_3$

(l) $CaSO_4 \cdot 2H_2O \longrightarrow CaSO_4 + 2H_2O$

(m) $(NH_4)_2S + CdCl_2 \longrightarrow CdS + 2NH_4Cl$

(n) $H_3PO_3 + 3NaOH \longrightarrow Na_3PO_3 + 3H_2O$

(o) $2NH_3 + H_2SO_4 \longrightarrow (NH_4)_2SO_4$

(p) $CS_2 + 3Cl_2 \longrightarrow S_2Cl_2 + CCl_4$

(q) $Ca_3(PO_4)_2 + 3H_2SO_4 \longrightarrow 2H_3PO_4 + 3CaSO_4$

(r) $2NH_3 + 3Cu_2O \longrightarrow N_2(g) + 3H_2O + 6Cu$

(s) $NH_4NO_3 \longrightarrow 2H_2O + N_2O(g)$

(t) $Mg(OH)_2 + Zn(NO_3)_2 \longrightarrow Zn(OH)_2 + Mg(NO_3)_2$

Combination: (o)

Decomposition: (g), (i), (l), (s)

Single displacement: (c), (j)

Metathesis: (a), (b), (d), (e), (h), (k), (m), (n), (q), (t)

None of these: (f), (p), (r)

10.29 (a) $2C + O_2 \longrightarrow 2CO$

(b) $S + O_2 \longrightarrow SO_2$

(c) $N_2 + 2O_2 \longrightarrow 2NO_2$

(d) $4Cs + O_2 \longrightarrow 2Cs_2O$

(e) $P_4 + 6H_2 \longrightarrow 4PH_3$

(f) $Mg + H_2 \longrightarrow MgH_2$

10.30 (a) Na_2O (b) SrO (c) MgO (d) Al_2O_3

10.31 (a) CO_2 (b) SO_2 (c) P_4O_{10} (d) SO_3

10.32 (a) $MgCO_3$ (b) $KClO_3$ (c) $KHCO_3$ (d) $Ba(NO_2)_2 \cdot H_2O$

(e) H_2O (f) $NaNO_3$ (g) SO_3

10.33 (a) $2Na + 2H_2O \longrightarrow 2NaOH + H_2$

(b) $Zn + 2HCl \longrightarrow ZnCl_2 + H_2$

(c) $Zn + Cu(NO_3)_2 \longrightarrow Cu + Zn(NO_3)_2$

(d) $Br_2 + 2RbI \longrightarrow I_2 + RbBr$

10.34 Acid and base

10.35 (a) $Al(NO_3)_3 + 3KOH \longrightarrow Al(OH)_3(s) + 3KNO_3$

(b) $K_2CO_3 + 2HCl \longrightarrow 2KCl + CO_2(g) + H_2O$

(c) $ZnCl_2 + Na_2CO_3 \longrightarrow ZnCO_3(s) + 2NaCl$

(d) $NH_4OH + HI \longrightarrow NH_4I(aq) + H_2O$

10.36 Soluble: (a), (e), (g)   Insoluble: (b), (c), (d), (f), (h)

10.37 (a) $8Zn(s) + S_8(s) \longrightarrow 8ZnS(s)$

(b) $4Al(s) + 3O_2(g) \longrightarrow 2Al_2O_3(s)$

(c) $2Na(s) + F_2(s) \longrightarrow 2NaF(s)$

(d) $3H_2(g) + N_2(g) \longrightarrow 2NH_3(g)$

(e) $Br_2(g) + I_2(g) \longrightarrow 2IBr(g)$

(f) $2CO(g) + O_2(g) \longrightarrow 2CO_2(g)$

(g) $3Mg(s) + N_2(g) \longrightarrow Mg_3N_2(s)$

(h) $2Ag(s) + Cl_2(g) \longrightarrow 2AgCl(s)$

10.38 (a) $PtO_2 \longrightarrow Pt + O_2$

(b) $CuSO_4 \cdot 5H_2O \longrightarrow CuSO_4 + 5H_2O$

(c) $2NaNO_3 \longrightarrow 2NaNO_2 + O_2$

(d) $SrCO_3 \longrightarrow SrO + CO_2$

(e) $2LiHCO_3 \longrightarrow Li_2CO_3 + H_2O + CO_2$

(f) $2H_2O \longrightarrow 2H_2 + O_2$

(g) $MgSO_3 \cdot 6H_2O \longrightarrow MgSO_3 + 6H_2O$

(h) $2H_2O_2 \longrightarrow 2H_2O + O_2$

10.39 (a) $Zn + Pb(NO_3)_2 \longrightarrow Pb + Zn(NO_3)_2$

(b) $Ba + 2H_2O \longrightarrow H_2 + Ba(OH)_2$

(c) $Ni + SnBr_2 \longrightarrow Sn + NiBr_2$

(d) $Hg + Fe(NO_3)_2 \longrightarrow NR$

(e) $KCl + I_2 \longrightarrow NR$

(f) $Cu(ClO_3)_2 + Mn \longrightarrow Mn(ClO_3)_2 + Cu$

(g) $Pb + H_2SO_4 \longrightarrow H_2 + PbSO_4$

(h) $2HC_2H_3O_2 + Zn \longrightarrow H_2 + Zn(C_2H_3O_2)_2$

10.40 (a) $NiCl_2 + Ca(OH)_2 \longrightarrow CaCl_2 + Ni(OH)_2(s)$

(b) $Hg(C_2H_3O_2)_2 + K_2CO_3 \longrightarrow 2KC_2H_3O_2 + HgCO_3(s)$

(c) $H_3PO_4 + 3AgNO_3 \longrightarrow Ag_3PO_4(s) + 3HNO_3$

(d) $3H_2SO_3 + 2Al(OH)_3 \longrightarrow Al_2(SO_3)_3(s) + 6H_2O$

(e) $(NH_4)_2S + BaI_2 \longrightarrow BaS(s) + 2NH_4I$

(f) $Cs_2CO_3 + 2HC_2H_3O_2 \longrightarrow 2CsC_2H_3O_2 + CO_2(g) + H_2O(l)$

(g) $Li_2SO_4 + Co(NO_3)_2 \longrightarrow 2LiNO_3 + CoSO_4$

(h) $NH_4CN + HBr \longrightarrow NH_4Br + HCN(g)$

10.41 (a) $2AgNO_3 + CuCl_2 \longrightarrow 2AgCl(s) + Cu(NO_3)_2$

(b) $Fe + H_2O \longrightarrow NR$

(c) $2KOH + H_2SO_4 \longrightarrow K_2SO_4 + 2H_2O$

(d) $4NH_3 + 5O_2 \longrightarrow 4NO + 6H_2O$

(e) $SO_3 + H_2O \longrightarrow H_2SO_4$

(f) $H_2SO_4 + Zn \longrightarrow H_2(g) + ZnSO_4$

(g) $2NO + O_2 \longrightarrow 2NO_2$

(h) $4Al + 3O_2 \longrightarrow 2Al_2O_3$

(i) $H_2 + Sn(NO_3)_2 \longrightarrow NR$

(j) $Na_2SO_4 \cdot 10H_2O \longrightarrow Na_2SO_4 + 10H_2O$

(k) $NH_3 + HCl \longrightarrow NH_4Cl$

(l) $NaNO_2 + HCl \longrightarrow NaCl + HNO_2$

(m) $NaC_2H_3O_2 + Pb(C_2H_3O_2)_2 \longrightarrow NR$

(n) $Au_2S_3 + 2Fe \longrightarrow Fe_2S_3 + 2Au$

(o) $2As + 3Cl_2 + 2AsCl_3$

(p) $2Ca(NO_3)_2 \xrightarrow{\Delta} 2CaO + 4NO_2 + O_2$

(q) $Li_2O + H_2O \longrightarrow 2LiOH$

(r) $SnCl_4 + 2Mg \longrightarrow 2MgCl_2 + Sn$
(s) $2HNO_3 + CaO \longrightarrow Ca(NO_3)_2 + H_2O$
(t) $2B + 3F_2 \longrightarrow 2BF_3$
10.43 (a) 48 kJ   (b) 36°C
10.44 Endothermic: (a), (b)   Exothermic: (c), (d)
10.45 (a) . . . produce *basic* solutions . . .
(b) *Most* inorganic . . .
(c) . . . can be used . . .
(d) . . . insoluble except those of alkali metals and ammonium.
(e) . . . into liquid water *does not react.*
(f) . . . reaction *can be* a precipitate.
(g) . . . undergo *exothermic* reactions . . .
(h) . . . bonds are *formed*, heat energy . . .
10.48 (a) $N_2 + 3H_2 \longrightarrow 2NH_3$
(b) $4NH_3 + 5O_2 \longrightarrow 4NO + 6H_2O$
(c) $2NO + O_2 \longrightarrow 2NO_2$
(d) $2NO_2 + H_2O \longrightarrow HNO_3 + HNO_2$
10.49 (a) $CaCO_3 \longrightarrow CaO + CO_2$
(b) $CO_2 + NH_3 + H_2O + NaCl \longrightarrow NaHCO_3 + NH_4Cl$
(c) $2NaHCO_3 \longrightarrow Na_2CO_3 + CO_2 + H_2O$
(d) $CaO + 3C \longrightarrow CaC_2 + CO$
(e) $CaC_2 + N_2 \longrightarrow CaCN_2 + C$
(f) $CaCN_2 + Na_2CO_3 + C \longrightarrow 2NaCN + CaCO_3$
(g) $2NaCN + H_2SO_4 \longrightarrow 2HCN + Na_2SO_4$

CHAPTER 11

11.3 0.58 mol NH_3
11.4 33.7 mol O_2
11.5 6.0×10^3 mol H_2O
11.6 26.3 mol SO_2
11.7 (a) 14.9 g P_4 (b) 66.3 g PCl_3
11.8 572 g F_2
11.9 14 g Cl_2
11.10 7.10×10^3 kJ
11.13 (a) Neither   (b) 1 mol C   (c) 1.0 g O_2 (d) 11 g C
(e) 0.0750 g C
11.14 (b) 1 mol O_2 (c) 0.052 mol C   (d) 0.18 mol O_2
(e) 0.0008 mol C
11.15 322.3 g NaBr

11.17

	N$_2$	+	O$_2$	⟶	2NO
Molecules:	1 molecule		1 molecule		2 molecules
Molecules:	6.02×10^{23}		6.02×10^{23}		1.20×10^{24}
Moles:	1 mol		1 mol		2 mol
Mass, g:	28 g		32 g		60 g
Mass, g:	2.8 g		3.2 g		6.0 g

11.18

	C$_5$H$_{12}$	+	8O$_2$	⟶	5CO$_2$	+	6H$_2$O
Molecules:	1 molecule		8 molecules		5 molecules		6 molecules
Molecules:	6.02×10^{23}		4.82×10^{24}		3.01×10^{24}		3.61×10^{24}
Mole:	1 mole		8 moles		5 moles		6 moles
Mole:	0.1 mole		1 mole		0.6 mole		0.8 mole
Mass:	72 g		2.6×10^2 g		2.2×10^2 g		1.1×10^2 g

11.19 (a) 1 mol CaF$_2$ (b) 78 g CaF$_2$ (c) 40 g Ca + 38 g F$_2$ ⟶ 78 g CaF$_2$

11.20 1 mol Br$_2$:   160 g
1 mol Cl$_2$: $\underline{71\ g}$
2 mol BrCl: 231 g

11.21 (a) 2 mol AlCl$_3$ (b) 2 mol AlCl$_3$ (c) 10 mol AlCl$_3$
(d) 8 mol AlCl$_3$ (e) 3.3 mol AlCl$_3$

11.22 $\dfrac{1\ \text{mol C}_3\text{H}_8}{3\ \text{mol CO}_2}$, $\dfrac{1\ \text{mol C}_3\text{H}_8}{4\ \text{mol H}_2\text{O}}$, $\dfrac{5\ \text{mol O}_2}{3\ \text{mol CO}_2}$, $\dfrac{5\ \text{mol O}_2}{4\ \text{mol H}_2\text{O}}$

11.23 (a) 0.33 mol Cl$_2$ (b) 1.5 mol H$_2$O   (c) 24 mol HNO$_3$
(d) 3.7 mol Cl$_2$, 11 mol KNO$_3$ (e) 0.025 mol Cl$_2$, 0.025 mol NOCl, 0.050 mol H$_2$O, 0.075 mol KNO$_3$

11.24 (a) 1 mol Fe$_2$O$_3$, 4 mol SO$_2$ (b) 3 mol Fe$_2$O$_3$, 12 mol SO$_2$
(c) 0.31 mol Fe$_2$O$_3$, 1.2 mol SO$_2$ (d) 4 mol Fe$_2$O$_3$, 16 mol SO$_2$ (e) 1.6 mol Fe$_2$O$_3$, 6.5 mol SO$_2$

11.25 (a) 96 g O$_2$ (b) 0.0122 mol O$_2$ (c) 1.02 mol KCl, 1.52 mol O$_2$ (d) 7.7×10^3 g KClO$_3$

11.26 (a) 36.5 g HCl   (b) 0.86 mol CO$_2$ (c) 2.1×10^2 g NaCl
(d) 57 g NaHCO$_3$

11.27 (a) 2×10^2 g Cu   (b) 2.3×10^3 g Cu   (c) 11 g Cu
(d) 2252 g Cu   (e) 8.27×10^4 g Cu   (f) 3.269 g Cu

11.28 (a) 0.931 g Ag   (b) 0.216 g Ag   (c) 8.50 g Ag
(d) 2.4×10^3 g Ag

11.29 (a) 55 g Pb(NO$_3$)$_2$ (b) 1.68×10^4 g Pb(NO$_3$)$_2$
(c) 2.85×10^3 g Pb(NO$_3$)$_2$ (d) 0.8670 g Pb(NO$_3$)$_2$

11.30 (a) 0.563 g CH$_4$ (b) 544 g Cl$_2$ (c) 7.1 g CH$_4$
(d) 1.040×10^4 g CH$_4$, 1.844×10^5 g Cl$_2$

11.31 (a) 8.60 g B   (b) 12.0 g H$_2$ (c) 1.30×10^2 g BCl$_3$
(d) 9.0×10^1 g H$_2$, 3.3×10^2 g B, 3.2×10^3 g HCl

11.32 (a) 2.78 g XeF$_4$ (b) 12.7 g XeF$_4$ (c) 2.96×10^{-4} g H$_2$O
(d) 4.27 g XeOF$_4$, 5.02 g Xe, 3.07 g HF, 0.920 g O$_2$

11.33 (a) 7.54 g NiCl$_2$·6H$_2$O   (b) 31.7 g NiCl$_2$ (c) 976 g NiCl$_2$·6H$_2$O   (d) 407 g NiCl$_2$, 338 g H$_2$O

11.34 (a) $2\ KNO_3 \longrightarrow 2\ KNO_2 + O_2$
(b) 2 mol KNO_3; 2 mol KNO_2; 1 mol O_2 (c) 252.6 g KNO_2
(d) 0.2287 g O_2

11.35 (a) $SiO_2 + 2\ C + 2\ Cl_2 \longrightarrow SiCl_4 + 2CO$
(b) 450 g $SiCl_4$ (c) 4.04×10^{19} molecules CO

11.36 (a) $Sb_2S_3 + 3\ Fe \longrightarrow 2\ Sb + 3\ FeS$
(b) 38.0 g Fe   (c) 3.27×10^3 g Sb   (d) 1132 g Sb_2S_3, 557.0 g Fe

11.37 (a) $2\ H_2 + O_2 \longrightarrow 2\ H_2O + 569\ kJ$
(b) 4.04 g H_2, 32.0 g O_2 (c) 68.0 kcal   (d) 4.25 kcal

11.38 (a) 25.7 g NO   (b) 22.8 kJ   (c) 4.28×10^{-3} mol O_2
(d) 60.0 g NO

11.39 (a) 12.0 kcal   (b) 83.1 g C_2H_4 (c) 125 kcal   (d) 0.02 g C_2H_4

11.40 (a) 3.2×10^3 kcal   (b) 1.97×10^3 kcal   (c) 4.9×10^2 kcal
(d) 2×10^1 kJ

11.41 (a) $CaCO_3 + 178\ kJ \longrightarrow CaO + CO_2$
(b) $CaCO_3 + 42.5\ kcal \longrightarrow CaO + CO_2$

11.42 5.43×10^3

11.43 (a) 123 g HI   (b) 1.22×10^3 g HI

11.44 (a) 1.94 g HCl   (b) 11 g HCl

11.45 (a) 121.5 g Na_2SO_4 (b) 32.8 g Na_2SO_4

11.46 (a) $Pb(NO_3)_2(aq) + 2KI(aq) \longrightarrow PbI_2(s) + 2KNO_3(aq)$ (b) 3.48 g PbI_2 (c) 2.50 g $Pb(NO_3)_2$

11.47 (a) 61.3 g NaCN   (b) 17.5 g N_2

11.48 $C_6H_6 + Cl_2 \longrightarrow C_6H_5Cl + HCl$
(a) 18.8 g C_6H_5Cl (b) 60.3%

11.49 86.8%

11.50 (a) 588 g $Ca(OH)_2$;   (b) 2.62×10^5 g $Ca(OH)_2$

11.51 (a) $2ZnS + 3O_2 \longrightarrow 2ZnO + 2SO_2$
$ZnO + C \longrightarrow Zn + CO$
(b) 5.30×10^5 g Zn   (c) 13.4 g Zn

11.52 29.2 g I_2

CHAPTER 12

12.4 (a) 0.988 torr   (b) 0.513 atm   (c) 1.4×10^2 kPa
(d) 1.64 atm

12.7 0.29 L

12.8 59.8 mL

12.9 2.1×10^2 L

12.11 (a) 0.0100 mol   (b) 4.60 mol   (c) 0.0033 mol

12.12 (a) 45 L   (b) 45 L   (c) 45 L

12.13 0.257 g

12.14 0.237 L

12.15 62.4 (torr·L)/(mol·K)

12.16 4.4 L

12.17 272°C

12.18 0.0423 mol

12.19 (a) 26 L NO_2 (b) 318 L NO   (c) 44 L O_2

12.20 12 L O_2

12.21 0.232 L O_2

12.24 2.92 atm

12.25 721 mm Hg

12.37 (a) 1.35×10^2 kPa   (b) 1.01×10^3 torr   (c) 1.33 atm
(d) 1.01×10^3 mm Hg

12.38 (a) 615 torr   (b) 8.20×10^4 Pa   (c) 8.20×10^1 kPa

12.39 (a) 1214 torr   (b) 1.597 atm   (c) 1.618×10^5 Pa
(d) 1.618×10^2 kPa

12.40 (a) 3.3×10^{-4} atm   (b) 3.3×10^1 Pa   (c) 3.3×10^{-2} kPa

12.41 (a) 4.88×10^4 mm Hg   (b) 1.5 Pa   (c) 5.4 atm
(d) 3.2×10^{-2} torr   (e) 971 torr

12.42 1.5×10^3 torr

12.44 (a) 0.33 L   (b) 0.066 L   (c) 0.66 L   (d) 0.0022 L
(e) 0.38 L

12.45 (a) 10 L   (b) 2.5 atm   (c) Approaches zero

12.46 (a) 4.8 L   (b) 9.6 L   (c) 0.38 L   (d) 51 L   (e) 8.9 L

12.47 2.33×10^4 mL

12.48 10.4 L

12.49 (a) 92 L   (b) 7.0×10^4 L   (c) 9.3×10^6 L   (d) 9.3×10^3 L

12.50 (a) 0.6 atm   (b) 2 atm   (c) 6×10^1 atm   (d) 0.1 atm

12.51 2.6×10^3 balloons

12.53 (a) 17 L   (b) 20 L   (c) 18 L   (d) 11 L   (e) 40 L

12.54 (a) 340 K   (b) 200 K   (c) 25 L

12.56 (a) 2.45 L   (b) 9.8 L   (c) 9.4 L   (d) 3.4 L

12.57 5.307 L

12.58 (a) 328 K   (b) 197 K   (c) 498 K

12.59 (a) 0.0084 L   (b) 3.0×10^3 atm

12.60 (a) Decrease   (b) Increase   (c) Increase   (d) Increase

12.61 (a) 5.0 L   (b) 289 mL   (c) 2.38 L   (d) 337 L

12.62 (a) 1.1 L   (b) 1.09×10^3 mL   (c) 0.785 L
(d) 9.5×10^3 mL

12.63 (a) 1.17 atm   (b) 79.8 kPa   (c) 838 torr
(d) 1.24×10^{-4} atm

12.64 2.32×10^4 mm Hg

12.65 273°C

12.66 (a) 9.02 mol   (b) 22.0 mol   (c) 0.0029 mol
(d) 1.5×10^{-5} mol   (e) 0.042 mol

12.67 (a) 5.05 L   (b) 0.012 L   (c) 0.836 L   (d) 0.29 L
(e) 26.7 L

12.68 (a) 933 g H_2O (b) 0.87 g HBr   (c) 15.0 g C_2H_2 (d) 64 g
UF_6 (e) 0.14 g Ar

12.69 (a) 73.9 L H_2 (b) 96.0 L ClF (c) 7.98×10^{-3} L SO_3
12.70 (a) 383 L He (b) 17.8 L O_2 (c) 0.244 L Cl_2 (d) 1.30 L H_2S
12.71 (a) 1.25 g/L N_2 (b) 1.25 g/L CO (c) 3.74 g/L Kr
(d) 1.34 g/L C_2H_6 (e) 5.94 g/L $C_2F_2Cl_2$
12.72 (a) 0.644 g/L O_2 (b) 0.672 g/L C_2H_6O (c) 2.86 g/L CF_4
12.74 (a) 82.1 (mL·atm)/(mol·K) (b) 6.24×10^4 (mL·mm Hg)/(mol·K) (c) 8.31 (L·kPa)/(mol·K)
12.75 (a) 727 L (b) 30.2 L (c) 39.1 L
12.76 (a) 0.0579 mol (b) 4.11 mol (c) 0.00572 mol
12.77 (a) 0.217 atm (b) 9.02 atm (c) 1.74 atm
12.78 19.7 g/L
12.79 34 g/mol
12.80 132 g/mol
12.81 (a) 36.9 L H_2 (b) 2.0 L H_2 (c) 0.14 L H_2 (d) 180.0 L H_2
12.82 (a) 3.68 L SO_3 (b) 0.654 L SO_3 (c) 49.4 mL SO_3
(d) 1.05×10^3 L SO_3
12.83 (a) 1.3 L O_2 (b) 4.91 L O_2 (c) 114 L O_2 (d) 0.0840 L O_2
12.84 (a) 7.7 g PbS (b) 1.66×10^{-3} g PbS (c) 2.42×10^3 g PbS
(d) 2.21 g PbS
12.85 (a) 1.75 L (b) 64.5 L (c) 14.1 L (d) 0.0281 L
12.86 (a) 17.1 L NO (b) 0.864 L NO (c) 35.2 L NO (d) 615 L NO
12.87 13.1 L
12.88 337 mm Hg CO
12.89 0.65 atm He, 1.2 atm Ar
12.90 80.8 mL
12.91 0.713 L
12.92 (a) 0.0355 mol (b) 0.428 g Mg 1.28 g HCl
12.93 0.223 g $KClO_3$
12.106 7.0 atm
12.107 (a) 3.89 g/L (b) 3.50×10^3

CHAPTER 13

13.10 Halogen molecules are nonpolar; therefore, only dispersion forces are involved. These forces are stronger in molecules with larger numbers of electrons. Fluorine is a gas at 25°C and astatine is a solid.
13.16 (a) Liquid (b) Liquid to gas (c) The molar heat of fusion
(d) The heat capacity of the gas
13.18 (a) Aluminum (b) NaCl

13.29 (a), (b) Weak intermolecular forces   (c), (d) Particles in liquid phase can move about each other but are held at a constant average distance.   (e) Liquid particles are close together, but gas particles are many diameters apart.

13.30 (a) Density of liquids is greater.   (b) Fluidity of gases is greater. (c) Intermolecular forces of liquids are greater.   (d) Compressibility of gases is greater.   (e) Shapes of gases and liquids vary.

13.32 The liquid that quickly evaporates has a higher vapor pressure due to weaker intermolecular forces.

13.33 The heat from the skin is used to supply the heat of vaporization of the alcohol, evaporating the alcohol and reducing the body temperature.

13.34 The gas particles move away from the liquid. Therefore, the chance of condensation of gas back into that container of liquid is very small.

13.35 (a) B   (b) D   (c) F   (d) H

13.36 (a) I   (b) 22°C   (c) I (lowest), II, III   (d) I, 17°C; II, 72°C; III, 98°C   (e) I, 11°C; II, 67°C; III, 95°C

13.37 Vapor of that liquid.

13.38 The temperature of the water did not change.

13.39 (a) It would take longer to hard-boil eggs in Denver. The water boils at a temperature <100°C because the atmospheric pressure in Denver is <1 atm. Therefore, the eggs require more time in the <100°C water to absorb the same amount of heat or cook to the same "hardness."   (b) There would be no difference. Frying does not depend on atmospheric pressure.

13.40 A pressure greater than that of the room is created above the water in the pressure cooker by not allowing steam to escape. Water must reach a temperature >100°C to boil when the pressure above the liquid is more than 1 atm. Therefore, the water gets hotter than it would in an open pan and cooks the food faster.

13.44 (a) Water molecules cohere to each other and adhere to the glass. (b) Mercury atoms cohere to each other and do not adhere to the glass.

13.45 An increase in temperature decreases viscosity. Warm syrup pours more easily than does cold syrup.

13.46 (a) K   (b) M

13.49 $\overset{\delta+}{:\ddot{B}r}\!\!-\!\!\overset{\delta-}{\ddot{C}l:}$ $\overset{\delta+}{:\ddot{B}r}\!\!-\!\!\overset{\delta-}{\ddot{C}l:}$

13.51 Q, hydrogen bonding; R, dispersion forces; S, dipole forces

13.53 (a) Ne, Ar, Kr   (b) C_5H_{12}, C_8H_{12}, $C_{10}H_{22}$

13.54 Additional mass.

13.57 Structure would change.

13.59 Bonds in the crystal lattice are broken; therefore, the ions in the liquid are free to migrate.

13.60 Covalent solids are composed of a network of atoms that are bonded by strong covalent bonds; thus, the bonds require a lot of energy to break them apart.

13.61 The formation of vapor involves overcoming dispersion forces among carbon dioxide molecules.

13.63 Metal atoms can move around each other because bonding electrons are not localized.

13.64 (a) (1) liquid   (2) gas   (3) solid
    (b) FP = $-70°C$ BP = $70°C$
    (c) (UV) Kinetic energy decreases
       (VW) Potential energy decreases
       (WX) Kinetic energy decreases
       (XY) Potential energy decreases
       (YZ) Kinetic energy decreases

13.66 The heat is being used to overcome intermolecular forces, not to increase kinetic energy.

13.67 22 kJ

13.68 5.78 kJ

13.69 27 kJ

13.70 Water would be nonpolar liquid if it were linear. This would cause the boiling point, the melting point, and heats of fusion and vaporization to be lower.

13.71 The animals would die.

13.72 (a) High fluidity, 100°C liquid range, high heat of vaporization, high specific heat   (b) all properties in (a) and its polarity, which make it an excellent solvent

13.74 (a) $BaCl_2 \cdot 2H_2O(s) \longrightarrow BaCl_2(s) + 2H_2O(l)$
    (b) $CuSO_4(s) + 5H_2O(l) \longrightarrow CuSO_4 \cdot 5H_2O(s)$

13.75 Temperature, humidity

13.78 Graphite crystals are soft, and crystal layers easily slide past each other.

13.79

	Graphite	Diamond
Melting point	very high	very high
Boiling point	very high	very high
Conductivity	good	very low
Hardness	soft	very hard
Density	2.3 g/mL	3.5 g/mL

13.81 Graphite has delocalized electrons.

13.82 High pressure is required.

13.84 Charcoal adsorbs the impurities in vodka.

13.86 The carbon dioxide is converted to carbonates and bicarbonates, and it is used by aquatic plants in photosynthesis.

13.87 Possibly, a large amount of the excess CO_2 dissolves in the oceans.

13.88 The steam contains more energy to release to your skin. The additional energy is the heat of vaporization.

13.89 4.9 g H_2O

13.90 There is a greater difference in electronegativity between hydrogen and fluorine than between hydrogen and oxygen, causing a greater dipole and greater force of attraction.

CHAPTER 14

14.5 (a) $2Rb + H_2O \longrightarrow 2RbH$ (b) $6Rb + N_2 \longrightarrow 2Rb_3N$
(c) $4Rb + O_2 \longrightarrow 2Rb_2O$

14.9 (a) $3Ba + N_2 \longrightarrow Ba_3N_2$
(b) $Ba + 2H_2O \longrightarrow Ba(OH)_2 + H_2$ (c) $2Ba + O_2 \longrightarrow 2BaO$

14.17 (a) $H_2 + Br_2 \longrightarrow 2HBr$ (b) $Sr + I_2 \longrightarrow SrI_2$
(c) $Cl_2 + CH_4 \longrightarrow CH_3Cl + HCl$

14.27 (a) Increase   (b) Decrease   (c) Decrease   (d) Decrease

14.28 (a) $1s^2 2s^2 2p^6 3s^1$ (b) $1s^2 2s^1$ (c) $1s^2 2s^2 2p^6 3s^2 3p^6 4s^1$

14.29 (a) Rb·   (b) $Na^+ \ \left[:\ddot{F}: \right]^-$ (c) $2K^+ \ \left[:\ddot{O}: \right]^{2-}$

(d) $3Cs^+ \ \left[:\ddot{N}: \right]^{3-}$

14.31 (a) 1145°C   (b) Greater

14.33 14.4 g Rb

14.34 5.43 kJ

14.35 (a) $2K + Cl_2 \longrightarrow 2KCl$ (b) $2Rb + H_2 \longrightarrow 2RbH$
(c) $4Li + O_2 \longrightarrow 2Li_2O$
(d) $2Cs + 2H_2O \longrightarrow 2CsOH + H_2$
(e) $2Na + 2HCl \longrightarrow 2NaCl + H_2$

14.36 (a) LiOH   (b) Rb_2SO_4 (c) K_2S (d) Na_2O_2 (e) K_3PO_4

14.38 (a) $2Li + Br_2 \longrightarrow 2LiBr$ (b) Combination   (c) 4.67 g LiBr

14.39 (a) 101 g K

14.40 43.4% Na

14.41 $HCl(aq) + NaHCO_3(aq) \longrightarrow NaCl(aq) + H_2O(l) + CO_2(g)$

14.42 3.1×10^{22} atoms K

14.45 26.3% Mg

14.46 33.3 kcal; 1.40×10^2 kJ

14.47 Melting point, boiling point, heat of vaporization, ionization energy

14.48 (a) Mg: (b) $Sr^{2+} \ \left[:\ddot{S}: \right]^{2-}$

(c) $Ca^{2+} \ \left[\ddot{O} = C - \ddot{O}: \right]^{2-}$ (d) $Mg^{2+} \ \left[:\ddot{O} - \overset{\displaystyle :\ddot{O}:}{\underset{\displaystyle :\ddot{O}:}{S}} - \ddot{O}: \right]^{2-}$

14.49 (a) $3Mg + N_2 \longrightarrow Mg_3N_2$ (b) $Ba + F_2 \longrightarrow BaF_2$
(c) $Sr + H_2 \longrightarrow SrH_2$ (d) $Ba + 2H_2O \longrightarrow Ba(OH)_2 + H_2$
(e) $Be + Cl_2 \longrightarrow BeCl_2$

14.52 (a) $BaO_2 + 2HCl \longrightarrow BaCl_2 + H_2O_2$ (b) 178 g H_2O_2

14.53 22.8 kg plaster of paris

14.54 (a) $CaO + H_2O \longrightarrow Ca(OH)_2$ (b) 4.7×10^3 kJ; 1.1×10^3 kcal

14.55 (a) 0.406 g HCl   (b) 0.141 g HCl

14.58 (a) $Al^{3+} \ \left[:\ddot{N}: \right]^{3-}$ aluminum nitride

(b) $Al^{3+} \ 3 \left[:\ddot{F}: \right]^-$ aluminum fluoride

(c) $2Al^{3+}$ $3\left[:\!\overset{..}{\underset{..}{O}}\!:\right]^{2-}$ aluminum oxide

(d) Al^{3+} $3\left[:\!\overset{..}{\underset{..}{O}}\!-H\right]^{-}$ aluminum hydroxide

14.59 $Al(OH)_3 + HCl \longrightarrow Al(OH)_2Cl + H_2O$
$Al(OH)_3 + 2HCl \longrightarrow Al(OH)Cl_2 + 2H_2O$

14.61 (a) $2Al + Fe_2O_3 \longrightarrow 2Fe + Al_2O_3$ (b) 1.51×10^3 kJ
(c) 4.8×10^3 g H_2O

14.62 The bonds in $AlCl_3$ are less polar than the bonds in AlF_3; thus, the bonds in $AlCl_3$ have more covalent character.

14.63 $Al_2O_3 + 6HF \longrightarrow 2AlF_3 + 3H_2O$

14.65 (a) $:\!\overset{\overset{\displaystyle :\overset{..}{\underset{..}{I}}:}{}}{\underset{\underset{\displaystyle :\overset{..}{\underset{..}{I}}:}{}}{:\!\overset{..}{\underset{..}{I}}\!:\!C\!:\!\overset{..}{\underset{..}{I}}\!:}}$ (b) $H\!:\!\overset{..}{\underset{..}{F}}\!:$ (c) $:\!\overset{..}{\underset{..}{Cl}}\!:\!\overset{..}{\underset{..}{F}}\!:$ (d) $Ca^{2+}2\left[:\!\overset{..}{\underset{..}{I}}\!:\right]^{-}$

(e) $:\!\overset{..}{Cl}\!-\!\overset{..}{\underset{..}{S}}\!-\!\overset{..}{Cl}\!:$

14.66 6.31 L Cl_2

14.67 Oxidizing agent, fluorine; reducing agent, astatine

14.69 (a) $8F_2 + S_8 \longrightarrow 8SF_2$ (b) $Ca + Br_2 \longrightarrow CaBr_2$
(c) $H_2 + I_2 \longrightarrow 2HI$ (d) $F_2 + MgCl_2 \longrightarrow MgF_2 + Cl_2$
(e) $CH_4 + Cl_2 \longrightarrow CH_3Cl + HCl$

14.70 0.741 g SO_2Cl_2

14.71 129 g HF

14.72 (a) 33%S   (b) 53%S   (c) 13%S

14.73 (a) $S_8 + 8Ca \longrightarrow 8CaS$ (b) $S_8 + 8H_2 \longrightarrow 8H_2S$
(c) $S_8 + 4C \longrightarrow 4CS_2$

14.74 13.9 kJ

14.75 0.24 kg H_2SO_4

14.76 1.12 L CO_2

14.77 (a) Phosphorus pentafluoride   (b) Phosphorous acid
(c) Tetraphosphorus hexoxide   (d) Sodium dihydrogen phosphate   (e) Aluminum phosphide

14.78 26.5% P

14.79 11.5 kg P_4

14.80 22.38 g H_3PO_4

14.81 (a) $Ca_3P_2 + 6H_2O \longrightarrow 2PH_3 + 3Ca(OH)_2$ (b) 8.6 L PH_3

14.82 (a) $4P_4 + 5S_8 \longrightarrow 4P_4S_{10}$ (b) $8P_4 + 3S_8 \longrightarrow 8P_4S_3$
(c) $8P_4 + 7S_8 \longrightarrow 8P_4S_7$

CHAPTER 15

15.10 Add 148 g acetone to 527 g H_2O.

15.11 3.76 g $CaCl_2$

15.12 Dissolve 22 g NaOH, then dilute to 375 mL.

15.13 277 g $Mg(NO_3)_2$

15.14 0.520 mol C_2H_6O

15.15 23.5 M formic acid

15.16 523 mL H_2O

15.19 (a) $H^+(aq) + NO_3^- + K^+(aq) + OH^-(aq) \longrightarrow$
$$H_2O(l) + K^+(aq) + NO_3^-(aq)$$
(b) $2H^+(aq) + SO_4^{2-}(aq) + 2K^+(aq) + 2OH^-(aq) \longrightarrow$
$$2H_2O(l) + 2K^+(aq) + SO_4^{2-}(aq)$$
(c) $3H^+(aq) + PO_4^{3-}(aq) + 3K^+(aq) + 3OH^-(aq) \longrightarrow$
$$3H_2O(l) + 3K^+(aq) + PO_4^{3-}(aq)$$

15.22 $T_b = 102°C$ $\quad$ $T_f = -7.4°C$

15.23 6.1 m

15.24 $T_b = 100.517°C$

15.27 (a) Evaporation of water $\quad$ (b) No

15.31 (a) Solid–liquid $\quad$ (b) Gas–gas $\quad$ (c) Solid–solid $\quad$ (d) Solid–liquid $\quad$ (e) Liquid–liquid

15.32 (a) H_2O $\quad$ (b) CH_3OH $\quad$ (c) CH_3OCH_3 $\quad$ KCl is more soluble in polar solvents.

15.33 (a) 15.5 g $NaNO_3$ + 435.5 g H_2O $\quad$ (b) 15.0 g KOH + 55.0 g H_2O $\quad$ (c) 45 g NH_4Cl + 8055 g H_2O

15.34 (a) 10 g HCl $\quad$ (b) 2.5 g HCl $\quad$ (c) 5.7×10^3 g HCl $\quad$ (d) 0.18 g HCl

15.35 (a) 2.6×10^2 g NH_3 $\quad$ (b) 15 g NH_3 $\quad$ (c) 0.0065 g NH_3

15.36 1.9×10^3 g KI solution

15.37 84.8 g H_2O

15.38 7.97%

15.39 (a) 7.58% NH_4NO_3 $\quad$ (b) Add 5.5 g NH_4NO_3

15.40 (a) 7.47 g $Ca(NO_3)_2$ diluted to 100 mL $\quad$ (b) 9.28 g NH_3 diluted to 500 mL $\quad$ (c) 535 g $C_6H_{12}O_6$ diluted to 3.25 L $\quad$ (d) 23.6 g H_3PO_4 diluted to 83.2 mL

15.41 (a) 4.6 M $\quad$ (b) 1.23 M $\quad$ (c) 0.152 M $\quad$ (d) 1.25×10^{-5} M

15.42 (a) 16 M $\quad$ (b) 6.3 M

15.43 (a) Add 70 mL H_2O. $\quad$ (b) Add 14.9 L H_2O. $\quad$ (c) Add 836 L H_2O.

15.44 (a) 0.0417 mol CsOH $\quad$ (b) 0.0159 mol HBr $\quad$ (c) 51.2 mol $NaNO_3$

15.45 0.694 M $AgNO_3$

15.46 3.3 M H_2SO_4

15.47 553 mL HI

15.48 (a) 103 mL 4.55 M NaOH $\quad$ (b) Dilute 103 mL 4.55 M to 935 mL.

15.49 (a) Bright $\quad$ (b) Dim $\quad$ (c) No light $\quad$ (d) No light $\quad$ (e) Dim

15.50 Large amount of HF(aq), small amounts of $H^+(aq)$ and $F^-(aq)$

15.51 (a) $KCl(s) \xrightarrow{H_2O} K^+(aq) + Cl^-(aq)$

(b) $NaHCO_3(s) \xrightarrow{H_2O} Na^+(aq) + HCO_3^-(aq)$

(c) $Pb(NO_3)_2(s) \xrightarrow{H_2O} Pb^{2+}(aq) + 2NO_3^-(aq)$

(d) $CaI_2(s) \xrightarrow{H_2O} Ca^{2+}(aq) + 2I^-(aq)$

(e) $MgCrO_4(s) \xrightarrow{H_2O} Mg^{2+}(aq) + CrO_4^{2-}(aq)$

15.52 (a) $Li^+(aq) + Br^-(aq) + Ag^+(aq) + NO_3^-(aq) \longrightarrow$
$$AgBr(s) + Li^+(aq) + NO_3^-(aq)$$
$Ag^+(aq) + Br^-(aq) \longrightarrow AgBr(s)$

(b) $6Na^+(aq) + 2PO_4^{3-}(aq) + 3Ni^{2+}(aq) + 6Cl^-(aq) \longrightarrow$
$$Ni_3(PO_4)_2(s) + 6Na^+(aq) + 6Cl^-(aq)$$
$3Ni^{2+}(aq) + 2PO_4^{3-}(aq) \longrightarrow Ni_3(PO_4)_2(s)$

(c) $2NH_4^+(aq) + S^{2-}(aq) + Co^{2+}(aq) + SO_4^{2-}(aq) \longrightarrow$
$$CoS(s) + 2NH_4^+(aq) + SO_4^{2+}(aq)$$
$Co^{2+}(aq) + S^{2-}(aq) \longrightarrow CoS(s)$

(d) $H^+(aq) + HSO_4^-(aq) + K^+(aq) + OH^-(aq) \longrightarrow$
$$H_2O(l) + K^+(aq) + HSO_4^-(aq)$$
$H^+(aq) + OH^-(aq) \longrightarrow H_2O(l)$

(e) $Hg_2^{2+}(aq) + 2NO_3^-(aq) + Ca^{2+}(aq) + 2Br^-(aq) \longrightarrow$
$$Hg_2Br_2(s) + Ca^{2+}(aq) + NO_3^-(aq)$$
$Hg_2^{2+}(aq) + 2Br^-(aq) \longrightarrow Hg_2Br_2(s)$

15.53 (a) $3Ba^{2+}(aq) + 2PO_4^{3-}(aq) \longrightarrow Ba_3(PO_4)_2(s)$

(b) NR

(c) $Fe^{2+}(aq) + CO_3^{2-}(aq) \longrightarrow FeCO_3(s)$

(d) $Ag^+(aq) + Cl^-(aq) \longrightarrow AgCl(s)$

(e) $Mg^{2+}(aq) + S^{2-}(aq) \longrightarrow MgS(s)$

15.54 (a) $0.12\ m$ (b) $0.250\ m$ (c) $1.36\ m$ (d) $0.099\ m$

15.55 (a) $0.20\ m$ (b) $6.9\ m$ (c) $10.6\ m$

15.56 (a) 15.7 g $Cu(NO_3)_2 + 893$ g H_2O (b) 4.0×10^3 g
$C_{12}H_{22}O_{11} + 6.91$ kg H_2O

15.57 (a) $0.0621\ m$ I_2 (b) 1.55% I_2

15.59 A solution of two volatile liquids.

15.60 (a) $100.17°C$ (b) $103.0°C$ (c) $101.95°C$

15.61 (a) $-0.61°C$ (b) $-11.0°C$ (c) $-7.09°C$

15.62 (a) $1.8°C$ (b) $11°C$ (c) $8.8°C$ (d) $18°C$ (e) $13°C$

15.63 (a) $97°C$ (b) $82.3°C$ (c) $100°C$

15.64 (a) $-28.5°C$ (b) $0.8°C$ (c) $-36°C$

15.65 $T_b = 101.10°C$ $T_f = -4.00°C$

15.66 $T_b = 83.2°C$ $T_f = -1.1°C$

15.67 Time is decreased

15.68 $T_b = 100.512°C$ $T_f = -1.86°C$

15.69 8.9×10^3 g $C_2H_6O_2$

15.70 $0.0767\ M$ Na^+ $0.0384\ M$ Ca^{2+} $0.153\ M$ Cl^-

15.71 126 g/mol

15.72 (a) $18\ M$ (b) $245\ m$ (c) 2.4 L H_2O (d) 1.18×10^3 g

CHAPTER 16

16.6 (a) Decreases   (b) Increases   (c) Decreases

16.9 (a) Step 1   (b) O

16.13 (a) $K = \dfrac{[SO_3][NO]}{[NO_2][SO_2]}$ (b) $K = \dfrac{[Cl_2]^2[H_2O]^2}{[HCl]^4[O_2]}$

16.16 (a) (c) (d) Products   (b) Reactants

16.17 (a) Reactants   (b) (c) Products   (d) No change

16.20 (a) (c) (d) High   (b) Moderate   (e) Low

16.24 (a) No   (b) Yes   (c) No

16.25 (a) Decrease   (b) Increase   (c) Increase

16.26 (a) High   (b) Low   (c) Low   (d) High

16.27 (a) > (b) > (c)

16.28 (a) (c) (d) (e) Increase   (b) Decrease

16.29 Low rate

16.34 Pressure, temperature, concentration

16.35 Oxidize nitrogen oxides and carbon monoxide

16.36 (b) HI   (c) Step 1

16.37 (a) $OCl^- + I^- \longrightarrow OI^- + Cl^-$ (b) HOCl, OH$^-$, HOI

16.38 (a) At equilibrium the rate of the forward reaction equation equals the rate of the reverse reaction.   (b) . . . are *stable* and *do not* undergo . . .   (c) . . . reactions *continue.*

16.39 Addition or removal of a reactant or product disturbs a chemical equilibrium.

16.40 Double arrows indicate that both reactions are occurring at equal rates.

16.42 (a) $K = \dfrac{[NO_2][SO_2]}{[NO][SO_3]}$ (b) $K = \dfrac{[SO_2][Cl_2]}{[SO_2Cl_2]}$

(c) $K = \dfrac{[CH_4][H_2S]^2}{[CS_2][H_2]^4}$ (d) $K = \dfrac{[NO]^4[H_2O]^6}{[NH_3]^4[O_2]^5}$

(e) $K = \dfrac{[Cl_2]^2[H_2O]^2}{[HCl]^4[O_2]}$ (f) $K = \dfrac{[NO][Cl_2]^{\frac{1}{2}}}{[NOCl]}$

16.43 If $K > 1$, then the products are in greater concentration at equilibrium. If $K < 1$, then the reactants are in greater concentration at equilibrium.

16.44 (a), (c) Increase products   (b), (d) Increase reactants

16.45 Increases forward reaction and decreases reverse reaction.

16.46 (a) Reactants   (b), (c) No change   (d) Products

16.47 The system increases pressure by producing more gas molecules, increasing the concentration of O_2.

16.48 (a) Products   (b) Reactants

16.49 A catalyst increases the rate of both the forward and reverse reactions.

16.50 (a) Increase reactants   (b) Increase products   (c) Increase reactants   (d) Increase reactants   (e) No change

CHAPTER 17

17.4 Acid, HCN; conjugate base, CN^- Base, NH_3; conjugate acid, NH_4^+

17.7 F^- is a stronger base than Cl^-.

17.8 (a) HNO_2 (b) NH_4^+ (c) H_3O^+ (d) H_2CO_3

17.9 (a) $H^+(aq) + ClO_4^-(aq) + Rb^+(aq) + OH^-(aq) \longrightarrow$
$$H_2O(l) + Rb^+(aq) + ClO_4^-(aq)$$
(b) $Na^+(aq) + OH^-(aq) + HNO_2(aq) \longrightarrow$
$$H_2O(l) + Na^+(aq) + NO_2^-(aq)$$
(c) $H^+(aq) + I^-(aq) + NH_3(aq) \longrightarrow NH_4^+(aq) + I^-(aq)$

17.11 (a) $1 \times 10^{-14}\ M$ (b) $1.2 \times 10^{-13}\ M$ (c) $5.3 \times 10^{-11}\ M$

17.12 (a) 0.040 M $HClO_4$

17.14 (a) 2.0   (b) 5.0   (c) 2.03

17.15 11.30

17.18 (a) 1.17×10^3 mL   (b) 107 mL   (c) 3.08 L

17.19 (a) 81 g HBr/equiv   (b) 33 g H_3PO_4/equiv   (c) 41 g H_2SO_3/equiv   (d) 45 g $H_2C_2O_4$/equiv

17.20 9.03 mL of 36.0 N diluted to 500 mL

17.21 0.780 N NaOH

17.24 Acids: (a), (d), (e)   Bases: (b), (c), (f)

17.25 (a) $HClO \xrightarrow{H_2O} H^+ + ClO^-$

(b) $Al(OH)_3 \xrightarrow{H_2O} Al^{3+} + 3OH^-$

(c) $RbOH \xrightarrow{H_2O} Rb^+ + OH^-$ (d) $HI \xrightarrow{H_2O} H^+ + I^-$

(e) $HC_2H_3O_2 \xrightarrow{H_2O} H^+ + C_2H_3O_2^-$

(f) $NH_3 + H_2O \xrightarrow{H_2O} NH_4^+ + OH^-$

17.26 (a) Acid, HCl; base, NH_3 (b) NO

17.27

	Acid	Base	Conjugate acid	Conjugate base
(a)	H_2O	CO_2	H_2CO_3	H_2CO_3
(b)	H_2SO_4	NaF	HF	$NaHSO_4$
(c)	HCN	KOH	H_2O	KCN
(d)	$HC_2H_3O_2$	NH_3	NH_4^+	$C_2H_3O_2^-$

17.28 (a) $H_2S(aq) + 2Na^+(aq) + 2OH^-(aq) \longrightarrow$
$$2H_2O(l) + 2Na^+(aq) + S^{2-}(aq)$$
(b) $2H^+(aq) + 2Br^-(aq) + Ca^{2+}(aq) + 2OH^-(aq) \longrightarrow$
$$2H_2O(l) + Ca^{2+}(aq) + 2Br^-(aq)$$
(c) $H_3PO_4(aq) + 3Li^+(aq) + 3OH^-(aq) \longrightarrow$
$$3H_2O(aq) + PO_4^{3-}(aq) + 3Li^+(aq)$$
(d) $2NH_3(aq) + 2H^+(aq) + SO_4^{2-}(aq) \longrightarrow 2NH_4^+(aq) + SO_4^{2-}(aq)$

17.29 $H_2O(l) + NH_3(aq) \longrightarrow NH_4^+(aq) + OH^-(aq)$
$H_2O(l) + HCl(aq) \longrightarrow H_3O^+(aq) + Cl^-(aq)$

17.30 (a) $H_2PO_4^-(aq) + H^+(aq) \longrightarrow H_3PO_4(aq)$
$H_2PO_4^-(aq) + OH^-(aq) \longrightarrow HPO_4^{2-}(aq) + H_2O$
(b) $NH_3(aq) + H^+(aq) \longrightarrow NH_4^+(aq)$
$NH_3 + OH^- \longrightarrow NH_2^- + H_2O$
(c) $HS^-(aq) + H^+(aq) \longrightarrow H_2S(aq)$
$HS^-(aq) + OH^-(aq) \longrightarrow H_2O(l) + S^{2-}(aq)$
(d) $HCO_3^-(aq) + H^+(aq) + H_2CO_3(aq)$
$HCO_3^-(aq) + OH^-(aq) \longrightarrow H_2O(l) + CO_3^{2-}(aq)$

17.31 (a) NO_3^- (b) SO_4^{2-} (c) O^{2-} (d) F^- (e) NH_2^-
(f) PO_4^{3-}

17.32 (a) HCN (b) HNO_2 (c) NH_4^+ (d) H_3O^+
(e) $H_2NO_3^+$ (f) $H_2PO_4^-$

17.33 (a) H_2SO_4 (b) NH_2^- (c) NH_3 (d) H_2O (e) HCO_3^-

17.34 (a) HCl (b) H_2O (c) H_2Te (d) AsH_3

17.35 (a) $HCl(aq) + RbOH(aq) \longrightarrow H_2O(l) + RbCl(aq)$
(b) $H_2SO_4(aq) + Ca(OH)_2(aq) \longrightarrow 2H_2O(l) + CaSO_4(aq)$
(c) $2HClO_4(aq) + Ba(OH)_2(aq) \longrightarrow 2H_2O(l) + Ba(ClO_4)_2(aq)$
(d) $NaOH(aq) + HNO_3(aq) \longrightarrow H_2O(l) + NaNO_3(aq)$
(e) $2HI(aq) + Sr(OH)_2(aq) \longrightarrow 2H_2O(l) + SrI_2(aq)$

17.36 (a) $H^+(aq) + Cl^-(aq) + Rb^+(aq) + OH^-(aq) \longrightarrow$
$$H_2O(l) + Rb^+(aq) + Cl^-(aq)$$
$H^+(aq) + OH^-(aq) \longrightarrow H_2O(l)$
(b) $2H^+(aq) + SO_4^{2-}(aq) + Ca^{2+}(aq) + 2OH^-(aq) \longrightarrow$
$$2H_2O(l) + Ca^{2+}(aq) + SO_4^{2-}(aq)$$
$H^+ + OH^-(aq) \longrightarrow H_2O(l)$
(c) $2H^+(aq) + 2ClO_4^-(aq) + Ba^{2+}(aq) + 2OH^-(aq) \longrightarrow$
$$2H_2O(l) + Ba^{2+}(aq) + 2ClO_4^-(aq)$$
$H^+(aq) + OH^-(aq) \longrightarrow H_2O(l)$
(d) $Na^+(aq) + OH^-(aq) + H^+(aq) + NO_3^-(aq) \longrightarrow$
$$H_2O(l) + Na^+(aq) + NO_3^-(aq)$$
$H^+(aq) + OH^-(aq) \longrightarrow H_2O(l)$
(e) $2H^+(aq) + 2I^-(aq) + Sr^{2+}(aq) + 2OH^-(aq) \longrightarrow$
$$2H_2O(l) + Sr^{2+}(aq) + 2I^-(aq)$$
$H^+(aq) + OH^-(aq) \longrightarrow H_2O(l)$

17.38 (a) $OH^-(aq) + HF(aq) \rightleftharpoons H_2O(l) + F^-(aq)$
(b) $OH^-(aq) + HNO_2(aq) \rightleftharpoons H_2O(l) + NO_2^-(aq)$
(c) $OH^-(aq) + H_2C_2O_4(aq) \rightleftharpoons H_2O(l) + HC_2O_4^-(aq)$

17.39 (a), (d) Acidic   (b), (c) Basic

17.40 (a) $1.0 \times 10^{-11} M$ (b) $1.9 \times 10^{-7} M$ (c) $1.0 \times 10^{-8} M$

17.41 (a) $1.0 \times 10^{-13} M$ (b) $1.0 \times 10^{-10} M$ (c) $1 \times 10^{-2} M$

17.42 (a) $0.0100 M\ H^+$; $1.00 \times 10^{-12} M\ OH^-$ (b) $0.0235 M\ H^+$;
$4.25 \times 10^{-13} M\ OH^-$ (c) $8.63 \times 10^{-3} M\ H^+$; $1.16 \times 10^{-12} M$
OH^-

17.43 (a) $5.6 \times 10^{-4} M\ OH^-$; $1.79 \times 10^{-11} M\ H^+$ (b) $0.096 M\ OH^-$;
$1.0 \times 10^{-13} M\ H^+$ (c) $9.64 \times 10^{-5} M\ OH^-$; $1.04 \times 10^{-10} M$
H^+

17.44 (a) 5.00   (b) 1.0   (c) 1.0

17.45 (a) 10.0   (b) 3.00   (c) 7.00

17.46 (a) 1.82   (b) 2.21   (c) 3.365   (d) 10.77   (e) 9.52

17.47 2.031   (b) 1.42   (c) 11.08   (d) 8.29   (e) 10.256

17.48 (a) 28.6 mL   (b) 1.79 L   (c) 1.43 mL   (d) 46.7 L

17.49 0.136 M

17.50 (a) 34.5 mL   (b) 379 mL   (c) 155 mL

17.51 (a) 0.466 L   (b) 265 mL   (c) 0.151 mL

17.52 (a) 0.00796 mol NaOH   (b) 0.291 g HCl   (c) 0.162 M
     (d) 0.790

17.53 2.90×10^{-13} M H^+; 0.0345 M OH^-; 0.0750 M ClO_4^-; 0.110 M
     K^+; pH 12.538

17.54 (a) 85.5 g   (b) 17 g   (c) 24 g   (d) 41 g

17.55 0.0538 equiv   (b) 0.725 equiv   (c) 0.241 equiv
     (d) 0.00453 equiv   (e) 0.536 eqiv

17.56 (a) 0.332 N   (b) 0.00456 N   (c) 0.00962 N   (d) 0.00038 N

17.57 (a) 2.2 g HNO_3 (b) 1.7 g H_2SO_4 (c) 1.1 g H_3PO_4 (d) 2.8 g
     HBr

17.58 (a) 7.85 L H_2SO_4 (b) 1.95 mL H_2SO_4 (c) 0.18 mL H_2SO_4

17.59 (a) 199.0 g/equiv   (b) 398 g/mol

17.60 11.3 L

17.61 (a) 0.2909 L   (b) 0.05284 M Na^+ (c) 0.1061 L NaOH

CHAPTER 18

18.3

	Oxidized/reducing agent	Reduced/oxidizing agent
(a)	H_2	I_2
(b)	CO	O_2
(c)	Mg	N_2
(d)	PCl_3	Cl_2
(e)	Fe^{2+}	$Cr_2O_7^{2-}$

18.4 (a) $PH_3 + 2I_2 + 2H_2O \longrightarrow H_3PO_2 + 4I^- + 4H^+$
     (b) $4Zn + NO_3^- + 10H^+ \longrightarrow 4Zn^{2+} + NH_4^+ + 3H_2O$
     (c) $4Zn + NO_3^- + 6H_2O + 7OH^- \longrightarrow 4Zn(OH)_4^{2-} + NH_3$

18.5 Cathode: $Sn^{2+} + 2e \longrightarrow Sn$ Anode: $Ni \longrightarrow Ni^{2+} + 2e$

18.12 Reaction would cease.

18.15 Electrons are transferred from reducing agent to oxidizing agent.

18.16 Oxidized: (a) Cu   (b) Fe   (c) Cu   (d) Zn
     (e) I^- (f) Cl^- (g) Br_2 Reduced: (a) Ag^+ (b) H^+
     (c) NO_3^- (d) MnO_2 (e) Cl_2 (f) H_2O (g) Br_2

18.17 Oxidizing agents: (a) F_2 (b) HNO_3 (c) MnO_4^- (d) PbO_2
      Reducing agents: (a) Al   (b) HCl   (c) Cl^- (d) Pb

18.18 (a) $14H^+ + Cr_2O_7{}^{2-} + 6Br^- \longrightarrow 2Cr^{3+} + 3Br_2 + 7H_2O$
      (b) $2H^+ + H_2O_2 + 2I^- \longrightarrow I_2 + 2H_2O$
      (c) $4Zn + 10H^+ + NO_3^- \longrightarrow 4Zn^{2+} + NH_4^+ + 3H_2O$
      (d) $H^+ + HOCl + 2NO_2 \longrightarrow 2NO_2^- + Cl^- + H_2O$
      (e) $6H_2O + 4AsH_3 + 24Ag^+ \longrightarrow 24Ag + As_4O_6 + 24H^+$
      (f) $12H^+ + 3Zn + 2H_2MoO_4 \longrightarrow 3Zn^{2+} + 2Mo^{3+} + 8H_2O$

18.19 (a) $H_2O + ClO^- + 2I^- \longrightarrow Cl^- + I_2 + 2OH^-$
      (b) $8OH^- + S^{2-} + 4I_2 \longrightarrow SO_4{}^{2-} + 8I^- + 4H_2O$
      (c) $2OH^- + 2Al + 6H_2O \longrightarrow 2Al(OH)_4^- + 3H_2$
      (d) $8OH^- + 4H_2O + 2P_4 \longrightarrow 4PH_3 + 4HPO_3{}^{2-}$
      (e) $H_2O + 6Fe_3O_4 + 2MnO_4^- \longrightarrow 9Fe_2O_3 + 2MnO_2 + 2OH^-$
      (f) $2OH^- + H_2O + Si \longrightarrow SiO_3{}^{2-} + 2H_2$

18.20 (a) $10OH^- + 2CrI_3 + 27H_2O_2 \longrightarrow 2CrO_4{}^{2-} + 6IO_4^- + 32H_2O$
      (b) $3H_2O + 9I^- + XeO_3 \longrightarrow Xe + 3I_3^- + 6OH^-$
      (c) $2OH^- + 2HXeO_4^- \longrightarrow XeO_6{}^{4-} + Xe + O_2 + 2H_2O$
      (d) $8H^+ + Pt + 6Cl^- + 4NO_3^- \longrightarrow PtCl_6{}^{2-} + 4NO_2 + 4H_2O$
      (e) $2H^+ + 3CN^- + 2MnO_4 \longrightarrow 3CNO^- + 2MnO_2 + H_2O$
      (f) $H_2O + 2CrO_4{}^{2-} + 3HSnO_2^- \longrightarrow$
                                      $2CrO_2^- + 3HSnO_3^- + 2OH^-$

18.24 (a) Ca, cathode   fluoride, anode
      (b) $Ca^{2+} + 2e^- \longrightarrow Ca$ $2F^- \longrightarrow F_2 + 2e^-$

18.26 (a) $2F^- \longrightarrow F_2 + 2e^-$ or $2H_2O \longrightarrow 4e^- + O_2 + 4H^+$
      (b) $K^+ + e^- \longrightarrow K$ or $2H_2O + 2e^- \longrightarrow H_2 + 2OH^-$

18.27 Use the impure Ag as the anode and electroplate silver onto a pure Ag cathode.

18.31 (a) $2e^- + PbO_2 + 4H^+ + SO_4{}^{2-} \longrightarrow PbSO_4 + 2H_2O$
      (b) $Cd + 2OH^- \longrightarrow Cd(OH)_2 + 2e^-$
      (c) $2e^- + 2NH_4 + 2MnO_2 \longrightarrow 2NH_3 + Mn_2O_3 + H_2O$

18.34 (a) Anode $PbSO_4 + 2H_2O \longrightarrow 2e^- + PbO_2 + 4H^+ + SO_4{}^{2-}$
      Cathode $PbSO_4 + 2e^- \longrightarrow Pb^{2+} + SO_4{}^{2-}$

CHAPTER 19

19.3 (a) 85   (b) 37   (c) 48

19.4 (a) 2, 8, 20, 50, 82, 126   (b) No

19.5 ^4_2He (b) $^{12}_6C$ $^{64+}_{30}Zn$

19.6 (a) α (b) β (c) α (d) γ (e) γ

19.8 (a) 1.3×10^9 yr   (b) 3.13 g

19.10 (a) $^{115}_{48}\text{Cd} \longrightarrow {}^{115}_{49}\text{In} + {}^{0}_{-1}\beta + \bar{\nu}$ (b) $^{233}_{92}\text{U} \longrightarrow {}^{229}_{90}\text{Th} + {}^{4}_{2}\alpha$

     (c) $^{198}_{79}\text{Au} \longrightarrow {}^{198}_{80}\text{Hg} + {}^{0}_{-1}\beta + \bar{\nu}$ (d) $^{89}_{38}\text{Sr}^* \longrightarrow {}^{89}_{38}\text{Sr} + \gamma$

19.13 (a) $^{40}_{18}\text{Ar}({}^{4}_{2}\alpha, {}^{1}_{1}p){}^{43}_{19}\text{K}$ (b) $^{59}_{27}\text{Co}({}^{1}_{0}n, {}^{4}_{2}\alpha){}^{56}_{25}\text{Mn}$ (c) $^{12}_{6}\text{C}({}^{2}_{1}\text{H}, {}^{1}_{0}n){}^{13}_{7}\text{N}$

19.14 (a) $^{238}_{92}\text{U} + {}^{1}_{0}n \longrightarrow {}^{239}_{93}\text{Np} + {}^{0}_{-1}\beta$

     (b) $^{242}_{96}\text{Cm} + {}^{4}_{2}\alpha \longrightarrow {}^{245}_{98}\text{Cf} + {}^{1}_{0}n$

     (c) $^{252}_{98}\text{Cf} + {}^{10}_{5}\text{B} \longrightarrow {}^{257}_{103}\text{Lr} + 5\,{}^{1}_{0}n$

19.22 3×10^{10} L

19.23 (a) 89 p^+, 138 n° (b) 33 p^+, 44 n° (c) 83 p^+, 127 n°

     (d) 24 p^+, 27 n° (e) 70 p^+, 105 n°

19.24 (a) $^{51}_{23}\text{V}$ (b) $^{97}_{43}\text{Tc}$ (c) $^{153}_{62}\text{Sm}$ (d) $^{192}_{77}\text{Ir}$

19.25 (a) $^{4}_{2}\text{He}$ (b) $^{16}_{8}\text{O}$ (c) $^{40}_{20}\text{Ca}$ (d) $^{207}_{82}\text{Pb}$

19.26 (a) $^{184}_{74}\text{W}$, most stable; $^{176}_{73}\text{Ta}$, least stable

     (b) $^{202}_{82}\text{Hg}$, most stable; $^{200}_{79}\text{Au}$, least stable

19.27 Larger number of neutrons in the nucleus help to diminish the electrostatic repulsions of the protons.

19.28 (a) Beta particles have less charge and mass.   (b) Alpha particles have a greater charge and mass.

19.29 Greater ionization of substances in biological cells.

19.30 (a) $^{89}_{38}\text{Sr}$ (b) 3.13 g

19.31 48 days

19.32 0.2 g

19.33 Alpha decay

19.34 (a) $^{222}_{86}\text{Rn} \longrightarrow {}^{218}_{84}\text{Po} + {}^{4}_{2}\alpha$ (b) $^{234}_{92}\text{U} \longrightarrow {}^{230}_{90}\text{Th} + {}^{4}_{2}\alpha$

     (c) $^{237}_{93}\text{Np} \longrightarrow {}^{233}_{91}\text{Pa} + {}^{4}_{2}\alpha$ (d) $^{253}_{99}\text{Es} \longrightarrow {}^{249}_{97}\text{Bk} + {}^{4}_{2}\alpha$

19.35 (a) $^{32}_{15}\text{P} \longrightarrow {}^{32}_{16}\text{S} + {}^{0}_{-1}\beta + \bar{\nu}$ (b) $^{144}_{58}\text{Ce} \longrightarrow {}^{144}_{59}\text{Pr} + {}^{0}_{-1}\beta + \bar{\nu}$

     (c) $^{76}_{33}\text{As} \longrightarrow {}^{76}_{34}\text{Se} + {}^{0}_{-1}\beta + \bar{\nu}$ (d) $^{191}_{76}\text{Os} \longrightarrow {}^{191}_{77}\text{Ir} + {}^{0}_{-1}\beta + \bar{\nu}$

19.36 Nuclear composition does not change during gamma emission.

19.38 (a) $^{232}_{90}\text{Th} \longrightarrow {}^{228}_{88}\text{Ra} + {}^{4}_{2}\alpha$

     (b) $^{228}_{88}\text{Ra} \longrightarrow {}^{228}_{89}\text{Ac} + {}^{0}_{-1}\beta + \bar{\nu}$

     (c) $^{228}_{89}\text{Ac} \longrightarrow {}^{228}_{90}\text{Th} + {}^{0}_{-1}\beta + \bar{\nu}$

19.39 (a) $^{131}_{54}\text{Xe}$ (b) $^{214}_{84}\text{Po}$ (c) $^{210}_{82}\text{Pb}$

19.40 (a) $^{0}_{-1}\beta$ (b) $^{247}_{98}\text{Cf}$ (c) $^{129}_{52}\text{Te}$

19.41 (a) $^{32}_{16}\text{S} + {}^{1}_{0}n \longrightarrow {}^{33}_{16}\text{S} + \gamma$

     (b) $^{130}_{52}\text{Te} + {}^{2}_{1}\text{H} \longrightarrow {}^{130}_{53}\text{I} + 2\,{}^{1}_{0}n$

     (c) $^{43}_{20}\text{Ca} + {}^{4}_{2}\alpha \longrightarrow {}^{46}_{21}\text{Sc} + {}^{1}_{1}p$

     (d) $^{9}_{4}\text{Be} + {}^{1}_{1}p \longrightarrow {}^{6}_{3}\text{Li} + {}^{4}_{2}\alpha$

19.42 (a) $^{238}_{92}\text{U}({}^{16}_{8}\text{O}, 5{}^{1}_{0}n){}^{249}_{100}\text{Fm}$ (b) $^{241}_{95}\text{Am}({}^{4}_{2}\alpha, 2{}^{1}_{0}n){}^{243}_{97}\text{Bk}$

     (c) $^{238}_{92}\text{U}({}^{22}_{10}\text{Ne}, 4{}^{1}_{0}n){}^{256}_{102}\text{No}$ (d) $^{242}_{94}\text{Pu}({}^{22}_{10}\text{Ne}, 4{}^{1}_{0}n){}^{260}_{104}\text{Unq}$

19.43 (a) $^{2}_{1}\text{H}$ (b) $^{2}_{1}\text{H}$ (c) $3{}^{1}_{0}n$ (d) $^{4}_{2}\alpha$ (e) γ

19.44 (a) $^{239}_{94}\text{Pu} + {}^{1}_{0}n \longrightarrow {}^{240}_{95}\text{Am} + {}^{0}_{-1}\beta$

19.45 (b) $^{246}_{96}\text{Cm} + {}^{12}_{6}\text{C} \longrightarrow {}^{254}_{102}\text{No} + 4{}^{1}_{0}n$

     (c) $^{249}_{98}\text{Cf} + {}^{15}_{7}\text{N} \longrightarrow {}^{260}_{105}\text{Unp} + 4{}^{1}_{0}n$

19.46 Conversion of matter into energy.

19.50 5700 yr

CHAPTER 20

20.3 (a) CH_4 (b) C_3H_8 (c) C_6H_{14} (d) $C_{10}H_{22}$ (e) $C_{10}H_{22}$

20.4 (a) CH_3CH_3 (b) $CH_3CH_2CH_2CH_3$
(c) $CH_3CH_2CH_2CH_2CH(CH_3)CH_3$

20.5 $CH_3CH_2CH_2CH_2CH_2CH_3$ $\quad$ $CH_3\overset{\displaystyle |}{C}HCH_2CH_2CH_3$
$\quad\quad$ Hexane $\quad\quad\quad\quad\quad\quad\quad\quad\quad$ CH_3
$\quad\quad\quad\quad\quad\quad\quad\quad\quad\quad\quad\quad$ 2-Methylpentane

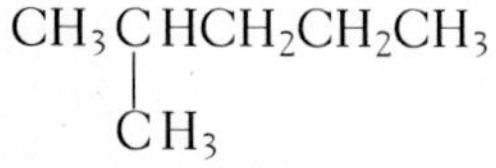

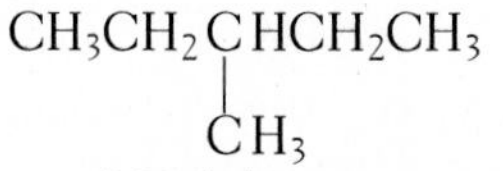

$CH_3CH_2\overset{\displaystyle |}{C}HCH_2CH_3$ $\quad$ $CH_3\overset{\displaystyle |}{C}H\overset{\displaystyle |}{C}HCH_3$ $\quad$ $CH_3\overset{\displaystyle |}{C}CH_2CH_3$
$\quad\quad CH_3$ $\quad\quad\quad\quad\quad CH_3$ $\quad\quad\quad\quad CH_3$
3-Methylpentane 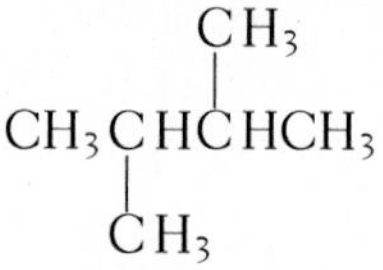 $\quad$ 2,3-Dimethylbutane 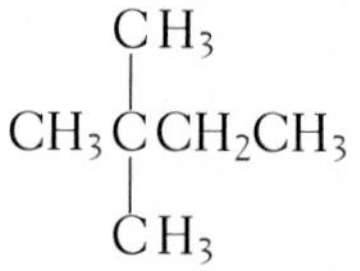 $\quad$ 2,2-Dimethylbutane

20.6 (a) 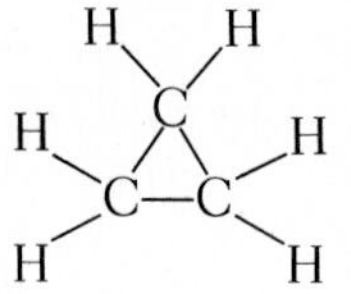 (b) Propane, $CH_3CH_2CH_3$, is composed of a chain.

20.8 (a) $HC{\equiv}CH$ (b)  (c) $CH_2{=}CH{-}CH{=}CH_2$

(d)

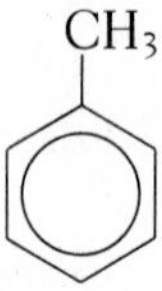

20.10

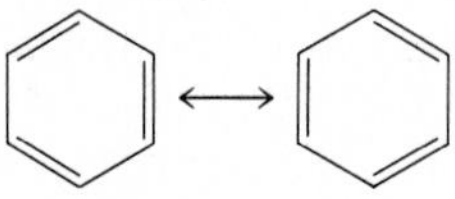

20.11

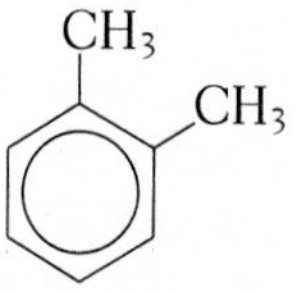

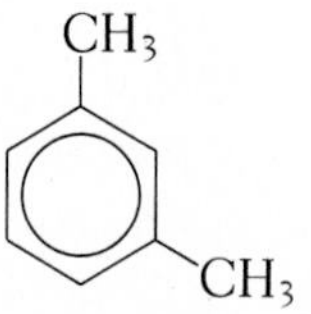

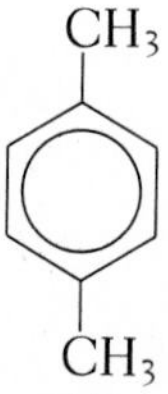

$\quad\quad$ o-Xylene $\quad\quad\quad\quad$ m-Xylene $\quad\quad\quad\quad$ p-Xylene

20.12 A polycyclic aromatic contains two or more fused benzene rings. Naphthalene is an example.

20.13 (a) Ether (b) Amine (c) Nitrile (d) Aldehyde (e) Acid

20.14 (a) RCOR (b) RCOOR (c) $RCONH_2$ (d) RX
(e) ROH

20.15 (a) CH_3OH (b) CH_3NH_2 (c) CH_3OCH_3 (d) HCOOH
(e) CH_3COCH_3

20.16 (a) Chlordane (b) Halothane (c) Phenol (d) Formalin
(e) Benzaldehyde

20.17 (a) Pyridine   (b) DDT   (c) Ethyl propanoate
(d) $CH_3CONHCH_3$ (e) Cyclohexanone

20.20 (a) Amino acids   (b) Monosaccharides   (c) Nucleotides

20.21 (a) Alcohols, aldehydes, ketones   (b) Amides, acids, amines
(c) Acids, esters

20.22 Sucrose   (b) Stearic acid   (c) Glycine   (d) Cholesterol
(e) DNA   (f) Starch

20.23 A simple protein is composed only of bonded amino acids. A conjugated protein contains some other group besides the protein chain.

20.24 (a) Glycerol + fatty acids   (b) Two monosaccharides
(c) Pentose + phosphate + heterocyclic amine   (d) Amino acids   (e) Acid + alcohols

20.28 The functional group

20.29 (a) 7   (b) 4   (c) 10   (d) 2   (e) 8

20.30

(a)
```
    H   H   H   H   H
    |   |   |   |   |
H — C — C — C — C — C — H
    |   |   |   |   |
    H   H   H   H   H
```

(b)
```
    H   H   H   H   H   H   H
    |   |   |   |   |   |   |
H — C — C — C — C — C — C — C — H
    |   |   |   |   |   |   |
    H   H   H   H   H   H   H
```

(c)
```
    H   H   H   H   H   H
    |   |   |   |   |   |
H — C   C — C — C — C — H
    |   |   |   |   |
    H   |   H   H   H
    H — C — H
        |
        H
```

(d)
```
            H
            |
        H — C — H
    H       |   H   H   H   H
    |       |   |   |   |   |
H — C       C — C — C — C — H
    |       |   |   |   |   |
    H       |   H   H   H   H
        H — C — H
            |
            H
```

20.31 (a) Pentane   (b) Heptane   (c) Heptane   (d) Octane

20.32 (a) 2,3-Dimethylbutane   (b) 2,8-Dimethylnenneane (or nonane)
(c) 3-Ethylhexane   (d) 2,3,4,5-Tetramethylheptane   (e) 5-Ethyl-3-methyloctane   (f) 2,2,3,3-Tetramethylbutane

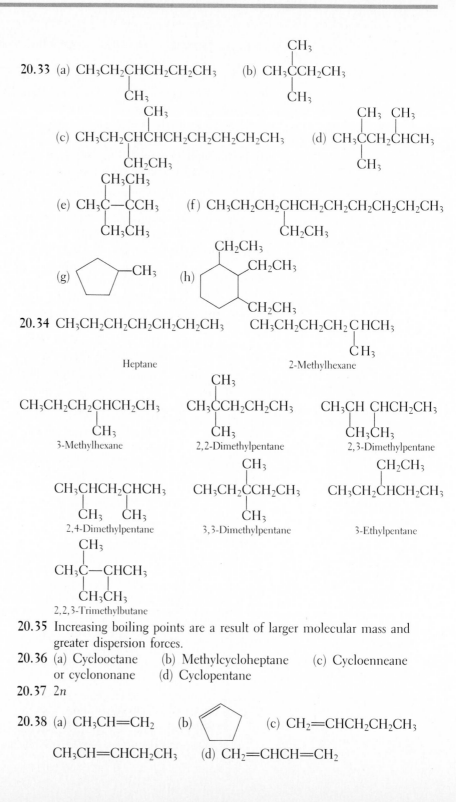

20.33 (a) $CH_3CH_2CHCH_2CH_2CH_3$

                                CH_3

$$CH_3$$

              |
(b) $CH_3CCH_2CH_3$

                  CH_3

                      CH_3

(c) $CH_3CH_2CHCHCH_2CH_2CH_2CH_3$

                    CH_2CH_3

                              CH_3 CH_3

(d) $CH_3CCH_2CHCH_3$

                             CH_3

           CH_3CH_3

(e) $CH_3C—CCH_3$

           CH_3CH_3

(f) $CH_3CH_2CH_2CHCH_2CH_2CH_2CH_2CH_3$

                          CH_2CH_3

(g)   ⬠—CH_3

(h)

          CH_2CH_3
             CH_2CH_3
          CH_2CH_3

20.34 $CH_3CH_2CH_2CH_2CH_2CH_2CH_3$ $CH_3CH_2CH_2CH_2CHCH_3$

                                                            CH_3

                    Heptane   2-Methylhexane

$CH_3CH_2CH_2CHCH_2CH_3$ $CH_3CCH_2CH_2CH_3$ $CH_3CH\ CHCH_2CH_3$

               CH_3 CH_3 CH_3CH_3

     3-Methylhexane   2,2-Dimethylpentane   2,3-Dimethylpentane

$CH_3CHCH_2CHCH_3$ $CH_3CH_2CCH_2CH_3$ $CH_3CH_2CHCH_2CH_3$

         CH_3 CH_3 CH_3 CH_2CH_3

   2,4-Dimethylpentane   3,3-Dimethylpentane   3-Ethylpentane

          CH_3

$CH_3C—CHCH_3$

         CH_3CH_3

  2,2,3-Trimethylbutane

20.35 Increasing boiling points are a result of larger molecular mass and greater dispersion forces.

20.36 (a) Cyclooctane   (b) Methylcycloheptane   (c) Cycloenneane or cyclononane   (d) Cyclopentane

20.37 $2n$

20.38 (a) $CH_3CH{=\!=}CH_2$ (b) ⬠   (c) $CH_2{=\!=}CHCH_2CH_2CH_3$

     $CH_3CH{=\!=}CHCH_2CH_3$ (d) $CH_2{=\!=}CHCH{=\!=}CH_2$

$$CH_3CH{=}C{=}CH_2 \qquad \text{(e)}$$

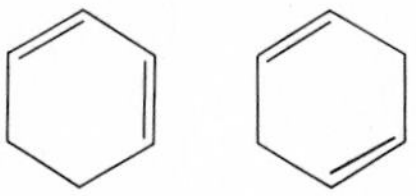

20.39 (a) 10   (b) 16   (c) 14   (d) 4   (e) 6   (f) 18

20.40 $CH_2{=}CHCH_2CH_2CH_3 \qquad CH_3CH{=}CHCH_2CH_3$

$$CH_2{=}\overset{CH_3}{\underset{}{C}}CH_2CH_3 \qquad CH_3\overset{CH_3}{\underset{}{C}}{=}CHCH_3 \qquad CH_3\overset{CH_3}{\underset{}{C}}HCH{=}CH_2$$

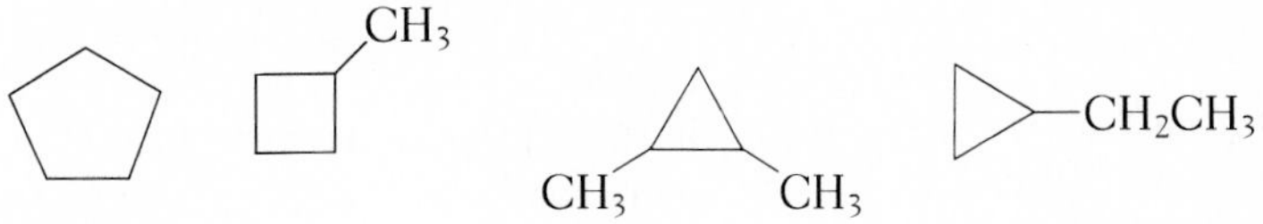

20.42 C_nH_{2n-2}

20.43

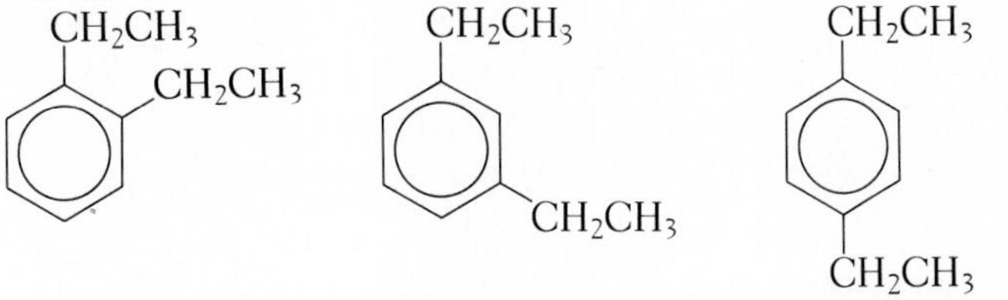

1,2-Diethylbenzene   1,3-Diethylbenzene   1,4-Diethylbenzene

20.44

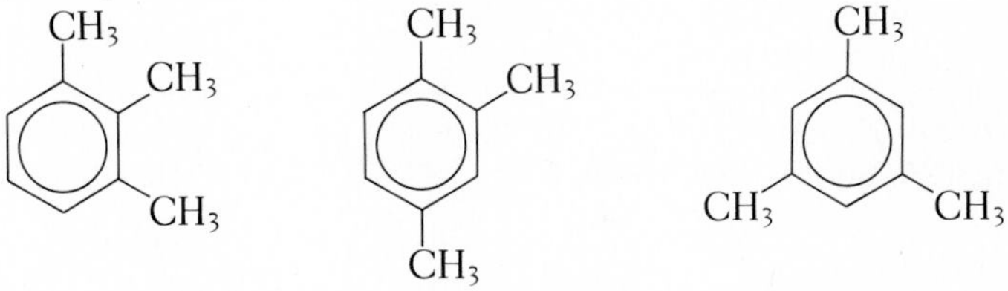

1,2,3-Trimethylbenzene   1,2,4-Trimethylbenzene   1,3,5-Trimethylbenzene

20.45

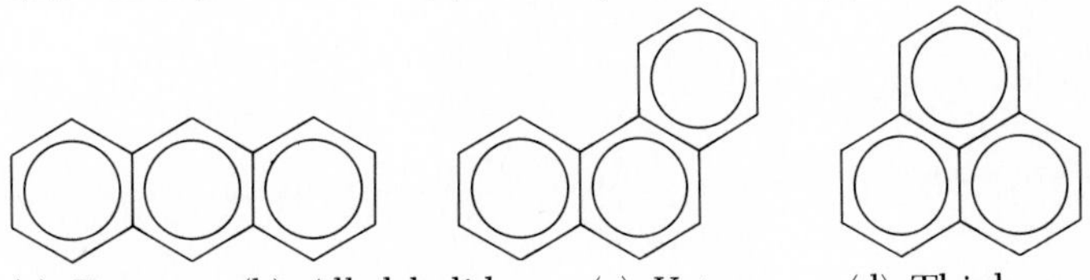

20.46 (a) Ester   (b) Alkyl halide   (c) Ketone   (d) Thiol
(e) Amide

20.47 (a) —CHO   (b) —COOH   (c) R—NH$_2$ (d) RSR

20.48 (a) Ester, alcohol   (b) Halides, amine, aldehyde
(c) Ether, ketone, amide, double bond

20.49 (a) CH_3CH_2OH (b) HCHO   (c) $CH_3CH_2CH_2CH_2COOH$
(d) CH_3COCH_3 (e) $CH_3CH_2CH_2CH_2NH_2$
(f)

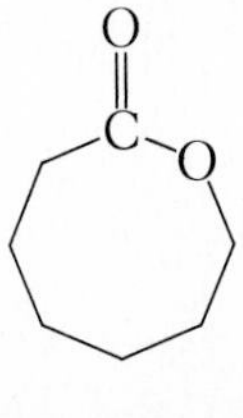

plus many others

20.50 (a) Ethanol   (b) Propanone   (c) Methanoic acid   (d) Butanoic acid
20.51 (a) Heterocyclic amine   (b) Aldehydes and ketones   (c) Acids (d) Organic halides
20.52 (a) Drinking   (b) Anesthetic   (c) Antifreeze (d) Preservative   (e) Flavoring agent   (f) Solvent
20.53 The carbonyl group is at the end of a chain in an aldehyde. The carbonyl group is bonded to the nonterminal carbon in a ketone.
20.54 (a) HOOCCOOH   (b) Contains two carboxyl groups instead of one.
20.55 (a) Distinctive, mainly pleasant, odors   (b) Methyl formate, ethyl butyrate
20.56 Aspirin, an analgesic   methyl salicylate, a muscle analgesic phenyl salicylate, a coating for pills
20.57 (a) CH_3CONH_2 (b) $CH_3CONHCH_3$ (c) $C_6H_5CON(CH_3)_2$
20.60 Heterocylic amines contain one or more nitrogen atoms within a ring structure.
20.61 $CH_4 + Cl_2 \longrightarrow CH_3Cl + HCl$
20.62 Liver damage and cancer production

APPENDIX 1

1. (a) 33   (b) 76   (c) −139   (d) 3.778
2. (a) 11   (b) 5   (c) $n/(m(a + b)) - 2$ (d) $b - (a + 2)/(n + 1)$
3. (a) 18   (b) 32   (c) 18   (d) 14
4. (a) 9   (b) 156   (c) 1.25   (d) 72   (e) 55
5. (a) $1/an - m/n$ (b) $((a + b)/n) - m$ (c) $(3a + b - n)/4$ (e) $4b/a$
6. (a) −2.10   (b) −17   (c) 9   (d) 2
7. (a) 4.9×10^{-4} (b) 1.2×10^8 (c) 1.35×10^{-9} (d) 1.0×10^{-20}
8. (a) 0.311   (b) 0.0000000290   (c) 7774.2 (d) 8,001,000,000,000,000,000
9. (a) 4.36×10^{-1} (b) 1.215×10^5 (c) 1.001×10^{-9} (d) 9.79×10^{10} (e) 6.361×10^{-8}
10. (a) 7.75×10^5 (b) 1.33×10^{-2} (c) 2.6575×10^9 (d) 3.073×10^{35} (e) 3.9433×10^{19}
11. (a) 3.663×10^5 (b) 9.8488×10^{-12} (c) -5.435×10^{46} (d) -1.6877×10^{-32}
12. (a) 4.27×10^6 (b) 3.47628×10^{137} (c) 4.41824×10^{41}
13. (a) 4.47529×10^{-17} (b) 5.40×10^{118} (c) 3.988995×10^5
14. (a) 1.2564×10^9 (b) 1.000675
15. (e) 3
16. (a) 0.56   (b) 37°C, 24°C, −32°C, 57°C

· GLOSSARY ·

Absolute zero the lowest possible temperature, 0 K or $-273.15°C$

Accuracy how close a measured value is to the actual value

Acid a substance that donates H^+ to water (the Arrhenius definition); a proton donor (the Brønsted-Lowry definition)

Acid anhydride a substance that reacts with water to produce an acidic solution; a nonmetal oxide

Acid-base indicator a dye that changes color depending on the pH of a solution

Actinide series the 14 elements that come after actinium, Ac, on the periodic table; elements with atomic numbers 90–103

Activated complex a high-energy intermediate species produced in chemical reactions as a result of the collision of the reactant molecules

Activation energy the minimum energy needed to produce the activated complex; the minimum energy required for a reaction to occur

Actual yield the mass of product obtained in a chemical reaction

Aerobic able to take place in the presence of oxygen gas, O_2

Alcohol an organic compound that has a —OH group attached to a hydrocarbon group

Aldehyde an organic compound that has a —CHO, a carbonyl group and hydrogen atom, attached to a hydrocarbon chain or ring

Alkali metals group IA elements: Li, Na, K, Rb, Cs, and Fr

Alkaline earth metals group IIA elements: Be, Mg, Ca, Sr, Ba, and Ra

Alkane a hydrocarbon that contains all carbon-carbon single bonds; a saturated hydrocarbon

Alkene a hydrocarbon that contains at least one carbon-carbon double bond; an unsaturated hydrocarbon

Alkyl group a group that results when one hydrogen is removed from a nonaromatic hydrocarbon, for example, methyl, CH_3—, ethyl, C_2H_5—, etc.

Alkyne a hydrocarbon that contains at least one carbon-carbon triple bond; an unsaturated hydrocarbon

Allotropes different forms of the same element, e.g., graphite and diamond, two distinct forms of carbon

Alloy a solution of metals

Alpha particle a high-energy helium nucleus, He^{2+}, that is emitted by some heavy nuclides when undergoing radioactive decay

Amalgam the solution that results when a metal solute is dissolved in liquid mercury

Amide an organic compound that contains a nitrogen bonded to a carbonyl carbon, $-CONH_2$, $-CONHR$, or $-CONR_2$

Amine an organic compound that contains one or more alkyl groups attached to a nitrogen; RNH_2, R_2NH, or R_3N

Amino acid an organic compound that contains both an amino group and a carboxylic acid group; amino acids combine together to produce proteins

Amorphous solid a solid whose structure lacks the long-range order of crystalline solids

Ampere (amp) a unit of electric current; the amount of coulombs of charge per second

Amphoteric a substance that can combine with either H^+ or OH^-, thus behaving as either an acid or base, depending on the conditions

Anaerobic able to take place without the presence of oxygen, O_2; the opposite of aerobic

Anion a negative ion; one with extra electrons

Anode the electrode where oxidation occurs in electrochemical cells

Antimatter a form of matter that has the opposite properties of "regular" matter; if antimatter contacts regular matter, the two annihilate each other and are transformed totally into energy

Aqueous solution a water solution

Aromatic hydrocarbon an organic compound that contains a benzene ring or has benzenelike properties

Atmosphere (atm) a unit of pressure equivalent to 101 kPa; the amount of pressure necessary to support a column of mercury 760 mm Hg high

Atom a tiny neutral particle composed of protons, neutrons, and electrons that is the smallest unit that retains the chemical properties of an element

Atomic mass the average mass of the naturally occurring isotopes compared with ^{12}C; traditionally called *atomic weight*

Atomic mass unit (see unified atomic mass unit)

Atomic number the number of protons in the nucleus of an atom; the positive charge on the nucleus of an atom

Atomic radius a measure of the relative size of an atom; the average distance from the nucleus to the outermost electron, normally measured as half the distance between two bonded nuclei

Atomic weight (see atomic mass)

Avogadro's number the number of particles in 1 mol, 6.022×10^{23}

Balance a laboratory instrument used to measure the mass of objects

Barometer a laboratory device used to measure atmospheric pressure

Base a substance that increases the OH^- concentration when dissolved in water (Arrhenius definition); a proton acceptor (Brønsted-Lowry definition)

Base units fundamental SI units of measurement from which all other SI units are obtained

Beta particle a high-energy electron that is emitted by a nucleus undergoing one type of radioactive decay

Binary acid an acid that contains hydrogen bonded to a nonmetal, HX

Binary compound a compound that is composed of two different elements

Boiling point the temperature at which the vapor pressure of a liquid equals the applied pressure; when the boiling point is reached, bubbles of the liquid's vapor form throughout the liquid

Boiling point elevation the increased boiling temperature of a solvent after a nonvolatile solute is added

Bond the primary force of attraction that holds atoms together in molecules and lattice structures

Bonding electrons electrons that are attracted by two nuclei; shared electrons

Bond length the average distance between the nuclei of two atoms that are bonded together

Calorie a unit of heat energy; 1 cal = 4.184 J

Calorimeter a laboratory instrument used to measure heat transfers in chemical and physical changes

Carbohydrate a class of biologically important compounds that includes sugars and starches

Carbonyl group a functional group in organic chemistry that consists of a carbon that is doubly bonded to an oxygen, $C=O$; aldehydes and ketones both contain carbonyl groups within their structures

Carboxyl group a functional group contained in organic acids; it consists of a carbon that has a doubly bonded oxygen and a hydroxy group attached, —COOH

Catalyst a substance that increases the rate of a chemical reaction by lowering its activation energy; generally, a catalyst is fully recovered after the reaction

Cathode the electrode where reduction occurs in electrochemical cells

Cation an ion with a positive charge; one that has lost electrons

Celsius temperature scale a temperature scale that is displaced 273 degrees from the absolute temperature scale; water's boiling point is 100°C (373 K), and its freezing point is 0°C (273 K)

Centi a prefix attached to units that decreases the magnitude by $\frac{1}{100}$

Chalcogen a name applied to group VIA elements: O, S, Se, Te, and Po

Chemical bond the force of attraction that holds atoms together in compounds

Chemical change a change in which the composition of a substance is altered; also called a chemical reaction

Chemical equation an expression of symbols, formulas, and coefficients that describes mass, volume, and mole relationships of specific chemical reactions; in chemical equations the reactants are written to the left of an arrow and the products are written to the right of the arrow

Chemical equilibrium a closed chemical system in which the forward and reverse reaction rates are equal

Chemical family another name for a group in the periodic table; a vertical column in the periodic table, usually denoted by a roman numeral and a letter (see chemical group)

Chemical formula a combination of chemical symbols with appropriate subscripts that indicate the ratio of atoms within molecules and formula units

Chemical group a vertical column of elements in the periodic table with similar outer electronic configurations

Chemical kinetics the study of the rates of chemical reactions

Chemical nomenclature a system of rules and guidelines for writing unique names for each element and compound

Chemical property a property that describes a chemical change that a substance undergoes

Chemical symbol either one, two, or three letters used to represent each element; normally, these letters are the beginning letters of the modern or classical name of the element

Chemistry the study of matter and its interactions

Coefficient a number or algebraic quantity preceding a variable, unknown quantity, or chemical formula

Colligative property a property of a solution that depends on the number of dissolved nonvolatile solute particles, rather than their type, e.g., freezing point, boiling point, and vapor pressure

Collision theory a theory that attempts to explain the rate of chemical reactions in terms of the number of effective collisions of reactants that take place in a specified time interval

Combination reaction a reaction in which two or more reactants are chemically combined to produce a singular product

Composition the amount and type of components in a sample of matter

Compound a pure substance composed of two or more different elements that have been chemically combined

Concentrated the term applied to describe solutions in which a large quantity of solute is dissolved in a solvent

Condensation the process in which a vapor changes to a liquid

Conversion factor a fraction that relates the value of one set of units to the value of another set of units, for example, 1 cal/4.184 J

Coordinate covalent bond a covalent bond that results when one atom contributes both electrons in the formation of the bond

Coulomb a unit of electric charge; the amount of charge that passes in an electric circuit when one ampere flows for one second

Covalent bond a chemical bond that results when electrons are shared between two nuclei; the overlap of atomic orbitals from two different atoms

Critical mass the minimum mass of a fissionable element, like U, that is necessary to sustain a nuclear fission reaction

Cryogenics the branch of physics that deals with the study of very low temperatures and their effects

Crystalline solid a solid with atoms, ions, and molecules arranged in an orderly, regular, three-dimensional pattern

Data the information collected when conducting an experiment

Decomposition reaction a reaction in which a single reactant is broken down to two or more products

Deliquescence the property of various solids to absorb moisture from the air, and then dissolve in this added water

Density the mass to volume ratio of a substance

Derived units SI units that are obtained from combinations of the seven base units

Diatomic molecule a molecule that contains two atoms, for example, Br_2, O_2, and HF.

Dilute the term applied to describe solutions with a small quantity of solute per solvent; the opposite of concentrated

Dipole (electric) case in which there is charge separation in a molecule; the positive center of charge does not correspond with the negative center

Dipole-dipole interactions attractive intermolecular forces that exist among polar covalent substances

Dispersion force the attractive intermolecular force existing in all molecules as a result of momentary induced dipoles; this force is most important in molecules that do not have other types of intermolecular forces

Dissociation the separation of a larger chemical species into smaller ones, generally the separation of ions in salts entering solution.

Distillation a chemical separation procedure in which one component is selectively vaporized and condensed to remove it from other substances

Double bond a covalent bond in which four electrons are shared between two nuclei

Dynamic equilibrium an equilibrium that results when the rates of two opposing processes are equal

Effective collision a collision between two reactant particles that results in the formation of the products; the colliding particles must possess the proper amount of energy and be properly oriented

Efflorescence the loss of water by hydrated salts

Electrolysis cell a container in which substances are decomposed by passing a direct electric current through them

Electrolyte a substance that exists as ions when dissolved in solution

Electron the low-mass, negatively charged particle found in atoms outside the nucleus; it has a mass of 0.000549 u

Electron dot formulas symbols and formulas that illustrate the number of outer electrons located in elements and molecules; commonly called Lewis structures or Lewis electron dot formulas

Electronegativity the property of atoms to attract electrons in chemical bonds

Electron energy levels regions of space about the nucleus where electrons reside; they are subdivided into smaller regions called sublevels and orbitals

Electronic configuration the arrangement and population of electrons in specific energy levels, sublevels, and orbitals in atoms

Electron spin the property of electrons to appear as if they are spinning on an axis

Element a pure substance that cannot be decomposed by chemical means

Empirical formula a formula that expresses the simplest ratio of atoms within a compound; also known as the simplest formula

Endothermic a term used to describe a chemical process in which heat flows from the surroundings to the observed system; applied to reactions in which heat is absorbed

End point the point at which the indicator changes color during a titration, indicating that the titration is completed

Energy the ability to do work or produce a change

Enthalpy a quantity that describes the heat content of substances; the difference in enthalpy of products and reactants is equal to the amount of heat liberated or absorbed

Enzyme a high-molecular-mass protein structure within living systems that catalyzes chemical reactions (see protein)

Equilibrium (see chemical equilibrium)

Equivalence point the point in an acid-base titration where the number of moles of H^+ from the acid equals the number of moles of OH^- from the base

Equivalent mass for acids and bases, the mass of a substance that gives up or takes in 1 mol of H^+ or electrons

Ester a class of organic compounds that results when an organic acid combines with an alcohol; esters have the general formula RCOOR′

Ether a class of organic compounds that contains two hydrocarbon groups bonded to an oxygen, R—O—R′

Evaporation the process whereby surface molecules of liquids break free of the intermolecular forces that hold them in the liquid and enter the gas phase

Exothermic term used to describe a chemical process in which heat flows from a system to the surroundings; applied to reactions in which heat is liberated

Faraday the quantity of charge possessed by 1 mol of electrons, 96,485 C

Fission (see nuclear fission)

Fluidity the property of substances to flow; the opposite of viscosity

Fluorescence the property of a substance to release light energy after being excited by other energy forms

Force a push or a pull

Formula an expression used to represent the type and number of atoms in a molecule or ion

Formula mass the sum of the atomic masses of all atoms in a particular formula unit

Freezing point the temperature at which a liquid changes state and becomes a solid

Freezing point depression the decrease in the freezing point of a solvent after the addition of a solute

Functional group a group of atoms in organic compounds that give a molecule its characteristic chemical and physical properties, e.g., carbonyl group, carboxyl group, and alcohol group

Fusion (see nuclear fusion)

Galvanic cell an electrochemical cell that produces an electric current from redox reactions; also called a voltaic cell or battery

Gamma ray a high-energy radiation form that is released by unstable nuclei during radioactive decay

Gas constant (R) the numerical constant that relates volume, pressure, temperature, and moles in the ideal gas equation, $PV = nRT$; the numerical value is 0.082056 $L \cdot atm/(mol \cdot K)$

Group a vertical column in the periodic table denoted by a roman numeral and either the letter A or B; sometimes periodic groups are called families

Half-life the amount of time required for one-half of the reactants in a chemical reaction to change to products or for one-half of the radioactive nuclei to decay to a new nuclide

Half-reaction a pseudoreaction that represents either the oxidation or reduction part of a redox reaction; half-equations are written to represent half-reactions, for example, $Cu^{2+} + 2e^- \rightarrow Cu$

Halide ion the negative ion produced when a halogen takes in an electron, for example, F^-, Cl^-, Br^-, and I^-

Halogen an element that belongs to group VIIA in the periodic table: F, Cl, Br, I, and At

Heat a form of kinetic energy that, when transferred to an object not undergoing a state change, increases its temperature

Heat capacity the amount of heat required to increase the temperature of a fixed amount of substance (usually 1 mol or 1 g) by 1 K

Heat of fusion the amount of heat needed to change a fixed amount of solid to liquid at a fixed temperature

Heat of vaporization the amount of heat required to change a fixed amount of liquid to vapor at a fixed temperature

Heterogeneous composed of two or more distinct components; applied to types of matter with more than one observable phase

Homogeneous mixture a mixture of pure substances that has the same composition throughout; a solution

Homologous series a group of compounds in which one member differs from the one preceding and the one following by a fixed amount

Hydrate a chemical species, generally a salt, that has attached water molecules, for example, $CuSO_4 \cdot 5H_2O$

Hydration addition of water to another substance

Hydration energy the energy released when solute particles are surrounded by water molecules in the solution process

Hydride an ionic or covalent binary compound of hydrogen; examples of ionic hydrides are LiH and CaH_2, and examples of covalent hydrides are NH_3 and SiH_4

Hydrocarbon an organic compound that contains only carbon and hydrogen; includes the alkanes, alkenes, alkynes, and aromatics

Hydrocarbon derivative an organic compound that contains at least one other atom besides carbon and hydrogen; each group of hydrocarbon derivatives are characterized by a functional group

Hydrogen bond the dipole-dipole interaction between molecules that have hydrogen bonded to F, O, or N; the strongest intermolecular force

Hydrolysis a chemical reaction in which the water molecule is split

Hydronium ion ion that results when a hydrogen ion combines with water, $H^+ + H_2O \rightarrow H_3O^+$ (hydronium ion); a hydrated proton

Hygroscopic a term used to describe salts that take up and retain moisture without dissolving

Hypothesis a tentative guess based on previously collected facts that is proposed to explain regularities in data

Ideal gas a nonexistent gas that behaves exactly as predicted by the ideal gas law; some real gases approach ideal gas behavior at low pressures and high temperatures

Ideal gas equation the equation that expresses the relationship of pressure, volume, temperature, and number of moles of an ideal gas, $PV = nRT$

Immiscible the term used to describe two or more liquids that are not soluble in each other; they are identified by observing two or more layers

Inert atmosphere an environment of stable gases such as helium, argon, or nitrogen that will not enter into a chemical reaction

Inert gases the old name for the noble gases

Inhibitor a substance that decreases the rate of chemical reactions by increasing the activation energy; a negative catalyst

Intermolecular forces attractive forces among molecules that are responsible for holding molecules in a particular physical state; the primary intermolecular forces are dispersion forces, dipole-dipole interactions, and hydrogen bonding

International System of Units (SI) a system of measurement units that is based on the metric system and is used by scientists throughout the world

International Union of Pure and Applied Chemistry Nomenclature System (IUPAC System) a systematic set of rules used to assign a unique name to chemical compounds

Ion a charged atom (see anion and cation)

Ionic bond a chemical bond in which electrons are transferred from a metal to a nonmetal or polyatomic ion, resulting in the formation of ionic species

Ionization a process by which a substance is changed to ions

Ionization energy the amount of energy required to remove the highest-energy electron from a neutral gaseous atom

Ionizing radiation the radiation that produces ions as it traverses matter

Isoelectronic a term used to describe different atomic species with the same electronic configuration; for example, Na^+ is isoelectronic with Ne

Isomers compounds with the same molecular formula but different structures

Isotopes atoms with the same atomic number but different mass numbers

IUPAC (International Union of Pure and Applied Chemistry) an organization that is responsible for establishing the rules and regulations for naming chemical substances consistently throughout the world

Joule (J) a unit of energy, equivalent to 0.239 calorie; 1 cal = 4.184 J

Kelvin temperature scale a temperature scale in which the zero point is absolute zero, the lowest possible temperature; each degree is 1/273.16 of the temperature of the triple point of water

Ketone an organic compound that contains a carbonyl group attached to two hydrocarbon groups, RCOR′

Kilo a prefix that is placed in front of units to increase their magnitude 1000 times.

Kinetic energy the energy possessed by moving bodies; it is equal to one-half the mass of an object times its velocity squared, $KE = \frac{1}{2}mv^2$

Lanthanides the fourteen elements in the periodic table that come after lanthanum, elements 58 through 71; also called the rare earths

Law of conservation of mass the law that states that mass cannot be created or destroyed in normal chemical changes

Law of constant composition the law that states that the mass ratios of elements within a compound are fixed

Le Chatelier's principle the principle that states that when a physical property of an equilibrium system is changed, the equilibrium attempts to absorb the change and return to a state of equilibrium

Lewis structure (see electron dot formula)

Limiting reagent the reactant that determines the maximum amount of products produced; when all of it is consumed, the reaction ceases even though the other reactants are still present

Lipids a class of biologically important compounds that include triglycerides, steroids, waxes, and prostaglandins

Liter (L) a non-SI unit of volume, equivalent to the SI unit $1 \; dm^3$

Logarithm of a number the exponent of 10 (common logarithms) that gives a quantity equal to the number; for example, log 1000 = 3, because $10^3 = 1000$

Malleable the property of substances that enables them to be hammered and shaped into different forms; a characteristic property of metals

Manometer a laboratory instrument used to measure gas pressures

Mass the measure of the quantity of matter in an object

Mass number the total number of protons and neutrons (nucleons) in an atom

Matter anything that has mass and occupies space

Mechanism (see reaction mechanism)

Melting point the temperature at which a solid changes to a liquid, and the solid and liquid exist in equilibrium; the same temperature as the freezing point

Metal a substance that is a good conductor of heat and electricity, readily loses electrons to form cations, is malleable, is ductile, and has a shiny metallic appearance

Metalloid an element with properties different from those of metals or nonmetals; examples of metalloids include B, Si, Ge, and As

Metathesis reaction a double replacement reaction

Meter (m) SI unit of length; 1 m = 39.37 in.

Metric system the decimal system of measurement units from which the International System (SI) was derived

Milli a prefix placed in front of a unit to diminish its size by $\frac{1}{1000}$

Millimeter of Hg, mm Hg (torr) a unit of pressure equal to $\frac{1}{760}$ of an atmosphere; 1 atm = 760 mm Hg

Miscible the property of two or more liquids to be mutually soluble in each other

Mixture a combination of two or more pure substances; two groups of mixtures exist: (1) homogeneous mixtures, or solutions, and (2) heterogeneous mixtures, those with more than one identifiable phase

Molality (m) a unit of solution concentration that relates moles of solute per kilogram of solvent, mol (solute)/kg (solvent)

Molarity (M) a unit of solution concentration that relates moles of solute per liter of solution, mol (solute)/L (solution)

Molar mass the mass of one mole of a substance

Molar volume the volume of one mole of a substance at a fixed set of conditions

Mole the SI unit for the amount of a substance; a mole is equivalent to 6.022×10^{23} particles

Molecular formula a formula that indicates the type and exact number of atoms within a molecule

Molecular geometry the three-dimensional shape of a molecule; it indicates the position of each atom relative to all other atoms within the molecule

Molecular mass the sum of all the atomic masses of atoms within a molecule

Molecule the most fundamental unit in a compound that retains the chemical properties of the compound; molecules are composed of atoms that are chemically combined

Monomer a basic repeating molecular structure(s) that combines together to produce polymers

Monosaccharides simple sugars that combine together to yield all other carbohydrates; most naturally occurring ones contain three to seven carbons

Multiple covalent bonds covalent bonds with more than two shared electrons; includes double and triple bonds

Neutralization the combination of an acid and base to yield a salt and water

Neutron a particle in the nucleus of an atom possessing no electric charge; its mass, 1.008665 u, is slightly larger than that of a proton

Noble gas group VIIIA elements, including He, Ne, Ar, Kr, Xe, and Rn

Nomenclature (see chemical nomenclature)

Nonelectrolyte a substance that does not ionize when dissolved in solution

Nonmetals elements on the right side of the periodic table that possess filled or nearly filled outer electronic configurations and have chemical and physical properties opposite to the metals

Nonpolar covalent bond a bond in which electrons are shared equally; there is no separation of charge

Normal boiling point the temperature at which the vapor pressure of a liquid equals 760 mm Hg

Normality (N) a unit of concentration that relates the number of equivalents of solute per liter of solution, equiv(solute)/L(solution)

Nuclear fission a nuclear change in which a high-mass nucleus breaks up into two or more smaller fragments; a large quantity of energy is released

Nuclear fusion a nuclear change in which two low-mass atoms are united to produce a higher-mass atom; a large amount of energy is released during the fusion

Nucleic acid a class of biologically important molecules that are composed of nucleotides, including deoxyribonucleic acids (DNA) and ribonucleic acids (RNA)

Nucleon a particle located in the nucleus, either a proton or a neutron

Nucleus the small, dense, positively charged region in the center of an atom; it contains the protons and neutrons

Orbital a region of space where electrons are found within an atom; it is the smallest subdivision of electron energy levels, holding a maximum of two electrons

Ore the rock or mineral from which elements, commonly metals, can be extracted

Organic compound any carbon compound except those that exhibit properties of inorganic compounds, such as carbonates, metallic cyanides, hydrogen carbonates, carbides, simple carbon oxides

Oxidation a chemical change in which electrons are released; the addition of oxygen or the loss of hydrogen by a substance

Oxidation number a number that is assigned to atoms to assist in predicting chemical changes and writing chemical formulas

Oxide a binary compound of oxygen

Oxidizing agent a substance that brings about the oxidation of another substance by accepting electrons from it

Oxyacid an inorganic acid that contains oxygen, for example, HNO_3, H_2SO_4, and $HClO_3$

Partial pressure the pressure exerted by an individual gas in a gaseous mixture

Parts per million (ppm) a unit of concentration that relates the number of parts of solute per million total parts, parts(solute)/1,000,000 total parts

Pascal the SI unit for pressure; 133.3 Pa = 1 mm Hg

Percent (mass to volume) a unit of concentration that relates the mass of solute per 100 mL of solution, mass(solute)/100 mL(solution)

Percentage yield the actual yield divided by the theoretical yield times 100; percentage of the theoretically calculated yield that is actually obtained

Percent by mass a unit of concentration that relates the mass of solute to 100 g of solution, mass(solute)/100 g(solution)

Percent by volume a unit of concentration that relates the volume of solute per 100 mL of solution, volume(solute)/100 mL(solution)

Period a horizontal row in the periodic table

Periodic properties the chemical and physical properties that recur regularly with increasing atomic numbers

Peroxide a compound that contains an oxygen-oxygen single bond

pH the negative logarithm of $[H^+]$

Phase a homogeneous region of matter with observable boundaries

Physical change a change in physical properties of a substance with no change in composition

Physical property a property associated with an individual substance that can be described without referring to any other substance, e.g., color, size, mass, and density

Physical states various forms in which substances exist, depending on temperature, pressure, and intermolecular forces; solid, liquid, and gas

pOH the negative logarithm of $[OH^-]$

Polar covalent bond a bond in which electrons are shared unequally; there is a separation of charge

Polyatomic ion an ion containing more than one atom

Polymer a high-molecular-mass compound that is composed of long chains of repeating small molecules (monomers) that are bonded together

Polyprotic acid an acid that has the capacity to donate more than one H^+

Polysaccharides polymers of monosaccharides found in living systems, e.g., starch, cellulose, and glycogen

Positron a positively charged electron, e^+; a form of antimatter

Potential energy stored energy as a result of an object's position, condition, or composition

Precipitation a process whereby a solid insoluble substance is produced in an aqueous reaction

Precision expresses how close repeated measurements are grouped; describes the reproducibility of measurements

Pressure force applied to an area; gas pressure is measured in units of kilopascals, atmospheres, millimeters of mercury, and torr

Product the end result of a chemical reaction, written to the right of the arrow in a chemical equation

Property a physical or chemical characteristic used to identify a sample of matter

Proteins a class of biologically important molecules that are major structural and controlling agents in cells; chemically, they are polymers of amino acids

Proton a positively charged particle within the nucleus of atoms; it has a mass of 1.007276 u

Rad a measure of the amount of energy absorbed per gram of living tissue as a result of being exposed to ionizing radiation

Radioactive element an element that emits various matter-energy forms at a measurable rate

Radioactivity the emission of particles and energy forms by unstable nuclei

Random errors unidentifiable errors associated with all measurements

Rare earth elements (see lanthanides)

Rate constant the proportionality constant relating the rate of a chemical reaction to the concentration of one or more of the reactants raised to a power; k is the symbol used to indicate rate constants

Reactant a substance that is initially present in a chemical reaction

Reaction mechanism a series of steps that occurs when the reactant molecules collide and form the products

Reaction rate the change in concentration, or pressure, of reactants or products over a unit time interval; how fast or slow a reaction proceeds

Real gas a gas that does not behave exactly in the manner predicted by the ideal gas laws

Redox a contraction meaning reduction and oxidation

Reducing agent a substance that brings about the reduction of another substance; a substance that undergoes oxidation

Reduction a chemical process whereby electrons are taken in; adding hydrogen or removing oxygen from a substance results in the reduction of the substance

Replacement reaction a reaction whereby an element combines with a compound and displaces one of the compound's components

Resonance the case where more than one Lewis electron dot structure can be written for a molecule; the actual molecule would most resemble the average of all the Lewis structures

Reversible reaction a reaction whereby the products can combine to re-form the reactants

Salt a substance that results when an acid is combined with a base; salts are ionic substances composed mainly of combinations of metals and nonmetals or metals and polyatomic ions

Saturated hydrocarbon hydrocarbons that contain only carbon-carbon single bonds

Saturated solution a solution in which the maximum amount of solute is dissolved in a solvent for a particular set of conditions; the dissolved solute particles would be in equilibrium with undissolved solute, if present

Scale a laboratory instrument used to measure an object's weight

Scientific exponential notation the expression used to write large and small numbers as the product of a decimal and exponential factor; the decimal factor always has a numerical value between 1 and 9.99, and the exponential factor is a power of 10

Significant figures measured digits plus one estimated digit that together indicate the uncertainty of a measurement

Simplest formula (see empirical formula)

Single bond a covalent bond with two electrons shared between two atoms

Solubility the amount of solute that is dissolved in a fixed amount of solvent at a specified temperature, normally measured in grams of solute per 100 mL of solvent

Solute the component of a solution that is present in smaller amount; the solid component in a solid-liquid solution

Solution a homogeneous mixture of pure substances

Solvent the component of a solution that is present in larger amount; the liquid component of a solid-liquid solution

Specific gravity the ratio of the density of a substance to the density of water; a unitless ratio

Specific heat the amount of heat required to raise a gram of substance by one kelvin

Spectator ion an ion that is not chemically changed in aqueous reactions

Stable a term used to describe substances that do not tend to undergo spontaneous changes

Standard temperature and pressure (STP) when applied to gases, the conditions of 1 atm and 273 K (0°C)

Stoichiometry the study of quantitative relationships in chemical reactions and formulas

STP (see standard temperature and pressure)

Strong acid an acid that dissociates nearly 100% in dilute aqueous solutions, adding a large quantity of H^+ to water

Strong base a base that dissociates nearly 100% in dilute aqueous solution, adding a large quantity of OH^- to water

Structure the three-dimensional arrangement of the components of matter

Sublevel a subdivision of electron energy levels; the four primary sublevels are designated by the letters s, p, d, and f

Subliming point the temperature at which a solid changes to a vapor; solid-vapor transition point

Surface tension a property of the surface of a liquid to act as if it has a membrane covering

Systematic errors correctable errors in measurement; they result from poor techniques and procedures, uncalibrated equipment, and human error

Temperature the measurement of the relative hotness of an object; it determines the direction of heat flow

Ternary compound a compound containing three different elements

Theoretical yield the maximum obtainable yield of a chemical reaction predicted from stoichiometric relationships

Theory a unified set of hypotheses that are consistent with one another and with experimentally observable phenomena

Thermodynamics the study of energy and its transformation

Titration a laboratory procedure that determines the volume of one chemical species needed to totally combine with another

Torr another name for the unit of pressure mm Hg

Transition metal a metal belonging to periodic groups with B designations

Transition state a transient complex that forms when reactant molecules collide together; once formed, it breaks apart as either the reactants or products

Transmutation the conversion of one nuclide to another nuclide

Transuranium element an element on the periodic table that comes after U within the actinide series; elements 93 through 103

Triglyceride an ester of three fatty acids and glycerol; the most common form of lipids

Triple covalent bond a bond in which six electrons are shared between two atoms

Unified atomic mass unit (u) a mass equivalent to $\frac{1}{12}$ the mass of a ^{12}C atom (1.666×10^{-24} g); it is used to express the mass of individual atoms

Unsaturated hydrocarbon an organic compound that contains double or triple bonds

Unsaturated solution a solution in which more solute can be dissolved; an equilibrium does not exist between dissolved and undissolved solute

Valence electron an electron in the outermost energy level of an atom

Vapor a substance in the gas phase

Vapor pressure of a liquid the pressure of a vapor above a liquid; this normally refers to the equilibrium vapor pressure

Viscosity the resistance of a substance to flow, directly related to the strength of the substance's intermolecular forces

Volatile a term used to describe a liquid that evaporates readily at relatively low temperatures.

Volt a unit of electromotive force; electrical potential difference

Voltaic cell (see galvanic cell)

Volume the space occupied by a mass

Weak acid an acid that dissociates to a small degree, producing few H^+ in solution

Weak base a base that dissociates to a small degree, producing few OH^- in solution

Weight the measure of the gravitational force of attraction on a mass

· I N D E X ·

USEFUL CONVERSION FACTORS

Mass

$1\ g = 10^{-3}\ kg$
$1\ lb = 0.453592\ kg = 453.592\ g$
$1\ kg = 2.205\ lb$
$1\ u = 1.66057 \times 10^{-27}\ kg$
$1\ metric\ tonne = 1000\ kg$

Length

$1\ cm = 10^{-2}\ m$
$1\ mm = 10^{-3}\ m$
$1\ nm = 10^{-9}\ m$
$1\ Å = 10^{-10}\ m$
$1\ Å = 0.10\ nm$
$1\ inch = 2.54\ cm$
$1\ mile = 1.609\ km$

Temperature

$0°C = 273.15\ K$
$0\ K = -273.15°C$

Volume

$1\ L = 10^{-3}\ m^3$
$1\ L = 1.057\ qt$
$1\ qt = 0.9463\ L$
$1\ cm^3 = 10^{-6}\ m^3$
$1\ mL = 1\ cm^3$
$1\ L = 1000\ mL$

Energy

$1\ cal = 4.184\ J$
$1\ erg = 10^{-7}\ J$
$1\ erg = 2.3901 \times 10^{-8}\ cal$

Pressure

$1\ atm = 101,325\ Pa$
$1\ atm = 760\ mm\ Hg$
$1\ mm\ Hg = 1\ torr$
$1\ mm\ Hg = 133.322\ Pa$